The Test Access Port and Boundary Scan Architecture

Colin M. Maunder and Rodham E.Tulloss

THE TEST ACCESS PORT AND BOUNDARY-SCAN ARCHITECTURE

Colin M. Maunder and Rodham E. Tulloss

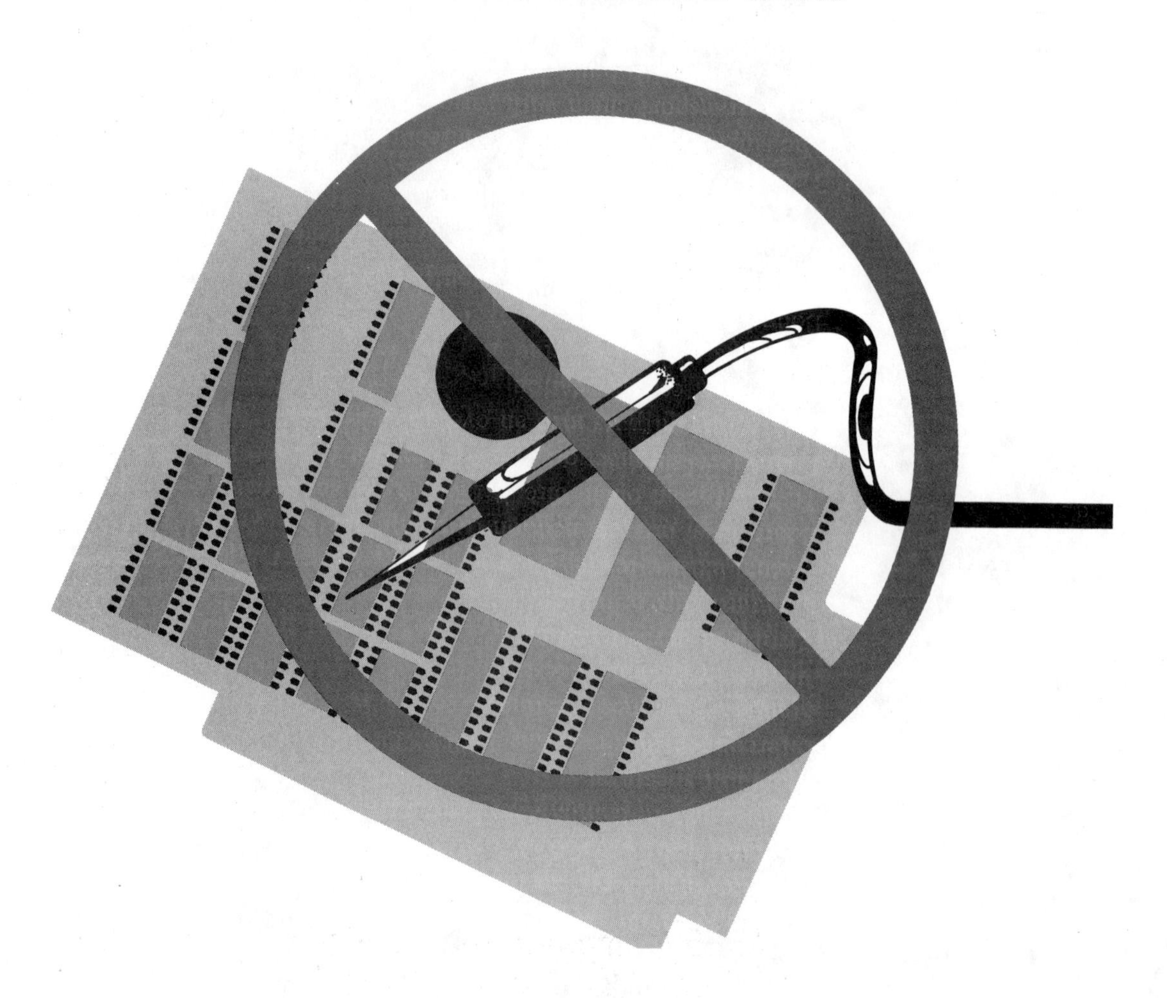

IEEE Computer Society Press Tutorial

The Test Access Port and Boundary Scan Architecture

Colin M. Maunder and Rodham E.Tulloss

IEEE Computer Society Press
Los Alamitos, California

Washington • Brussels • Tokyo

Library of Congress Cataloging-in-Publication Data

The test access port and boundary-scan architecture /
[edited by] Colin M. Maunder and Rodham E. Tulloss.
p. cm.
Includes bibliographical references and index.
ISBN 0-8186-9070-4
1. Electronic circuits — Testing — Data processing. 2. Computer architecture.
I. Maunder, Colin M. II. Tulloss, Rodham E.
TK7867.T39 1990
621.381—dc20
90-39682
CIP

Published by the
IEEE Computer Society Press
10662 Los Vaqueros Circle
PO Box 3014
Los Alamitos, CA 90720-1264

Cover design from *The Institute,* November 1989, page 1
Sponsored by IEEE Test Technology Technical Committee

IEEE Computer Society Press Order Number 2070
Library of Congress Number 90-39682
IEEE Catalog Number EH0321-0
ISBN 0-8186-9070-4 (case)
ISBN 0-8186-6070-8 (microfiche)

Additional copies can be ordered from

IEEE Computer Society Press
Customer Service Center
10662 Los Vaqueros Circle
PO Box 3014
Los Alamitos, CA 90720-1264

IEEE Service Center
445 Hoes Lane
PO Box 1331
Piscataway, NJ 08855-1331

IEEE Computer Society
13, avenue de l'Aquilon
B-1200 Brussels
BELGIUM

IEEE Computer Society
Ooshima Building
2-19-1 Minami-Aoyama
Minato-ku, Tokyo 107
JAPAN

Second printing, 1992

Technical Editor: Pradip K. Srimani
Production Editor: Lisa O'Conner

Printed in the United States of America by Victor Graphics, Inc.

THE INSTITUTE OF ELECTRICAL AND ELECTRONICS ENGINEERS, INC.

The individuals presenting information on IEEE Standards are participants in the development of the standards under discussion, and they are presumed to have expert knowledge on the subject. However, it must be understood that an IEEE Standard represents a consensus developed from many differing viewpoints. Comments of any individual may represent a somewhat different perspective from that of the IEEE and should be considered as the views of an individual expert, rather than the formal position taken by the IEEE.

Extract from the IEEE Standards Manual

Foreword

During a visit to an exhibition on the history of Chinese technology, I learned of the existence of the very first tester. Two centuries after the birth of Christ, Chinese farmers had developed a hand-operated mill to separate chaff from grain. Amazingly enough, this equipment was not introduced into Europe for more than 16 centuries! There could be several reasons for this. Perhaps the agricultural industry in Europe was not sufficiently developed to require the equipment, or perhaps the ubiquitous "not invented here" syndrome was already hampering the progress of technological change. More likely, however, is that there was no communication on the development and transfer of technology between workers within the same industry -- in this case, between Chinese and Europeans.

In today's world, testers are needed to separate the bad from the good. However, the complexity of our electronic circuit designs and the continuing miniaturization of the finished product have made the difference between good and bad more subtle and more difficult to detect. Now, testing can be an expensive process, but fortunately we have the freedom to design the product to improve its testability. The Chinese farmers didn't.

Recently, we have come to realize the value of discussing approaches to design-for-testability between companies and countries and, as a result, the standard described in this book has been created. The development of the *IEEE Standard Test Access Port and Boundary-Scan Architecture* began in 1985 when representatives from a small group of European electronics companies met in The Netherlands to discuss problems caused by the increased use of surface-mount technology and very large-scale integration (VLSI). At that first meeting, a consensus was reached about the problems and a willingness emerged to cooperate in solving them. More meetings were organized and, to identify the activity, a name was chosen: the Joint European Test Action Group. Later, as companies from North America joined the group, the name was changed to the Joint Test Action Group (JTAG).

JTAG started to define a test methodology that would address the foreseen problems and to describe the methodology in a technical proposal. This proposal, which became the JTAG Test Access Port and Boundary-Scan Architecture, was promoted at technical conferences and workshops to raise the interest and awareness of other companies, especially the integrated circuit manufacturers and the automatic test equipment vendors. The reaction from the electronics industry was very enthusiastic, with support coming from the test community and the management of many companies. Letters of endorsement were provided by the senior management of major electronics companies, demonstrating the benefit that adoption of the JTAG proposal would have for their businesses. This, in turn, increased the motivation of the JTAG members involved in the technical development.

By the summer of 1988, the JTAG proposal had matured into a specification that met many requirements of the electronics industry, and the support of the IEEE was sought to convert the ad–hoc JTAG proposal into a formal standard. Also at that time, the designers of the companies involved in JTAG began to develop the first integrated circuit designs for production and inclusion in their products. Commercially–available integrated circuits (ICs) and application–specific integrated circuit (ASIC) cells followed shortly afterwards.

Looking back, it is surprising that so much interest in boundary–scan techniques developed so rapidly. This achievement was only possible through the cooperation and support of all the companies involved, and through the significant contributions made by those involved in the technical development activities.

I hope that the examples contained in this book will give you some idea of the range of applications of the standard that JTAG created, as well as the potential value of the standard for your business. If you have a need for the solutions described in this book, don't wait 17 centuries before you use them!

Harry Bleeker
JTAG Chairman
Philips Telecommunications and Data Systems
Hilversum, The Netherlands

TABLE OF CONTENTS

Chapter		Page
6	The Boundary-Scan Register	59
6.1	The Provision of Boundary-Scan Cells	59
6.2	The Minimum Requirement	63
6.3	The INTEST Instruction	70
6.4	The RUNBIST Instruction	75
Part III	**Applications to Loaded-Board Testing**	79
7	Taking Advantage of Boundary-Scan	81
7.1	Loaded-Board Testability Problems and Traditional Test Techniques	81
7.2	100 Percent Boundary-Scan Testing	84
7.3	Test-Access Strategies for Mixed-Technology Boards	87
7.4	Conclusion	94
7.5	References	95
8	A Test Program Pseudocode	97
8.1	Introduction	97
8.2	Initialization	100
8.3	Test Circuitry Check	101
8.4	Interconnect Check	107
8.5	BIST Part Check	110
8.6	The Remaining Chips	111
8.7	Comments on Diagnosis	111
8.8	Conclusion	112
8.9	Acknowledgment	112
8.10	References	112
9	Diagnosing Faults in the Serial Test Data Path	115
9.1	Objective	115
9.2	A Basic Path Test	115
9.3	Use of the Device Identification Register	116
9.4	More Complex Methods	117
9.5	Reference	121
10	In-Circuit Testing	123
10.1	Mixed In-Circuit and Boundary-Scan Testing	123
10.2	Method 1	125
10.3	Method 2	126
10.4	Method 3	126
10.5	Conclusions	126

Chapter **Page**

LIST OF ILLUSTRATIONS

LIST OF TABLES

Preface

...denn da is keine Stelle,
die dich nicht sieht. Du musst dein Leben aendern.
[...for there is no place at all
that isn't looking at you. You must change your life.]
R. M. Rilke, *"Archaischer Torso Apollos"*

Pasteur's was the most enviable life I had yet encountered. It was his privilege to do things until they were done.
A. Dillard, *An American Childhood*

We are able to witness achievements in the arts and everything else, not because of those who adhere to the established order, but because of the innovators, who dare to change or move things that need change or correction.
Isocrates, *Evagore*

It has been our pleasure, as well as that of our colleagues, to witness changes in the art of electronic testing. We know that it is not given to everyone to witness and participate in such things. We know we have been party to a rare experience -- something bound to be matched by few other experiences in our professional lives. We witnessed and served during the birth and development of an international standard for testing -- IEEE Std 1149.1. We had the opportunity to work with a set of international, expert volunteers on a critical task, and, like Pasteur, to work on our job until it had been completed -- until a standard was successfully described and promulgated.

There is no question that the situation in electronic testing is in need of change; indeed, significant change is inevitable whether or not it is promulgated with an accompanying, technically sound, supportive test technology. We believe that such change is being forced on us today caused (at least in large part) by the following factors:

- the constant pressures for greater integration;
- the widespread adoption of surface mount technology (SMT) employed on both one-sided and two-sided printed wiring boards (PWBs);
- the shrinking feature sizes of these PWBs;
- the decreasing distances between pins of SMT devices;
- the consequent difficulties of continuing to test PWBs via physical contact by spring-loaded nails;

- the growing gap in speed between product and automatic test equipment (ATE);
- the increasing cost of acquiring capital equipment such as ATE and the increasing cost of developing associated test fixtures;
- the significant difficulty of rapidly developing accurate, automated diagnoses for loaded boards and systems;
- the desire to have a test methodology compatible with assembly processes that are rapidly reconfigurable through software and aimed at lot-size-of-one manufacturing;
- the continuing, if not increasing, consumer demand for high reliability and maintainability; and
- the need for generic solutions that can be repeatedly reused in a variety of digital products.

The engineers who formed the ad-hoc group known as the Joint Test Action Group (JTAG) were all aware of the impact on product quality that would arise if solutions to these needs were not found.

The continuous process of increasingly greater integration never gives a process engineer time to congratulate himself/herself on nearing perfection before process and product change again. We cannot expect perfect processes -- process engineers have to expect to be working on process improvement and alteration throughout their careers. What is the best source of guidance to them? It is the carefully analyzed results of testing. The widely accepted concept of quality through continuous improvement is not possible if one cannot assess causes of failure and their frequency. The results of testing and subsequent failure mode analysis form a treasure trove for the engineer concerned with quality. When testing is threatened, quality is threatened. And this is a threat to an electronics firm's ability to manufacture -- a threat to its existence.

IEEE Std 1149.1 was not only developed to contain testing costs. Basically, it was developed because the ability to perform tests and to learn from test results was perceived to be under dire threat.

Once IEEE Std 1149.1 was well along the way in its development, it became clear that application notes and other supportive information that could not be properly considered part of the standard document were going to be needed. More than a dozen application notes were sketched to some degree of completion by members of JTAG, but there was a need for much more than that. There was a need to provide a teaching vehicle that would provide motivation, history, and theory as well as application suggestions and records of successful use. The result of evaluation of these needs is this book.

The book is composed of five parts:

I. *"Background."* Chapter 1 describes the situation giving rise to the development of the *IEEE Standard Test Access Port and Boundary–Scan Architecture*. Chapter 2 introduces the boundary–scan technique and shows how it can provide a solution to the problems identified in Chapter 1. The technology that was available in the literature when JTAG first set to work and the steps in the development of the standard are reviewed in Chapter 3.

II. *"Tutorial."* Chapters 4 to 6 contain a tutorial introduction to the circuitry defined by the standard.

III. *"Applications to Loaded Board Testing."* In Parts III and IV, we have gathered material for this tutorial book especially written or rewritten by authors from companies that contributed to the creation of the standard. The chapters in Part III discuss the application of IEEE Std 1149.1 to the testing of loaded boards -- that is, applications in the problem area originally targeted by JTAG. The topics discussed in this part of the book include:

- the structure of a typical board test program;
- testing and diagnosis of the standardized test logic; and
- the testing of boards containing components that are incompatible with IEEE Std 1149.1.

IV. *"Implementation Examples and Further Applications."* In Part IV, we discuss the implementation of IEEE Std 1149.1 and give a view of the range of applications of the standard beyond board testing. A sampling of topics includes:

- silicon implementations and related costs;
- interfacing to scan design and built–in self–test;
- analog and mixed–signal applications;
- applications to systems debugging and emulation; and
- testing throughout the assembly hierarchy: integrated circuit (IC) to system.

V. *"Bibliography and Reprints."* The final part of the book contains an extensive annotated bibliography and reprints of selected papers. The papers selected describe key steps in the development of boundary–scan prior to IEEE Std 1149.1 and continue the discussion of applications for the standard.

IEEE Std 1149.1 was developed for your use. As more engineers and more firms use it, it will become more valuable. The more expertise in ATEs, circuit design, catalog ICs, application–specific ICs (ASICs), etc. that is developed collectively, the more we all can benefit from reuse of generic solutions to common technological problems. As Harry Bleeker did in his foreword to this book, we urge you to use the standard, we urge you to participate in its further evolution, and we urge you to do so in the superbly constructive and cooperative spirit that has infused JTAG.

Guard the mysteries.
Constantly reveal them.
L. Welch, *"Course College Courses: Religion"*

Colin M. Maunder
British Telecom Research Labs
Design Technology Division
Martlesham Heath, Ipswich, UK

Rodham E. Tulloss
AT&T Bell Labs
Engineering Research Center
Princeton, New Jersey, U.S.A.

Part I: Background

In Part I, the trends in product and test technology that motivated the development of IEEE Std 1149.1 are discussed and the concept of standardized test-support features at the integrated circuit and loaded printed wiring board levels is introduced.

Chapter 1 outlines "traditional" test techniques for loaded boards and examines the effects of trends in design-for-test and circuit miniaturization. Readers familiar with this material may want to move to Chapter 2 where the boundary-scan technique is introduced. Chapter 3 concludes Part I with an overview of the work of the Joint Test Action Group and, subsequently, of the IEEE P1149.1 Working Group.

Chapter 1. Test Technology Prior to IEEE Std 1149.1

1.1: Test Technology for Loaded Boards

Over the years, the automatic test equipment (ATE) used to test electronic products has evolved to cope with continued increases both in the number of integrated circuit packages used on, say, a printed wiring board (PWB) and in the complexity of the integrated circuits (ICs) themselves. Typically, manufacturers of loaded boards will use high pin count in-circuit and functional† board test systems, either separately or in sequence, to detect defects and to enable high quality levels to be achieved in shipped products.

Using the in-circuit test technique, tests are applied directly to individual components by backdriving their connections from other devices in the product. The objective is to apply an appropriate test sequence for the component type regardless of the environment in which it is used. Direct access is made to the component's outputs to monitor the test results, enabling the function of each component in the circuit and the interconnections between the various components to be checked. This method reduces the expense of test development for each circuit design since, as long as an IC's functionality is not modified by externally wired connections (e.g., by direct connection to power or ground), the same test can be applied irrespective of where the IC is used. Clearly, the process requires extensive access to the circuit, because every connection must be driven and monitored directly to apply the tests to the individual components. This access is provided through a bed-of-nails interface in which spring-loaded probes are used to make contact with the interconnections on the PWB (Figure 1-1).

In the functional test technique, the principal interface for applying test stimuli and for observing circuit responses is that provided by the board's normal terminations -- for example, the edge connector (Figure 1-2). Access may also be made to connections internal to the loaded board, but this is on a more limited scale than that required by an in-circuit test system; frequently such access is limited to monitoring, rather than to driving, the connection. In contrast with in-circuit testing, the functional test technique is able to confirm that the various components used to construct the product interact correctly and that the overall required function is achieved. In the process, the correctness of both the components in the circuit and their interconnections is verified. However, the achievement of a thorough test is a difficult task since tests must be generated separately for each board design. This task can be both time-consuming and extremely expensive, sometimes prohibitively so [1].

† The term "functional" is used to describe test systems that do not require the use of backdriving. This includes edge-to-edge functional testing, structural testing, or a modular test approach.

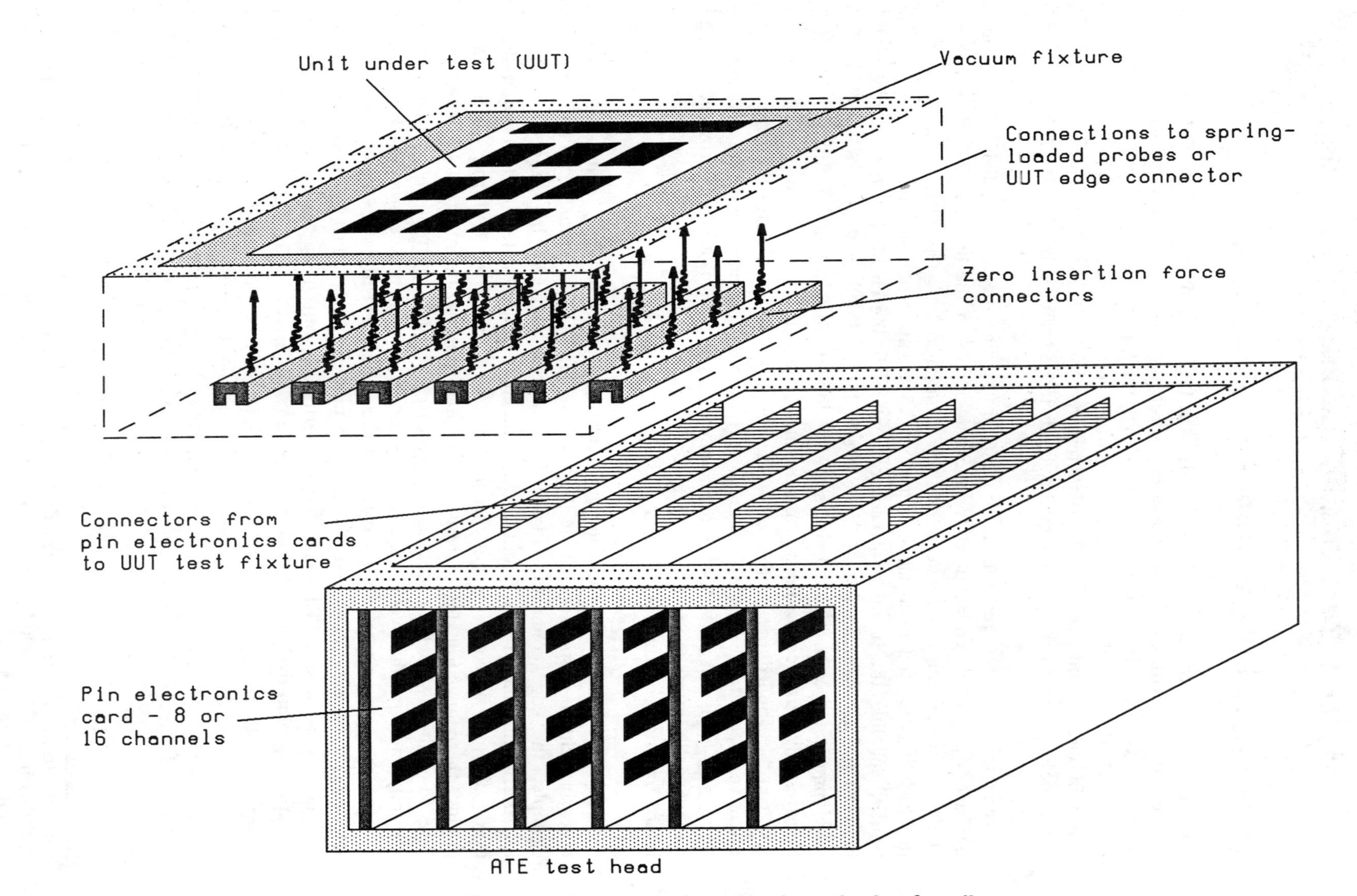

Figure 1–1: In–circuit test using a bed–of–nails.

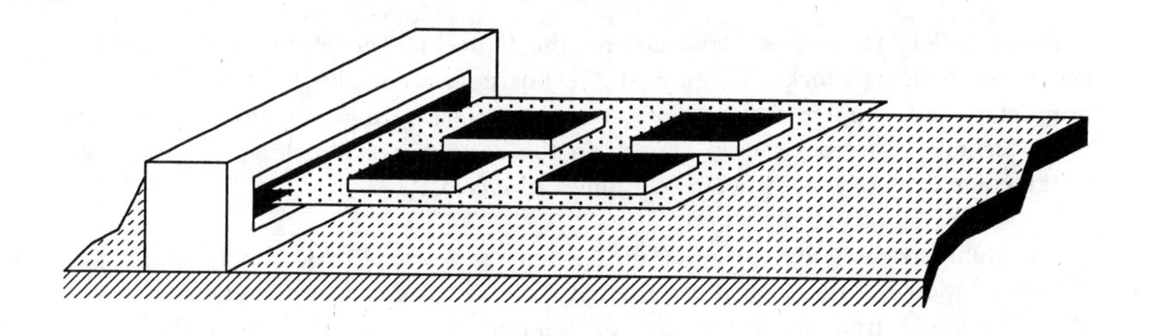

Figure 1–2: Functional test using the board connector.

Due to the differences in operation and failure detection capability between the in–circuit and functional test techniques, a common approach is to use the two techniques in sequence to achieve a high–quality test (Figure 1–3). Initial product screening is performed by using an in–circuit test system since this is able to rapidly detect and diagnose the most common failures in newly assembled boards -- for example, those due to soldering errors or to incorrect or wrongly–inserted components [2]. Once a loaded board has passed the screening test, it is passed forward to a functional test system where checks are made for more complex (and less frequent) failures caused by faulty interaction between components. To allow the mix between the two test techniques to be more easily optimized for a given product, test equipment that supports both techniques within a single system has more recently become available.

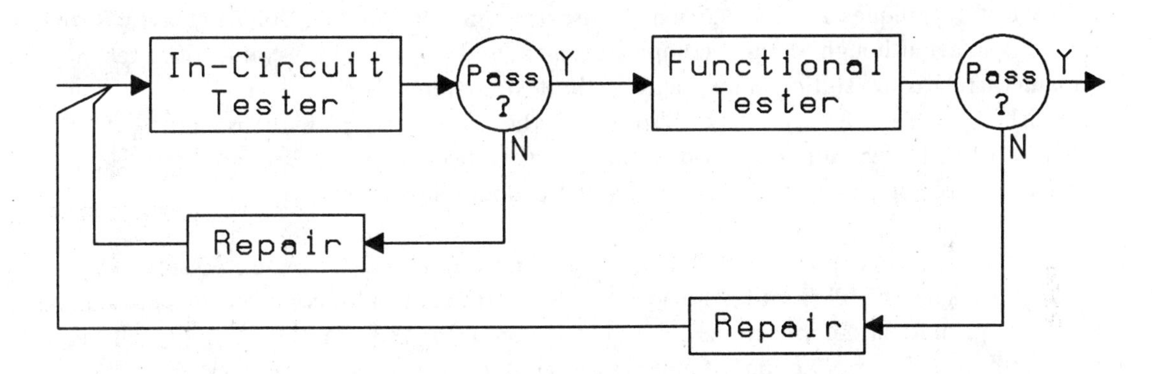

Figure 1–3: Sequential use of in–circuit and functional test.

1.2: Trends in Design–for–Testability

We have already remarked on the expense of generating tests, especially for use on functional test systems. Because of this, the past decade has seen the development of many circuit structures and design techniques that can be used to improve the testability of digital circuits, thus reducing the cost of the various test tasks [3]. Perhaps the most prominent among these have been scan–design [4], data generation and compaction circuits based on linear–feedback shift–registers (e.g., [5,6]), and the development of built–in self-test (BIST) techniques based around the built-in logic block observer (BILBO) [7] and other building blocks. For use at the board level, families of components offering proprietary test-support features have become available (e.g., [8, 9]). Today, these and other techniques are being used to restrain the costs of test development and application as the complexity of loaded boards continues to increase.

Unfortunately, however, many of the techniques are applicable only in situations where an organization has the capability to adopt a consistent design-for-testability approach at all design levels, from IC to system. For example, the scan-design technique can be used at the board level if a complete set of scan-design components is available from which the board can be constructed. Typically, however, board designs are constructed from catalog ICs obtained from a variety of sources on the open market. Almost without exception, these ICs fail to offer the board designer facilities that will assist him to develop tests for his circuit.

For this reason design-for-testability at the board level has been a less structured activity than for many complex IC designs. Also, there has been less need for structured approaches because of the extensive use of in-circuit test techniques in industry. At the board level, the ease of making access through a bed-of-nails fixture has obviated or, at least, has significantly delayed the need for more structured techniques to gain access to circuit nodes such as those essential in IC testing.

1.3: The Effect of Miniaturization

The test techniques for loaded boards discussed in Section 1.1 evolved steadily during a period when, although circuit complexity increased rapidly, interconnection technology remained relatively static. Thus, automatic test systems began to rely heavily on the established dual-in-line package (DIP) and the associated plated-through-hole (PTH) PWB. DIP/PTH technology provided the extensive access to component interconnections needed by bed-of-nails fixtures or for guided probing during fault diagnosis.

Recently, however, there has been an increasing trend toward the use of surface-mount package designs and PWBs that no longer rely on through-hole connections between their layers of printed interconnections [9] and it is now clear that such technology will become the norm for the production of complex digital boards in the 1990s. This trend is the result of a number of factors, among them the need to produce packages that can accommodate high pin-count components and the pressures for continued product miniaturization.

The new interconnection technology has had a considerable impact on current loaded-board test techniques primarily due to reduced pin spacings on packages, the fact that package pins may no longer be directly accessible during test, and the increased density with which packages may be mounted onto the host PWB. As a result, the cost of bed-of-nails fixtures for surface-mount boards is high and probing of component interconnections can be impossible where components are densely packed. Further, the test heads of current board test systems are optimized toward an assumed even distribution of test contacts over a large area on one side of a board as required for DIP/PTH technology, whereas surface-mount products may require contact to both sides through "toaster" and "clam-shell" style fixtures (e.g., as described in [11]).

So far, the established test techniques have succeeded in meeting the challenges of surface-mount technology (SMT). For example, test fixtures can be constructed to permit the use of in-circuit test techniques for surface-mount boards if care is taken in the design of the PWB artwork and if components are spaced sufficiently apart from one another [12]. However, such fixtures are extremely expensive and the technical problems in producing them are increasing as SMT continues to develop. Further, the need to design the loaded boards so that they can be probed acts against the area-conserving thrust of SMT.

Looking to the future, the lack of a test methodology that can be applied cost-effectively to products formed by surface-mount interconnection of complex functions will be a major obstacle to the adoption of the very high-density packaging techniques currently under development. Examples of high-density packaging techniques include silicon-on-silicon and direct-chip-attach (e.g., [13-16]).

1.4: The Need for a New Approach

To summarize, two key factors are having an increasingly adverse effect on the cost of testing loaded boards.

First, the ICs used in such products are becoming increasingly complex and this complexity contributes significantly to the difficulty in testing the loaded board. Generally, ICs available on the open market do not offer the test support facilities that the board producer needs, although some do contain design features (such as self-test capability) that could be of considerable interest to the purchaser (e.g., [17,18]). For in-circuit testing, it is difficult to perform a comprehensive test of the function of a complex IC due to the need to keep the test length sufficiently short that surrounding ICs will not be damaged by the backdriving techniques employed during test application [19]. Also, board wiring may tie together component inputs -- severely restricting the set of usable test patterns.

Second, increasing use is being made of surface-mount interconnection technology, where access to connections is considerably more limited than for the established dual-in-line technology. It is clear that existing test techniques -- particularly in-circuit test -- will be faced with increasing difficulties as this technology continues to develop. In effect, use of SMT is increasing the similarity between ICs and loaded boards from the test viewpoint; access to connections is becoming increasingly difficult to achieve. Therefore, loaded-board testing must be done increasingly through the normal input and output connections in a "functional" manner as is already the case for ICs.

While the functional test technique is better able to cope with the results of advanced surface-mount technology, the technique carries the penalty of requiring generation of comprehensive test programs for each separate design. This process is extremely expensive for complex boards due to the complexity of controlling and observing individual components through the others on the board. For ICs, where functional testing is the only technique that can be used, structured design-for-test techniques (e.g., scan design, BIST) are often used to make all parts of a design sufficiently testable by improving either controllability or observability or both at critical circuit connections.

Arguably, therefore, the way forward is to use a structured technique similar to scan design or self-test at the board level, rather than through continued evolution of existing board test techniques. As we will see in Chapter 2, a version of scan design called boundary-scan provides the functionality that would be required. In fact, the boundary-scan technique has been used for some time by several companies to solve precisely the problems highlighted in this chapter. The advantage that these companies had was that they designed their boards entirely (or predominantly) from application-specific ICs (ASICs) developed to their own requirements. They were able to design features into these ASICs to help solve their board test problems.

The opportunity of being able to design every board entirely out of ASICs is, however, comparatively rare. In most companies, boards are designed primarily using the off-the-shelf ICs advertised in vendor catalogs. Therefore, most companies will be able to move to use of structured board-level design-for-test techniques only when both ASICs *and* off-the-shelf ICs include the facilities that this requires. Further, it is essential that ICs offered by different manufacturers can interact with each other appropriately and predictably during the testing of the loaded board.

A widely-supported standard is therefore essential if the electronics industry at large is to make progress in solving the increasing test problems that it faces. The prime objective of IEEE Std 1149.1 is to meet this requirement.

1.5: References

[1] P. Goel, "Test Costs Analysis and Projections," *IEEE Design Automation Conference Proceedings*, IEEE Computer Society Press, Los Alamitos, Calif., 1980, pp. 77-82.

[2] Factron Schlumberger, *The Primer of High-Performance In-Circuit Testing*, Factron Schlumberger, Wimborne, Dorset, UK, 1985.

[3] T.W. Williams and K.P. Parker, "Design for Testability -- A Survey," *IEEE Transactions on Computers*, Vol. C-31, No. 1, January 1989, pp. 2-15.

[4] E.B. Eichelberger and T.W. Williams, "A Logic Design Structure for LSI Testability," *Journal of Design Automation and Fault-Tolerant Computing*, Vol. 2, No. 2, May 1978, pp. 165-178.

[5] R.A. Frohwerk, "Signature Analysis: A New Digital Field Service Method," *Hewlett Packard Journal*, Vol. 28, No. 9, May 1977, pp. 2-8.

[6] P.H. Bardell and W.H. McAnney, "Self-Testing of Multi-Chip Logic Modules," *IEEE International Test Conference Proceedings*, IEEE Computer Society Press, Los Alamitos, Calif., 1982, pp. 200-204.

[7] B. Konemann et al., "Built-In Logic Block Observation Techniques," *IEEE Test Conference Proceedings*, IEEE Computer Society Press, Los Alamitos, Calif., 1979, pp. 37-41.

[8] Advanced Micro Devices Inc., *On-Chip Diagnostics Handbook*, Advanced Micro Devices Inc., Sunnyvale, Calif., 1985.

[9] J. Turino, "Enhancing Built-In Test on SMT Boards," *Evaluation Engineering*, June 1985.

[10] C. Maunder, D. Roberts, and N. Sinnadurai, "Chip Carrier Based Systems and Their Testability," *Hybrid Circuits*, No. 5, Autumn 1984, pp. 29-36.

[11] R.N. Barnes, "Fixturing for Surface-Mounted Devices," *IEEE International Test Conference Proceedings*, IEEE Computer Society Press, Los Alamitos, Calif., 1983, pp. 72-74.

[12] M. Bullock, "Designing SMT Boards for In-Circuit Testability," *IEEE International Test Conference Proceedings*, IEEE Computer Society Press, Los Alamitos, Calif., 1987, pp. 602-613.

[13] R. Keeler, "Chip-on-Board Alters the Landscape of a PC Board," *Electronic Packaging and Production*, Vol. 25, July 7th 1985, pp. 62-67.

[14] G.L. Ginsberg, "Chip and Wire Technology: The Ultimate in Surface Mounting," *Electronic Packaging and Production*, Vol. 25, August 8th 1985, pp. 78-83.

[15] K. Gilleo, "Direct Chip Interconnect Using Polymer Bonding," *Electronic Components Conference Proceedings*, IEEE, New York, 1989, pp. 37-44.

[16] C.J. Bartlett, J.M. Segelken, and N.A. Teneketges, "Multichip Packaging Design for VLSI Based Systems," *Electronics Components Conference Proceedings*, IEEE, New York, 1987, pp. 518-525.

[17] J.R. Kuban and J.E. Salick, "Testing Approaches in the MC68020," *VLSI Design*, Vol. 5, No. 11, 1984, pp. 22-30.

[18] P.P. Gelsinger, "Design and Test of the 80386," *IEEE Design and Test of Computers*, June 1987, pp. 42-50.

[19] L.J. Sobotka, "The Effects of Backdriving Digital Integrated Circuits During In-Circuit Testing," *IEEE International Test Conference Proceedings*, IEEE Computer Society Press, Los Alamitos, Calif., 1982, pp. 269-286.

Chapter 2. An Introduction to Boundary-Scan

This chapter provides an introduction to boundary-scan. It shows how the technique can provide an answer to the problems identified in Chapter 1.

2.1: Scan Testing at the Board Level

At the chip level, the scan-design technique can be used to guarantee testability and to permit use of automatic test pattern generation (ATPG) tools [1]. Many companies have used the technique, some to the extent that every chip on a board is scan testable. In such cases, the board can be made scan testable by daisy-chain interconnection of the scan paths in the individual integrated circuits (ICs) (Figure 2-1).

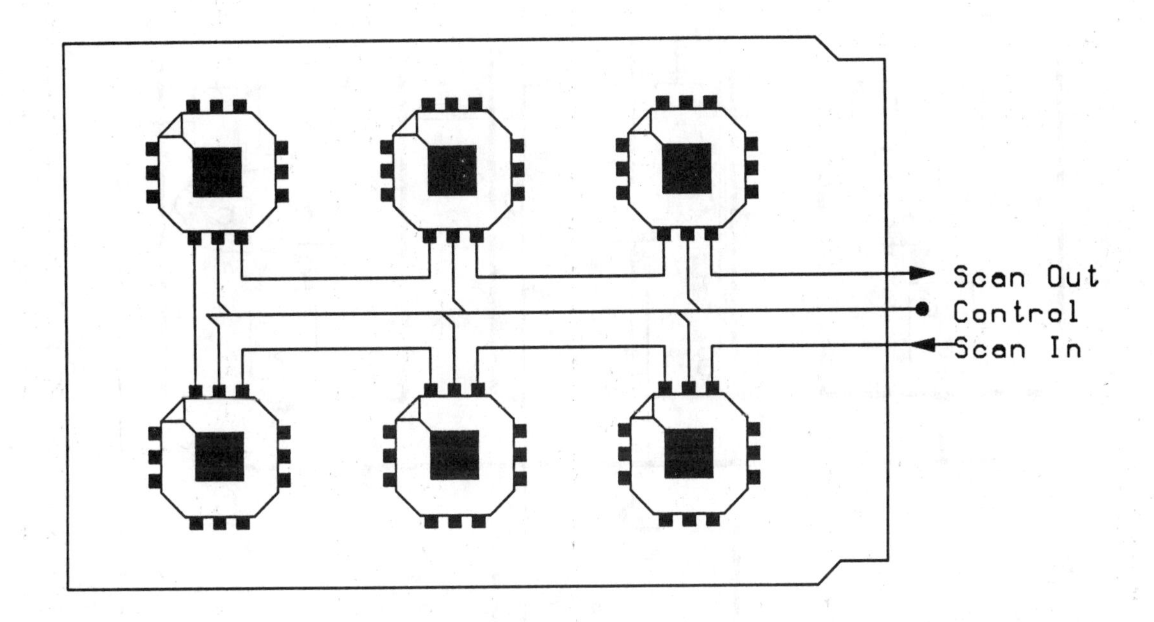

Figure 2-1: Scan design at the board level.

In these cases, the board design has the same structure as each individual chip -- it is formed from a combinational logic block and one or more shift-register paths. Test generation for the loaded board can, in principle, be approached in exactly the same way as for a chip.

Several problems arise, however. First, the combinational logic block will be many times larger than for any individual chip. This will result in increased test generation costs and, as the complexity of the product increases, may cause the capacity of the ATPG software to be exceeded. Second, test generation costs for the loaded board would be considerably reduced if the tests created for stand-alone chip testing could be reused in the board environment. Unfortunately, however, this is far from straightforward. Whereas for the chip a test may require that a logic 1 is applied at a package pin, at the board level this logic value must be applied by scanning appropriate patterns into the chips that drive the signal. Consider, for example, the process of applying a logic 1 to the input of chip D in Figure 2-2. This requires that the board-level bus is set to 1, which can be achieved by setting the output of one of the driving chips (A, B, or C) to 1 while the others are set to high-impedance. These conditions can be achieved by shifting appropriate 1s and 0s into the scan paths of chips A, B, and C. The precise patterns to be shifted in can be determined by analysis of the combinational logic network that controls the output of each IC.

Similar problems arise when trying to observe an output from one IC on the board. When the IC is tested using automatic test equipment (ATE), the output can be observed directly; however, to apply the same test when the chip is on the board the output signal must be observed using the scan paths in the ICs that receive the signal. This requires that a path is set up through the combinational logic between the inputs of the receiving ICs and their internal scan paths, such that the signal can be examined by loading the scan path and then scanning the contents out of the board to the ATE.

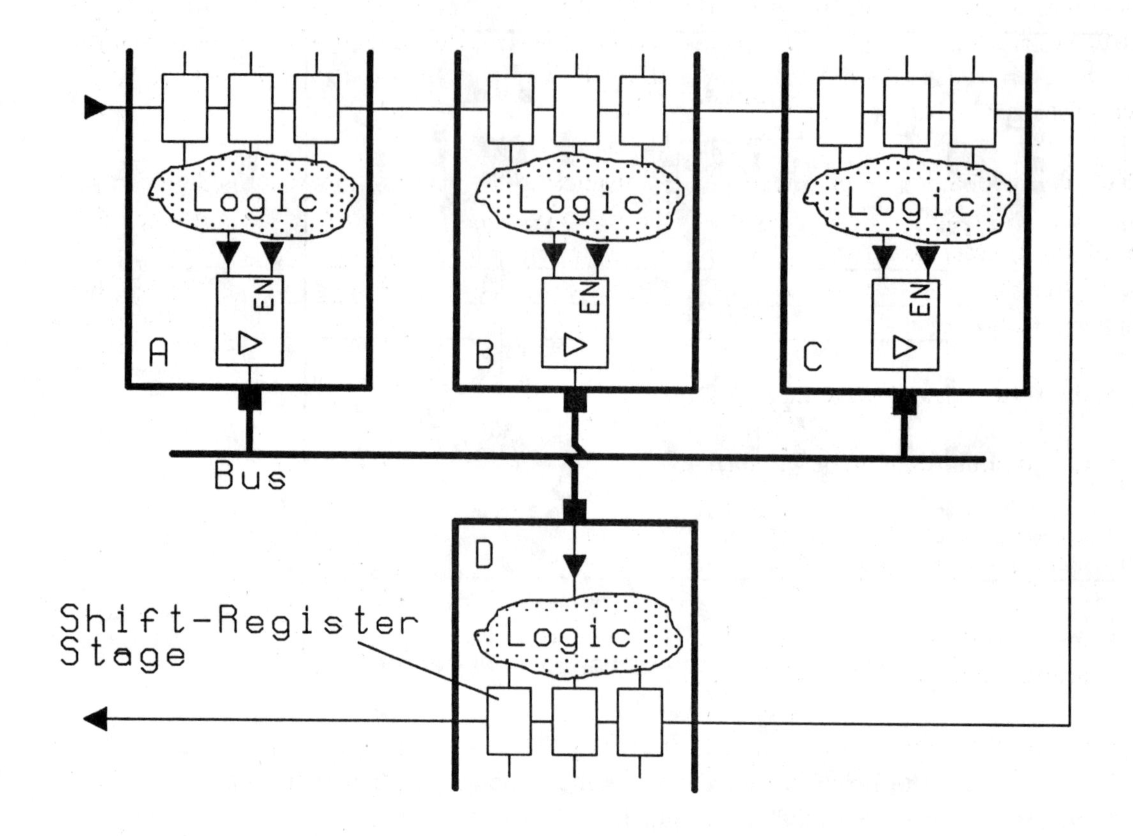

Figure 2-2: Testing a board-level bus by using embedded scan paths.

Consider, for example, the case where a test of component A in Figure 2-2 produces a 0 at its output. The following actions are necessary to allow the board-level ATE to check this result:

1. All other ICs that can drive the bus must have their outputs set to high-impedance. In this case, this is achieved by controlling the scan paths in components B and C.

2. The signal received at the input to component D must be observed using the component's scan path. The combinational logic between the input and the scan path must therefore be controlled such that a change at the input from the bus produces a corresponding change in the value loaded into one or more scan path stages (i.e., so that a sensitive path is set up between the input pin and the scan path). This may require conditions to be established at the outputs of other components on the board or in the scan path within component D.

The consequence of the problems just discussed is that a chip test cannot be used directly once the IC has been assembled onto a board. A significant amount of computation is required to compute the values that must be scanned into and out of the other ICs on the board to apply the test. As shown in Figure 2–2, there may be a significant amount of logic in the combinational logic networks that drive or are driven by any chip pin, particularly where board–level bus structures are involved, such as on a microprocessor board.

The final problem is a practical one -- diagnosing the cause of any failures detected so that repair is possible. That is, what is the cause of a particular error in the data scanned out of the board? Referring to Figure 2–2 again, if an incorrect value is scanned out of one of the shift–register stages in chip D, this could be caused by a fault in one of the following locations:

- the shift–register stage in chip D;
- the combinational logic in chip D;
- the chip–to–board connections of one of the chips connected to the bus (e.g., an open–circuit joint);
- the board level bus (e.g., a short–circuit to another signal or a broken printed–circuit track);
- the combinational logic in a chip that drives the bus (A, B, or C); or
- one of the shift–register stages in a chip that drives the bus.

A more accurate diagnosis can be achieved only by careful analysis of the data scanned out of chip D in response to a number of tests. In some cases, further tests may need to be generated to achieve an acceptable level of diagnostic accuracy.

2.2: The Value of Boundary-Scan

The problems reviewed above can be overcome by placing a scan shift-register stage adjacent to every input or output pin of each chip -- that is, at the component boundaries. To achieve this, specialized test circuitry may need to be added to an IC design between the pin and the logic to which it is connected, as shown in Figure 2-3. These test circuits, called boundary-scan cells, are connected into a shift-register path around the periphery of the IC. This is called the boundary-scan path.

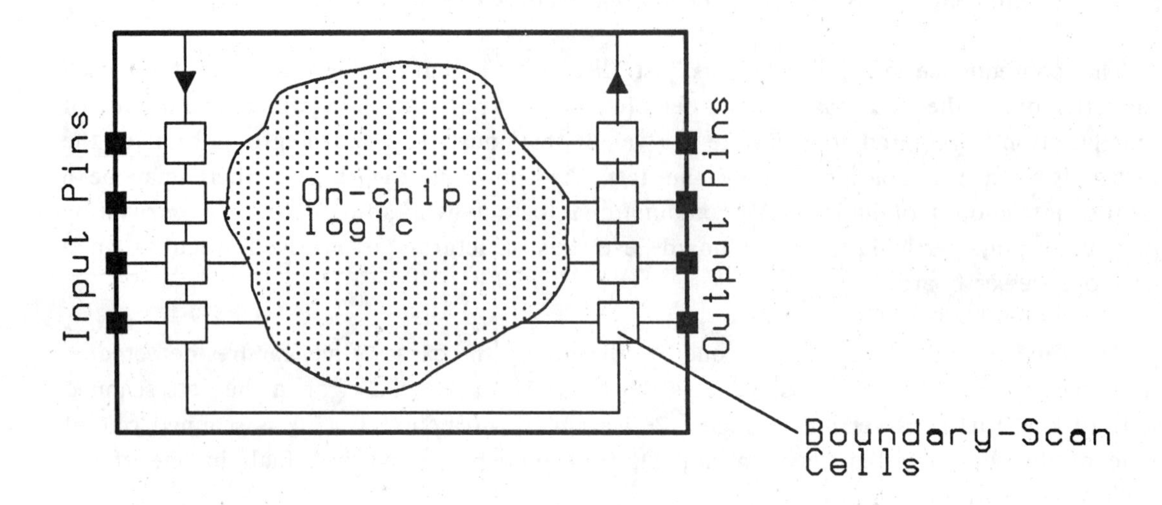

Figure 2-3: Inclusion of boundary-scan cells in an IC.

An example design for a boundary-scan cell is shown in Figure 2-4. Note that the boundary-scan cells defined by IEEE Std 1149.1 are more complex than the cell shown here. The simplified cell designs used in this chapter only illustrate the processes involved in boundary-scan testing. (Chapter 6 describes the constraints within which boundary-scan cells compatible with the standard must be designed.) Note also that throughout this book (as in IEEE Std 1149.1), signal names that end in an asterisk (e.g., Load*) are active-low, while others (e.g., Shift) are active-high.

Data can flow directly through the boundary-scan cell (from Data_In to Data_Out) when normal operation of the component is required. During testing, the cells at output pins can be used to drive signal values onto the external network, while those at the input pins can capture the signals received.

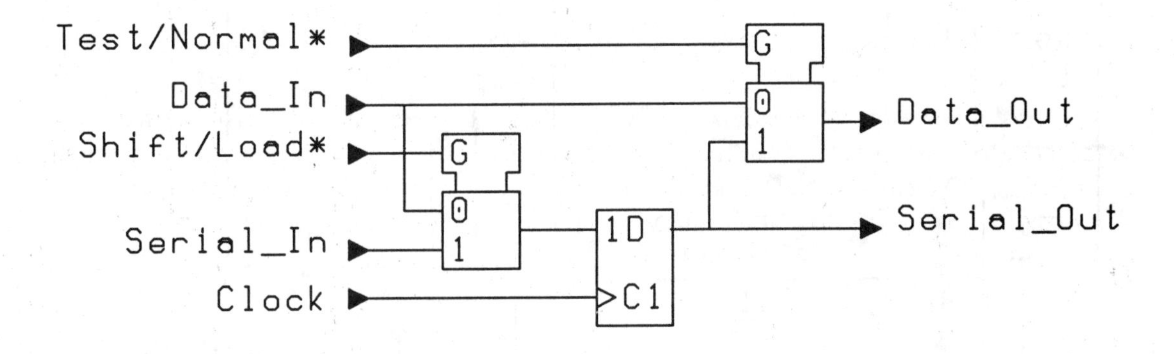

Figure 2–4: A basic boundary–scan cell.

2.3: Testing a Board with Boundary–Scan

With boundary–scan cells of the form shown in Figure 2–4, testing could proceed in two stages:

1. *Testing interconnections between chips:* Test patterns are shifted into the boundary–scan cells at component output pins and driven onto the board–level interconnections by setting their Test/Normal* inputs to 1. The responses that arrive at chip input pins are loaded into their boundary–scan cells (while Shift/Load* is 0) and shifted out for examination (while Shift/Load* is 1). By careful selection of the test patterns, the interconnections can be tested for stuck–at, short–circuit, open–circuit, and other fault types. Figure 2–5 shows a circuit that contains a short–to–ground (stuck–at–0) fault and a wire–OR short–circuit fault in the board interconnect (e.g., a solder bridge). Table 2–1 shows some test vectors for these faults.

2. *Testing the chip:* Figure 2–6 shows a simple IC that contains a NAND gate. To apply tests to this gate, the Test/Normal* control signals for the cells at input pins (i.e., pins that drive into the on–chip logic) would be held at 1, while those at output pins are held at 0. Test vectors are shifted into the boundary–scan path and applied to the gate. The result is then loaded into the cell at the output pin (Shift/Load* = 0) and shifted out for examination (Shift/Load* = 1). For the NAND gate, test vectors would be as shown in Table 2–2.

If the target chip is scan testable, then operation of its internal scan path can be synchronized to that of the surrounding boundary–scan path during application of the chip test. Note that, in contrast to the situation without boundary–scan, the process for converting the stand–alone chip tests into a test that can be used on the loaded board is simple. It requires only that the correct sequences of 1s and 0s are scanned through the boundary–scan path.

Because the board–level interconnections can be tested independently of the circuitry within any chip, the problem of fault diagnosis is eased considerably .

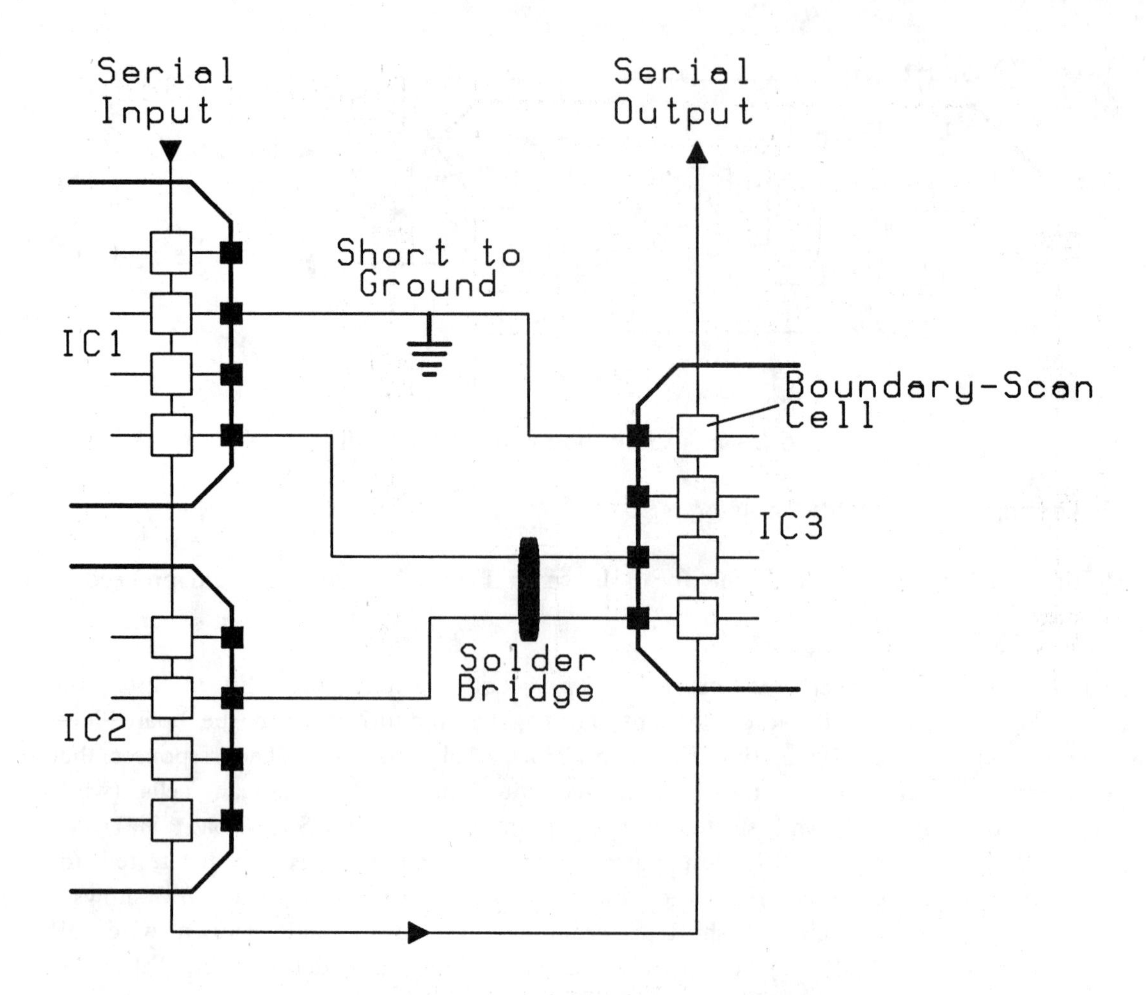

Figure 2–5: Testing for interconnect faults.

Table 2–1: Example tests for interconnect faults.

Input	Output	
	Expected	Actual
x1x1x0xxxxxx	xxxxxxxx01x1	xxxxxxxx**1**1x**0**
x0x0x1xxxxxx	xxxxxxxx10x0	xxxxxxxx1**1**x0

NOTE: The rightmost bit of the above data values is shifted into the serial input, or out of the serial output, first. Bold type is used to highlight the output data bits that are changed by the faults.

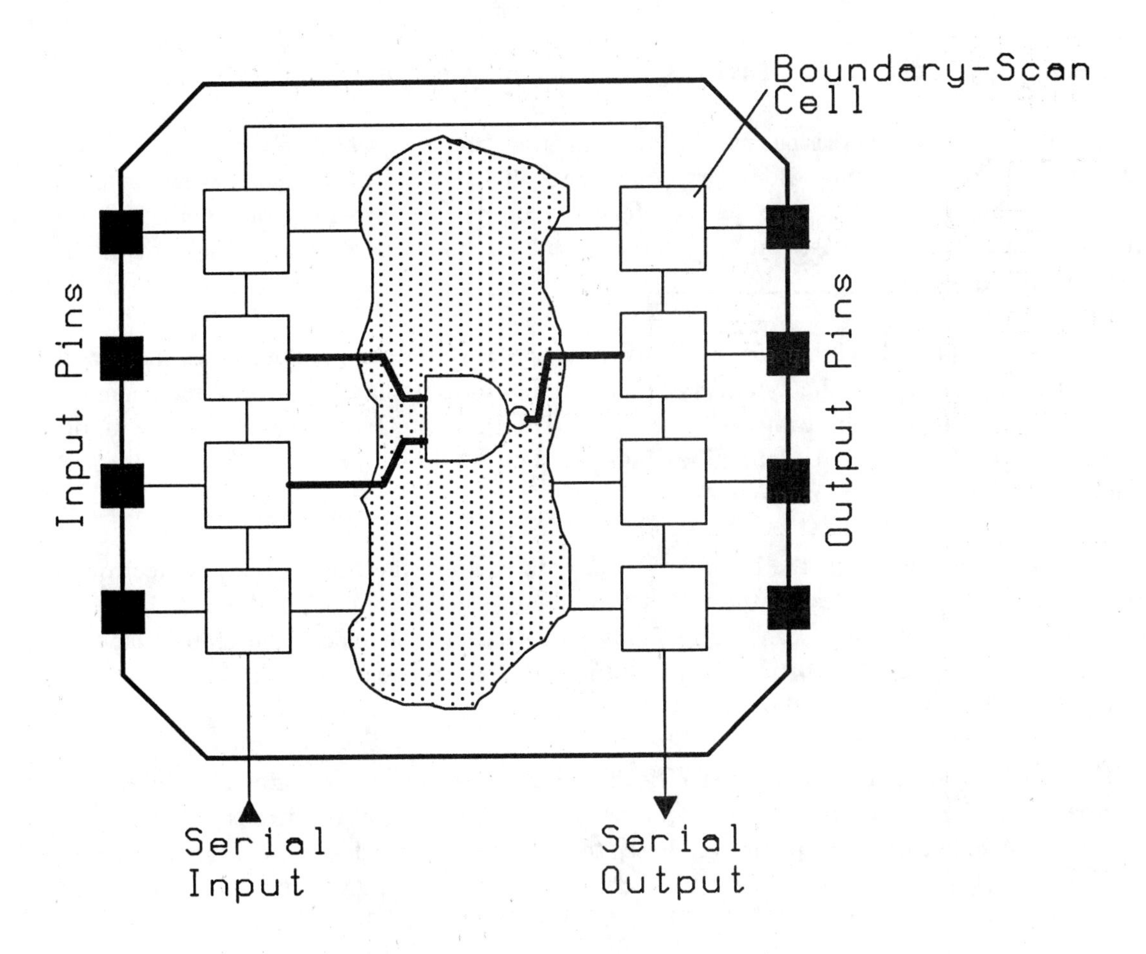

Figure 2–6: Testing on–chip logic.

Table 2–2: Example tests for the NAND gate.

Input	Expected Output
x10xxxxx	xxxxx1xx
x01xxxxx	xxxxx1xx
x11xxxxx	xxxxx0xx

NOTE: The rightmost bit of the above data values is shifted into the serial input, or out of the serial output, first.

2.4: Boundary–Scan for ICs That Are Not Themselves Scannable

There will be an increasing number of ICs that are *themselves* too complex to test efficiently via scan testing. Boundary–scan can still handle interconnect testing in such situations, but a different chip test strategy is required -- hopefully one that produces tests that are able to be used at chip, board, and system levels of assembly, both in the factory and in the field.

Testing of board–level interconnections can proceed in exactly the same manner as previously described -- the boundary–scan cells at output pins apply the test stimulus, while those at input pins capture the results. To allow the boundary–scan cells at the component's output pins to determine the signals driven from the IC, the Test/Normal* controls for those cells are set to 1.†

For the test of the chip, the boundary–scan path assumes the role of the pin electronics on a chip tester. Each test pattern that would have been applied to the IC's inputs is shifted into the boundary–scan path. When the pattern is in place, the chip is clocked once. The test response is then captured into the boundary–scan cells at the IC's output pins and shifted out for examination.

There is a problem that may be significant. The test is applied at a greatly reduced rate compared to the stand–alone chip test because of the need to shift patterns and responses through the boundary–scan paths. At best, the speed will be reduced by a factor close to the number of non–test signal pins on the chip under test; typically, the speed will be tens or hundreds of times slower than the maximum possible during chip testing.

This significant reduction in test application rate can make it impossible to test certain types of logic. Even when a test is possible using this approach, the test length may be undesirable. Consider the following cases:

1. *An IC that does not contain dynamic logic:* ‡ In this case, a slow–speed static test can be applied. Static faults (e.g., stuck–at faults or short–circuits) will be detected, while other faults that require "at–speed" testing will not be found. As already mentioned, the run time for a high coverage test may be significant for a complex, high pin–count chip and, as a result, the amount of testing that can be achieved economically may be limited. In practice, it may only be possible to apply an "are–you–alive?" test of limited fault coverage.

---

† Note that it may be advisable to ensure that the signals arriving at the IC's input pins during the test do not place the on–chip system logic in a state where damage to the IC might occur. For example, if inputs to the on–chip logic are set to conditions that would not arise normally, several drivers within the IC may be enabled simultaneously onto a single bus. Such problems can be avoided in a number of ways, for example by disabling the clock.

‡ Dynamic logic circuits contain stored–state logic elements (e.g., latches, flip–flops, etc.) that do not hold their state indefinitely. Typically, a clock must be applied at a specified minimum frequency to prevent the stored–state elements from "forgetting" their state.

2. *A self–testing IC:* Here, the surrounding boundary–scan paths can be used to trigger execution of the self–test, apply any required starting patterns at the chip's input pins, and examine the test results. The chip is tested to the same extent as when the self–test is executed; there need be no reduction in test quality. The relatively low test throughput of the boundary–scan path is not a problem because data are shifted only at the beginning and end of the test.

   As we will see from the application examples in Parts III and IV of this book, boundary–scan and self–test together provide an excellent solution to chip and loaded–board testing. The boundary–scan path isolates the on–chip logic from neighboring ICs while the self–test runs in addition to allowing chip–to–chip interconnections to be tested. Self–test can provide a high–quality test of the on–chip system logic. A standard boundary–scan architecture and protocol provides a gateway to reusable self–test and, by providing added leverage for system–house purchasers, encourages development and use of self–test technology by IC suppliers.

3. *An IC that contains dynamic logic:* Due to the low test application rate, it is not practical to use the boundary–scan path to test a chip that contains dynamic circuitry unless self–test features are available. The operation of dynamic circuitry depends on the ability to store a charge on internal chip connections. After a relatively limited period of time, this charge will decay, resulting in incorrect operation of the component. Therefore, a minimum clock rate is generally specified for dynamic logic circuits, and it might be impossible to achieve this clock rate where test patterns are being shifted in and out by using the boundary–scan path. An exception would be where the chip could be placed in a "hold" mode while each test was shifted such that the clock could continue to be applied to the on–chip logic without changing its state.

In summary, boundary–scan can be used to test board interconnections whether or not the chips themselves are designed to be scan testable. Self–test and scan testable ICs can be tested on the board by using their boundary–scan paths just as effectively as they can be stand–alone tested. Without scan or self–test, some limited tests can be performed on static logic designs, but, in such cases, on–board testing of dynamic logic might be impractical.

2.5: Boundary–Scan Compared to In–Circuit and Functional Test

As was discussed in Chapter 1, the motivation for producing a standard form of boundary–scan was to address the problems of increasing IC complexity and of reducing product size. We have seen that boundary–scan techniques can be used to apply tests to digital circuit boards without the necessity of extensive physical access (e.g., using a bed–of–nails), but how effective are these tests? What is the fault coverage and diagnostic resolution of these tests?

Figure 2–7 illustrates the region tested by using an in–circuit test system. Typically, the loaded board is tested for shorts between interconnections (i.e., between bed–of–nails probes of which there is often only one per *net*) before power is applied. For complex interconnections, the number of potential faults that is tested during each test will be quite large, and may include faults in segments of interconnect that are provided solely for test purposes (e.g.,, branches leading only to test pads). When power has been applied, tests are applied by using backdriving techniques on a chip-by-chip basis to the various ICs on the board. These tests detect many faults in the board interconnections (open-circuits, stuck-ats, etc.) and some defects in the chips. The precise coverage will depend on the quality of the test applied, and the speed of application.

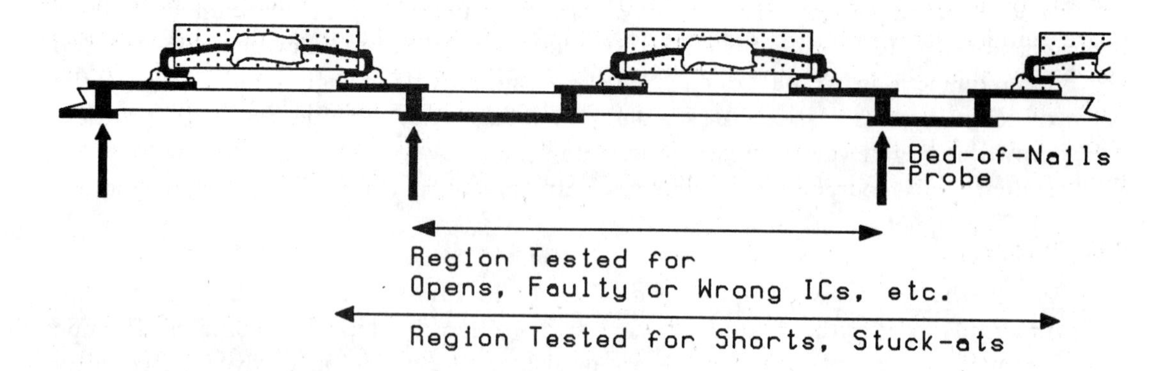

Figure 2–7: Test coverage using an in-circuit tester.

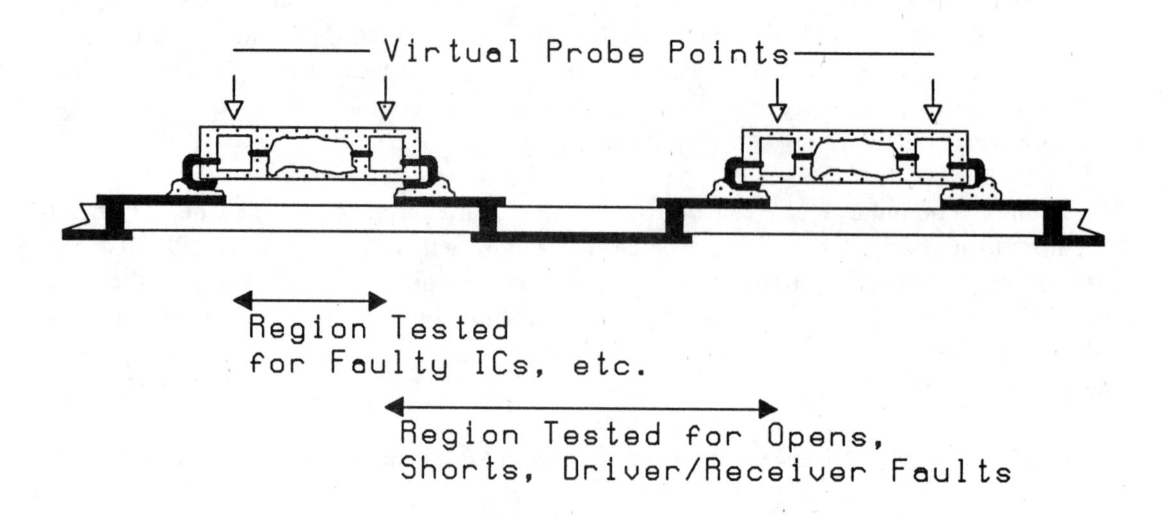

Figure 2–8: Test coverage using boundary-scan.

Figure 2–8 shows the regions tested by the interconnect and component boundary-scan tests. As for the in-circuit test, the quality of the test performed on each component will vary -- in this case, depending on the type of chip concerned. For example, ICs that offer a self-test facility will probably be tested more thoroughly than those tested by shifting patterns in and out through the boundary-scan path. Note that the chip-to-chip interconnect test will detect faults both in the interconnection itself and in the drivers and receivers of the chips at each end, covering those parts of the chips and the board that are most likely to be incorrectly manufactured or damaged during either chip or board assembly, or later in the product's life.

Regardless of whether in-circuit test or boundary-scan is used, errors in "at speed" interactions between chips will not be thoroughly tested -- each IC on the loaded board is tested in isolation from all others. Therefore, it might be necessary to follow both in-circuit and boundary-scan tests with a further test that exercises the complete loaded board in its normal operating mode. This test could be applied by using a functional test system, or it could be a board-level self-test (e.g., applied by a microprocessor on the board running some specially-designed test firmware). Note, however, that this functional test can normally be accomplished without the necessity for extensive bed-of-nails contact with the board; contact through the board connector, etc. is usually sufficient. In cases in which it is possible, another option is to design ICs so that they form groups that can be treated as self-testing "meta-components" or clusters during board test.

2.6: Reference

[1] E.B. Eichelberger and T.W. Williams, "A Logic Design Structure for LSI Testability," *Journal of Design Automation and Fault-Tolerant Computing*, Vol. 2, No. 2, May 1978, pp. 165-178.

Chapter 3. The Development of IEEE Std 1149.1†

The effort of establishing IEEE Std 1149.1 began with the creation of an ad-hoc group of systems electronics companies. This group became the Joint Test Action Group (JTAG) and, subsequently, the core of the IEEE Working Group that developed IEEE Std 1149.1.

In this chapter, we will review the steps in the technical development of IEEE Std 1149.1, from the formation of the JTAG through publication of the IEEE Standard. As shown in Figure 3–1, the technical activity developed in four key steps and each will be reviewed in turn in the following sections. Note that the development of the standard has continued since its approval, with the aim of extending the functionality of the circuitry described and of improving the clarity of the document.

3.1: The Joint Test Action Group

JTAG was set up following a paper by Frans Beenker of Philips Research Labs in 1985 [1,2]. He discussed the need for a structured approach to loaded-board testing and considered the value of boundary-scan as a solution to the problems he identified.

The initial JTAG meeting was attended by representatives from several major European electronics companies. By the end of 1986, however, JTAG had become an international group involving both European and North American companies, all of whom were seeking solutions for the test problems in hybrid and loaded-board products created by the combination of complex integrated circuits (ICs) and surface-mount technology. During 1986, JTAG members decided the problems they were facing could be solved if a standardized form of boundary-scan was available that allowed correct test interaction between various vendor's ICs.

3.2: JTAG Version 0

The initial JTAG proposal [3] was created by Frans Beenker (Philips Research Labs, The Netherlands), Chantal Vivier (Bull Systemes, France), and Colin Maunder (British Telecom Research Labs, UK) in June 1986, based on their understanding of work done with boundary-scan in their respective companies and of other material published internationally. Among the developments reviewed were the following:

- IBM: Chip partitioning aid (CPA) [4].
- IBM: Electronic chip in place test (ECIPT) [5]
- Control Data Corporation: On-chip maintenance system (OCMS) [6]

---

† The text of this chapter is derived from a segment of the IEEE Satellite Seminar *Chip–to–System Testability* transmitted March 1, 1989.

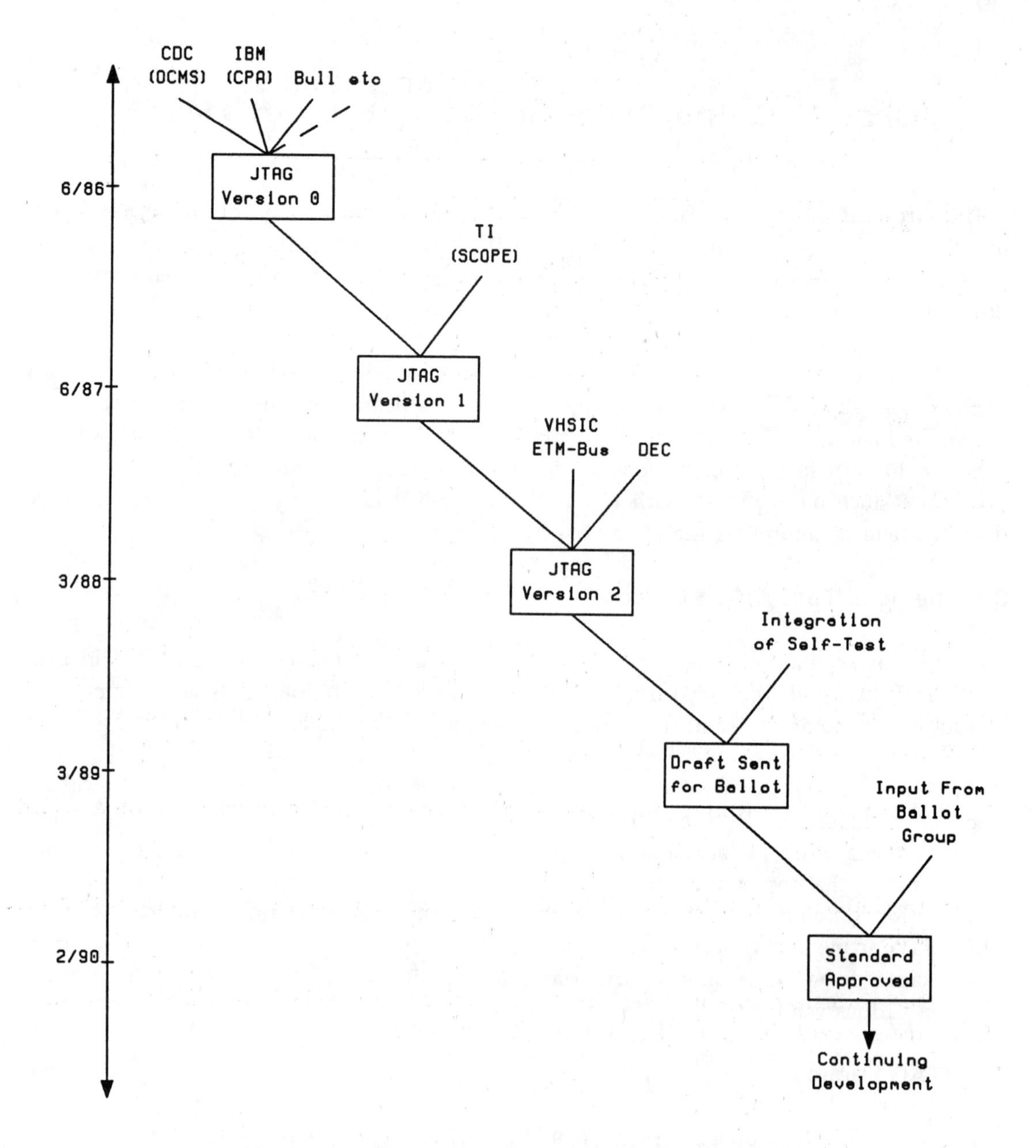

Figure 3–1: The development of IEEE Std 1149.1.

- STC Computer Research Corporation: Shift test control logic (STCL) [7].
- Bull Systemes [8].
- Control Data Corporation: Built–in evaluation and self–test (BEST) [9].
- Hewlett–Packard [10].

The proposal was for an architecture based on a single serial shift–register path and was targeted solely at boundary–scan testing, as shown in Figure 3–2.

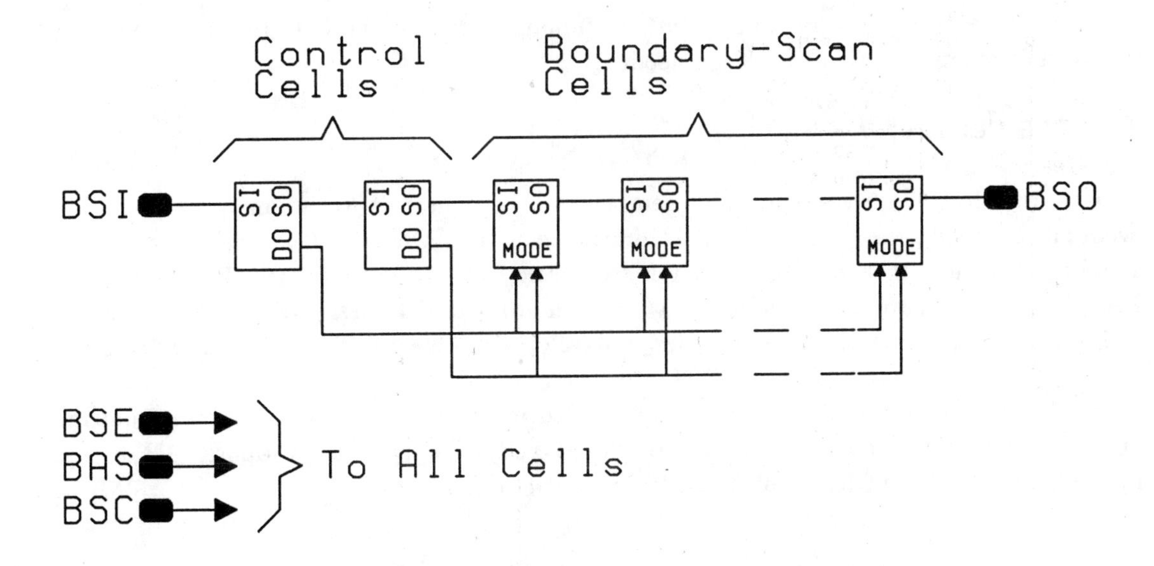

Figure 3–2: JTAG version 0.1 architecture.

The serial path was constructed from two control register cells and a number of boundary–scan register cells -- one for each system input or output of the chip. The control cells allowed the boundary–scan cells to be set into three operating modes:

- the exterior test mode that allowed the interconnections between chips on a loaded board to be tested;
- the interior test mode that allowed slow–speed static testing of the logic within the chip; and
- the normal operation mode where the boundary–scan cells were configured to allow the system function of the chip to occur unimpeded.

Five pins were required for this architecture:

- A test mode control, boundary–scan enable (BSE), that enables the boundary scan circuitry.
- A test clock, boundary–scan clock (BSC).
- A signal to select between loading and shifting of the boundary–scan path -- boundary apply/scan (BAS).
- A serial data input, boundary scan input (BSI).
- A serial data output, boundary–scan output (BSO).

This architecture was very simple, but its functionality was limited. It was soon clear that a more complex design would be needed.

3.3: JTAG Version 1.0

Over the following year, JTAG members worked on this straw-man proposal -- eventually forming the Technical Sub-Committee to focus on the technical development activities. During this period, Lee Whetsel from Texas Instruments (TI) joined the Technical Sub-Committee, bringing with him the initial designs for TI's SCOPE architecture [11] -- a boundary-scan design developed in TI's Military Products Division.

The JTAG proposal was improved and extended to include a number of inputs from the SCOPE design and other sources, resulting in the JTAG version 1.0 proposal [12] -- the first document to be widely mailed in Europe and North America.

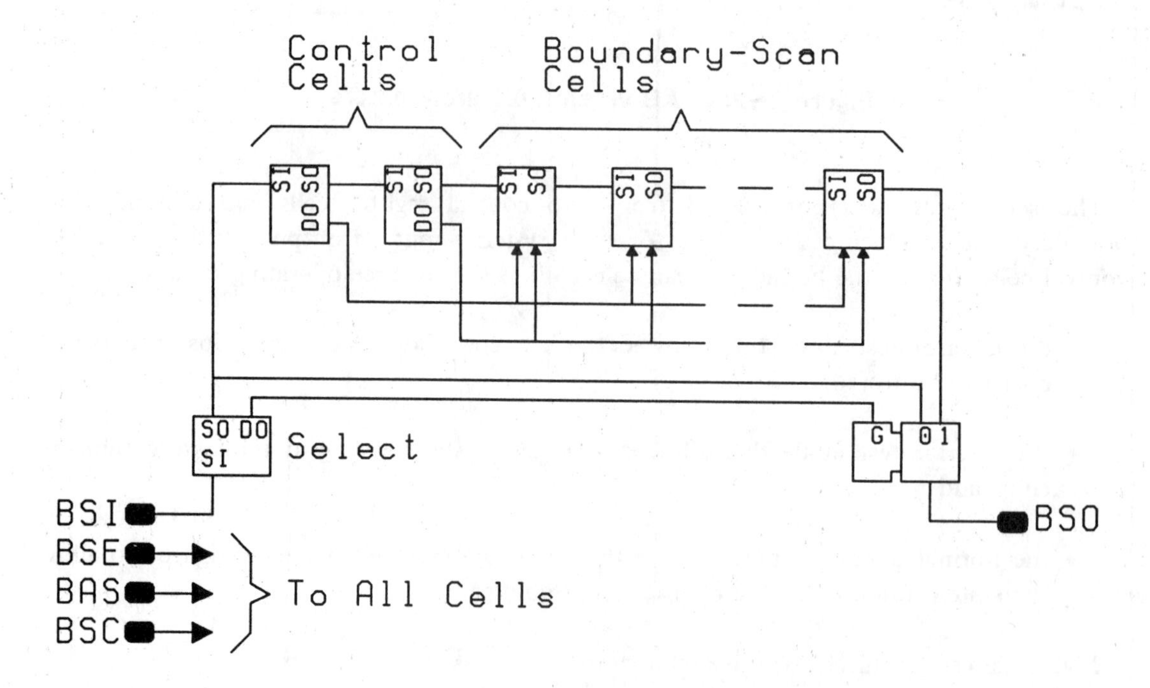

Figure 3-3: JTAG version 1.0 architecture.

The version 1.0 architecture (Figure 3-3) included two key features in addition to those of the initial design:

1. The design allowed the serial path through the chip to be short-circuited under control of one or more select bits placed at the head of the path. This feature allows a chip to be bypassed when it is not involved in a particular test, with the result that the volume of test data can be significantly reduced. The control and boundary-scan segments of the path are accessed only when necessary.

2. The design allowed the serial path to be extended by adding shift–register stages at its tail end, between the boundary–scan cells and the multiplexer shown in Figure 3–3. This feature allows design–for–test features other than the boundary–scan register to be accessed, increasing the scope and value of the proposal considerably. For example, access to embedded self–test features in a design is now possible by using the same pins as those provided for boundary–scan.

3.4: JTAG Version 2.0

The version 1.0 proposal was discussed widely both in Europe and North America and was the subject of a paper at the 1987 IEEE International Test Conference (ITC) [13]. Also, at ITC in 1987 an evening meeting was arranged to allow discussion of the JTAG proposal. This was attended by more than 100 engineers from many electronics companies.

At that evening meeting, and at working meetings hastily arranged later during the conference week, a number of key suggestions for improvements and extensions to the proposed design were made. These included input from Digital Equipment Corporation (DEC) and from a number of people involved in the development of the United States Department of Defense's VHSIC† Element Test and Maintenance Bus -- the ETM–Bus [14].

The principal suggestion was that the design should be altered to allow efficient access to *any* serial design–for–test circuitry embedded in a chip. Simply, given that a number of package pins need to be dedicated to test to provide access to the boundary–scan cells, the objective is to exploit these pins to the fullest extent possible.

The JTAG version 2.0 architecture [15,16,17] (Figure 3–4) is structurally identical to the design embodied in IEEE Std 1149.1. Since the detail of the standard is presented in Part II, the discussion here is intended only to highlight the changes between JTAG version 1.0 and JTAG version 2.0.

In contrast to the earlier architectures, the JTAG version 2.0 design was based on parallel instruction and test data registers located between common serial input and output pins.

The instruction register provides the functions of the select and control registers of the earlier designs and is also extensible to meet the particular needs of any chip. The alternative path consists of a parallel bank of test data registers, each of which can be accessed when an appropriate instruction is loaded into the instruction register. The bank of test data registers can support a whole range of test, maintenance, and other functions embedded in the chip design -- in addition to the boundary–scan test capability that was the prime focus of JTAG activity from the outset.

† VHSIC – Very high–speed integrated circuit

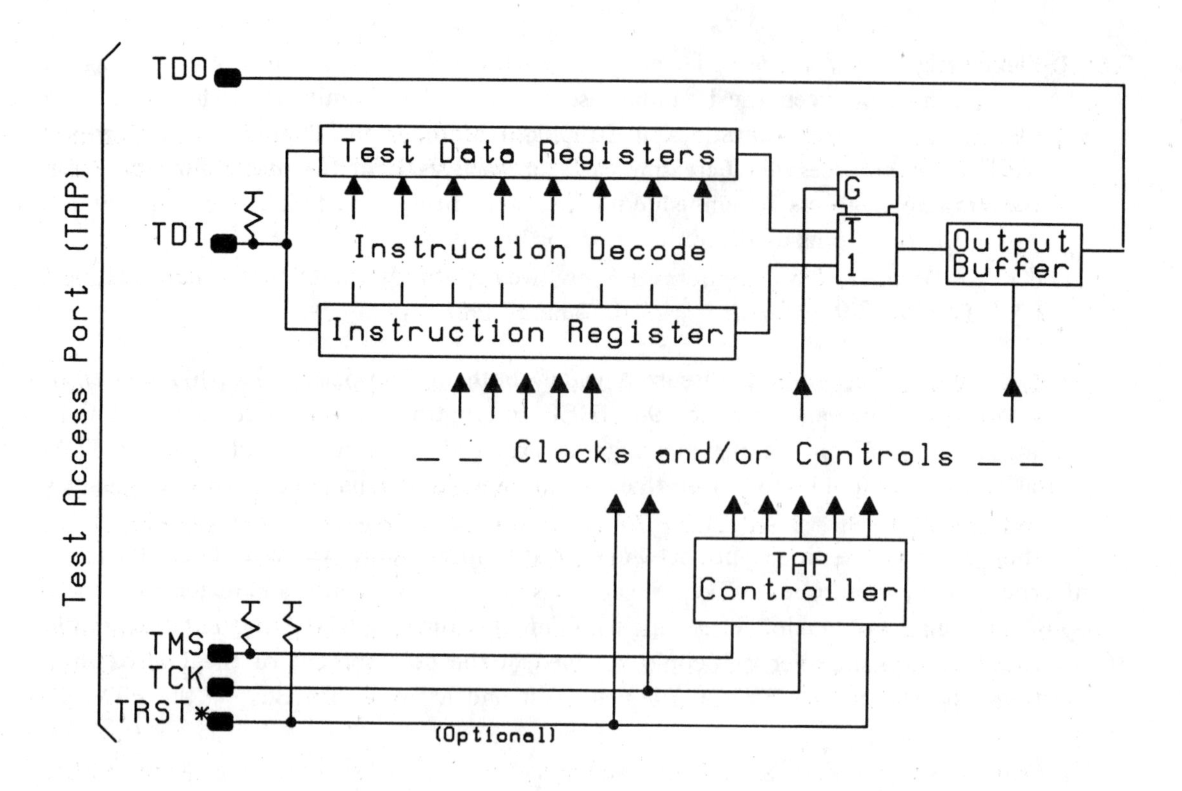

Figure 3–4: JTAG version 2.0 architecture.

Also, the minimum requirement for dedicated test pins has been reduced from five to four compared to the earlier designs. This change reflects the widely–held view that the number of pins dedicated to test must be kept to the absolute minimum.

3.5: IEEE Std 1149.1

Late in 1987, JTAG decided to approach the IEEE to discuss the possibility of formalizing their technical proposal as an IEEE Standard. As a result of this approach, the JTAG proposal became one of a range of testability approaches being developed by the IEEE Testability Bus Standards Committee. At the same time, the JTAG Technical Sub–Committee became the core of the working group responsible for the further development of the JTAG proposal as IEEE Draft Standard P1149.1.

Technical changes made by the Working Group prior to approval of the standard concentrated on the detailed design of the boundary–scan register, the instruction set, the device identification register, and on the integration of built–in self–test features within the overall design.

3.6: References

[1] F.P.M. Beenker, "Systematic and Structured Methods for Digital Board Testing," *IEEE International Test Conference Proceedings*, IEEE Computer Society Press, Los Alamitos, Calif., 1985, pp. 380-385.

[2] F.P.M. Beenker, "Systematic and Structured Methods for Digital Board Testing," *VLSI Systems Design*, Vol. 8, No. 1, January 1987, pp. 50–58.

[3] The Joint Test Action Group, *A Standard Boundary–Scan Architecture –– Draft 3,* September 1986.

[4] S. DasGupta et al, "Chip Partitioning Aid: A Design Technique for Partitionability and Testability in VLSI," *IEEE Design Automation Conference Proceedings*, IEEE Computer Society Press, Los Alamitos, Calif., 1978, pp. 203–208.

[5] P. Goel and M.T. McMahon, "Electronic Chip–in–Place Test," *IEEE International Test Conference Proceedings*, IEEE Computer Society Press, Los Alamitos, Calif., 1982, pp. 83–90.

[6] D.R. Resnick, "Testability and Maintainability With a New 6K Gate Array," *VLSI Design*, Vol. 4, No. 2, March/April 1983, pp. 34–38.

[7] J.J. Zasio, "Shifting Away From Probes for Wafer Test," *IEEE Compcon Proceedings*, IEEE Computer Society Press, Los Alamitos, Calif., 1983, pp. 395–398.

[8] D. Laurent, "An Example of Test Strategy for Computer Implemented with VLSI Circuits," *IEEE International Conference on Circuits and Computers Proceedings*, IEEE Computer Society Press, Los Alamitos, Calif., 1985, pp. 679–682.

[9] R. Lake, "A Fast 20K Gate Array with On–Chip Test System," *VLSI Systems Design*, Vol. 7, No. 6, June 1986, pp. 46–55.

[10] D. Weiss, "VLSI Test Methodology," *Hewlett-Packard Journal*, September 1987, pp. 24/5.

[11] L. Whetsel, communication to JTAG, 1986.

[12] The Joint Test Action Group, *A Standard Boundary–Scan Architecture –– Version 1.0,* June 1987.

[13] C. Maunder and F. Beenker, "Boundary–Scan: A Framework for Structured Design–for–Test," *IEEE International Test Conference Proceedings*, IEEE Computer Society Press, Los Alamitos, Calif., 1987, pp. 714–723.

[14] L. Avra, "A VHSIC ETM-Bus-Compatible Test and Maintenance Interface," *IEEE International Test Conference Proceedings,* IEEE Computer Society Press, Los Alamitos, Calif., 1987, pp. 964-971.

[15] M.M. Pradhan, R.E. Tulloss, H. Bleeker and F.P.M. Beenker, "Developing a Standard for Boundary-Scan Implementation," *IEEE International Conference on Computer Design: VLSI in Computers and Processors,* IEEE Computer Society Press, Los Alamitos, Calif., 1987, pp. 462-466.

[16] L. Whetsel, "A View of the JTAG Port and Architecture," *ATE and Instrumentation Conference*, January 1988, pp. 385-401.

[17] The Joint Test Action Group, *A Test Access Port and Boundary-Scan Architecture -- Version 2.0,* March 1988.

Part II: Tutorial

Part II provides a tutorial introduction to the circuitry defined by IEEE Std 1149.1. The material is a considerably reduced description compared to that given in the standard itself. It is therefore strongly recommended that readers intending to build an integrated circuit that conforms to the standard consult a copy of IEEE Std 1149.1 before doing so.

Copies of the standard may be obtained from: IEEE Standards Department, P.O. Box 1331, 445 Hoes Lane, Piscataway, New Jersey 08855–1331, U.S.A.

Acknowledgment

Acknowledgment is made to the IEEE Standards Department for permission to use several figures from IEEE Std 1149.1 in this part of the book.

Chapter 4. IEEE Std 1149.1: The Top–Level View

This chapter provides an introduction to the test circuitry defined by IEEE Std 1149.1 and shows how it can be used to perform a number of basic test operations. The chapter also indicates how further test circuitry can be added to that specified by the standard to allow access to test functions beyond the minimum required.

4.1: The IEEE Std 1149.1 Architecture

The top–level schematic of the test logic defined by IEEE Std 1149.1 includes three key blocks (Figure 4–1):

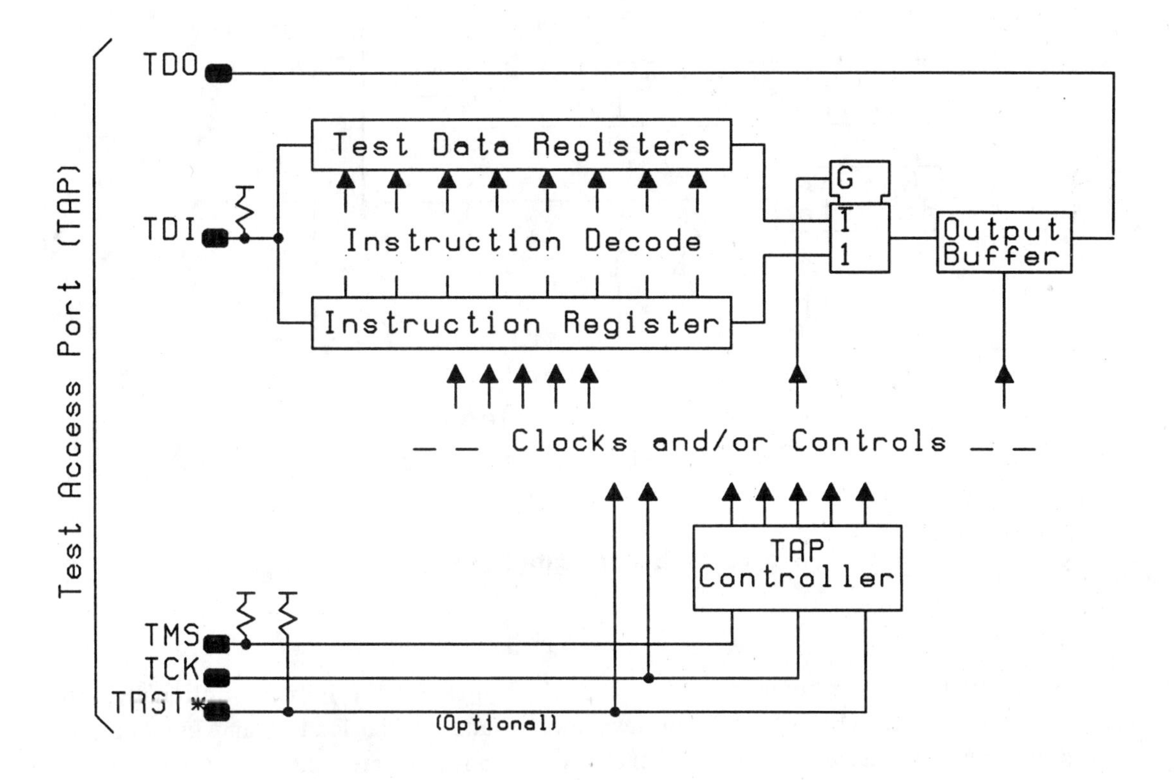

Figure 4–1: IEEE Std 1149.1 test logic.

- *The TAP controller:* This responds to the control sequences supplied through the test access port (TAP -- see below) and generates the clocks and control signals required for correct operation of the other circuit blocks.

- *The instruction register:* This shift–register–based circuit is serially loaded with the instruction that selects a test to be performed.

- *The test data registers:* This is a bank of shift–register based circuits (Figure 4–2). The stimuli or conditioning values required by a test are serially loaded into the test data register selected by the current instruction. Following execution of the test, the results can be shifted out for examination.

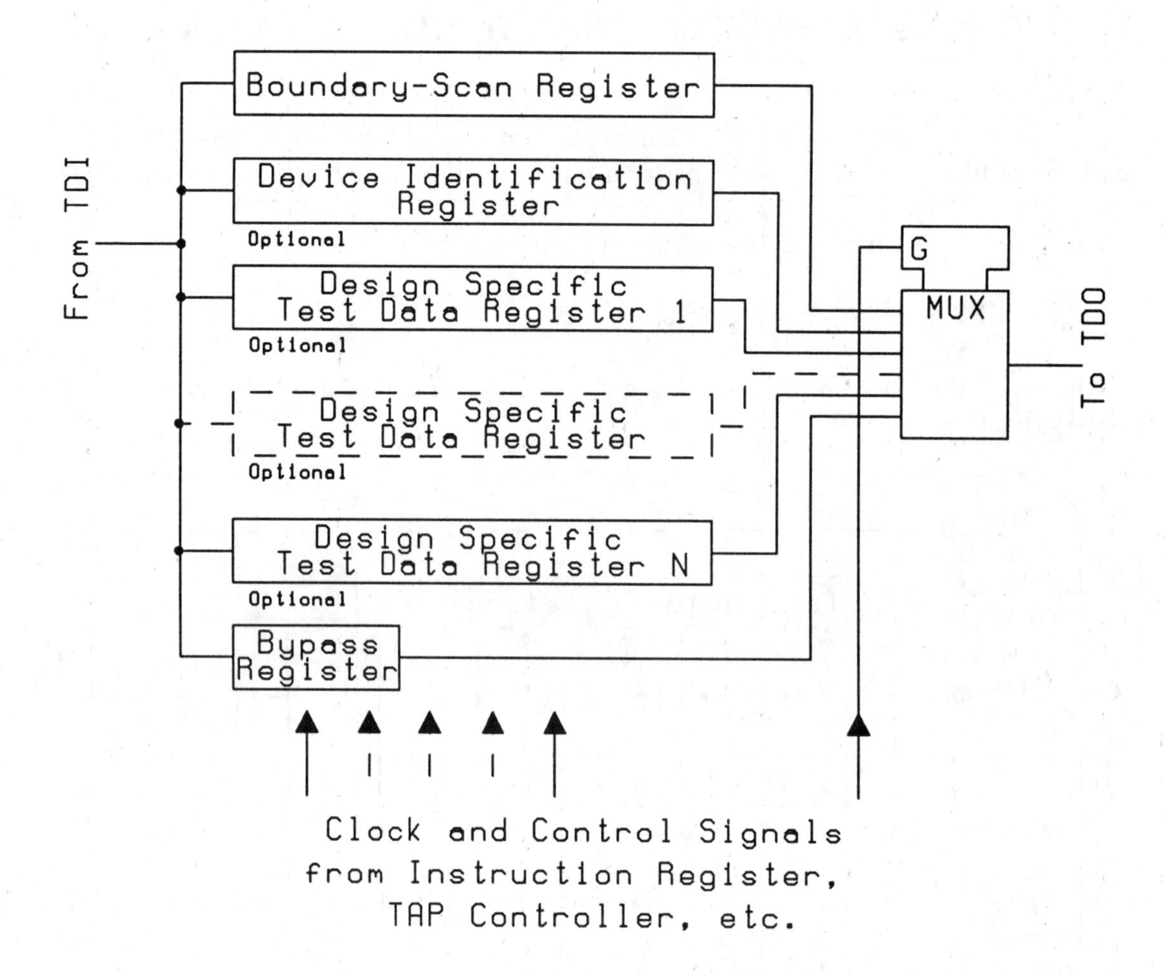

Figure 4–2: Test data registers.

These circuit blocks are connected to a TAP which includes the four or, optionally, five signals used to control the operation of tests and to allow serial loading and unloading of instructions and test data. The role of the TAP on an integrated circuit (IC) is directly analogous to the "diagnostic" socket provided on many automobiles -- it allows an external test processor to control and to communicate with the various test features built into the product.

In addition, the test data registers can be connected to the system circuitry within the chip (i.e., the circuitry that performs the particular function, other than test, for which the chip was designed) or to the pins that are connected to the system circuitry. These connections allow tests of the system circuitry to be performed. The operation of the test data register is described in Section 4.5.

The following sections discuss the TAP and the main circuit blocks in more detail.

4.2: The TAP

The TAP contains four or, optionally, five pins. These are:

- *The test clock input (TCK):* This is independent of the system clock(s) for the chip so that test operations can be synchronized between the various chips on a printed wiring board. Both the rising and falling edges of the clock are significant: the rising edge is used to load signals applied at the TAP input pins (test mode select(TMS) and test data input (TDI)), while the falling edge is used to clock signals out through the TAP test data output (TDO) pin. As will be discussed in Chapter 6, the boundary-scan register defined by the standard is controlled such that data is loaded from system input pins on the rising edge of TCK while data are driven through system output pins on the falling edge.

- *The test mode select input (TMS):* The operation of the test logic is controlled by the sequence of 1s and 0s applied at this input, with the signal value typically changing on the falling edge of TCK. As will be discussed in Section 4.3, this sequence is fed to the TAP controller which samples the value at TMS on each rising edge of TCK and which uses this information to generate the clock and control signals required by the other test logic blocks. TMS is either equipped with a pull-up resistor or otherwise designed such that, when it is not driven from an external source, the test logic perceives a logic 1.

- *The test data input (TDI):* Data applied at this serial input are fed either into the instruction register or into a test data register, depending on the sequence previously applied at TMS. Typically, the signal applied at TDI will be controlled to change state following the falling edge of TCK, while the registers shift in the value received on the rising edge. Like TMS, TDI is either equipped with a pull-up resistor or otherwise designed such that, when it is not driven from an external source, the test logic perceives a logic 1.

- *The test data output (TDO):* This serial output from the test logic is fed either from the instruction register or from a test data register depending on the sequence previously applied at TMS. During shifting, data applied at TDI will appear at TDO after a number of cycles of TCK determined by the length of the register included in the serial path. The signal driven through TDO changes state following the falling edge of TCK. When data are not being shifted through the chip, TDO is set to an inactive drive state (e.g., high-impedance).

- *The optional test reset input (TRST*):* The need to be able to initialize a circuit to a known starting state (the "reset" state) is crucial in testing. As will be discussed in Section 4.3, the TAP controller is designed so that this state can be quickly entered under control of TCK and TMS. The standard also requires that the test logic can be initialized at power-up independently of TCK and TMS. This can be achieved either by building features into the test logic itself (e.g., a power-up reset circuit) or by adding the optional TRST* signal to the TAP. Application of a 0 at TRST* asynchronously forces the test logic into its reset state. Note that, in this state, the test logic cannot interfere with the operation of the on-chip system logic, so TRST* can also be viewed as a "test mode enable" input.

By loading the signals applied to the test logic through chip input pins (e.g., through TMS and TDI) on the rising edge of TCK, while using the falling edge to clock signals out through chip output pins (such as TDO), operation of the IEEE Std 1149.1 test logic can be made race-free. For example, when chips compatible with the standard are serially connected (e.g., as in Figure 4-3) data are applied to TDO by the first chip one half cycle of TCK prior to the time when they are loaded from the TDI input of the second. This allows time to account for delays in the serial path, skew between the clocks fed to the neighboring ICs, and other factors.

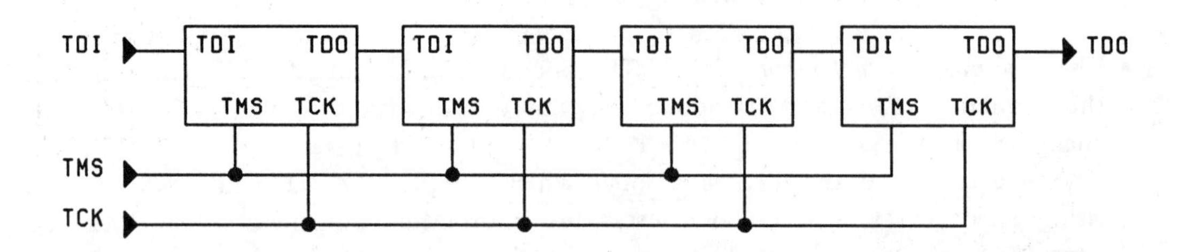

Figure 4-3: Simple serial connection of IEEE Std 1149.1-compatible ICs.

Since TDO is set to an inactive drive state when no data are being shifted, the TAPs of individual chips can, if required, be connected to give parallel serial paths at the board level (e.g., as shown in Figure 4-4). In such cases, a different TMS signal is required for each serial path. These signals should be controlled such that no two paths attempt to shift data simultaneously.

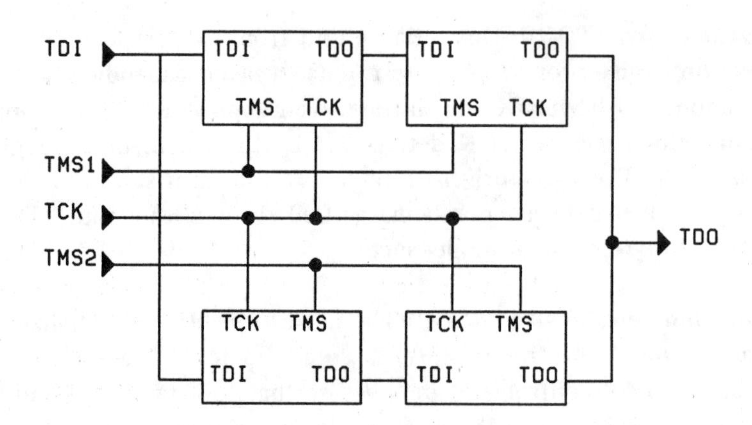

Figure 4-4: Hybrid serial/parallel connection of IEEE Std 1149.1-compatible ICs.

At the board level, the test signals can be controlled either by external automatic test equipment (ATE) or by an on-board bus-master chip. In the latter case, the bus-master chip might provide an interface between the interface defined by the IEEE Std 1149.1 TAP and some higher level test and maintenance messaging system (Figure 4-5) [e.g., [1]).

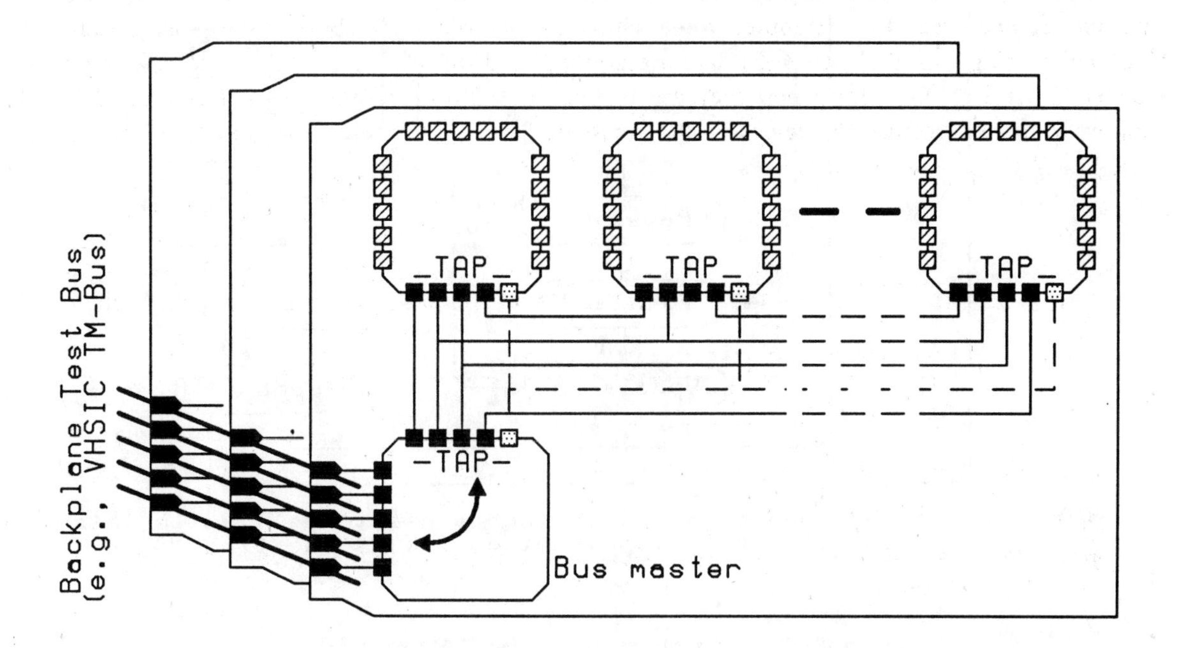

Figure 4-5: Use of a bus-master chip to control ICs compatible with IEEE Std 1149.1.

4.3: The TAP Controller

A key goal during the development of IEEE Std 1149.1 was to keep the number of pins in the TAP to a minimum, based on the knowledge that many ICs are pin- (rather than silicon-) limited. As test engineers are only too aware, designers are always reluctant to allocate pins for test purposes.

The TAP controller allows us to meet this goal. It is a 16-state finite state machine that operates according to the state diagram shown in Figure 4-6. Note that in the states whose names end "*-DR*" the test data registers operate, while in those whose names end "*-IR*" the instruction register operates. A move along a state transition arc occurs on every rising edge of TCK. The 0s and 1s shown adjacent to the state transition arcs show the value that must be present on TMS at the time of the next rising edge of TCK for the particular transition to occur.

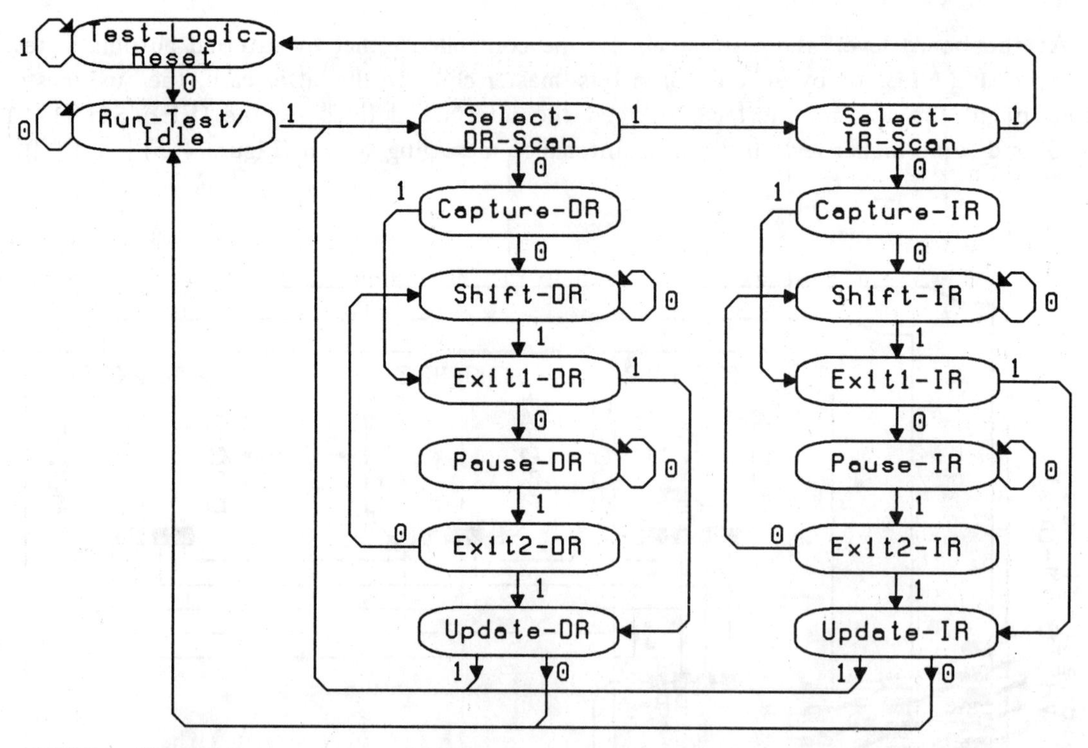

NOTE: The value shown adjacent to each state transition in this figure represents the signal present at TMS at the time of a rising edge at TCK.

Figure 4–6: State diagram for the TAP controller.

Eight of the 16 controller states determine operation of the test logic, allowing the following test functions to be performed:

- *Test–Logic–Reset*: In this controller state, all test logic is reset. As mentioned earlier, when the test logic is reset, it is effectively disconnected from the on–chip system logic, allowing normal operation of the chip to occur without interference. Regardless of the starting state of the TAP controller, the *Test–Logic–Reset* controller state is reached by holding the TMS input at 1 and applying five rising edges at TCK. Further, this controller state must be entered automatically when power is applied to a chip that does not have the optional TRST* input. Alternatively, where TRST* is provided, it can be used to force the controller asynchronously into the *Test–Logic–Reset* controller state both at power–up and at any desired point during circuit operation.

- *Run–Test/Idle:* The operation of the test logic in this controller state depends on the instruction held in the instruction register. When the instruction is, for example, one that activates a self-test, then the self-test will be run when the controller is in this state.† In another case, if the instruction in the instruction register is one that selects a data register for scanning, then the test logic is idle in the *Run-Test/Idle* controller state.

- *Capture-DR:* Each instruction must identify one or more test data registers that are enabled to operate in test mode when the instruction is selected. In this controller state, data are loaded from the parallel input of these selected test data registers into their shift-register paths on the rising edge of TCK.

- *Shift-DR:* Each instruction must identify a single test data register that is to be used to shift data between TDI and TDO in the *Shift-DR* controller state. Shifting allows the previously captured data to be examined and new test input data to be entered. Shifting occurs on the rising edge of TCK in this controller state. In the *Shift-DR* controller state, the TDO output is active (it is inactive in all other controller states except the *Shift-IR* state).

- *Update-DR:* This controller state marks the completion of the shifting process. Some test data registers may be provided with a latched parallel output to prevent signals applied to the system logic, or through the chip's system pins, from rippling while new data are shifted into the register. Where such test data registers are selected by the current instruction, the new data is transferred to their parallel outputs on the falling edge of TCK in this controller state.

- *Capture-IR, Shift-IR,* and *Update-IR:* These controller states are analogous to *Capture-DR, Shift-DR,* and *Update-DR* respectively but cause operation of the instruction register. By entering these states, a new instruction can be entered and applied to the test data registers and/or other specialized circuitry. This instruction becomes "current" on the falling edge of TCK in the *Update-IR* controller state.

The actions of the instruction and test data registers in each of these controller states will be described in more detail in the following sections of this chapter. Figure 4-7 shows where the actions described occur in each controller state.

In the remaining eight controller states, no operation of the test logic occurs - that is, the test logic is "idle." The "pause" states (*Pause-DR* and *Pause-IR*) are provided to allow the shifting process to be temporarily halted, for example while an ATE or other equipment controlling the test logic fetches more test data from backup memory (e.g., disc).

† Note: An important goal in the development of IEEE Std 1149.1 was to allow built-in self-test (BIST) functions to be integrated within the test logic. As was discussed in Chapter 2, the combination of BIST and boundary-scan is especially powerful -- allowing effective testing of ICs once they have been mounted on a board.

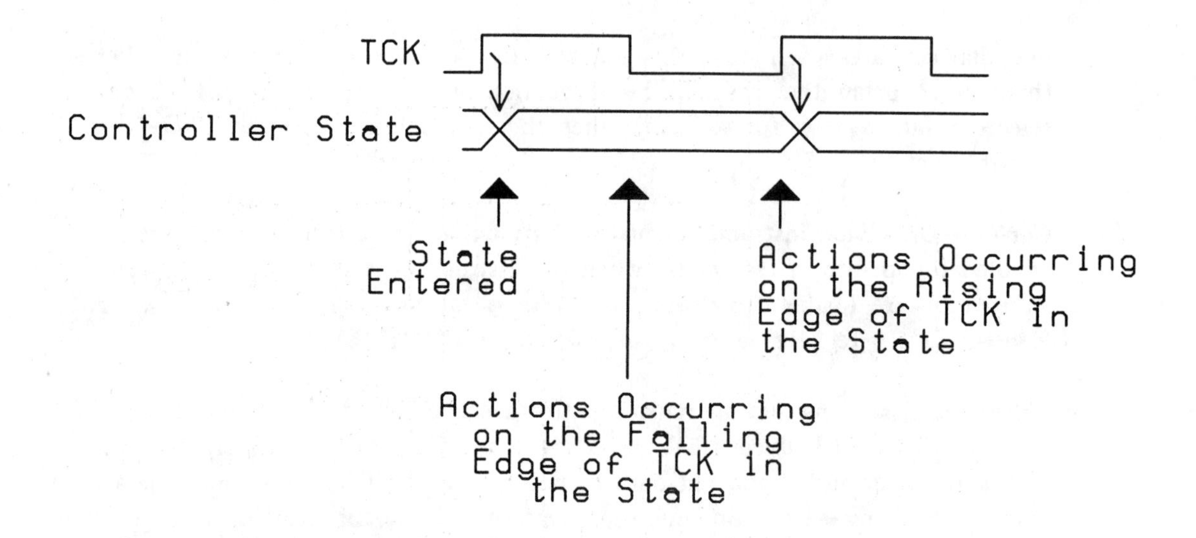

Figure 4–7: The timing of events within a controller state.

The final six controller states (*Select–DR–Scan, Select–IR–Scan, Exit1–DR, Exit1–IR, Exit2–DR,* and *Exit2–IR*) are decision points that allow choices to be made as to the route to be followed around the controller's state diagram. For example, in the *Exit1–DR* controller state a choice is made, depending on the signal applied at the TMS input, between entry into the *Pause–DR* state or entry into the *Update–DR* state.

Without the TAP controller, the nine functions fulfilled by the states previously described (*Test–Logic–Reset, Run–Test,* and *Idle* plus *Capture, Shift,* and *Update* for the two register types) would need to be selected by using at least four control inputs. With the TAP controller, only one control input (TMS) is required. The penalties are that a certain amount of logic must be built into every component to decode the signals received at TMS and that the ability to move between the functions is slightly constrained. Neither of these penalties is severe, however. As shown by the example controller implementation in Figures 4–8 and 4–9, construction of the controller requires only approximately 80 2–input NAND gates.† This is a small cost in the context of a complex very–large scale integration (VLSI) IC that can contain upwards of 250,000 gates.

The restriction in the freedom to move arbitrarily between test operations is similarly not a significant one since freedom would, in many cases, be removed as a result of simplification of the software written to control the test logic.

The encoding of the controller states for the example controller implementation is shown in Table 4–1.

---

† For the remainder of this part of the book, implementation examples will be given that are compatible with the TAP controller implementation included here.

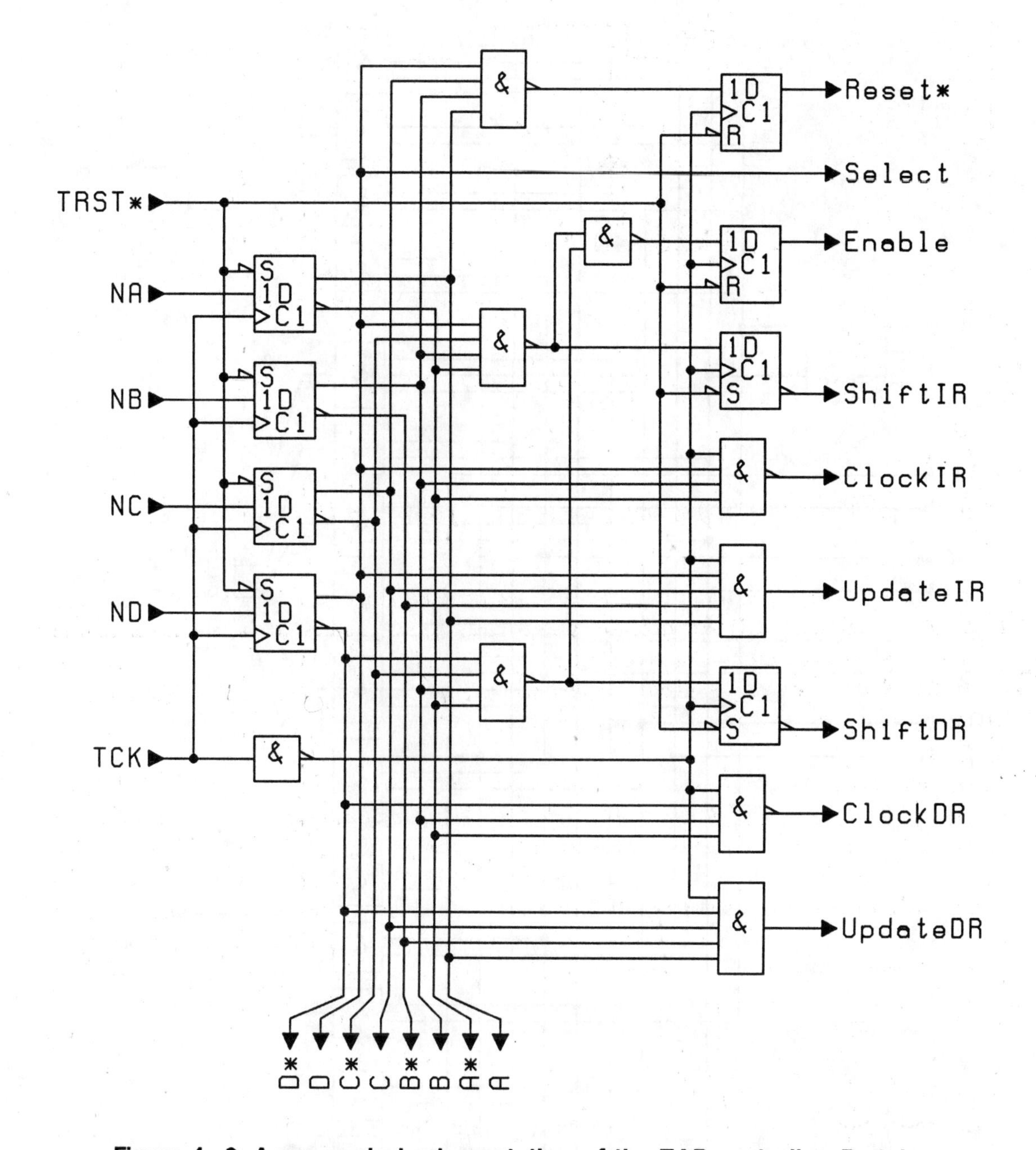

Figure 4–8: An example implementation of the TAP controller: Part 1.

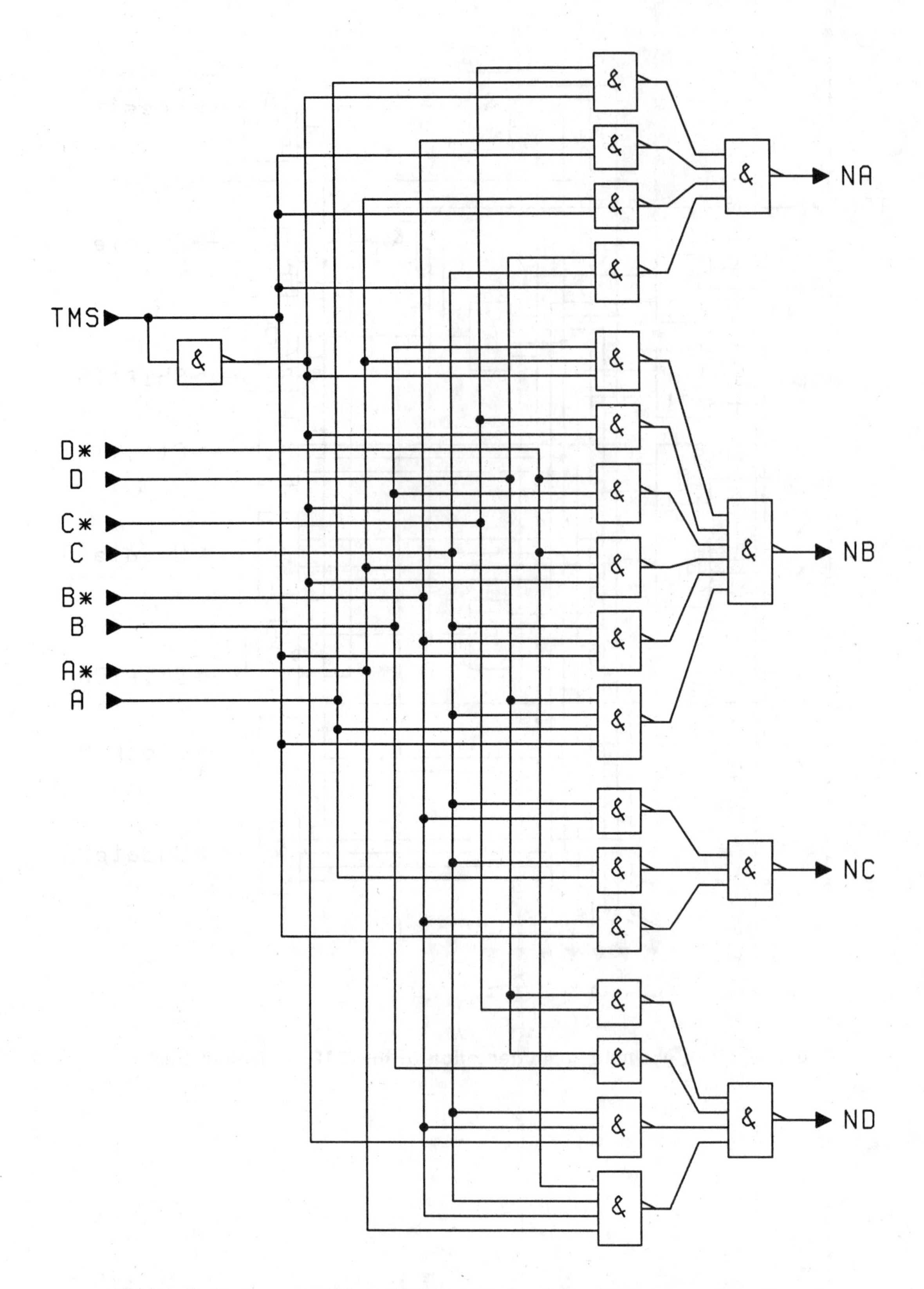

Figure 4–9: An example implementation of the TAP controller: Part 2.

Table 4-1: State assignments for the example TAP controller.

Controller state	DCBA (hex)
Exit2-DR	0
Exit1-DR	1
Shift-DR	2
Pause-DR	3
Select-IR-Scan	4
Update-DR	5
Capture-DR	6
Select-DR-Scan	7
Exit2-IR	8
Exit1-IR	9
Shift-IR	A
Pause-IR	B
Run-Test/Idle	C
Update-IR	D
Capture-IR	E
Test-Logic-Reset	F

4.4: The Instruction Register

The instruction register provides one of the alternate serial paths between TDI and TDO. It operates when the instruction scanning portion of the controller state diagram is entered (i.e., the portion where state names end "*–IR*").

The instruction register allows test instructions to be entered into each component along the board-level path. The instruction registers are daisy-chained together at the board level in the *Shift–IR* controller state (Figure 4-10), so a different instruction can be loaded into each chip on the path if required. Although it is unnecessary for each IC to be executing the same instruction at any given time, because instructions are shifted into all ICs on a single serial path at the same time, loading and execution of the instructions for each IC must be synchronized. For example, all ICs controlled by a single TMS signal must be simultaneously in the *Shift–IR* controller state.

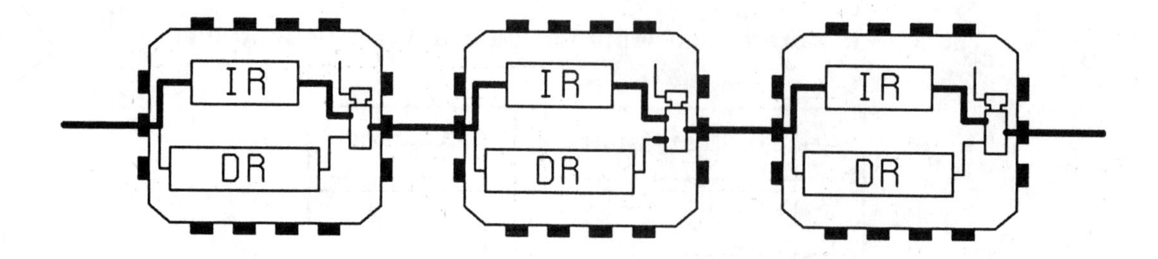

Figure 4–10: Daisy–chain connection of instruction registers.

4.4.1: Instruction Register Design

At the core of the instruction register's design is a shift register that must contain at least two stages (shown cross–hatched in Figure 4–11). No maximum length is defined, since this will be determined by the number of test instructions provided by the particular chip.

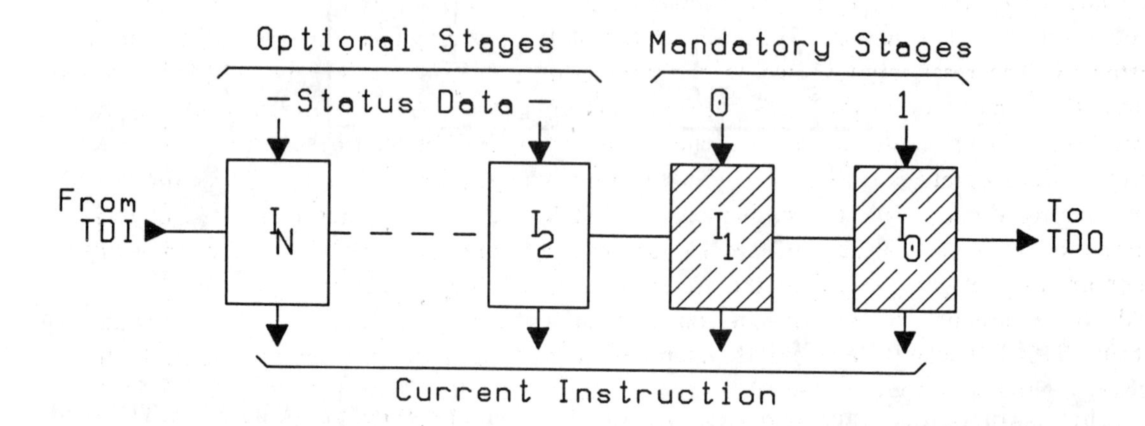

Figure 4–11: The instruction register.

The standard requires that stages I_1 and I_0 † must be set to 0 and 1 respectively on the rising edge of TCK in the *Capture–IR* controller state. These fixed values assist in detecting and locating faults in the serial path through chips on a board, as will be discussed in Chapter 9. Instruction register stages numbered I_2 or greater are optional and can have a parallel input from which data (typically, status information) are loaded.

Each shift–register stage in the instruction register might be designed as shown in Figure 4–12.

† Note that, within IEEE Std 1149.1, the convention is used that the least significant bit is that written or read from the shift–register stage closest to TDO. In addition, the least significant bit is numbered 0. For example, if the instruction register is named I the least significant stage is named I_0 and a minimum instruction register design must have stages I_1 and I_0.

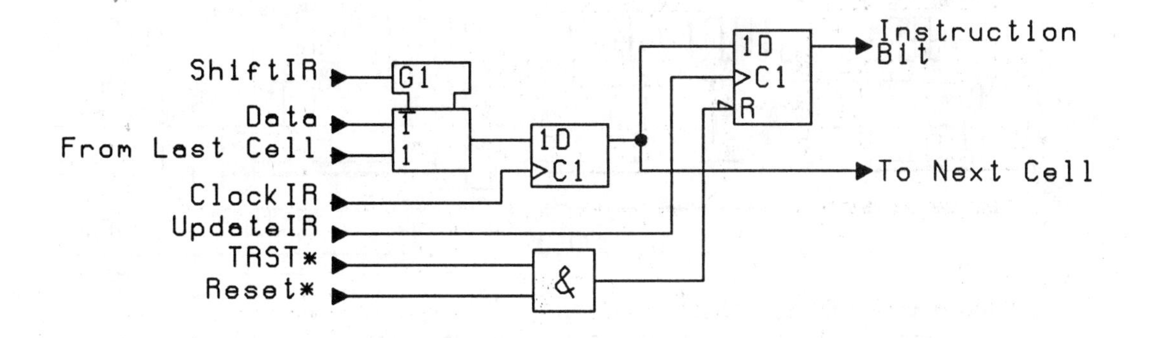

Figure 4–12: An example instruction register cell.

Each stage has a latched parallel output to which instructions are transferred when they are valid (i.e., on the falling edge of TCK in the *Update–IR* controller state -- at this time, the example TAP controller changes the UpdateIR signal from 0 to 1). The provision of a latched output means that the remaining test logic receives only valid instructions -- it does not see the changing contents of the shift–register while the new instruction is shifted in. The reset input shown to the parallel output register in Figure 4–12 forces a 0 onto the instruction register's output when the TAP controller enters the *Test–Logic–Reset* controller state (when the example TAP controller applies a 0 to Reset*). If this state is entered as a result of signals received at the TCK and TMS inputs, then the reset occurs on the falling edge of TCK. If, on the other hand, the state is entered through use of the optional TRST* input (or on power–up), then the reset will occur immediately on entry into the state. Note that some instruction register cells might need to be designed to have preset, rather than reset, capability for the latched parallel output. This is necessary because the standard requires that the instruction present at the register's parallel output in the *Test–Logic–Reset* controller state must be the *IDCODE* or, if the optional device identification register is not provided, the *BYPASS* instruction (see Chapter 5).

4.4.2: Instruction Register Operation

Figure 4–13 gives a view of the sequence of events involved in loading a new instruction into the test logic, starting from the *Test–Logic–Reset* controller state. This figure shows the signals applied to and generated by the example TAP controller design included in Section 4.3. The hexadecimal characters shown for signal "State" show the movement between certain of the 16 TAP controller states as represented by the states of the four state flip–flops in Figure 4–8 and summarized in hexadecimal encoding in Table 4–1.

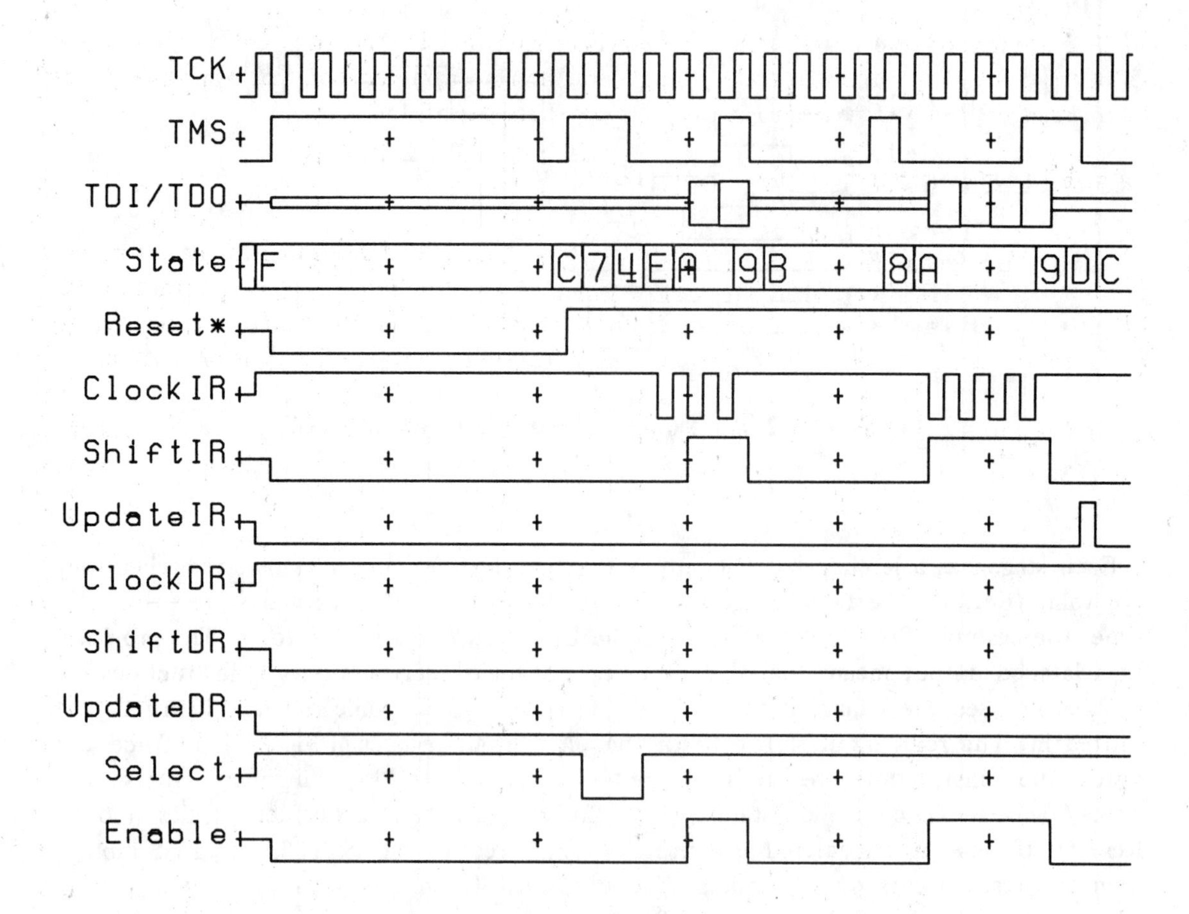

Figure 4–13: Loading a new instruction.

In the example of Figure 4–13, the circuit begins in the *Test–Logic–Reset* controller state. Instruction register scan is selected by manipulation of the signal applied to TMS. The scanning is interrupted by a pause and then continued. (Note the two periods of activity of ClockIR separated by a quiescent period.) Finally, instruction register scanning is completed and the TAP controller is taken to the *Run–Test/Idle* controller state.

Note that the new instruction becomes current on the rising edge of the UpdateIR signal from the TAP controller (i.e., on the falling edge of TCK in the *Update–IR* controller state).

4.5: The Test Data Registers

The test logic design provides for a bank of test data registers as shown in Figure 4–2. IEEE Std 1149.1 specifies the design of three test data registers, two of which must be included in the design. The mandatory test data registers are the bypass and boundary–scan registers. The provision of a device identification register is optional and further design–specific test data registers can be added as appropriate to a given design. The design–specific registers can be a part of the on–chip system logic and can have both system and test functions.

The design of the three test data registers specified by the standard is discussed in Chapters 5 and 6. In this section, the general design characteristics that apply to all test data registers (including design–specific registers) are described.

4.5.1: The Control of Test Data Registers

The operation of the various test data registers is controlled according to the instruction present at the output of the instruction register. An instruction can place several test data registers into their test mode of operation, but it might select only one register for connection as the serial path between TDI and TDO in the *Shift–DR* controller state.

IEEE Std 1149.1 requires that each named test data register must have a defined length (number of shift–register stages) and a defined set of operating modes. Thus, it will appear the same whenever it is accessed.

In practice, several test data registers can be constructed out of the same circuitry, for example, as shown in Figure 4–14. This circuit contains three test data registers:

1. a six stage register formed by enabling shifting through all six stages;

2. a three stage register formed from stages 2, 1, and 0; and

3. a three stage register formed from stages 5, 4, and 3.

This is acceptable provided the three test data registers are given unique names and each individually meets all the requirements of the standard. Therefore, some test data registers within an IC might appear as identifiable, dedicated circuit blocks while others might be "virtual" –– that is, they only exist when they are required by the current instruction.

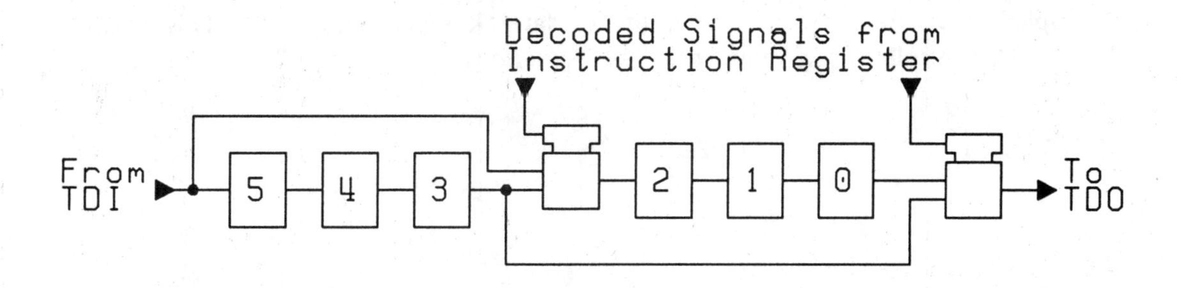

Figure 4–14: Sharing of circuitry between test data registers.

4.5.2: Test Data Register Operation

All test data registers operate according to the same principles:

- Registers that are not enabled for test operation by the current instruction are configured so that they do not interfere with operation of the on–chip system logic. Where a register can operate in either system or test mode, the system mode will be selected whenever the register is not required by the current test instruction. Because test data registers might not actually exist as distinct circuit blocks when they are not enabled (they can share circuitry with each other or with the system logic), they should be considered to have been left in an undefined, but safe (with respect to the system logic), state.

- The registers enabled for test operation by the current instruction will load data from their parallel inputs (if any) on the rising edge of TCK in the *Capture−DR* controller state, and will make any new data available at their latched parallel outputs (if any) on the falling edge of TCK in the *Update−DR* controller state. In other words, the results of a test are sampled in the *Capture−DR* controller state and the new test stimulus is available, at the latest, in the *Update−DR* controller state. Where test execution is required between the *Update−DR* and *Capture−DR* controller states (e.g., execution of a self−test), this occurs in the *Run−Test/Idle* state.

- The register selected by the instruction to be the serial path between TDI and TDO will shift data from TDI towards TDO in the *Shift−DR* controller state. Other test data registers enabled for test operation will hold their state while shifting occurs.

Figure 4−15 gives a view of the sequence of events involved in loading new test data into a selected test data register. We might imagine that Figure 4−15 is simply a continuation of Figure 4−13 which left the TAP controller in the *Run−Test/Idle* controller state after an instruction had been entered to select a data register. As in the earlier example, the shifting is done in two parts separated by a pause. (Note the activity on ClockDR.) At the completion of the shifting process, the UpdateDR signal goes active. This example ends with the controller being returned to the *Test−Logic−Reset* controller state.

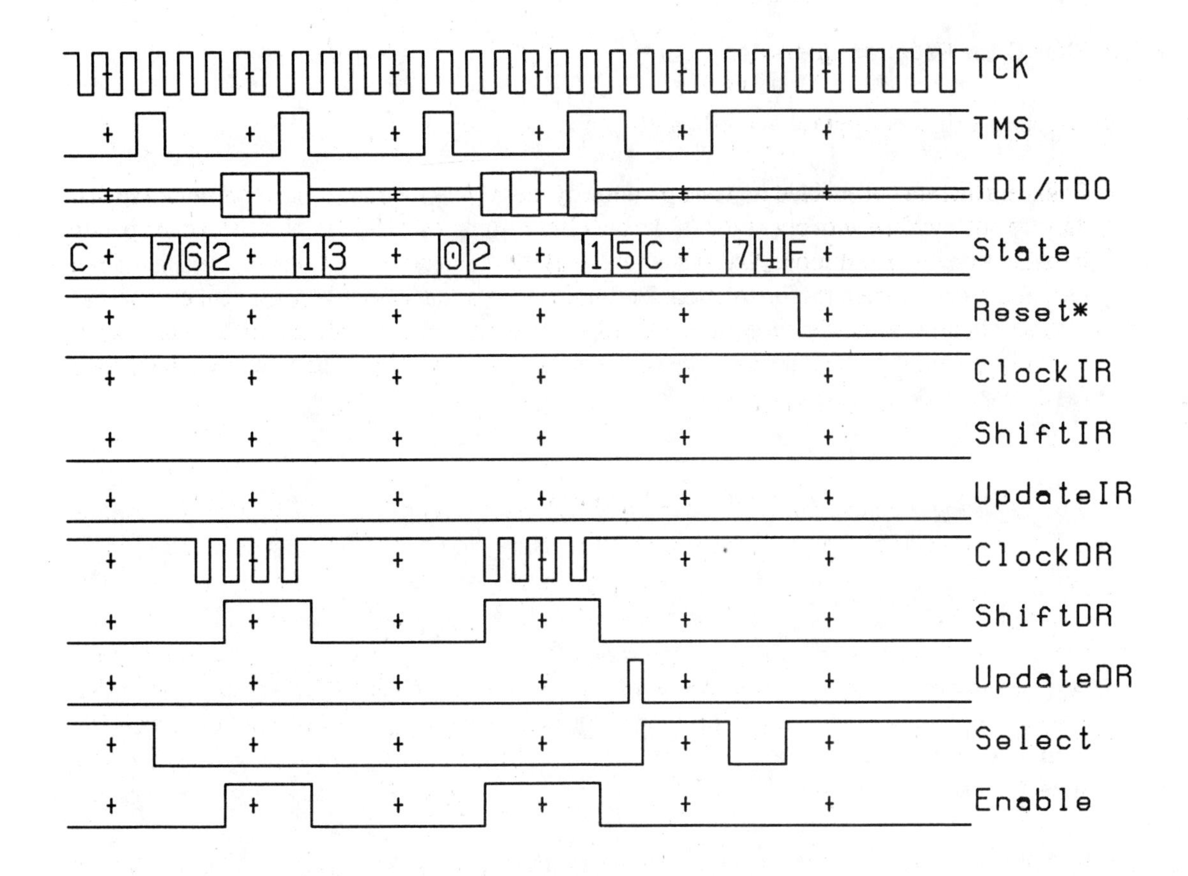

Figure 4–15: Loading new test data.

4.6: Reference

[1] IBM, Honeywell, and TRW, *VHSIC Phase 2 Interoperability Standards: TM–Bus Specification -- Version 3.0*, November 9, 1987 (available from J.P. Letellier, Naval Research Laboratory, Code 5305, Washington DC 20375, U.S.A).

Chapter 5. The Bypass and Device Identification Registers

This chapter describes two of the test data registers defined by IEEE Std 1149.1: the mandatory bypass register and the optional device identification register. The standard also defines three instructions for these registers: *BYPASS, IDCODE,* and *USERCODE*. These instructions are discussed below.

5.1: The Bypass Register

The bypass register must be present in all chips that conform to the standard. It provides a minimum length path between the test data input (TDI) and test data output (TDO) pins and can be accessed when there is no requirement to use another test data register in the chip. This allows data to be shifted through the chip without interfering with its system operation.

The bypass register consists of a single shift-register stage that loads a constant logic 0 in the *Capture-DR* controller state when the *BYPASS* instruction is selected. IEEE Std 1149.1 defines the binary code for the *BYPASS* instruction to be "all-1s" (i.e., a logic 1 entered into each stage of the instruction register).

The bypass register might be implemented as shown in Figure 5-1.

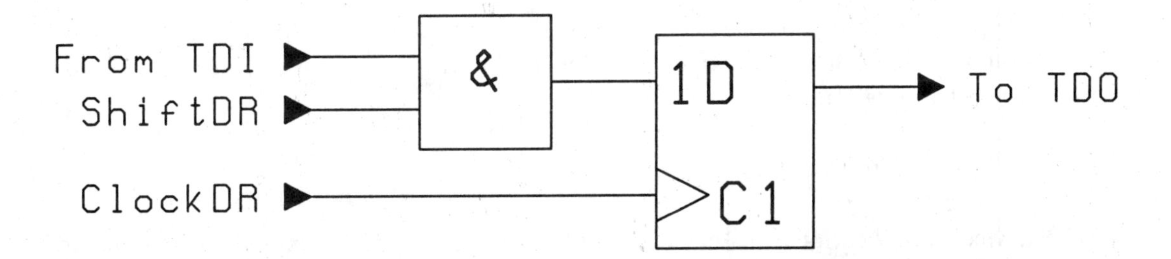

Figure 5-1: An example design for the bypass register.

The bypass register does not have a parallel data output so there is no significance to the data present in the register when shifting is completed. Its operation cannot interfere with that of the on-chip system logic.

5.1.1: Use of the Bypass Register

As an example of an occasion when the bypass register might be used, consider a board containing 100 integrated circuits (ICs), all with boundary-scan and connected into a single serial chain, a small part of which is shown in Figure 5-2. Assume that a need arises to access a test data register located in IC57, but that it is desired not to interfere with the operation of the remaining 99 ICs. (An example of such a situation might be when the target chip includes a "shadow" test data register that permits the state of its key internal registers to be read.)

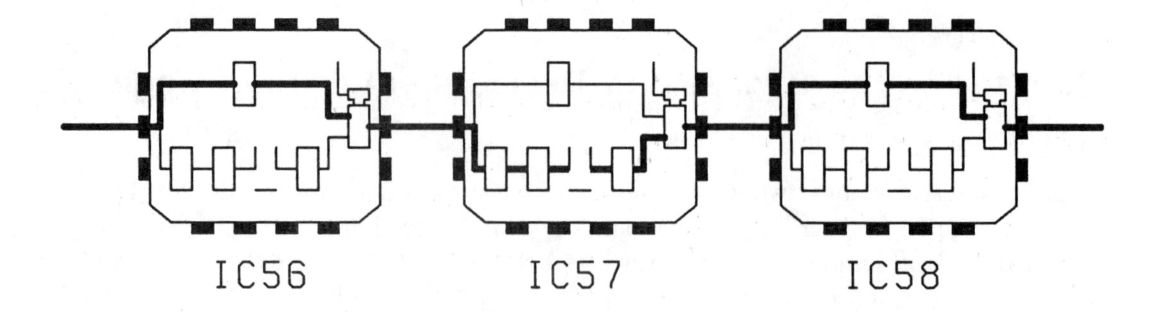

Figure 5–2: Use of the bypass register.

In this case, the required instruction would be loaded into IC57, with the *BYPASS* instruction being loaded into the other ICs. The serial bit stream shifted into TDI during the instruction scanning cycle would be:

111.......1111CCC...CCC1111.........111

where CCC...CCC is the instruction to be loaded into IC57. As a result of use of the "all–1s" value for the *BYPASS* instruction, the complexity of the bit stream input to the serial path is considerably reduced. This is an important consideration, since it reduces the data storage requirement for the automatic test equipment (ATE) or bus master chip that control the operation of the board during test.

Once the instructions are loaded, a minimum length serial path to and from the target chip is set up. This allows access to the chip of interest in the minimum possible time, increasing test throughput.

5.2: The Device Identification Register

The device identification register is an optional feature of the standard. Where included in the test logic, it allows a binary data pattern to be read from the chip that identifies the manufacturer, the part number, and the variant.

During testing, this information might be used to:

- adjust test program execution, depending on the source and/or variant of each chip present on the board;
- verify that the correct IC has been mounted in each board location; or
- establish which member of a plug–compatible family of boards is being tested.

5.2.1: Construction

The register contains 32 parallel-in, serial-out shift-register stages, each of which might be constructed as shown in Figure 5-3.

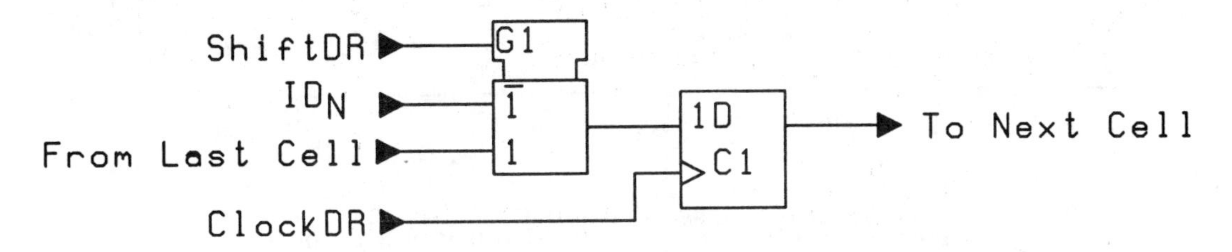

Figure 5-3: An example implementation of a device identification register cell.

Where a chip is programmed off-line (e.g., by blowing fuses or through some other nonreversible process), it is useful if the programmed state can also be observed via the device identification register. Therefore, where the function of the chip can be programmed by the user, each cell must have a pair of alternative data inputs so that two different 32-bit codes can be loaded -- one to identify the device and one to identify its programming. The former is loaded when the *IDCODE* instruction is selected, while the latter is loaded when the *USERCODE* instruction is selected.

When the register is addressed from the instruction register, the data pattern at its parallel input is loaded on the rising edge of the test clock (TCK) in the *Capture-DR* controller state. (At this time, the example TAP controller generates a rising edge on ClockDR while holding ShiftDR = 0.) These data are shifted toward TDO on the rising edge of TCK in the *Shift-DR* controller state, while data are shifted in from the TDI pin. (The example TAP controller changes ShiftDR to 1 and continues to generate clock edges on ClockDR.)

The device identification register has no parallel output and cannot interfere with the operation of the system logic in the chip. Therefore, when shifting is completed, the data present in the register have no significance.

5.2.2: The IDCODE Instruction

The structure of the data loaded into the device identification register in response to the *IDCODE* instruction is shown in Figure 5-4. As discussed previously, the data presented are loaded into the register from inputs ID_{31}-ID_0 in the *Capture-DR* controller state.

There are four separate fields:

1. *The header:* ID_0 loads a constant logic 1. Recall that the bypass register loads a constant 0 in the *Capture-DR* controller state. Later in this chapter, the advantage of this in determining the IC sequence for a given board will be explained.

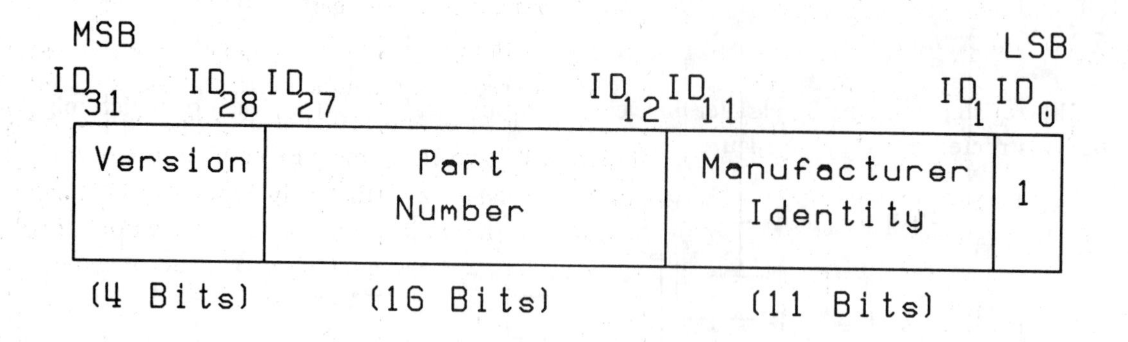

Figure 5–4: Structure of the device identity code.

2. *The manufacturer code:* ID_{11}–ID_1 load an 11–bit manufacturer code. This code is derived from a scheme managed by the Joint Electron Device Engineering Council (JEDEC) [1].

In the JEDEC scheme, each manufacturer is allocated a code consisting of one or more 8–bit bytes. The most significant bit in each byte ensures odd parity, so a maximum of 128 available manufacturers can be distinguished by a 1–byte JEDEC code. Clearly, however, there are more than 128 manufacturers of integrated circuits. To cater to those who cannot be allocated 1–byte codes, the code Hex 7F is reserved as a continuation character. One hundred and twenty–seven manufacturers are thus given codes consisting of just one byte, 127 are given 2–byte codes (the first byte being Hex 7F), a further 127 get 3–byte codes (the first two bytes being Hex 7F), and so on.

The scheme used in IEEE Std 1149.1 is a compressed form of this code containing a fixed number of bits (11) and is better suited to a serial environment. The 11 bits are derived from the JEDEC code as follows:

- Bits ID_7–ID_1 are the same as the seven data bits of the final byte of the JEDEC code.

- Bits ID_{11}–ID_8 contain a count of the number of continuation bytes in the JEDEC code (i.e., the total number of bytes in the JEDEC code minus one).

This scheme can uniquely identify up to 2032 manufacturers, since the pattern Hex 7F cannot occur in bits ID_7–ID_1. Section 5.3 will show how the 16 "invalid" manufacturer codes can be used to advantage during board testing. If more than 2032 manufacturer codes are issued by JEDEC, then the scheme will result in reuse of some code values within the manufacturer code field. However, the chance that a component from an incorrect manufacturer will have the same code *and* the same test functionality is acceptably low.

3. *The part number code:* $ID_{27}-ID_{12}$ provide a 16 bit part number, chosen by the manufacturer to distinguish a chip from the others that the company sells. In cases where more than 2^{16} chip types are offered by a manufacturer, part number codes might have to be reused. The objective is to minimize the chance that an incorrect chip in a given position on a board will have the same part number as the correct chip type. Given that 2^{16} codes are available, and that chip types will be further distinguished by the number of pins and the position of the test access port (TAP) pins, the chance of falsely receiving the expected part number code is extremely small.

4. *The version number code:* For chips that are manufactured in several different versions through their lives, bits $ID_{31}-ID_{28}$ can be used to distinguish up to 16 variants. As a minimum, the version code should distinguish variants of a chip that exhibit differences in the operation of the test logic -- e.g., different behavior in response to instructions or in the data to be sent or received through the TAP.

5.2.3: The USERCODE Instruction

In response to the *USERCODE* instruction, data are loaded from the alternative data input to the register -- $USER_{31}-USER_0$. Unlike the data presented to $ID_{31}-ID_0$, these data can be programmed by the user at the same time (and in the same way) that the function of the chip is programmed.

$USER_0$ must load a constant logic 1, while the structure of the data presented at $USER_{31}-USER_1$ could be identical to that of the device identification code (i.e., variant, part number, and company).

Note that this second data input is required only for chips whose function is "one-time" programmed off-line (e.g., by blowing fuses or through some other irreversible process) and cannot be modified through use of the test logic (e.g., by sending programming instructions through the TAP). For "soft" programmable chips whose programmed function is determined by instruction and data sequences entered through the TAP, the *USERCODE* instruction is not required. In such cases, the chip can be set arbitrarily to perform any desired function at the start of a test. Therefore, knowledge of the previously-programmed function is not required.

5.3: Learning the Structure of an Unknown Board

There are occasions when it would be useful to be able to access the device identification registers of chips to learn more about the precise mix of chips mounted on a particular board. For example, a board can be configured to perform one of a range of functions by including a different chip in some particular location.

There are two problems that have to be solved to permit this kind of "blind" interrogation:

- the device identification register is optional, so not every chip will include one; and

- because the value of the *IDCODE* instruction will vary from chip to chip -- indeed, the length of the instruction register can vary from component to component -- there is no way to know in advance what sequence of instructions to enter to select the device identification registers.

Two features are included in IEEE Std 1149.1 to allow these problems to be solved.

First, it is required that the instruction register's latched parallel output is initialized in the *Test–Logic–Reset* controller state to:

- the value of the *IDCODE* instruction if a device identification register is included in the chip; or

- the value of the *BYPASS* instruction if the device identification register is not provided.

Therefore, by moving from the *Test–Logic–Reset* controller state directly into the test data register scan sequence (starting with the *Capture–DR* controller state) all available identification codes on the board will be shifted out for examination. Referring to the TAP controller state diagram in Figure 4–6, the application of the sequence "111110100...0" to the TMS of all chips on a serial board–level path (one bit per cycle of TCK) will cause all available identification codes to be output, regardless of the starting states of the TAP controllers.

Second, because the standard requires that all identification codes have a logic 1 in the least significant bit (the header bit) while the bypass register is required to load a logic 0, it is possible to locate identification codes in the output bit stream. Consider, for example, the output sequence shown in Figure 5–5. A flow chart for decoding such an output sequence received is shown in Figure 5–6.

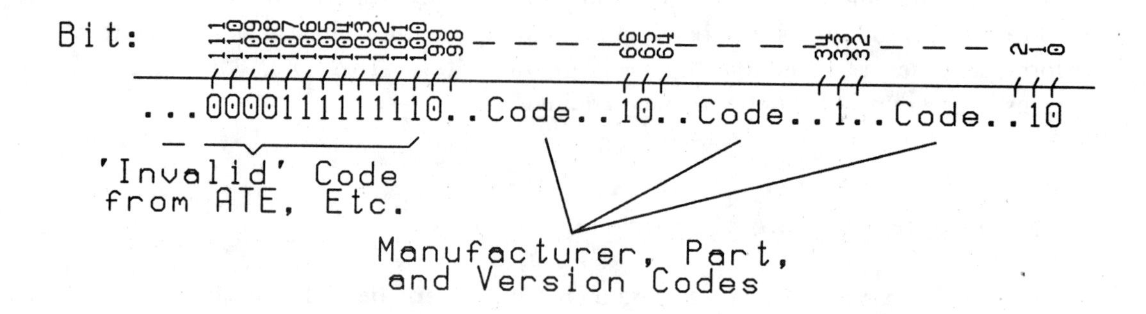

Figure 5–5: Output sequence following 'blind' access.

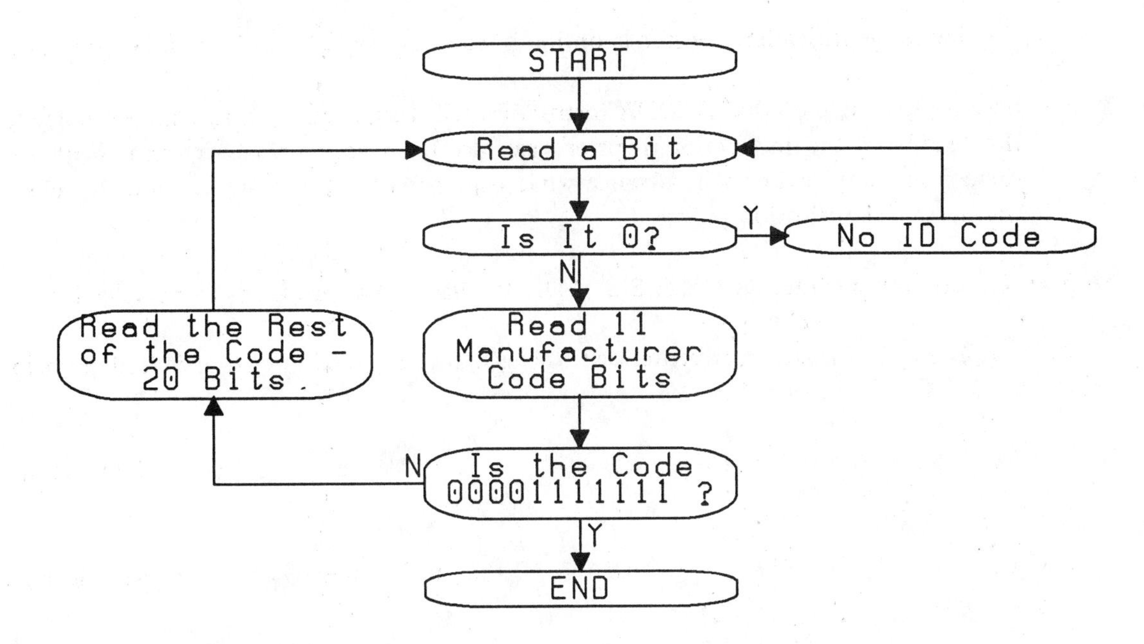

Figure 5–6: Flow chart for decoding output identity code sequence.

For the example sequence, this gives the result shown in Table 5–1. Note that by injecting at the board's serial input an identification code containing an invalid manufacturer code, it is possible to determine when the end of the sequence has been reached.

Table 5–1: Result of decoding the received sequence.

Bit(s)	Component	Comment
0	1	No ID code
1-32	2	ID code available
33-64	3	ID code available
65	4	No ID code
66-98	5	ID code available
99	6	No ID code
100-111	-	Invalid manufacturer - end of sequence

5.4: Reference

[1] Joint Electron Device Engineering Council, "Standard Manufacturer's Identification Code," *JEDEC Publication 106–A*, July 1986. (Obtainable from JEDEC, 2001 Eye Street. N.W., Washington, D.C. 20006, U.S.A)

Chapter 6. The Boundary-Scan Register

Every integrated circuit (IC) that complies with IEEE Std 1149.1 must include a boundary-scan register, which can be used to allow interconnections between ICs to be tested (the interconnect test described in Chapter 2). Optionally, it can also be used to support testing of the logic within the component -- either in conjunction with self-test or by shifting patterns and results on a test-by-test basis (again, as described in Chapter 2).

While many different implementations of boundary-scan are possible that would provide this level of functionality, the standard imposes a number of particular requirements. These ensure that boundary-scan paths included in chips obtained from two or more different vendors can be used reliably in concert to perform board interconnect testing.

Later in this chapter, we will describe the operation of the boundary-scan register and will illustrate how it might be designed through a series of example circuits. As we did in the introduction to Part II, we again stress that there are several features of the standard that we will not be able to discuss in this tutorial. Readers are therefore strongly recommended to consult the standard itself before implementing an IC design.

6.1: The Provision of Boundary-Scan Cells

Before discussing the provision of boundary-scan cells in an IC, two terms must be defined:

1. *The on-chip system logic:* This is the circuitry contained in the IC to allow it to perform the required "normal" function. For example, if the chip is intended to operate as a counter, then the on-chip system logic would comprise all the necessary circuitry to construct a counter.

2. *The test logic:* This is the circuitry built into the IC to assist either in testing of the on-chip system logic (e.g., confirming that the counter is indeed able to count) or in testing off-chip circuitry (e.g., board level interconnections).

Where design-for-test features are built into the on-chip system logic, these are regarded as a part of the test logic in their test mode of operation; otherwise, they are a part of the on-chip system logic.

To comply with IEEE Std 1149.1, an IC must contain boundary-scan cells at all off-chip system inputs and outputs†, as shown in Figure 6-1. That is, cells should be located:

† Cells are not required at connections between the test logic and the on-chip system logic or as the test access port (TAP) pins.

- between each system input pin (clock or data) and the corresponding input to the on–chip system logic;

- between each output from the on–chip system logic and the corresponding system output pin; and

- between each 3–state enable or direction control output from the on–chip system logic and the corresponding system pin output driver.

Note that, for chips that contain some analog circuitry between the on–chip logic and the system pins, the connections to and from the analog circuit block are treated exactly as if they were off–chip digital connections. This topic will be discussed further in Chapter 19.

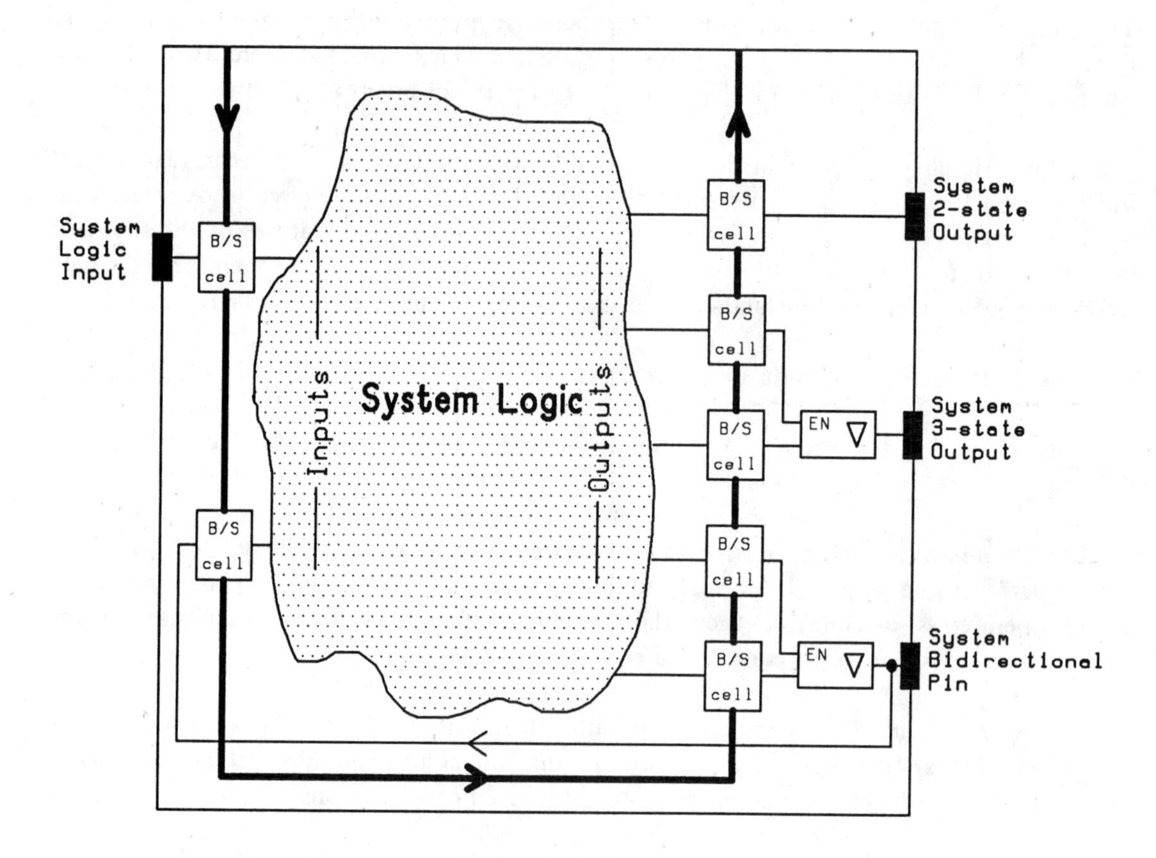

Figure 6–1: Provision of boundary–scan cells.

Of particular note are the cells located at the output enable and direction control outputs from the on–chip system logic to 3–state output and bidirectional pins, respectively. Operating in conjunction with the cells at the data connections of the on–chip system logic, these cells allow the state of the output driver (active or inactive), as well as the data value driven when the driver is active, to be controlled. The reason for the inclusion of these cells is illustrated in Figure 6–2.

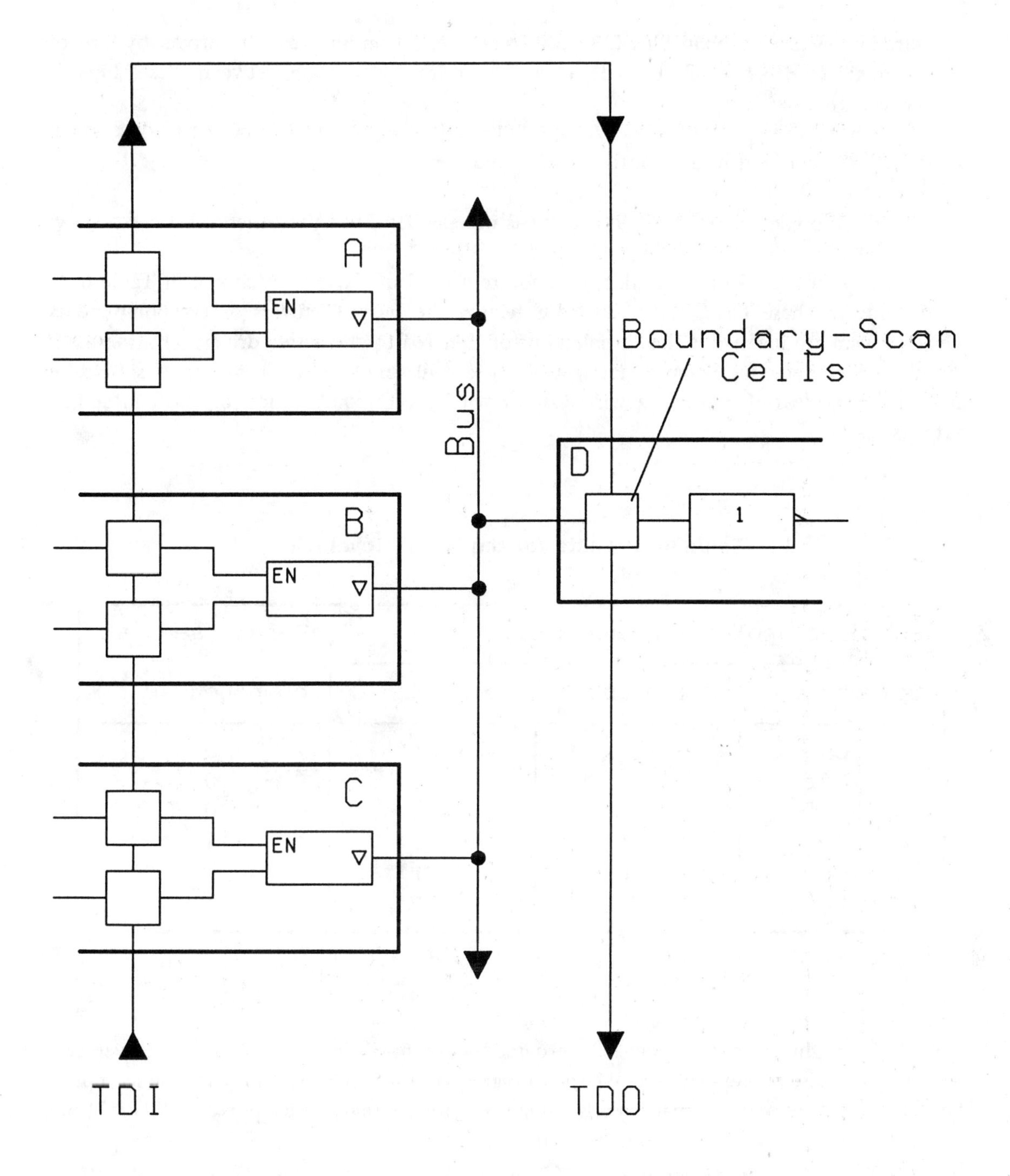

Figure 6–2: A board–level bus connection.

Figure 6–2 shows a board-level 3-state bus connection that can be driven by one of three chips: A, B, or C. To provide a test of the interconnection between these chips, it is necessary to check that:

- the bus can be driven to both 0 and 1; and
- each chip can drive signals onto the bus independently of the others.

For the circuit in Figure 6–2, this will require a total of six tests as shown in Table 6–1. Note that, in these tests, the data value fed to the output buffers of the components whose drivers are inactive is the complement of that fed to the active driver. This increases the chance of detecting a fault that would cause a driver to be active when it should be inactive, regardless of whether a wire-OR or wire-AND combination of the contending outputs results.

Table 6–1: Tests for the board–level bus.

Stimulus applied to bus from:			Result seen at
Component A	Component B	Component C	component D
0/on	1/off	1/off	0
1/off	0/on	1/off	0
1/off	1/off	0/on	0
1/on	0/off	0/off	1
0/off	1/on	0/off	1
0/off	0/off	1/on	1

While it might seem that the cells provided to control the activity of the driver at a 3-state or bidirectional pin might form a significant fraction of those in the complete boundary-scan register -- particularly where a chip has many such pins -- this will not normally be the case. The reason is that chips often have groups of 3-state outputs or bidirectional pins that are controlled from a single source. In such cases, all the outputs that form an address bus would be active or inactive simultaneously. It would be a design error if two or more such pin groupings were connected at the board level; therefore, it is only necessary to provide one output enable or direction control cell for each group of pins. Figure 6–3 provides an example to illustrate this point.

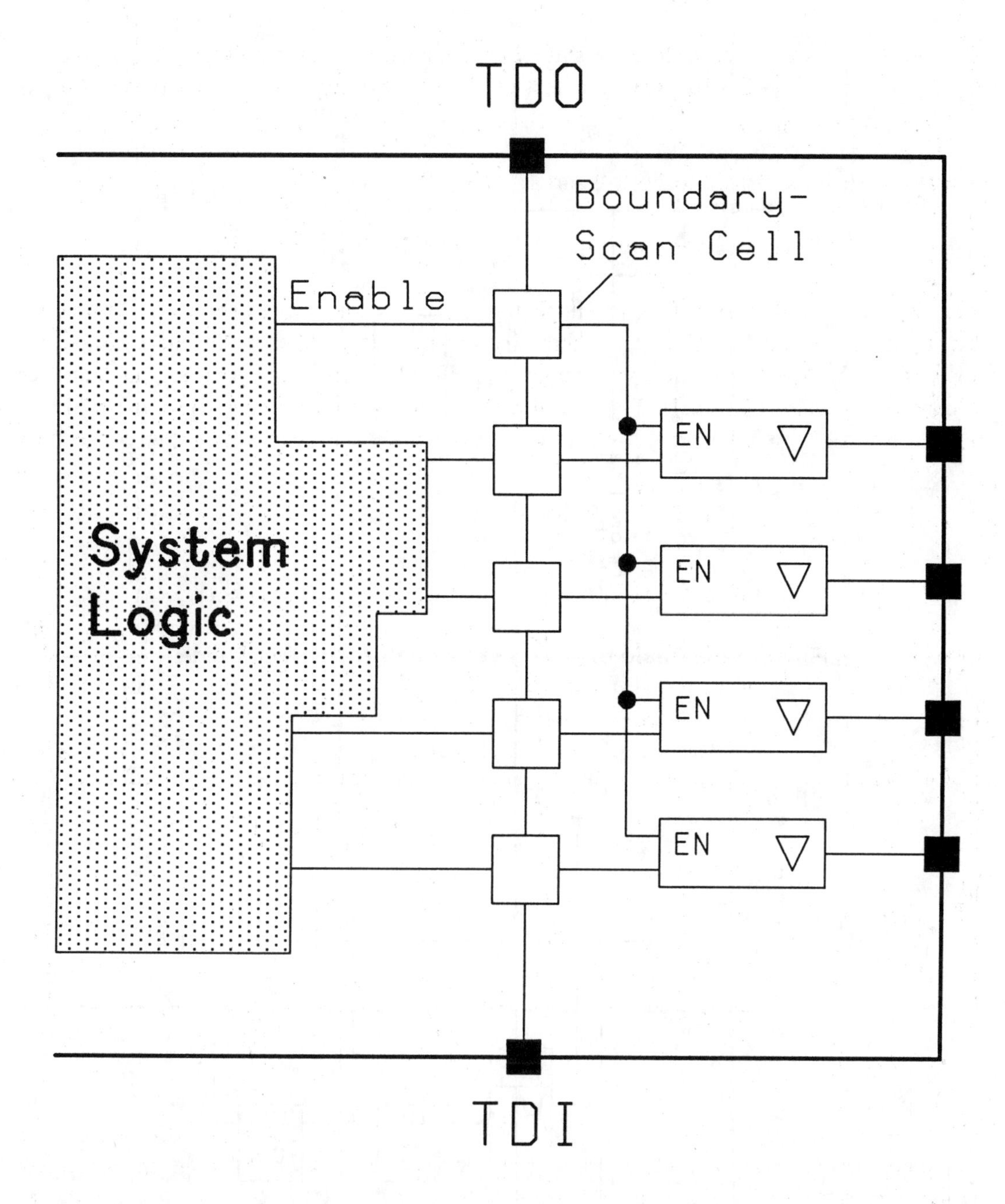

Figure 6–3: Control of multiple 3–state outputs from a single source.

6.2: The Minimum Requirement

Figures 6–4 and 6–5 show boundary–scan cell designs that meet the minimum requirements of the standard for input and output pins, respectively. In these cell designs, the signals ShiftDR, ClockDR, and UpdateDR are those generated by the example TAP controller (see Figures 4–8 and 4–9).

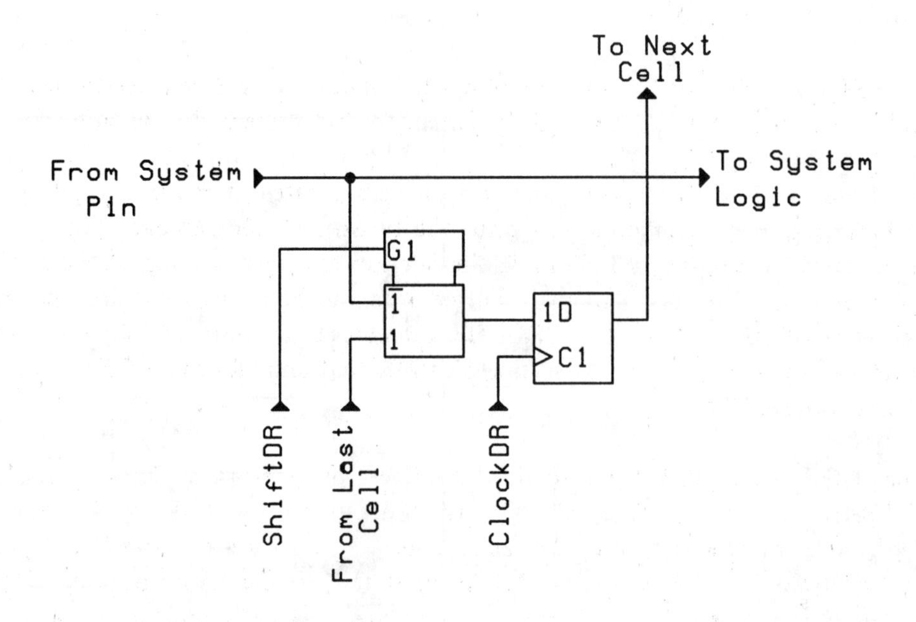

Figure 6–4: Basic boundary–scan cell for an input pin.

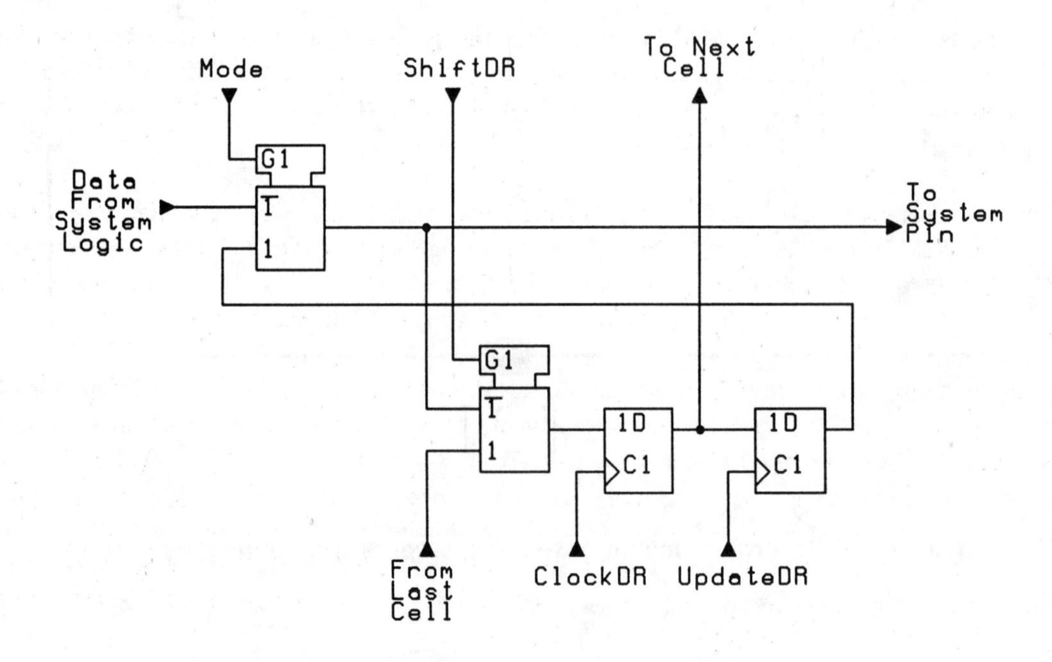

Figure 6–5: Basic boundary–scan cell for an output pin.

These boundary–scan cells allow an IC to support the two mandatory instructions defined by the standard: *EXTEST* and *SAMPLE/PRELOAD*. The Mode signal in Figure 6–5 is generated by decoding the current instruction and should be 1 when the *EXTEST* instruction is present; otherwise it should be 0.

6.2.1: EXTEST

The *EXTEST* (external test) instruction allows the boundary-scan register to be used for board-level interconnect testing in a similar manner to that presented in Chapter 2:

- Test stimuli shifted into the boundary-scan cells located at system output pins are driven through the connected pins onto the board interconnections. This process is started by first entering the *EXTEST* instruction and then moving to the *Shift-DR* controller state. One bit of data is shifted into the boundary-scan register on each rising edge of the test clock (TCK). The example TAP controller shown in Figures 4-8 and 4-9 enables shifting by setting ShiftDR to 1 and allowing TCK to propagate through to ClockDR.

- When entry of stimuli is concluded, the shifting process is completed by moving to the *Update-DR* controller state. On the falling edge of TCK in this state, the stimuli are transferred from the shift-register stages onto the latched parallel outputs of each cell. Because the Mode input to the cells at system output pins is set to 1 by the *EXTEST* instruction, the test is applied to the board interconnections at this time. The example TAP controller generates a rising edge on UpdateDR to cause the latched parallel outputs of the example boundary-scan cells to be updated from the associated shift-register stages.

- The test results are captured in the cells at the system input pins. This occurs on the rising edge of TCK in the next *Capture-DR* controller state. The example TAP controller causes data to be captured by holding ShiftDR at 0 and allowing TCK to propagate through to ClockDR.

- The test results are examined by moving back to the *Shift-DR* controller state. The data held in the boundary-scan register move one stage towards the test data output (TDO) on each rising edge of TCK. The data in cell number 0 (the cell nearest to TDO) appear at TDO on the falling edge of TCK *after* it reaches the cell.

Note that the output pin cell contains an additional register between the shift-register stage and the output to the connected system pin.† In Figure 6-5, this additional register is clocked by the signal UpdateDR, generated by the example TAP controller in the *Update-DR* controller state. This allows the data present in the shift-register stage to be latched onto the parallel output of the cell when shifting has been completed. It is held there until the next test stimulus has been completely shifted into the boundary-scan path, ensuring that the data driven from the cell when Mode is 1 changes cleanly from one serially-supplied stimulus to the next.

Provision of this latched output to the connected system pin allows the boundary-scan cells at system output pins to be used to apply test stimuli to circuitry external to the chip

† The standard permits use of either an edge-triggered register or a level-operated latch to fulfill this requirement. We have chosen to use an edge-triggered flip-flop in the examples contained in this book.

in a carefully–controlled manner. For example, clocks or inputs to asynchronous circuits can be included among the signals that feed into the external logic, as illustrated in Figure 6–6. These signals (as well as others -- for example, see the discussion in Chapter 19 regarding signals that feed into analog circuits) must not change state between one test pattern and the next. Any intervening changes will cause misoperation of the circuit under test. Therefore, it is necessary to prevent the data from being applied to the external circuitry as they pass along the boundary–scan path during shifting–in. The latched parallel output is included to meet this requirement.

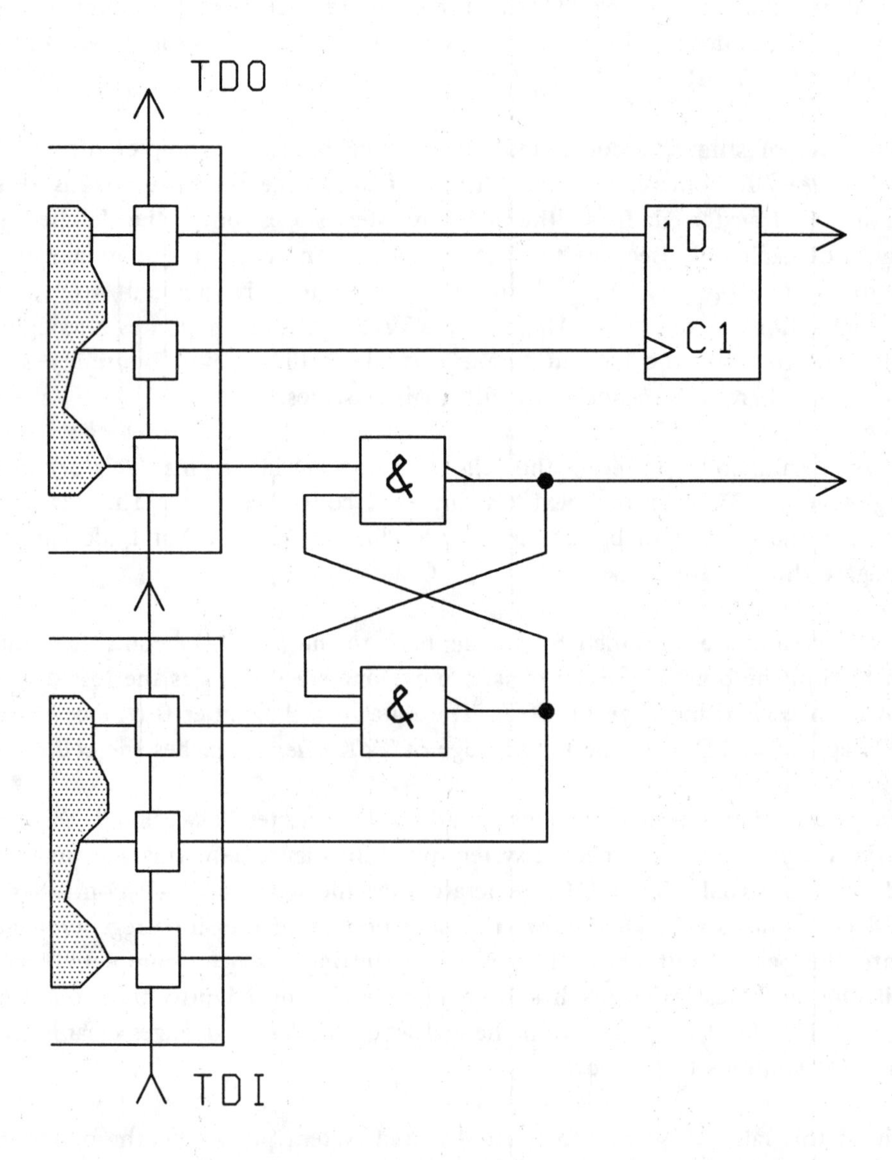

Figure 6–6: Using the boundary–scan path to test external logic.

6.2.2: SAMPLE/PRELOAD

While the *SAMPLE/PRELOAD* instruction is selected, the Mode input to the cells at system logic outputs is held at 0 -- allowing the chip to continue its normal operation without interference.

The instruction supports two distinct test operations.

In the first instance (*SAMPLE*), the boundary-scan cells at both inputs and outputs load the state of the signal flowing through them between the system pin and the on-chip logic:

- A snap-shot of the data flowing through the chip's system input and output pins is taken by first selecting the *SAMPLE/PRELOAD* instruction and then moving to the *Capture-DR* controller state. Data are sampled on the rising edge of TCK in this state. (At this time, the example TAP controller holds ShiftDR at 0 and applies a rising edge to ClockDR.)

- The captured data can be shifted out for examination in the *Shift-DR* controller state. On each rising edge of TCK, the data held in the boundary-scan register move one stage towards TDO. A data bit that arrives in cell number 0 (the cell nearest TDO) is driven through TDO on the following falling edge of TCK. (The example TAP controller holds ShiftDR at 1 and generates a rising edge on ClockDR for each rising edge of TCK.)

Applications of the *SAMPLE* test include debugging of prototype boards and a contactless form of the guided-probing process common on functional board testers.

In the second instance (*PRELOAD*), data can be shifted into the boundary-scan cells without interfering with the normal flow of signals between the system pins and the on-chip logic. This allows the latched parallel outputs in boundary-scan cells to be primed with data before another boundary-scan instruction is selected:

- The desired data are shifted into the boundary-scan register by first selecting the *SAMPLE/PRELOAD* instruction and then moving to the *Shift-DR* controller state. On each rising edge of TCK, one data bit is shifted into the register. (The example TAP controller generates clock transitions on ClockDR. ShiftDR is held at 0 for one clock cycle and then changed to 1.)

- When all data have been entered, shifting is halted by moving to the *Update-DR* controller state. On the falling edge of TCK, the data in each shift-register stage is shifted onto the cell's latched parallel output. (At this time, the example TAP controller generates a rising edge on UpdateDR.)

By loading suitable data when *PRELOAD* is selected, the user can ensure that all signals driven out of the chip are defined as soon as the *EXTEST* instruction is selected. The Mode input would change to 1 in response to the instruction change, allowing the data held in the boundary–scan cell (rather than the data generated by the on–chip logic) to be driven from the chip.

6.2.3: Cells for 3–state and Bi–directional Pins

Figures 6–7 and 6–8 show boundary–scan cells that could be used at 3–state output and bidirectional system pins, respectively, of an IC. These figures include the additional cell required to control the activity of the output driver. Both figures contain two shift–register stages -- one for data and one for output driver control. The signal CHIP_TEST* is 0 when the *INTEST* or *RUNBIST* instruction is selected (see Sections 6.3 and 6.4).

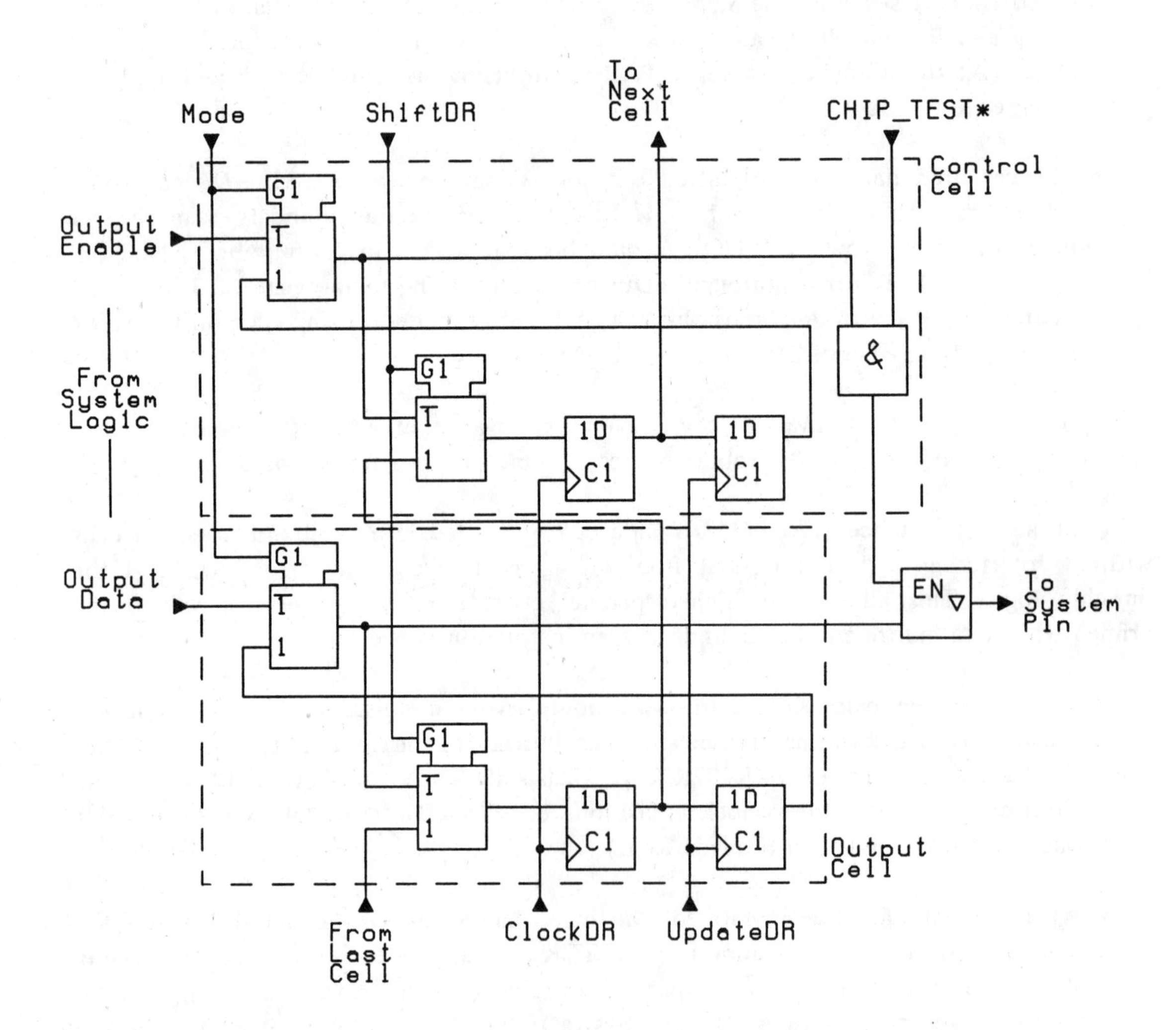

Figure 6–7: Basic boundary–scan cells for a 3–state output pin.

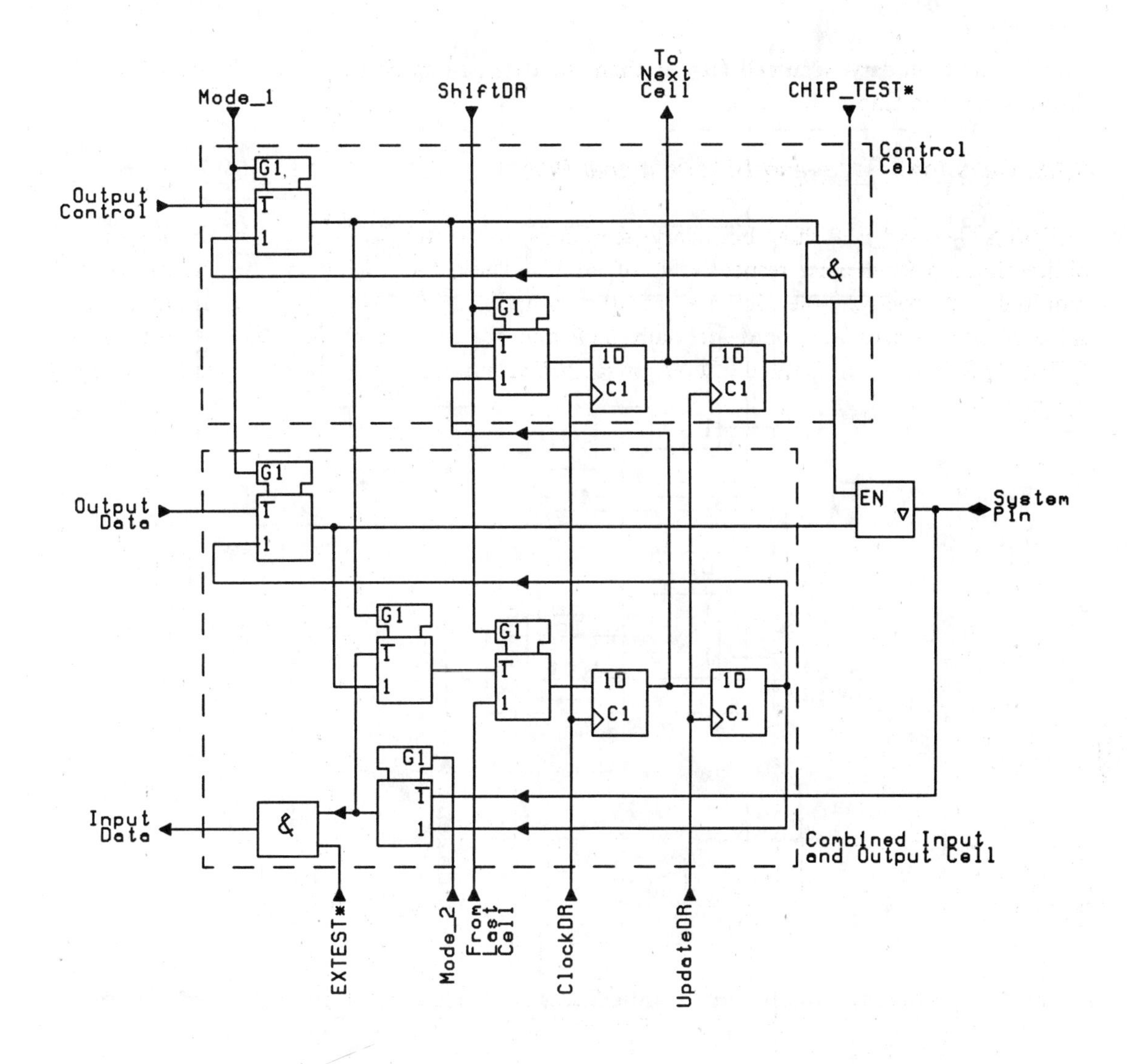

Figure 6–8: Basic boundary–scan cells for a 3–state bidirectional pin.

The design in Figure 6–8 is, in effect, a merging of those in Figures 6–4 and 6–7. It functions in the same way as the cell in Figure 6–7 when "output" operation is required and as the cell in Figure 6–4 when "input" operation is required. In Figure 6–8, the assumption is made that the bidirectional pin is either input or output at a given instant -- but never both simultaneously. This allows one shift–register stage to be used to convey the data value for the pin; two stages would be necessary were the pin to always be used as an input, allowing data to be driven out of the pin to be determined and data received at the pin to be monitored.

Figure 6–9 shows how a boundary–scan cell might be constructed for a 2–state open–collector bidirectional pin.

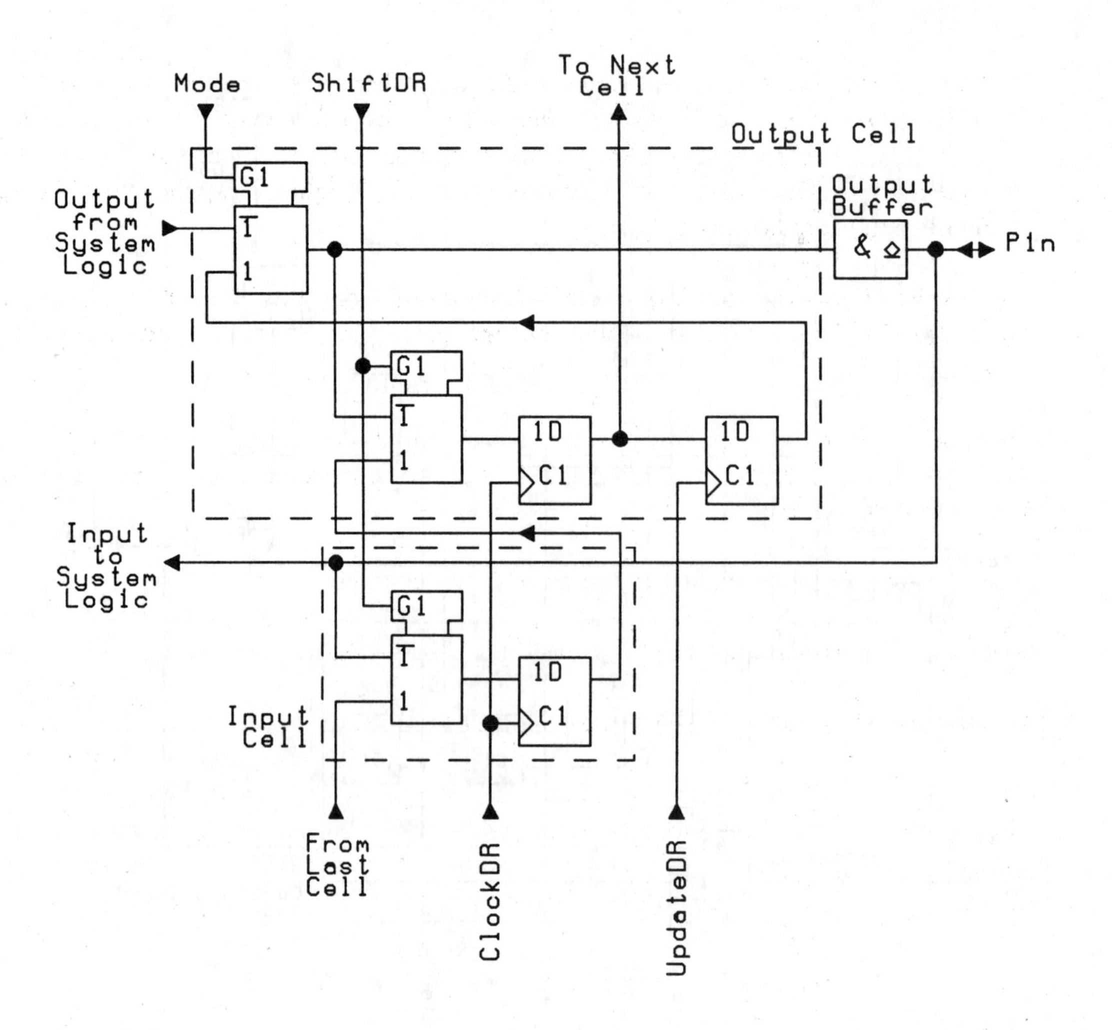

Figure 6–9: Basic boundary–scan cells for a 2–state open–collector bidirectional pin.

Boundary–scan cells for other types of pins can be constructed in a similar manner by correct combination of the cells for input and output pins.

6.3: The INTEST Instruction

The standard defines two optional instructions that can be used to perform tests of the on–chip system logic. The first of these is the *INTEST* instruction. The operation of the boundary–scan register when the *INTEST* instruction is selected is similar to that described for internal logic testing in Chapter 2:

1. Test stimuli for the on–chip logic are shifted into the cells at system input pins. Following the falling edge of TCK in the *Update–DR* controller state, the test stimulus is in place and is applied to the inputs of the on–chip system logic.

2. Between the *Update–DR* and *Capture–DR* controller states, the test is applied. For stored–state system logic designs, this will require entry into the *Run–Test/Idle* controller state where appropriate clock transitions will be applied to the on–chip system logic. This might require control of the clock signal(s) supplied to the clock input pin(s) (see Section 6.3.2).

3. On the rising edge of TCK in the *Capture–DR* controller state, the results are loaded into the cells at system output pins prior to being shifted out for examination.

Because of the slow test application rate, the chip must be able to support single–step operation where the *INTEST* instruction is offered. This requirement can be met in several ways, for example, where:

- no dynamic logic is included in the on–chip system logic; or
- the on–chip system logic can be placed in a "hold" state between tests.

6.3.1: Boundary–Scan Cell Designs That Support INTEST

Input Pins: To support this instruction, the design of the boundary–scan cells at non–clock system input pins must be extended beyond that of the cell shown in Figure 6–4. This is necessary to allow the data shifted into the cell to be driven to the connected system logic input. Figures 6–10 and 6–11 show two options for doing this.

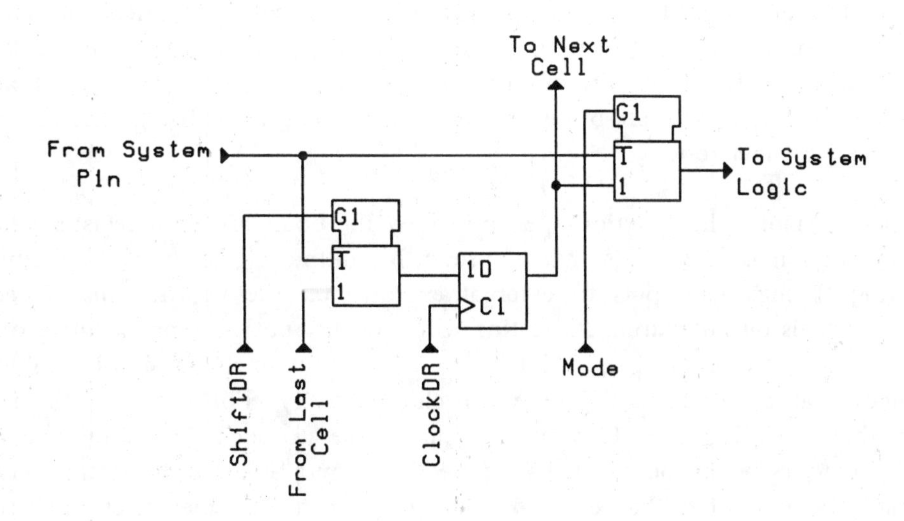

Figure 6–10: Enhanced boundary–scan cell for an input pin: Example 1.

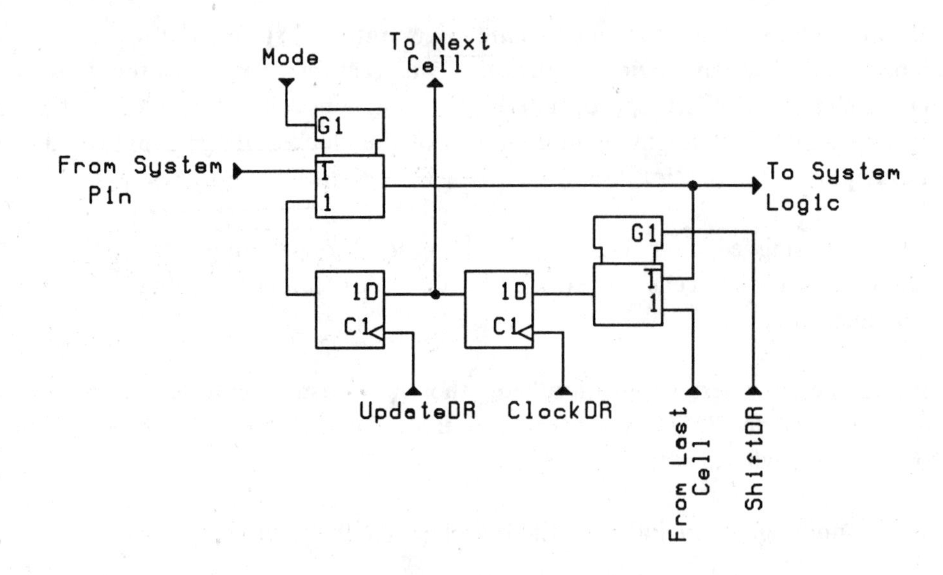

Figure 6–11: Enhanced boundary–scan cell for an input pin: Example 2.

In the design in Figure 6–10, the data in the shift–register stage are applied directly to the on–chip logic. This is acceptable provided the on–chip logic does not respond to the data that is shifted through the cell as each test is loaded and each set of results is examined. For example, the cell might feed the data input to a flip–flop whose clock was constrained not to change state during the shifting process. Therefore, the data applied from the boundary–scan cells become significant only when the flip–flop is clocked.

The design in Figure 6–11 is better than that in Figure 6–10 in cases where the circuitry fed by the cell will respond to the shifting data. This cell design is identical to that shown earlier for a system output pin (Figure 6–5) because it is targeted at the same problem. As when the cell is used at a system output pin, the added register (or latch) holds the stimulus data while new data are being shifted in, preventing the shifting data values from reaching the logic under test.

Output pins: Among the functions performed by the boundary–scan register when the *INTEST* instruction is selected is that of preventing output signals of the on–chip logic from flowing through chip pins to external circuitry on the board. This is necessary because the signals output during IC testing will probably not be representative of those generated as a result of normal operation. They might contain illegal signal combinations or sequences that cause damage to the off–chip circuitry. For example, the memory controller shown in Figure 6–12 would normally operate such that only one of the connected memories would be enabled to drive the output bus. During testing, however, signals might be generated that enabled two or more of the memories onto the bus simultaneously. The resulting contention between output drivers might cause damage to either memory chip.

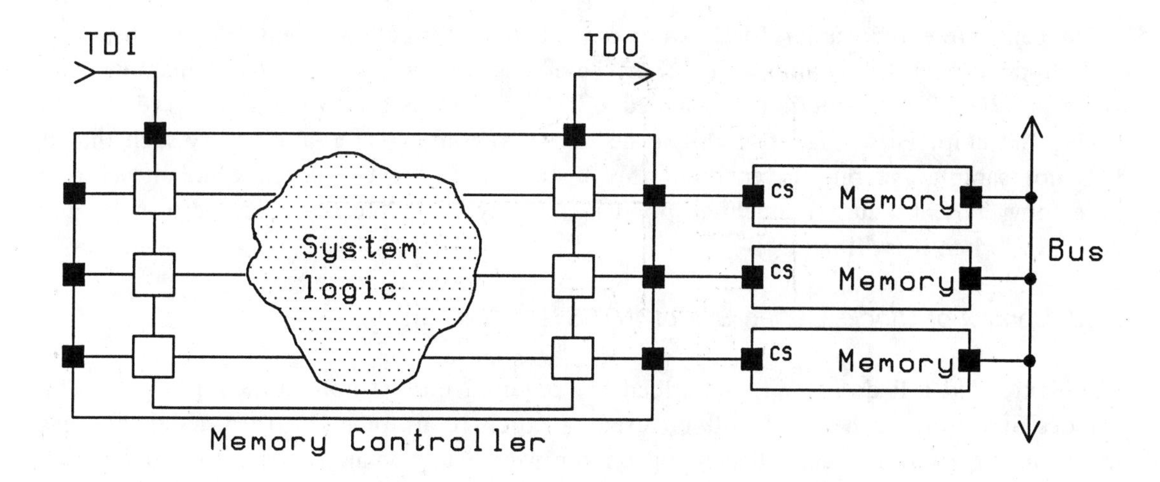

Figure 6–12: A circuit where bus contention might occur.

This problem is overcome by enhancing the design of boundary–scan cells for output pins when the *INTEST* instruction is to be supported. As shown in Figure 6–13, the design is changed so that data can be fed off–chip independently of that received from the on–chip system logic. This is not possible with the cell design in Figure 6–5, because of the feedback loop through the cell. If the data received from the on–chip system logic is to be captured into the cell, it will also be driven off–chip. The cell in Figure 6–13 is a feed–forward design that allows the user to define the chip's output, independently of the operation of the on–chip system logic, while the on–chip system logic test is in progress.

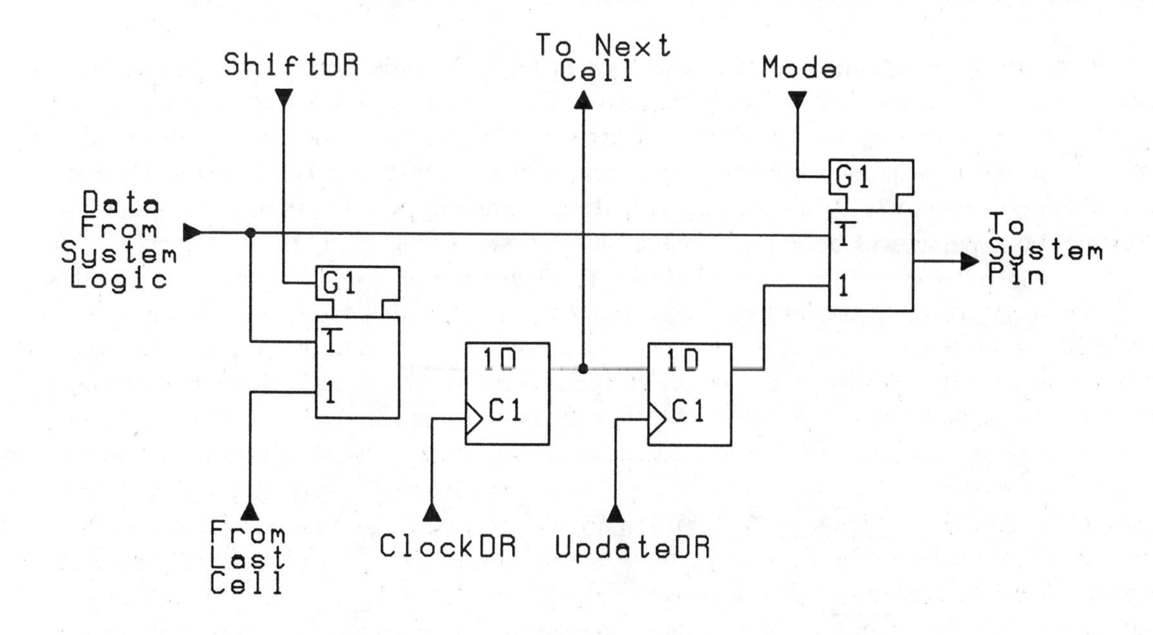

Figure 6–13: Enhanced boundary–scan cell for an output pin.

The cells presented earlier for 3–state output and bidirectional pins (Figures 6–7 and 6–8) support the *INTEST* instruction. For these cells, the output is set to an inactive state while the *INTEST* instruction is selected (CHIP_TEST* is set to 0). This prevents data leaving the chip. Note that it might be necessary to control external circuitry such that it does not sample the bus driven from the 3–state or bidirectional pin while the chip is undergoing test, because it might respond incorrectly when the bus is not driven by any chip (i.e., when it is "floating").

6.3.2: Control of Clocks During Use of INTEST

The extended cell designs just described are required only at non–clock input pins. The cell design of Figure 6–4 can still be used at clock input pins. Further design changes might be required, however, depending on the way that clocking of the on–chip system logic is to be controlled during testing. The following are three possibilities:

1. The system clock signal supplied to the chip can be externally controlled such that action–causing transitions will occur only in the *Run–Test/Idle* controller state, for example, as shown in Figure 6–14.

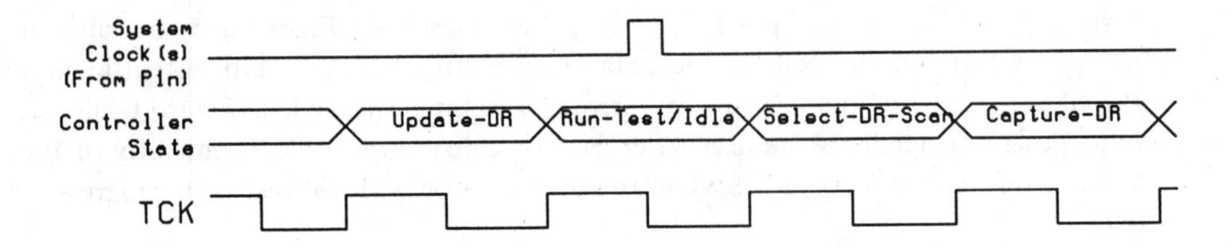

Figure 6–14: Control of the signal supplied to a clock input during INTEST.

2. A signal generated from TCK can be used in place of the externally–supplied signal while the *INTEST* instruction is selected. This signal must be controlled so that TCK pulses will be applied to the on–chip system logic only in the *Run–Test/Idle* controller state. The example shown in Figure 6–15 provides a positive edge clock to the on–chip system logic.

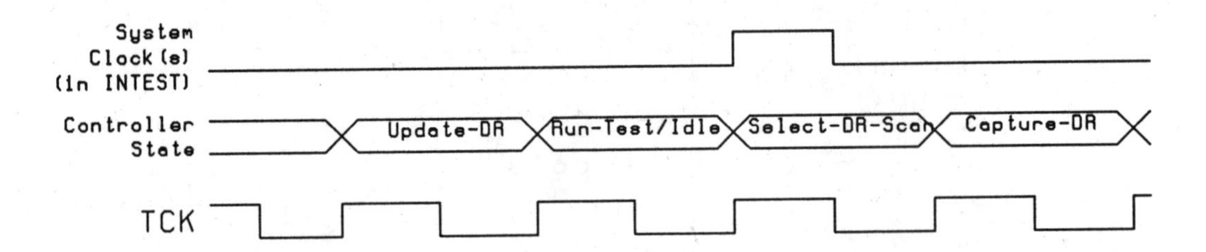

Figure 6–15: Generation of a system logic clock from TCK during INTEST.

3. A free-running clock could be supplied to the component and fed through to the on-chip system logic. In this case, the system logic must be placed in a "hold" state so that clock transitions received other than in the *Run-Test/Idle* controller state will not change the state of any of the stored-state devices contained in the on-chip system logic. Where a component has a HOLD* input (e.g., as is common on microprocessors to allow single-step operation), the signal fed to the on-chip system logic while the *INTEST* instruction is selected can be modified to be pulsed following entry into the *Run-Test/Idle* controller state. An example of how this could be achieved is shown in Figure 6-16. In this figure, the RT/I signal is 1 when the test logic is in the *Run-Test/Idle* state. The INTEST signal is true when the *INTEST* instruction is selected.

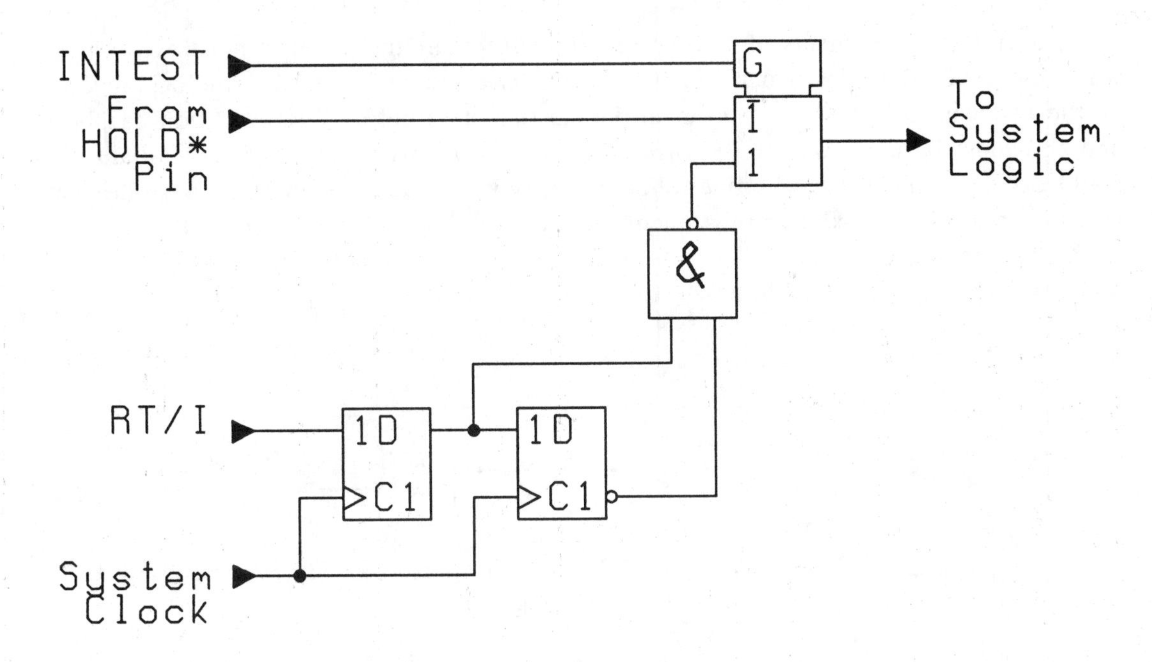

Figure 6-16: Generation of a "HOLD*" pulse.

6.4: The RUNBIST Instruction

The purpose of the optional *RUNBIST* instruction is to provide a consistent, straight-forward means of verifying the health of an IC through using embedded self-test facilities. The objective is to allow a health check to be run simultaneously in every chip on a board that supports the instruction without the need for complex control and/or data sequences. In effect, the *RUNBIST* instruction allows the user to ask the chip "Are you healthy?" and to receive the component's reply. As we will see in Section 6.4.2, when the *RUNBIST* instruction is selected it is necessary for output pins to be set to defined states independent of the operation of the on-chip system logic.

6.4.1: Execution of the Health Check

There are many different ways of building self-test features into an IC design. For example, self-test can be based on the inclusion of linear-feedback shift-registers (LFSRs), signature analyzers, or built-in logic block observers (BILBOs). The approach taken for any particular chip will depend on the nature of the circuit, on the preference of the circuit designer, and on many other factors.

The objective of the *RUNBIST* instruction is to provide users of ICs with a consistent means of accessing self-test features that is independent of the type of self-test offered by a chip and that requires only a very limited amount of data to be stored on the ATE system, on-board bus-master chip, or other unit in control of the board-level test bus.

To meet the requirements of the *RUNBIST* instruction, the self-test must execute only while the TAP controller remains in the *Run-Test/Idle* state. Typically, the logic involved in the test will need to be set to an initial starting state before test execution can begin and this must occur automatically within the chip. As shown in Figure 6-17, initialization could occur in the first clock cycle following entry into the *Run-Test/Idle* controller state and the test could execute in subsequent clock cycles.† In the figure, the RT/I signal is 1 while the test logic is in the *Run-Test/Idle* controller state; the RUNBIST signal is 1 when the *RUNBIST* instruction is selected.

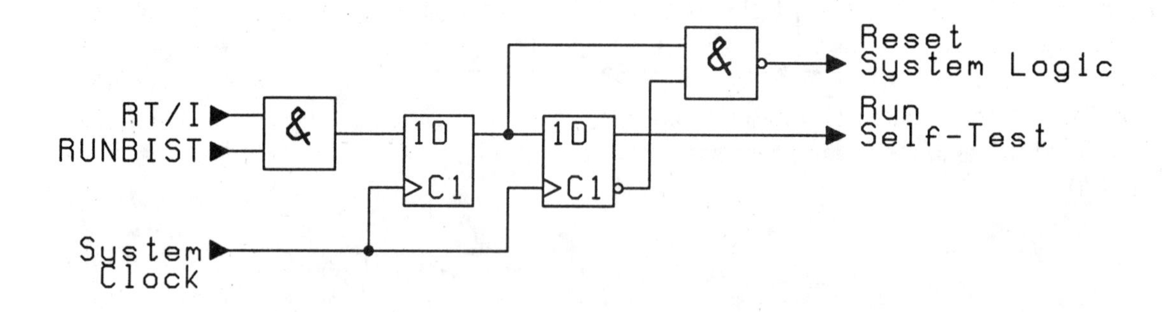

Figure 6-17: Control of on-chip system logic during RUNBIST.

The self-test will run to completion provided the TAP controller remains in the *Run-Test/Idle* state for a specified minimum period, for example, as measured by the number of clock cycles applied to the on-chip system logic. By moving to the *Capture-DR* controller state following this period, the result of the self-test can be loaded into the test data register selected by the *RUNBIST* instruction and then shifted out for examination.

To allow self-tests of different lengths to be run simultaneously in two or more chips on a board, the standard requires that, as long as the TAP controller remains in the

† Note that, as in the case of the *INTEST* instruction, the clock(s) for the on-chip system logic can be fed either from TCK or by an externally-generated clock source while the *RUNBIST* instruction is selected.

Run–Test/Idle state for more than the manufacturer–specified minimum period, the result loaded into the selected test data register must be invariant no matter how long the controller remains in this state. To illustrate, consider a board containing two ICs, one of which must receive 100 TCK cycles to complete its self–test and the other 1000 cycles. Once 100 cycles have been applied, the test on the first IC will have been completed and its result will be ready for inspection. After 1000 cycles have been applied, the results from both ICs will be ready for inspection. Therefore, by entering the *RUNBIST* instruction, moving to the *Run–Test/Idle* controller state for 1000+ clock cycles, and then moving through the *Capture–DR* controller state into the *Shift–DR* state, a test on the health of both ICs can be performed.

An additional benefit of this feature of the *RUNBIST* instruction is that it removes the need to maintain one version of the board test program for each variant of a chip used on the board. Should the length of the self–test change between variants, a board test program, which allows at least the maximum specified number of clocks to be applied to the on–chip system logic, will meet the requirements of both chip variants.

6.4.2: Control of the Boundary–Scan Register

While self–test execution is in progress, the boundary–scan register is used to hold the component's outputs at fixed values. This prevents the signals generated by the on–chip system logic during the test from propagating to neighboring components where they might cause unwanted or hazardous operation. For 2–state outputs, the value to be driven can be defined by the user. For 3–state outputs, some components might also allow the value to be user–defined; alternatively, the output might be set to the high–impedance state while the *RUNBIST* instruction is selected. Note that, in contrast to the other instructions described in this chapter, the boundary–scan register does not have to be selected by the *RUNBIST* instruction to form the serial path between TDI and TDO (although this is an option).

Typically, the values to be placed on the component's output pins will be shifted into place by use of the *SAMPLE/PRELOAD* instruction before the *RUNBIST* instruction is entered. Once the *RUNBIST* instruction has been entered, the Mode inputs of the cells connected to the chip's system output pins will change to 1, allowing the data held at the latched parallel outputs of the cells to be driven onto the board interconnections.† The latched parallel outputs of boundary–scan cells at system output pins are not updated in the *Update–DR* controller state while the *RUNBIST* instruction is selected; their state is held throughout the period for which the instruction is selected.

In some designs, the boundary–scan register can participate in the application of the self–test and can, if required, be the test data register enabled to shift data between TDI and TDO. For example, while the test is executing in the *Run–Test/Idle* controller state, the shift–register stages within the boundary–scan register cells could be configured to behave as LFSRs, multiple–input signature registers (MISRs), and other functions.

---

† In cases where the pin state cannot be programmed by the user, the output will be set to high–impedance.

Part III: Applications to Loaded–Board Testing

Part III contains application examples to illustrate the use of the *IEEE Standard Test Access Port and Boundary–Scan Architecture* in testing loaded boards. These examples show how boards composed purely of chips compatible with the standard can be tested and how the provision of boundary–scan facilities in some chips can help in the application of tests to others.

Further material on the application of boundary–scan techniques to loaded–board testing is contained in the reprinted papers in Part V.

Chapter 7. Taking Advantage of Boundary–Scan in Loaded–Board Testing

Peter Hansen
Teradyne Inc
321 Harrison Avenue
Boston, MA 02118, U.S.A.

Until recently, design–for–test (DFT) circuitry built onto chips was the province of large, vertically–integrated systems manufacturers. But that monopoly is fast disappearing now that the *IEEE Standard Test Access Port and Boundary–Scan Architecture* has been defined, as commercial parts incorporating that standard are being developed, and as application–specific integrated circuit (ASIC) technology gives more and more designers control over their own silicon.

Much of the drive towards DFT will focus on boundary–scan, which is implemented at the chip level and which can ease and simplify board–level testing. Boundary–scan offers test engineers a way around increasingly thorny testability problems that stem from advances in very large–scale integration (VLSI) integrated circuit (IC) processing and packaging technologies.

7.1: Loaded–Board Testability Problems and Traditional Test Techniques

VLSI processing advances have escalated IC gate counts; therefore, the number and complexity of test patterns needed for IC and board–level testing have also escalated. Meanwhile, device packaging advances such as surface–mount technology (SMT), tape automated bonding (TAB), and high pin–count IC packages have increasingly restricted the physical accessibility of device leads to fixtures and hand–held probes traditionally used to test and diagnose faults on printed wiring boards (PWBs).

7.1.1: The Fault Spectrum

The faults present on a board can be categorized as either structural or performance defects. A structural fault is created by a physical defect in a device or in an interconnect on the board, and can be detected at low test speeds. Test coverage is often simulated or thought about in terms of "stuck–at" faults measured at either the gate or device–pin level. The detection of performance faults is much more demanding, as test speed and operating modes might need to be close (if not identical) to actual system behavior.

Most faults that exist in manufacturing are structural faults. Performance defects are a much smaller, although a very troublesome, class. Manufacturers usually report that performance problems account for as much as 5 percent, to as little as a fraction of 1 percent, of all board failures. Unfortunately, however, performance faults require a disproportionately large amount of time and effort to diagnose and repair. [1]

The distribution of faults between analog and digital circuitry depends primarily on the make-up of the board. Good IC testing typically results in few bad chips being on boards, with the exception of devices that are grossly damaged during assembly. The latter class of faults is nearly always detected by a test that provides good coverage of pin-level stuck-at faults.

Figure 7-1 shows a rough representation of the frequency with which various fault classes occur on a predominantly digital VLSI board. Actual relative proportions depend on the types of components used on the board, as well as the design and quality practices used. Of utmost importance is the fact that structural faults dominate, and that structural faults occur mainly at device pins. Even in field returns, structural faults far outweigh performance failures, although a higher frequency of internal device faults would be expected.

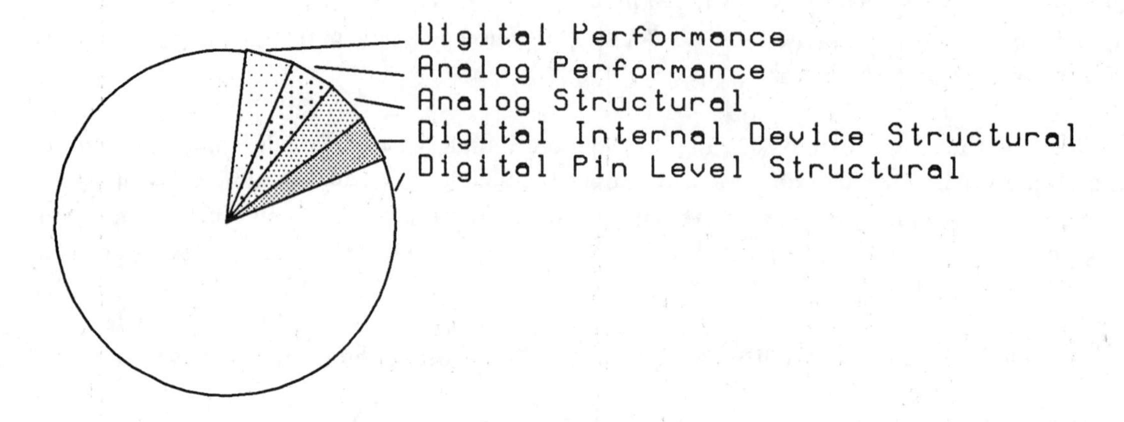

Figure 7-1: The fault spectrum.

To ship quality products, board manufacturers need to screen out both structural and performance faults. Since performance tests and their associated diagnostic techniques are far more expensive than structural testing, it is most important to eliminate virtually all structural defects prior to performance test.

7.1.2: In-Circuit Testing

In-circuit board testing for structural faults traditionally has offered three major benefits:

- fast, automated test generation;
- straightforward fault diagnosis; and
- relatively low capital equipment costs.

Escalating VLSI and ASIC complexity, however, is eroding these advantages.

Test generation, for example, is becoming more difficult as increasing gate counts demand more patterns. At the same time, the custom nature of ASICs means that engineers cannot pull in-circuit test sets ready-made from a pattern library.

Some patterns might be pulled directly from device test, although the ease and success of this practice usually depend on whether these patterns were developed specifically taking the target tester into account. Otherwise, test engineers might find that the chip-test patterns are too numerous to be handled efficiently by the in-circuit system's available pattern memory, and that some of the patterns conflict with the ASIC's wiring constraints in the board environment [2].

As a result, manufacturing test engineers often bear most of the burden of test development. Automatic test pattern generation (ATPG) tools, which can't handle circuits of large sequential depth, are only of limited assistance.

Backdrive and access restrictions also hamper in-circuit testing. Large ASICs, along with many advanced logic families, can be difficult or even impossible to backdrive, complicating the task of isolating neighboring components for in-circuit tests. Large pin-count ICs also can make board testing a highly channel-intensive proposition, thus driving up the cost of in-circuit test equipment.

Meanwhile, dense SMT and TAB packaging restricts the accessibility of component leads to conventional bed-of-nails fixtures. In some cases, the use of fine-pitch probes can overcome this problem, but only at the expense of more costly and less reliable fixtures.

Probe point density on boards can make it impossible to use vacuum-based fixturing techniques without causing excessive board flexing. And two-sided boards, packages with completely inaccessible leads, or fixtures that deny manual access all prevent the use of hand-held probes and make it impossible to diagnose even simple problems like open etches or bad solder joints.

7.1.3: Functional Testing

Entailing far more difficult program generation and diagnostics than in-circuit testing, functional testing typically is used to find simple structural faults only as a last resort. Functional techniques are best reserved for performance testing in critical applications.

Defense-related programs, for example, frequently use functional testing in instances where contractual agreements prohibit overdriving or conformal coatings preclude in-circuit access to boards under test. Commercial manufacturers might also adopt a functional strategy when device packaging so restricts access that in-circuit testing becomes impractical.

The pattern-generation and accessibility issues that affect in-circuit testing, however, impact functional test even more. Very large numbers of very complex patterns are needed to test full-board functionality; and these patterns must be generated anew, using logic and fault simulation, for each board design. Since structural faults are abundant and concentrated at device pins, very high stuck-at pin-fault coverage is called for; attaining this level of coverage with functional test patterns is extremely expensive.

Moreover, increasingly complex VLSI and ASIC components have sent board-level modeling, pattern generation, simulation, and diagnostic costs soaring. Finally, the same packaging technologies that restrict in-circuit access to a board also might block the hand-held guided probes traditionally used for functional fault diagnosis.

As these problems grow more acute, many boards will become impractical to test using either in-circuit or functional techniques. The way out of this dilemma is boundary-scan.

7.2: 100 Percent Boundary-Scan Testing

When boards incorporate boundary-scan components, the shift paths of these components are connected to form a larger shift path on the board. Through this path, a tester can access individual device leads, which serve as "virtual channels" providing control and visibility that otherwise would have to come from physical ATE channels. Backdriving is eliminated, test channel requirements are reduced, and access requirements for fixturing are simplified. Boundary-scan also decreases or eliminates the need for hand probing of a board to isolate faults, easing the diagnostic chore.

Boundary-scan, moreover, simplifies test development. By increasing the board's controllability and observability, boundary-scan makes it possible to partition the board test program to simplify test generation. The board test applications discussed in this chapter use one of the boundary-scan instructions defined by the standard: *INTEST*, *RUNBIST*, or *EXTEST*.

7.2.1: Checking Internal Logic with INTEST

Board test applications that require the highest possible fault coverage -- such as system test or field return testing -- include a comprehensive check for defects in the internal logic of the board's components. The *INTEST* instruction serves this function, allowing a tester to use the boundary-scan path to check the structural integrity of internal device logic. The tester can control the internal logic at device inputs and observe the results at device outputs.

The *INTEST* instruction usually cannot provide a complete gate-level test, however, because the quantity of data that would have to be clocked through the shift path to test a complex sequential IC would bog down test times. *INTEST* therefore must be augmented by other test-oriented circuitry in large, complex devices.

Sometimes the obvious choice is partial or full internal scan, which works very well for the static logic structures embedded in gate arrays. An IC having both internal scan and boundary-scan can be tested in the *INTEST* mode from the edge of the board, using patterns from incoming inspection or device test to provide nearly perfect gate-level fault coverage.

Boundary and internal scan techniques are not good for dynamic logic used in microprocessors and their peripherals because patterns can't be applied fast enough to keep dynamic devices alive. Moreover, adding scan capability in a highly repetitive logic structure such as a memory chip would double or triple the size of the device. An emerging alternative for these types of devices is built-in self-test (BIST), which designs test circuitry into the chip itself.

7.2.2: Implementing Chip Self-Test Using RUNBIST

In a boundary-scan IC containing BIST, provision of the *RUNBIST* instruction allows the test access port (TAP) to become the tester's means of accessing the BIST circuitry. The tester instructs the BIST circuitry on how to initialize the self-test, which typically uses pseudo-random pattern generation to create stimuli, and signature analysis for checking device response. The results of the signature analysis then are read from the shift path by the tester.

The internal fault coverage provided by BIST varies greatly. If coverage is very high, no supplemental testing by the automatic test equipment (ATE) is required. If coverage is reasonably good, a tester might supplement BIST with external patterns. Another possibility is that the BIST circuitry is only intended to test pieces of logic buried deep in the chip (logic that would otherwise be difficult for a tester to get at from the I/O pins), leaving the rest of the chip to be tested by more conventional means.

7.2.3: Verifying Board Interconnects Using EXTEST

Production board testing usually assumes that incoming inspection already has screened out nearly all components with internal defects and, therefore, concentrates on detecting the most common process faults: shorts and opens in device interconnections and stuck-at pin faults. If boards were composed entirely of boundary-scan parts, the *EXTEST* instruction could be used to do all this.

Faults detectable when using *EXTEST* interconnect testing occur between boundary-scan devices, and between these devices and primary inputs or outputs -- which must be connected to ATE channels. For digital portions of a board, *EXTEST* interconnect testing can provide fault coverage and diagnostic resolution far superior to that achieved through using manufacturing defect analyzers (MDAs) and in-circuit test systems.

In contrast with the in-circuit approach, moreover, *EXTEST* doesn't require the tester to have direct physical contact with individual device leads. Instead, a test system can control and observe boundary-scan device leads by clocking data to and from their associated shift-register cells (Figure 7-2). Thus, device pins along the boundary-scan path become the tester's "virtual channels" on the board. The test system can apply test patterns and capture response data through these virtual channels, much as it does through conventional ATE channels.

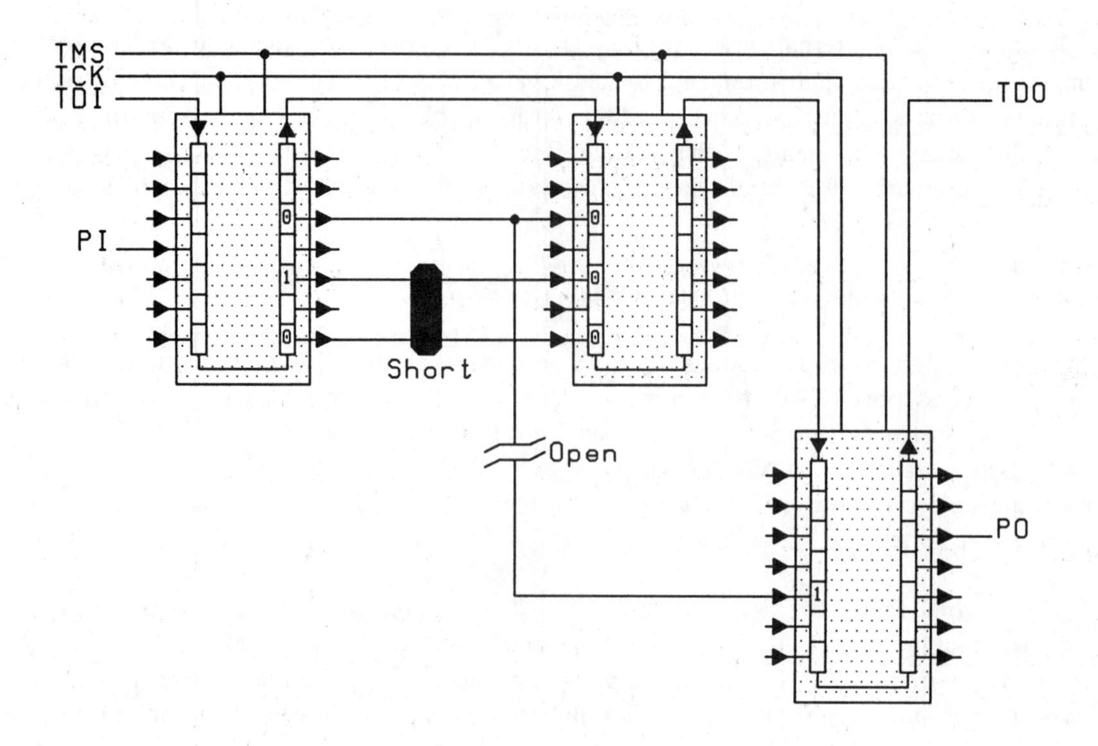

Figure 7–2: Interconnect testing using EXTEST.

Pattern generation also is simplified when boundary–scan testing is used to verify board interconnects, because failures don't have to be propagated through complex chips. Pattern generation algorithms have been developed that provide 100 percent fault coverage with minimal data size for both opens and shorts by using what are known as "counting" patterns [3]. These patterns can be generated automatically by using information extracted from netlist and boundary–scan configuration databases.

Since all boundary–scan device pins must be tested in both the logic–0 and logic–1 states to test the interconnect, 100 percent device–pin fault coverage is achieved during these tests. Thus, much more than the board interconnect is being tested. Each IC is shown to be basically functioning and the various interconnections –– from silicon to lead bonds, from solder bonds to the circuit board itself –– are shown to be intact.

While the counting patterns are fast and efficient, they do not provide the basis for good diagnosis. The reason is that many faults can cause tests to fail in an identical manner, a phenomenon known as fault aliasing and confounding [4].

The best means of dealing with this limitation is to use the counting patterns to identify failing networks and then to apply additional "walking" patterns (so called because they "walk" through the circuit, testing it by setting all networks except the failing one to logic–1 or logic–0) to provide information from which a precise diagnosis can be drawn. These patterns are also called "adaptive" patterns [4,5] because they are generated and applied by the ATE to the board immediately after a failure occurs, based on the specific nets that have failed.

A boundary-scan interconnect-fault diagnosis algorithm correlates the mass of serial response data shifted out of the boundary-scan path with topological data to identify physical defects and their locations on the board. Properly done, this method achieves diagnostic resolution comparable to a bed-of-nails tester for identifying shorts. It also gives superior resolution in pinpointing opens, because of the improved board visibility achieved by having boundary-scan cells at each chip pin.

EXTEST will be the most extensively used of boundary-scan's various modes for loaded-board testing -- not only because it deals with the interconnect faults described above, but also because it can be used to test mixed-technology boards containing both boundary-scan and conventional components. This application will be detailed in the remainder of this chapter.

7.3: Test-Access Strategies for Mixed-Technology Boards

Structural testing for an ideal board -- one implemented exclusively by using boundary-scan components -- is a fairly simple matter. Such boards are and will likely continue to be rare, however. Although boundary-scan is proliferating rapidly in gate array and standard-cell ASICs, it will advance only gradually and over a period of years in commercial components. In fact, the extra silicon or device leads required to implement boundary-scan may preclude its ever being used in chips such as small logic devices and some memory chips.

In consequence, mixed-technology boards populated by both boundary-scan and conventional components will predominate for the foreseeable future. Testing of these boards will combine boundary-scan testing with traditional in-circuit or functional testing of the conventional circuitry.

Analog components are not generally applicable to boundary-scan, so their board networks would require physical access if analog in-circuit testing is desired. Often, the extra power of having the full capability of ATE channels available at specific points on the board can make significant improvements in programming time, in test and diagnostic throughput, and in quality.

For most mixed-technology boards, having physical access to some of the board's networks will be the most critical factor in determining the economics of structural testing. The decision on which networks have fixture access must be made with a particular test strategy in mind during physical layout of the board.

Circuit designers must therefore convey to layout people both a knowledge of the virtual access provided by the leads of boundary-scan chips on the board, and the access requirements imposed by the specific test strategy to be implemented. This "design-for-access" methodology guarantees full access to a board through a combination of physical and virtual test channels.

The range of access strategies available for testing mixed-technology boards are listed below, in order from simplest to most difficult. It should be noted that the more complex strategies can contain one or more elements of simpler ones.

1. A standard in–circuit strategy is used when all the board's networks are fully accessible via bed–of–nails fixturing, providing contact to boundary–scan components and conventional logic on the board.

2. A virtual interconnect strategy combines nail–less testing of pure boundary–scan networks with standard in–circuit testing of conventional components.

3. A virtual in–circuit strategy tests individual non–scan components one by one; the leads of the device under test (DUT) are connected either to physical ATE channels via a fixture or to the virtual channels provided by the I/O pins of neighboring boundary–scan parts.

4. A standard cluster–test strategy groups non–scan devices together and tests them functionally through test nails at the cluster's periphery.

5. A virtual cluster–test strategy is applied when nails cannot be placed around a cluster; instead, the virtual channels associated with boundary–scan devices are used to test the cluster.

7.3.1: Standard In–Circuit Testing

If all board networks are accessible to traditional fixturing, a full bed–of–nails in–circuit test approach can be employed, as illustrated in Figure 7–3 (as in all the diagrams that follow, the Xs in Figure 7–3 represent physical tester–access points). While this might appear to be an overly conservative strategy (since a primary advantage of boundary–scan is to permit nail–less networks), a full bed–of–nails environment provides significant advantages. For one thing, all shorts testing can be done prior to powering up the board. Further, device patterns that achieve 100 percent pin–fault coverage for complex boundary–scan devices are made easy when ATE channels are used in conjunction with the boundary–scan DUT through the *EXTEST* instruction.

Stimulus applied by ATE channels is captured by the boundary–scan input cells. Stimulus shifted into boundary–scan device output cells is captured by ATE channels.

This procedure can be performed without requiring ATE channels to be simultaneously presented to the DUT. Taking advantage of multiplexing, DUT leads can be split into subsets that are tested independently, dramatically decreasing the channel count that would otherwise be required; the benefits of this approach are most significant when boundary–scan is present on the largest components on the board.

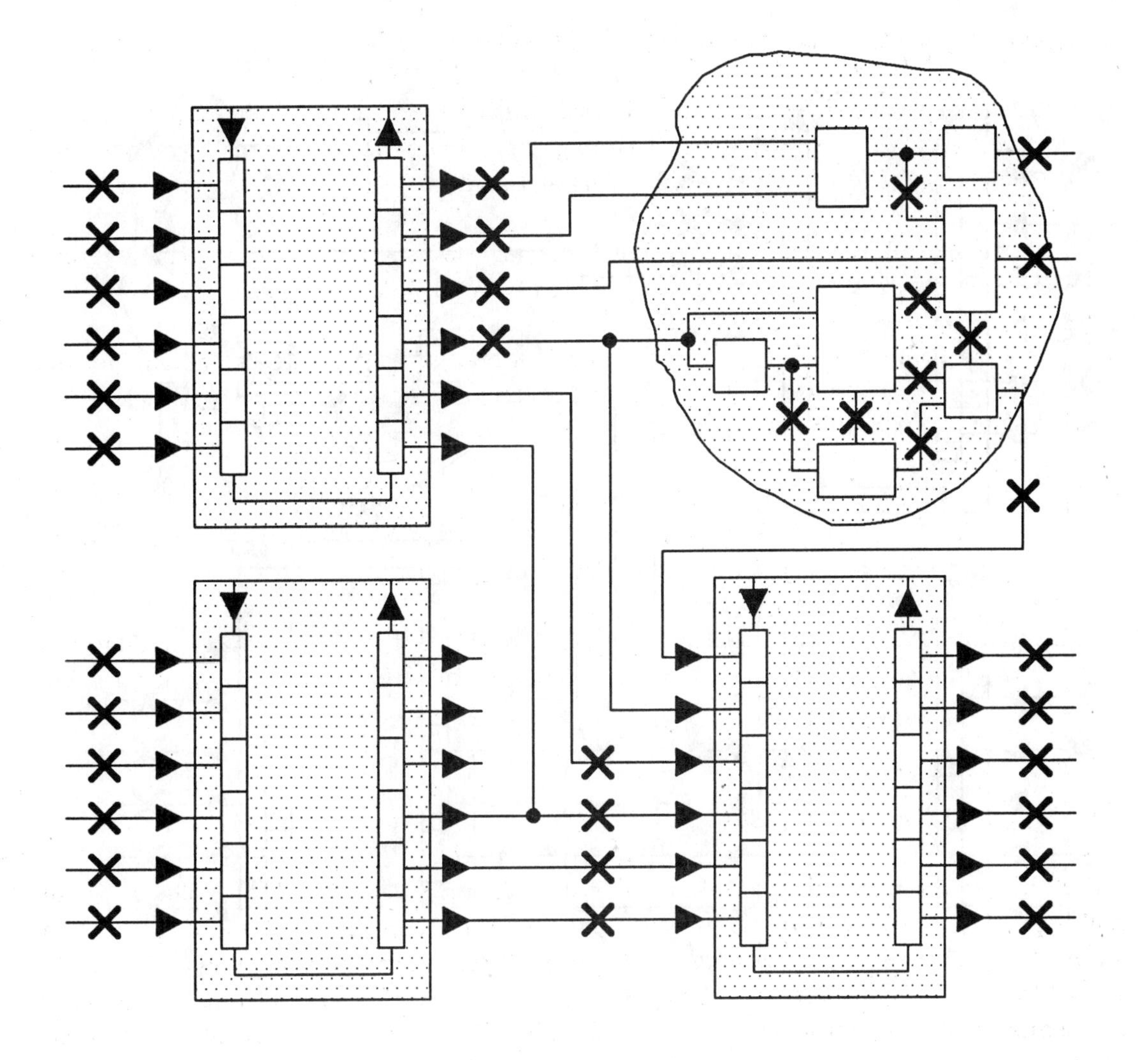

Figure 7–3: Standard in–circuit testing.

Diagnosis of open–circuit faults using boundary–scan devices is much better than with conventional ones in a bed–of–nails environment. With boundary–scan, opens can be isolated down to a single network without the aid of manual probing. Without boundary–scan, an open that causes a device failure can be properly identified only through manual probing. On double–sided SMT boards, this can be prohibited by device packaging and by fixture designs that do not allow access for hand–held probes.

7.3.2: Virtual Interconnect Testing

Leaving probes off networks that consist only of boundary–scan device interconnects can simplify in–circuit fixturing. In this virtual interconnect strategy, the only new element is that the *EXTEST* mode is used to test boundary–scan networks via the virtual channels provided by boundary–scan IC leads. Because all conventional chips are surrounded by physical test channels, they are tested by using existing in–circuit techniques and tools (Figure 7–4).

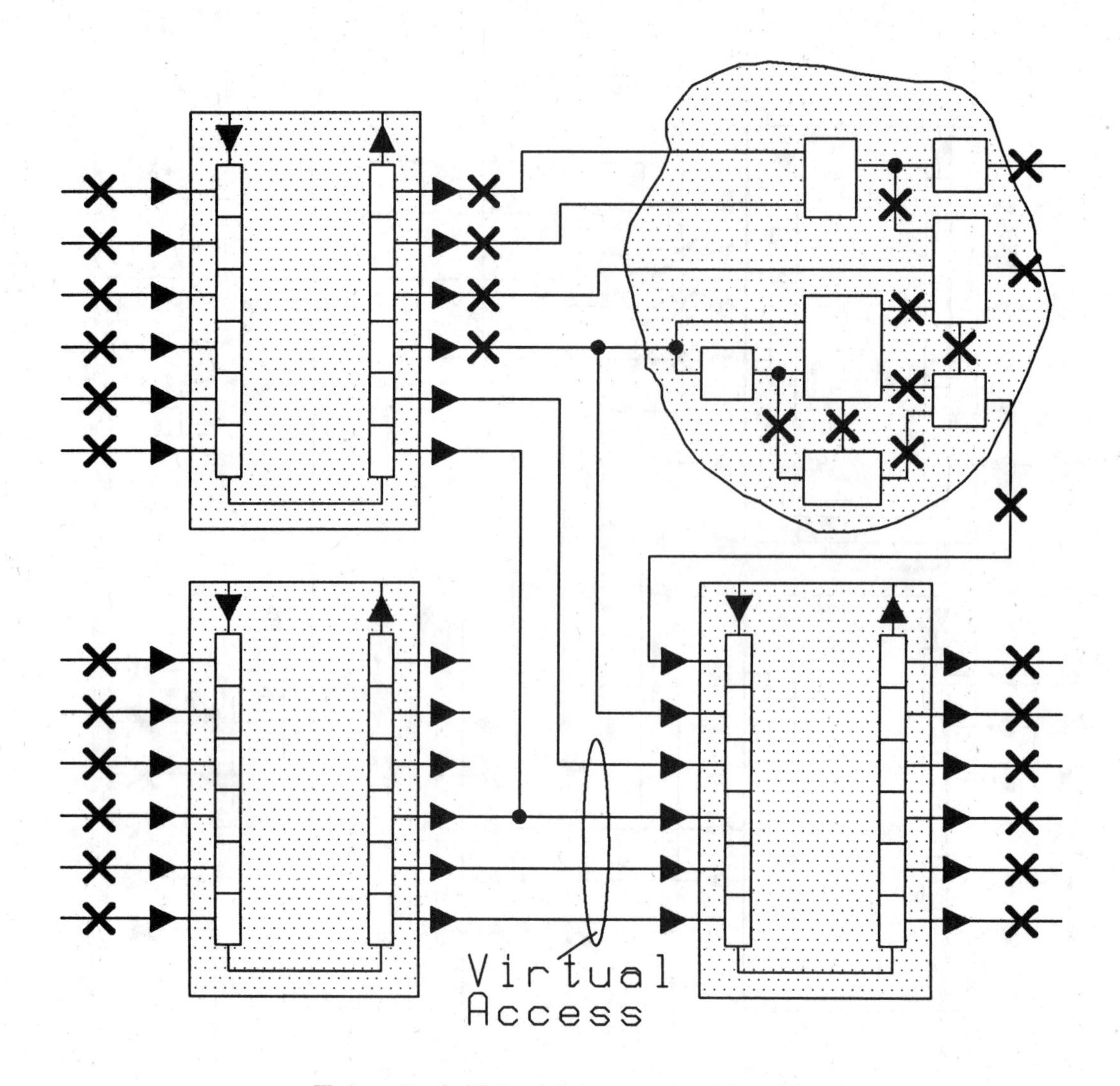

Figure 7–4: Virtual interconnect testing.

7.3.3: Virtual In–Circuit Testing

Virtual in–circuit testing, which accesses some or all of a DUT's leads via virtual channels provided by boundary–scan device I/O pins (Figure 7–5), retains the programming and diagnostic advantages of conventional in–circuit testing. Because it further reduces nail counts, however, this approach eases in–circuit testability problems stemming from physical access and overdrive restrictions.

A typical non–scan device tested by virtual in–circuit testing would be a small–, medium–, or large–scale integration (SSI, MSI, or LSI) IC whose test can be pulled from an in–circuit pattern library associated with the tester being used. These patterns are in parallel format because that's the way in which they have traditionally been applied.

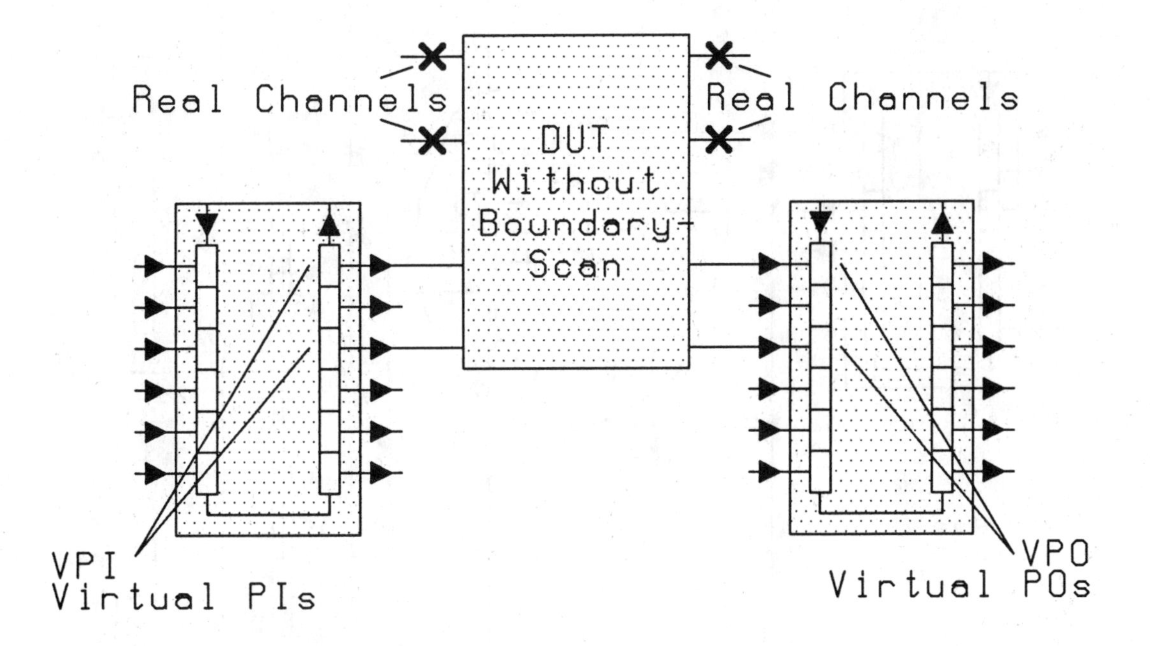

Figure 7–5: Virtual in–circuit testing.

While DUT pins connected to normal ATE channels are controlled conventionally using these parallel–format channels, the DUT pins actually serviced by virtual channels must be tested by using serial data. Many combinational testers can handle this requirement; thus, special scan–test hardware might not be needed.

Special data serializing software, though, is needed both for pattern serialization and fault diagnosis. Serialized patterns become a normal part of the automatic in–circuit programming, test, and diagnostic process.

Virtual in–circuit testing consumes local memory rapidly. While a conventional, parallel–format test pattern takes up only one location in a tester's channel memory, a serialized pattern requires many memory locations because of the repetitive shifting operations involved in testing through the boundary–scan path. Judicious use of hardware looping capabilities is required to compress consecutive "don't–care" states into one memory location to conserve space.

7.3.4: Standard Cluster Testing

In conventional cluster testing, chips that are inaccessible by using in–circuit techniques can be handled by grouping them together with other chips and placing nails or special test points around the cluster's periphery (Figure 7–6). A combinational test system can handle these clusters by using current functional test techniques and existing functional diagnostic tools such as guided probing, fault dictionary, or a combination of the two. Other parts of the board can be tested using elements of access strategies already described.

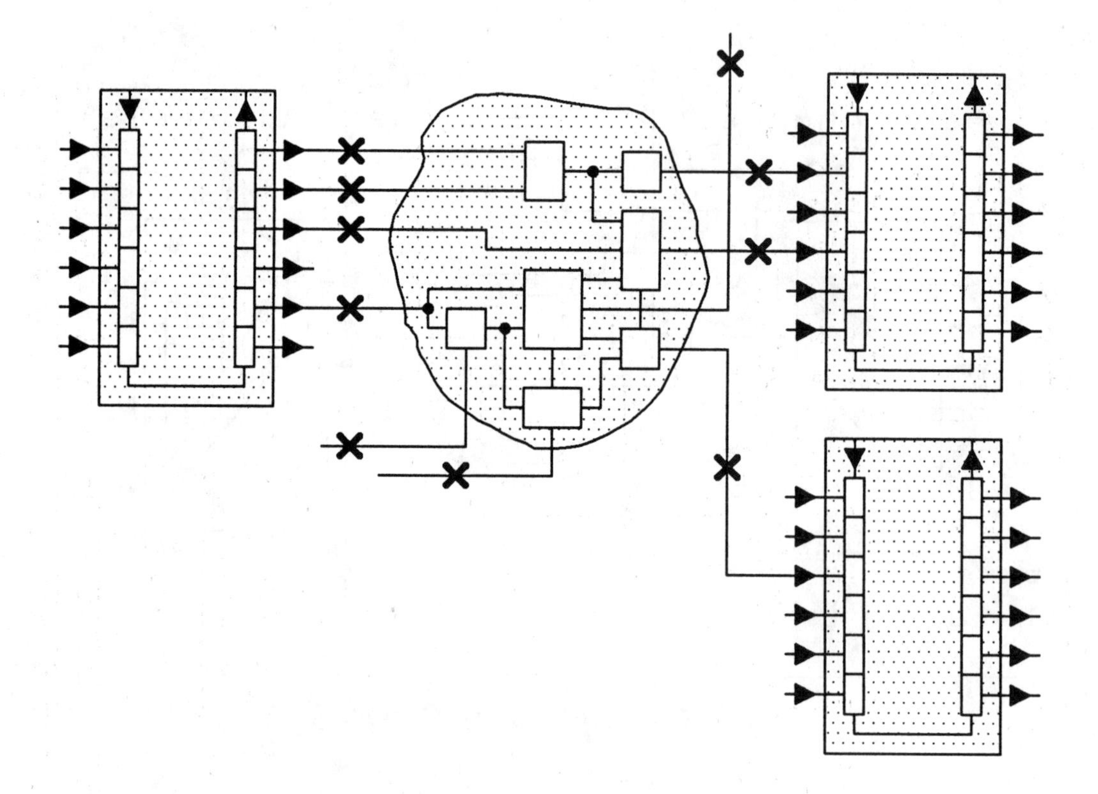

Figure 7–6: Standard cluster testing.

Standard cluster testing is valid for mixed–technology boards as long as the clusters can be accessed through nails or special test points. But a growing number of test applications effectively block all physical access to a board. In such cases, test engineers can once again resort to the board's boundary–scan path to get virtual access.

7.3.5: Virtual Cluster Testing

As is the case in the virtual in–circuit approach, virtual cluster testing uses a mixture of real test channels and virtual access through the boundary–scan path. Employing virtual access to test non–scan chips and device clusters as well as device interconnects (Figure 7–7), this strategy will be called upon often when edge–connector–based functional testing is the only available option.

As in standard cluster testing, the program for a virtual cluster test is generated by using logic and fault simulation. The test engineer creates a simulation model of the cluster, by using netlist–editing tools to extract the requisite device and interconnect data from a full–board netlist.

The engineer then writes test stimuli, which are applied to the cluster model in logic and fault simulation. The virtual channels, supplied by boundary–scan device I/O pins, are modeled as static ATE channels, which eliminates the need to simulate repetitive shifting operations.

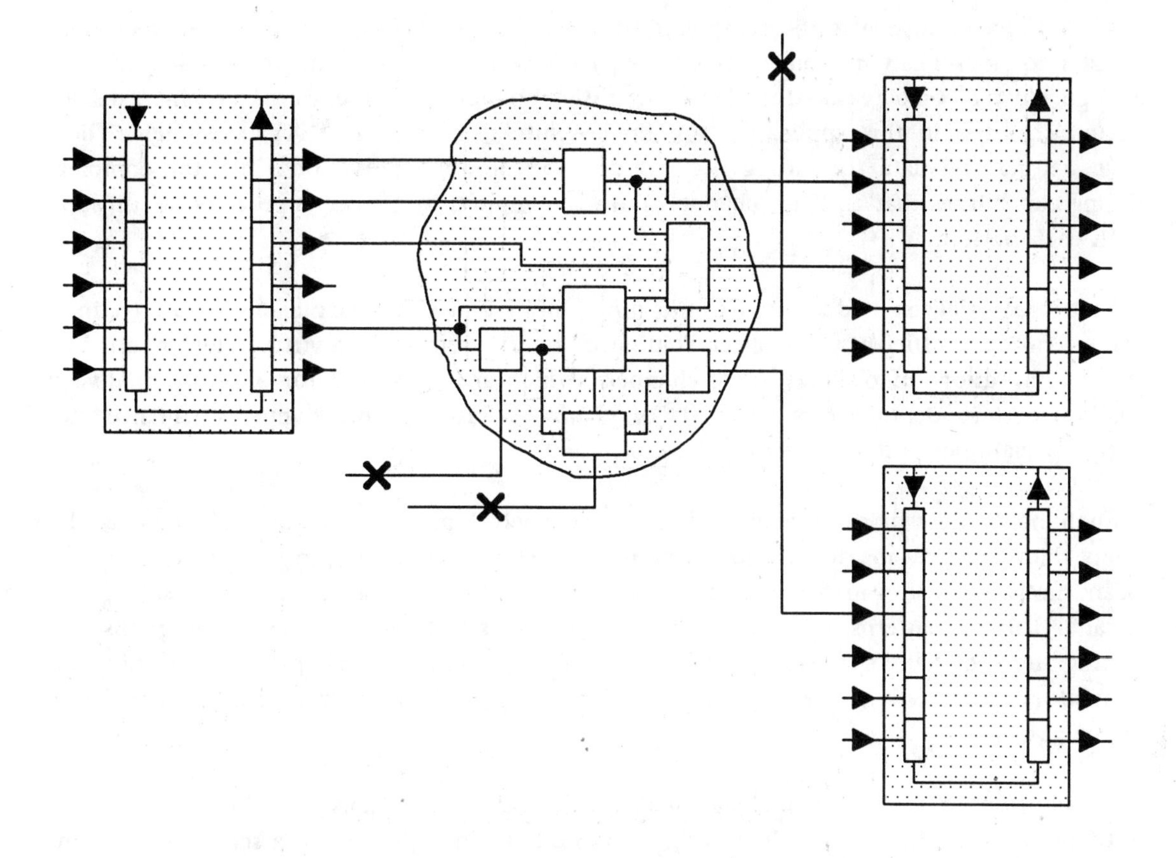

Figure 7–7: Virtual cluster testing.

Fault simulation grades the fault coverage provided by the test patterns written by the programmer. Near–perfect coverage of stuck–at pin faults should be the goal here, because faults that slip through this test will be more expensive to pinpoint in later test stages. With tomorrow's VLSI clusters likely to rival today's VLSI boards in complexity, achievement of this level of fault coverage will be an increasingly time–consuming task.

Board designers can help counter this trend by minimizing the size and sequential depth of non–scan clusters on the board. One way of breaking up a large block of sequential logic is to intersperse it with more boundary–scan components or physical access points to improve cluster controllability and observability. Designers also must take into account the basic static rate of virtual channels and must add physical access points or board–level BIST for dynamic logic.

When fault simulation indicates that the desired level of fault coverage has been reached, the stimulus and response data from the simulator is postprocessed for use by the target board test system. Converting the simulator's parallel–format test patterns into the serial data required for testing through the boundary–scan path, however, will require more than the software serializer described earlier.

Virtual cluster testing is the most pattern-intensive test strategy yet discussed, requiring a tester to clock many millions of bits through the boundary-scan path to functionally test a large cluster. One reason for this is that the boundary-scan device I/O pins used as virtual channels in this application are far less intelligent than real ATE channels. They have neither the data formatting (i.e., return-to-zero, return-to-one) nor the complex timing capabilities that an ATE channel routinely employs to convey large amounts of data with each pattern.

Software serializers, adequate for the more limited needs of virtual in-circuit testing, cannot produce efficiently the sheer volume of data required for a virtual cluster test due to the limitations of ordinary ATE channels, which are optimized for applying relatively shallow pattern depth across hundreds of channels. Instead, test engineers will need to apply special scan-test hardware to the task.

Such hardware would be able to handle extremely long serial data streams and could apply data-compression techniques to minimize testing times and storage requirements [6]. Scan hardware also can serialize a variety of test data for use in virtual cluster tests: parallel-format patterns or truth tables output by simulators or manual programs for testing most digital logic and possibly algorithmic patterns created by hardware number generators for testing memory devices (including signature analyzers for response compaction).

In addition to the problems associated with serializing patterns for go/no-go testing, automated diagnosis of cluster failures is important. Guided probing in a scan environment has been used for some time on boards with devices built by using internal-scan techniques such as level-sensitive scan design (LSSD) [7] and is applicable to virtual cluster testing through the boundary-scan path.

Guided probing can be integrated with fault dictionary diagnosis, which identifies likely fault locations without requiring physical access to the board [8,9] and can therefore be used even in situations where direct access to internal nodes of the loaded board is limited. To supply virtual cluster test diagnostics, the guided probe and fault dictionary tools must be adapted to accept serial response data clocked out of the boundary-scan path, just as they now accept parallel response data from conventional ATE channels.

7.4: Conclusion

On boards implemented exclusively with components designed according to IEEE Std 1149.1, boundary-scan testing allows automated generation of patterns with 100 percent coverage of digital structural pin-level faults, which account for the overwhelming majority of board failures. But because most boards in the foreseeable future will mix boundary-scan devices with conventional ICs, boundary-scan testing typically will be used in conjunction with current in-circuit and functional cluster test techniques.

This chapter has described five strategies for accessing, testing, and diagnosing mixed-technology boards. Where restricted physical access hampers traditional test methodologies, virtual access provided by boundary-scan device leads might offer the only means of assuring a comprehensive test of a complex board.

When a test application demands both physical and virtual access, these requirements must be taken into account during board design and layout. The designer must concentrate on breaking up the board's conventional circuitry as much as possible into isolated individual ICs or into relatively small clusters, which might be interspersed with boundary-scan chips for further improved visibility and controllability. Information about both design (i.e., data about which chips are boundary-scan ones and which are conventional) and test (i.e., the physical/virtual access requirements of the test strategy) must be factored into board layout to guarantee successful implementation of the test.

7.5: References

[1] D. Hebert and J. Arabian, "Implications of the Technique for Dynamic High Speed Functional Testing," *IEEE International Test Conference Proceedings,* IEEE Computer Society Press, Los Alamitos, Calif., 1982, pp. 548–556.

[2] P. Hansen, "Converting Device Test Vectors to an In-Circuit Board Test Environment," *IEEE International Test Conference Proceedings,* IEEE Computer Society Press, Los Alamitos, Calif., 1985, pp. 972–978.

[3] P.T. Wagner, "Interconnect Testing with Boundary-Scan," *IEEE International Test Conference Proceedings,* IEEE Computer Society Press, Los Alamitos, Calif., 1987, pp. 52–57.

[4] C. W. Yau and N. Jarwala, "A New Framework for Analyzing Test Generation and Diagnosis Algorithms for Wiring Interconnects," *IEEE International Test Conference Proceedings,* IEEE Computer Society Press, Los Alamitos, Calif., 1989, pp. 63–70.

[5] P. Goel, M.T. McMahon, "Electronic Chip-in-Place Test," *IEEE International Test Conference Proceedings,* IEEE Computer Society Press, Los Alamitos, Calif., 1982, pp. 83–90.

[6] P. Hansen, "Testing Conventional Logic and Memory Clusters Using Boundary-Scan Devices as Virtual ATE Channels," *IEEE International Test Conference Proceedings,* IEEE Computer Society Press, Los Alamitos, Calif., 1989, pp. 166–173.

[7] P. Hansen, "New Techniques for Manufacturing Test and Diagnosis of LSSD Boards," *IEEE International Test Conference Proceedings,* IEEE Computer Society Press, Los Alamitos, Calif., 1983, pp. 40–45.

[8] J. Richman, K.R. Bowden, "The Modern Fault Dictionary," *IEEE International Test Conference Proceedings,* IEEE Computer Society Press, Los Alamitos, Calif., 1985, p. 696.

[9] V. Ratford, P. Keating, "Integrating Guided Probe and Fault Dictionary: An Enhanced Diagnostic Approach," *IEEE International Test Conference Proceedings,* IEEE Computer Society Press, Los Alamitos, Calif., 1986, pp. 304–311.

8. A Test Program Pseudocode†

Rodham E. Tulloss and Chi W. Yau
AT&T Bell Laboratories
Engineering Research Center
Princeton, NJ 08540, U.S.A.

A product utilizing the IEEE Std. 1149.1 boundary-scan method, architecture, and protocol [1] is hypothesized. The product is assumed to contain a significant number of large- and very large-scale integration (LSI and VLSI) integrated circuits (ICs) equipped with the standard architecture and test access port (TAP). The parts equipped with boundary-scan are assumed to provide a single command activating all built-in self-test (BIST) capability which is available to the purchaser of the part in question. The pseudocode of a circuit board test program is laid out demonstrating:

- the initialization of the board for testing;
- the necessary steps to validate the test circuitry in those ICs equipped with boundary-scan;
- the verification of board-level interconnect circuitry;
- the activation of self-test features in parts equipped with the TAP and included on the boundary-scan path; and, very briefly,
- the testing of non-boundary-scan parts on the board.

The data available for diagnosis and its use in diagnosing the board are discussed briefly.

8.1: Introduction

This chapter describes pseudocode for a test program of a circuit board containing a significant number of chips designed with BIST and boundary-scan. This format has been chosen in order to present an operational view of the meaning of IEEE Std 1149.1.

Imagine that the "lines" of pseudocode become comments in the completed test program. In order to highlight the portions of pseudocode, they are presented indented in the COURIER font.

The boundary-scan standard provides for a 4-wire interface: test data input (TDI), test data output (TDO), test mode select (TMS), and test clock (TCK). An additional test logic reset (TRST*) line is optional and is predominantly used to provide protection against bus conflict during power up. There is no restriction in the standard regarding whether a single serial path is made up by connecting TDO and TDI lines of the chips

† An earlier version of this paper was presented at the First European Test Conference, Paris, April 1989.

EH0321-0/90/0000/0097$01.00 © 1990 IEEE

that have boundary–scan on a given board. A multiple ring configuration or a star configuration can also be designed by using chips with the features of the standard.

The pseudocode below is written:

- as if there were a single 5–wire port on the board under test; and
- as if there were a single serial path through all the chips on which boundary–scan is implemented.

For diagnostic purposes, we have assumed a duplication of the TMS line as shown in Figure 8–1. (This structure is also used in Chapter 9).

One of the TMS lines (TMS1) should be used to apply the TAP protocol to the even–numbered chips; and the other (TMS2), to apply the protocol to the TAPs of the odd–numbered chips. The presence of the optional TRST* line in an actual implementation is probable.

At the board level, one might assume that the first TDI on the ring, the last TDO on the ring, TCK, and test mode select lines are all available at board connectors. In multi–board systems, it is more likely that concerns over clock distribution and test mode select line coordination would lead to a master bus–controller. This device might be a stand–alone entity or it might be a peripheral to a programmable device that had other system (non–test) functions. It would source and sink the boundary–scan signals and provide protocol conversion between:

- commands arriving over a back–plane from a test and diagnostic processor; and
- the ICs constituting the boundary–scan ring.

In this paper it is not assumed that there is a master bus controller. If a master bus controller is present, the program should be read as providing instructions to the master bus–controller that then would generate the protocol necessary to do the various tasks. In case there is more than one serial path (ring) or even more than one test port, the program will become more complicated, especially in the case of interconnect testing where coordination of events between rings with separate TCK and TMS lines will be of great concern.

We assume that before this test begins, transistors, capacitors and other analog components have been checked by a process tester or have such high in–coming and process/assembly quality levels that such testing can be rationally eliminated. Alternative methods of testing these components are not excluded: They could be included as part of a lower level package and tested at that level rather than at the board level; or an in–circuit tester could be fixtured in such a way as to do in–circuit testing on non–boundary–scan parts and also to provide the necessary resources to carry out the portion of the test that is dependent on boundary–scan.

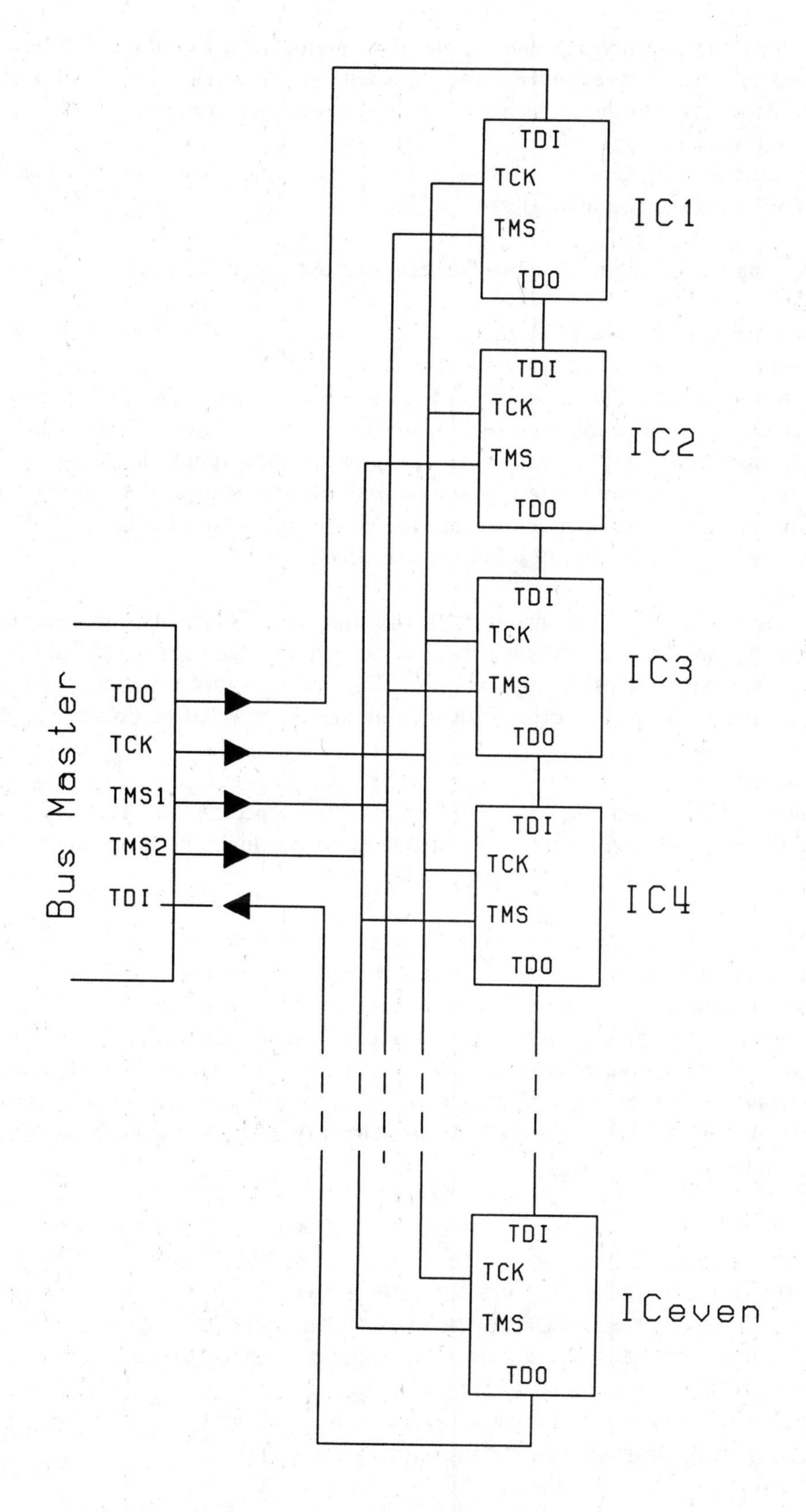

Figure 8–1: Board level interconnection of components.

In the electronic industry as a whole, the time period in which there will be mixing of boundary-scan and non-boundary-scan product on a single assembled unit is of unpredictable length. On the other hand the introduction of products:

- implemented in chip-on-board, double-sided surface mount boards, and silicon-on-silicon technologies;

- requiring encapsulation in controlled environment chambers; or

- requiring extensive, post-installation, diagnostic support features

are likely to require boundary-scan on all ICs involved -- even today. Of course, it is one goal of the engineers and firms who participated in the effort to establish IEEE Std 1149.1 that lower cost test facilities will be made possible as a result of successful standard promulgation. In other words, part of the desired effect of standardization is to provide a sufficiently simplified test interface so that the cost of board-level ATE could be reduced by an order of magnitude without a loss of test effectiveness.

The following sections contain the pseudocode and its explanation. A number of definitions of symbols and terms will be needed and these are defined in Table 8-1. By "odd(even)-numbered chips" we mean those chips in the odd (even) numbered positions on the boundary-scan path starting with the one nearest the TDI board-level input as the first.

You should become familiar with IEEE Std 1149.1, especially the TAP controller state diagram and the general operation of the standardized test logic, before reading further.

8.2: Initialization

First, the board is placed in the test fixture and powered up [2]. The parts, which include the boundary-scan standard architecture and TAP, will offer protection against damage caused by bus conflicts that might occur momentarily during the power-up [1]. This is done by a required reset capability that can be achieved by the use of an optional reset line (TRST*) or by internal chip design features. Whichever method is used, the finite state machine (FSM) of the TAP control circuitry is forced to the *Test–Logic–Reset* controller state.

```
1. POWER UP ON BOARD. BOUNDARY SCAN TEST-CIRCUITRY GOES INTO
RESET STATE. WHERE POWER-UP RESET REQUIRES THE USE OF THE
BOUNDARY-SCAN OPTIONAL RESET LINE, THIS LINE IS ACTIVATED BY
TOGGLING FROM THE INACTIVE (HIGH) TO THE ACTIVE (LOW) STATE. IN
ADDITION, BEFORE RUNNING THE FOLLOWING TESTS, AN L_BSR BIT LONG
SAFE VECTOR SHOULD BE LOADED INTO THE BOUNDARY-SCAN DATA
REGISTERS USING THE SAMPLE/PRELOAD INSTRUCTION. THIS WILL
ALLEVIATE POTENTIAL BUS CONTENTION PROBLEMS.
```

Table 8–1: Terms used in this chapter.

L_{BSR}	Total length of all boundary-scan data registers on the board
L_{EBS}	Total length of all boundary-scan data registers in even-numbered chips
L_{OBS}	Total length of all boundary-scan data registers in odd-numbered chips
L_{IR}	Total length of all instruction registers on the board
L_{EIR}	Total length of instruction registers in even-numbered chips
L_{OIR}	Total length of instruction registers in odd-numbered chips
M_{DR}	Maximum number of data registers per chip on the board
N	Mumber of chips with boundary-scan on the board
P	Period (in TCK clock cycles) of single entry into the Pause-DR or Pause-IR controller state
T_x	Number of interconnect tests required by the board

8.3: Test Circuitry Check

A known bit pattern, available in the ICs equipped with boundary-scan, assists in verifying that the serial path through the devices is intact. This technique was in use prior to the development of IEEE Std 1149.1 [3,4]. In the case of IEEE Std 1149.1, the pattern consists of a '01' in the two lowest order bits of the instruction register. This pattern is required to be loaded automatically by the test circuitry when in the *Capture–IR* controller state. When configured through the instruction registers of all the chips in this controller state, the serial path will pass through a long composite register having '01' at intervals known to the test programmer or to a test development tool used by the test programmer. This is the case because it is a further requirement of the standard that documentation of boundary-scan parts provide the lengths of all instruction and data registers in the boundary-scan test logic.

2. USING THE BOUNDARY-SCAN CLOCK (TCK) AND THE 2 TEST MODE SELECT LINES (TMS1 and TMS2), ALL BOUNDARY-SCANNABLE PARTS ARE PLACED IN THE INSTRUCTION REGISTER SELECT CONTROLLER STATE (SELECT-IR-SCAN).

Beginning with the test circuitry of all ICs in the *Test–Logic–Reset* controller state, this is accomplished in three cycles of TCK. Both TMS lines must provide the sequence '011' -- one bit is shifted in during each clock cycle.

3. NOT CONSUMING ANY TCK CLOCK PULSES, CONFIGURE RECEIVER ON TDO TO RECEIVE (L_{IR}+2) BITS. EXPECTED VALUES WILL BE THE CONCATENATED CONTENTS OF THE INSTRUCTION REGISTERS (EACH IR WILL CONTAIN A VECTOR OF THE FORM {X...X01}) FOLLOWED BY {01}. CONFIGURE DRIVER ON TDI TO SEND (L_{IR}+2) BITS CONSISTING OF THE SERIAL VECTOR {01} CONCATENATED WITH L_{IR} 1S.

There are a number of steps that we have expressed in the form used for step 3. Our intent is to indicate that preparation for serial input/output of vectors can be done as an ATE background activity. It is simply the case that the vectors must be ready when the next shifting activity is set to begin.

Note that the trailing '01' pattern detected by the TDO receiver will check the TDI edge connector pin for common defects such as opens, shorts, and stuck-ats.

4. USING 2 TCK CLOCK CYCLES PROCEED TO THE SHIFT-IR CONTROLLER STATE: BOTH TMS LINES TAKE ON THE VALUE 0 DURING BOTH CYCLES. [NOTE: 1 SHIFT CYCLE IS ACCOMPLISHED BY THIS OPERATION.]

5. APPLY 0 ON BOTH TMS LINES FOR (L_{IR}+1) TCK CYCLES THUS SHIFTING THE DRIVER SUPPLIED SERIAL INPUT VECTOR INTO THE BOUNDARY-SCAN PATH AND DUMPING THE PREVIOUS CONTENTS OF THE PATH FOLLOWED BY {01}.

/* THIS ALLOWS A CHECK FOR A BREAK IN THE SCAN PATH AND ALSO TESTS MANY STUCK-AT FAULTS IN THE SCAN PATH. THE CONNECTIONS IN THE CLOCK AND TEST MODE SELECT DISTRIBUTION NETS WILL ALSO BE CHECKED. IN CASE OF FAILURE OF THIS TEST, THE POINT IN THE SERIAL SCAN AT WHICH WRONG VALUES ARE FIRST DETECTED IS CRITICAL TO DIAGNOSIS. */

The loading of '1...1' into all instruction registers preload the instruction that will force selection of the bypass data registers in preparation for the next step of the test.

6. WITHOUT CONSUMING TCK CYCLES, CONFIGURE RECEIVER ON TDO TO RECEIVE A SERIAL VECTOR COMPOSED AS FOLLOWS: CONCATENATE (INTERLEAVED IN SEQUENCE OF CHIP POSITION ON THE BOUNDARY-SCAN PATH) THE EXPECTED CONTENTS OF THE INSTRUCTION REGISTERS (AS IN STEP 3) OF THE ODD-NUMBERED CHIPS AND A 0 FOR EACH EVEN-NUMBERED CHIP; FOLLOW THIS SEQUENCE WITH {01}. SIMULTANEOUSLY CONFIGURE THE DRIVER ON TDI TO SUPPLY THE SERIAL VECTOR COMPOSED BY CONCATENATING {01} WITH ($L_{OIR}+\lfloor N/2 \rfloor$) 1S.

Starting with Step 6, the trailing `01' pattern is included mainly for program simplicity. By this time, the integrity of the TDI edge connector pin has been checked. Therefore, a trailing pattern is needed only when the last IC on the boundary-scan path is in the bypass configuration. In this case, the minimum trailing pattern is a single-bit `1,' which is needed to detect the stuck-at-0 fault associated with the bypass data register of the last IC.

```
7. SELECT THE INSTRUCTION REGISTER IN CHIPS OCCUPYING THE
ODD-NUMBERED POSITIONS IN THE SERIAL PATH. SIMULTANEOUSLY SELECT
THE BYPASS DATA REGISTER IN THE EVEN-NUMBERED POSITION CHIPS.

/* STEP 7 REQUIRES 2 COORDINATED TEST MODE SELECT LINES. */
```

This action requires a minimum of 4 TCK cycles. The values required on TMSn (n = 1 or 2) are given in Table 8-2.

Table 8-2: Coordination of TMS lines.

TCK Cycle	TMS1	TMS2	Next TAP States	
			In Even ICs	In Odd ICs
1	1	1	Exit1-DR	Exit1-DR
2	1	1	Update-DR	Update-DR
†3	0	0	RunTest/Idle	RunTest/Idle
4	0	1	RunTest/Idle	Select-DR-Scan
5	1	1	Select-DR-Scan	Select-IR-Scan

When the FSMs of the test logic of the even-numbered chips enter the *Select-DR-Scan* controller state, the selected data register will be the bypass data register because all instruction registers contain vectors of the form `1...1.'

```
8. USING 2 TCK CYCLES, TAKE ALL ODD (EVEN) CHIPS TO THE
SHIFT-IR(DR) CONTROLLER STATE BY APPLYING 0 ON BOTH TMS LINES
DURING BOTH CYCLES. [NOTE: 1 SHIFT CYCLE IS ACCOMPLISHED BY THIS
OPERATION.]
```

† This cycle is optional. However, if time is needed to prepare an input/output vector for shifting, the third clock cycle in the above table can be stretched into many cycles because after the third clock cycle, the boundary-scan FSMs of all the chips will be in the *Run-Test/Idle* controller state. Maintaining `0' on both TMS lines will keep all FSMs in that controller state until the ATE is ready to proceed. This technique can be used in many similar steps below.

9. SHIFT SERIALLY THE PREPARED INPUT VECTOR INTO THE BOUNDARY-SCAN PATH BY SUPPLYING ($L_{OIR}+\lfloor N/2 \rfloor+1$) TCK PULSES WHILE HOLDING BOTH TMS LINES AT 0.

/* THE CONTINUITY THROUGH THE BYPASS REGISTERS IN THE EVEN-NUMBERED CHIPS HAS NOW BEEN CHECKED. IN CASE OF FAILURE OF THIS TEST, THE POINT IN THE SERIAL SCAN AT WHICH WRONG VALUES WERE FIRST DETECTED IS CRITICAL TO DIAGNOSIS. */

Note that the instruction registers in odd-numbered chips have again been pre-loaded with the instruction that will cause the bypass data register to be selected in those chips when the FSMs of their test logic enter the *Select-DR-Scan* state.

10. WITHOUT CONSUMING TCK CYCLES, CONFIGURE RECEIVER ON TDO TO RECEIVE A SERIAL VECTOR COMPOSED AS FOLLOWS: CONCATENATE (INTERLEAVED IN SEQUENCE OF CHIP POSITION ON THE BOUNDARY-SCAN PATH) THE EXPECTED CONTENTS (AS IN STEP 3) OF THE INSTRUCTION REGISTERS OF THE EVEN-NUMBERED CHIPS AND A 0 FOR EACH ODD-NUMBERED CHIP; FOLLOW THIS SEQUENCE WITH {01}. SIMULTANEOUSLY CONFIGURE THE DRIVER ON TDI TO SUPPLY THE SERIAL VECTOR COMPOSED BY CONCATENATING {01} WITH ($L_{EIR}+\lceil N/2 \rceil$) 0S.

Loading the instruction registers in even-numbered chips with '0...0' preloads them with the instruction that will cause selection of the boundary-scan data register for continuity checking once the bypass data register continuity is confirmed in all chips. Note that we must be careful about the values to be shifted into the boundary-scan data register so that potential problems such as bus contention are avoided (see Step 14).

11. SELECT THE INSTRUCTION REGISTER IN CHIPS OCCUPYING THE EVEN-NUMBERED POSITIONS IN THE SERIAL PATH. SIMULTANEOUSLY SELECT THE BYPASS DATA REGISTER IN THE ODD-NUMBERED POSITION CHIPS.

/* STEP 11 REQUIRES 2 COORDINATED TMS LINES. */

This action requires 4 TCK cycles. The sequences of values in Table 8-2, Step 7 (above) on TMS1 and TMS2 are swapped -- the sequence previously applied to TMS1 is applied, this time, to TMS2 and vice-versa.

12. USING 2 TCK CYCLES, TAKE ALL EVEN (ODD) CHIPS TO THE SHIFT-IR(DR) CONTROLLER STATE BY APPLYING THE 0 ON BOTH TMS LINES DURING BOTH CYCLES. [NOTE: 1 SHIFT CYCLE IS ACCOMPLISHED BY THIS OPERATION.]

13. SHIFT THE PREPARED SERIAL INPUT VECTOR INTO THE BOUNDARY-SCAN PATH BY SUPPLYING ($L_{EIR}+\lceil N/2 \rceil+1$) TCK PULSES WHILE HOLDING BOTH TMS LINES AT 0.

/* NOW CONTINUITY OF THE TEST PATH THROUGH ALL BYPASS REGISTERS HAS BEEN CHECKED. */

14. WITHOUT CONSUMING TCK CYCLES: CONFIGURE RECEIVER ON TDO TO RECEIVE A SERIAL VECTOR COMPOSED AS FOLLOWS: CONCATENATE (INTERLEAVED IN SEQUENCE OF CHIP POSITION ON THE BOUNDARY-SCAN PATH) THE EXPECTED CONTENTS (AS IN STEP 3) OF THE INSTRUCTION REGISTERS OF THE ODD-NUMBERED CHIPS AND SEQUENCES OF {X...X} OF THE LENGTH OF THE BOUNDARY-SCAN DATA REGISTERS IN THE EVEN-NUMBERED CHIPS; FOLLOW THIS SEQUENCE WITH {01}. SIMULTANEOUSLY CONFIGURE THE DRIVER ON TDI TO SUPPLY THE SERIAL VECTOR COMPOSED BY CONCATENATING {01} WITH A ($L_{OIR}+L_{EBS}$) BIT VECTOR WHICH, WHEN SHIFTED INTO THE BOUNDARY-SCAN PATH, WILL LOAD 0S INTO THE SELECTED INSTRUCTION REGISTERS, AND THE SAFE VECTORS INTO THE SELECTED BOUNDARY-SCAN DATA REGISTERS (SEE STEP 1).

15. SELECT THE INSTRUCTION REGISTER IN CHIPS OCCUPYING THE ODD-NUMBERED POSITIONS IN THE SERIAL PATH. SIMULTANEOUSLY SELECT THE DATA REGISTER IN THE EVEN-NUMBERED POSITION CHIPS.

/* STEP 15 REQUIRES 2 COORDINATED TMS LINES. */

This action requires 4 TCK cycles with TMS line signals as in Step 7, above.

16. USING 2 TCK CYCLES, TAKE ALL ODD (EVEN) CHIPS TO THE SHIFT-IR(DR) CONTROLLER STATE BY APPLYING 0 ON BOTH TMS LINES DURING BOTH CYCLES. [NOTE: 1 SHIFT CYCLE IS ACCOMPLISHED BY THIS OPERATION.]

The placing of '0...0' in all instruction registers of odd-numbered chips prepares them to cause selection of the boundary-scan data register the next time the FSMs of the odd-numbered chips go into the *Select–DR–Scan* controller state. Note that we must be careful about the values that are shifted into the boundary-scan data register to avoid potential problems such as bus contention (see Step 18).

17. SHIFT THE PREPARED SERIAL INPUT VECTOR INTO THE BOUNDARY-SCAN PATH BY SUPPLYING ($L_{OIR}+L_{EBS}+1$) TCK PULSES WHILE HOLDING BOTH TMS LINES AT 0.

/* CONTINUITY IN THE BOUNDARY-SCAN DATA REGISTERS OF THE EVEN-NUMBERED CHIPS HAS NOW BEEN CHECKED. THE FIRST BIT IN THE

SERIAL STREAM AT WHICH A WRONG VALUE OCCURS IS CRITICAL TO DIAGNOSIS. */

18. WITHOUT CONSUMING TCK CYCLES: CONFIGURE RECEIVER ON TDO TO RECEIVE A SERIAL VECTOR COMPOSED AS FOLLOWS: CONCATENATE (INTERLEAVED IN SEQUENCE OF CHIP POSITION ON THE BOUNDARY-SCAN PATH) THE EXPECTED CONTENTS (AS IN STEP 3) OF THE INSTRUCTION REGISTERS OF THE EVEN-NUMBERED CHIPS AND SEQUENCES OF {X...X} OF THE LENGTH OF THE BOUNDARY-SCAN DATA REGISTERS IN THE ODD-NUMBERED CHIPS; FOLLOW THIS SEQUENCE WITH {01}. SIMULTANEOUSLY CONFIGURE THE DRIVER ON TDI TO SUPPLY THE SERIAL VECTOR COMPOSED BY CONCATENATING {01} WITH A ($L_{EIR}+L_{OBS}$) BIT VECTOR WHICH, WHEN SHIFTED INTO THE BOUNDARY-SCAN PATH, WILL LOAD 0S INTO THE SELECTED INSTRUCTION REGISTERS, AND THE SAFE VECTORS INTO THE SELECTED BOUNDARY-SCAN DATA REGISTERS (SEE STEP 1).

The input vector used here is somewhat arbitrary because we do not plan to give an example of further checking of continuity through optional, user-defined data registers. If user-defined data registers are present, then the input vector of Step 18 would be slightly more complex. The positions in the vector representing values that would be in instruction registers at the end of the shifting sequence would be manufacturer-defined instructions selecting some set of user-defined data registers. The values that would end up in boundary-scan data register cells should still be "safe" (i.e., contention-free).

19. SELECT THE INSTRUCTION REGISTER IN CHIPS OCCUPYING THE EVEN-NUMBERED POSITIONS IN THE SERIAL PATH. SIMULTANEOUSLY SELECT THE DATA REGISTER IN THE ODD-NUMBERED POSITION CHIPS.

/* THIS REQUIRES 2 COORDINATED TMS LINES AND FOLLOWS THE METHOD OF STEP 11. */

20. USING 2 TCK CYCLES, TAKE ALL EVEN (ODD) CHIPS TO THE SHIFT-IR(DR) CONTROLLER STATE BY APPLYING 0 TO BOTH TMS LINES DURING BOTH CYCLES. [NOTE: 1 SHIFT CYCLE IS ACCOMPLISHED BY THIS OPERATION.]

21. SHIFT THE PREPARED SERIAL INPUT VECTOR INTO THE BOUNDARY-SCAN PATH BY SUPPLYING ($L_{IR}+L_{OBS}+1$) TCK PULSES WHILE HOLDING BOTH TMS LINES AT 0.

/* BY THIS OPERATION YOU WILL HAVE CONFIRMED CONTINUITY THROUGH THE BOUNDARY-SCAN REGISTERS OF ALL PARTS IN THE SERIAL PATH. THE FIRST BIT IN THE SERIAL STREAM AT WHICH A WRONG VALUE OCCURS IS CRITICAL TO DIAGNOSIS. */

To save space in this discussion, no additional data registers that might be in chips on the board will be checked for continuity in this pseudocode. It would be wise to check those paths in a real board should any exist. Such paths might contain registers read in determining the results of self-testing of ICs. They might also be used in deterministic scan testing of the on-chip logic of an IC or of groups of ICs. The method of checking them is analogous to that used to confirm continuity through the boundary-scan data register and the bypass data register.

The number of TCK cycles that have been employed in our test to this point can be obtained from the following formula:

$$L_{IR} + N + L_{BSR} + 35$$

In all probability, the dominating value will be L_{BSR}. It is worth noting that L_{BSR} is not computed merely by counting the number of boundary-scan input and output cells on all the chips with boundary-scan. On chips with 3-state and bidirectional leads, there will be cells added to the boundary-scan data register to provide control of those leads.

8.4: Interconnect Check

A set of deterministic patterns of verification of the interconnections between chips on the board is assumed to have been prepared in advance. Note that if no optional data registers exist, Steps 22–25 can be skipped.

22. SELECT THE INSTRUCTION REGISTERS IN ALL PARTS, CONSUMING 4 TCK CYCLES.

23. WITHOUT CONSUMING TCK CLOCK CYCLES, PREPARE A SERIAL INPUT STREAM OF L_{IR} 0S TO BE LOADED AT TDI. SET OUTPUT BUFFER ON TDO TO DONT CARE.

24. GO TO THE SHIFT-IR CONTROLLER STATE CONSUMING 2 TCK CYCLES. [NOTE: 1 SHIFT CYCLE IS ACCOMPLISHED BY THIS OPERATION.]

25. LOAD THE INSTRUCTION THAT WILL SELECT THE BOUNDARY SCAN DATA REGISTER IN ALL PARTS BY LOADING THE INPUT PREPARED IN STEP 23. THIS REQUIRES (L_{IR}-1) TCK CYCLES.

26. WITHOUT CONSUMING TCK CYCLES, PREPARE TO LOAD THE FIRST BOUNDARY-SCAN INTERCONNECT TEST VECTOR VIA TDI. THE LENGTH OF THIS VECTOR AND ALL SUBSEQUENT INTERCONNECT TEST VECTORS IS L_{BSR} BITS. AT THE SAME TIME, SET THE OUTPUT EXPECT BUFFER ON TDO TO DONT CARE. IT IS THE SAME LENGTH AS THE INPUT VECTOR.

27. PROCEED TO THE SHIFT-DR STATE. THIS CONSUMES 5 TCK CLOCK CYCLES. [NOTE: 1 SHIFT CYCLE IS ACCOMPLISHED BY THIS OPERATION.]

28. LOAD THE PREPARED INTERCONNECT TEST VECTOR. THIS CONSUMES (L_{BSR}-1) TCK CYCLES.

29. PROCEED TO THE UPDATE-DR STATE TO PLACE THE LOADED VALUES ON THE INTERCONNECT LINES. SIMULTANEOUSLY, PLACE ANY VALUES REQUIRED ON PRIMARY CIRCUIT PACK EDGE CONNECTOR LEADS ON THOSE LEADS. THIS REQUIRES 2 TCK CYCLES.

30. PROCEED TO THE CAPTURE-DR STATE TO COLLECT THE TEST RESULTS. THIS REQUIRES 2 TCK CYCLES.

31. WITHOUT CONSUMING TCK CYCLES, PREPARE THE NEXT INTERCONNECT TEST VECTOR FOR LOADING ON TDI. SIMULTANEOUSLY PREPARE THE EXPECTED OUTPUT BUFFER ON TDO WITH THE EXPECTED OUTPUT OF THE INTERCONNECT TEST JUST CARRIED OUT.

32. PROCEED TO THE SHIFT-DR STATE TO UNLOAD THE TEST RESULTS AND SIMULTANEOUSLY LOAD THE NEXT TEST. THIS REQUIRES 1 TCK CYCLE. [NOTE: 1 SHIFT CYCLE IS ACCOMPLISHED BY THIS OPERATION.]

/* UNLOAD RESULTS OF $TEST_{t-1}$ AND LOAD INPUT OF $TEST_t$ SIMULTANEOUSLY */

33. APPLY THE INTERCONNECT TEST PREPARED IN STEP 31 AND READ AND COMPARE OUTPUT OF PREVIOUS TEST TO EXPECTED OUTPUT PREPARED IN STEP 31. THIS REQUIRES (L_{BSR}-1) TCK CYCLES.

34. PROCEED TO THE UPDATE-DR CONTROLLER STATE TO PLACE THE LOADED VALUES ON THE INTERCONNECT LINES AND CONDITION PRIMARY CIRCUIT BOARD INPUTS -- AS IN STEP 29. THIS REQUIRES 2 TCK CYCLES.

35. PROCEED TO THE CAPTURE-DR CONTROLLER STATE AS IN STEP 30 (2 TCK CYCLES).

36. WHILE THERE IS ANOTHER INTERCONNECT TEST TO DO, GO TO STEP 31.

37. WHEN LAST INTERCONNECT TEST HAS BEEN INPUT, WITHOUT CONSUMING TCK CYCLES, PREPARE A SAFE/CONTENTION-FREE INPUT VECTOR FOR TDI; AND, SIMULTANEOUSLY, PREPARE AN EXPECTED OUTPUT VECTOR FOR THE LAST INTERCONNECT TEST FOR THE TDO OUTPUT BUFFER.

38. PROCEED TO THE SHIFT-DR CONTROLLER STATE. (1 TCK CYCLE.) [NOTE: 1 SHIFT CYCLE IS ACCOMPLISHED BY THIS OPERATION.]

39. UNLOAD THE LAST INTERCONNECT TEST RESULTS. (L_{BSR}-1 TCK CYCLES.)

All output data must be saved for use by the diagnostic program that will evaluate the interconnect test results. This evaluation can be done during the testing or following the interconnect testing section of the test program. Condensation of test results in this part of the test by a signature analysis approach is not acceptable in diagnostic situations. It is possible that a GO/NOGO situation might use a compacted signature; however, for efficiency of the test process, it might be better not to have to run a test twice to get the full data needed for diagnosis. The actual error bit position is used in the diagnosis.

If there are small- and medium-scale integration (SSI and MSI) ICs between chips on which boundary-scan is implemented, deterministic vectors can be applied to SSI/MSI clusters in a manner identical to that done in applying the interconnect test. In fact, if the diagnostic package is properly developed, the two tasks can be carried out with some (possibly considerable) overlap.

The number of TCK cycles required to carry out the interconnect test is:

$$L_{IR} + (T_x+1)(L_{BSR}+4) + 5$$

If SSI and MSI parts are to be tested using the boundary-scan path, then replace T_x in the above equation with $T_x+T_{SSI}+T_{MSI}$ in which T_{SSI} is the number of tests beyond T_x needed to get adequate fault coverage of SSI parts and T_{MSI} is defined analogously for the number of tests needed for MSI parts.

The expected values of T_x have been computed. The worst case situation requires 2^k tests for open and stuck-at faults (where k is the maximum number of boundary-scan data register output cells on a given net) and $2\log_2(n+2)$ tests for bridging faults including diagnosis (where n is the number of nets on the circuit board).

It is very clear that test length is dominated by the value of L_{BSR}. For this reason, it is very probable that future interconnect testing could be complicated by requiring synchronization of test application on more than one boundary-scan serial path per circuit board.

It might be the case that ATE input and output buffers on TDI and TDO can be set up for many vectors in advance. IEEE Std 1149.1 provides for a pause state even in the middle of shifting should ATE test vector buffers require reloading. If ATE buffers need to be reloaded r times during shifting, then the time of test application will be extended by a period of r(P+4) TCK cycles.

8.5: BIST Part Check

At this point, the integrity of the boundary-scan path, the integrity of the interconnect, and the possibility of faults in the MSI and SSI logic that is surrounded by boundary-scan paths will have been checked. There remains the task of testing the non-surrounded MSI and SSI chips and the LSI and VLSI parts on the board. First, we do all we can with the boundary-scan interface by checking the parts that have it and have BIST. Sometimes a group of parts might be considered as a single part for these purposes. This means that the capability of seeing parts individually and as part of a single "cluster" is important in the ATE system software.

In the following, we assume that a "safe" vector has been loaded into the boundary-scan data register through the last scan operation of the interconnect check (Section 8.4, Step 37). This alleviates many potential problems, such as bus contention, during execution of the BIST program.

```
40. SELECT THE INSTRUCTION REGISTER IN ALL ICS AND LOAD THE
RUNBIST COMMAND IN ALL ICS THAT HAVE BIST.

/* EVERY IC WITH PUBLICLY-ACCESSIBLE BIST IS REQUIRED TO HAVE A
COMMAND THAT RUNS (SERIALLY OR IN PARALLEL) ALL BIST FEATURES
AVAILABLE TO THE PURCHASER OF THE PART. */

41. SELECT THE RUN-TEST/IDLE STATE IN ALL ICS.

42. SUPPLY THE NUMBER OF SYSTEM CLOCK PULSES EQUAL TO THE
LARGEST NUMBER REQUIRED FOR SELF-TESTING OF ANY IC IN THE
BOUNDARY-SCAN CHAIN.

/* THE TEST LOGIC IS REQUIRED TO BE OF STATIC DESIGN SO THAT
PARTS THAT COMPLETE THEIR SELF-TESTS WILL HOLD THE RESULTS OF
THOSE TESTS UNTIL THEY ARE POLLED. */

43. SCAN THE RESULTS OF SELF-TEST OUT OF THE DATA REGISTERS
(WHICH ARE SIGNATURE REGISTERS AUTOMATICALLY SELECTED BY THE
RUNBIST INSTRUCTION).

/* THE TEST SOFTWARE MUST PARSE THIS TEST SEQUENCE SO THAT THE
SIGNATURES FOR EACH CHIP CAN BE READ AND CHECKED SEPARATELY BY
THE TEST/DIAGNOSTIC SYSTEM. */
```

It is important to note that the expected values of the self-test signatures that are scanned out in the last step are required to be supplied to IC purchasers in the data sheet of the IC.

It is possible that for some reason (e.g., power dissipation) not all the chips can be self-tested at once or that the self-test is carried out on a group of chips excluding others. If this is the case, Steps 40 through 43 would be modified to test groups of chips while others have their bypass data registers selected to reduce the length of the scan path when unloading the self-test results.

8.6: The Remaining Chips

At this point, the parts on the board that have not been tested are the parts without boundary-scan and those SSI and MSI parts not surrounded by boundary-scan. In early implementations, it might be s simplification to test these parts on an in-circuit tester while checking the passive components. However, if these parts have very good quality histories, it might be acceptable to test them only as part of a functional test following the test represented above in pseudocode. In the future, octal parts containing the boundary-scan TAP, boundary-scan path, and eight bits of signature analysis register or pseudorandom pattern generating register will be available from companies such as Texas Instruments [9]. These parts would allow the surrounding of all SSI and MSI parts with boundary-scan, thus reducing the problem of lack of coverage if in-circuit test were to be used only to check passive components on a partially assembled board.

8.7: Comments on Diagnosis

Full implementation of BIST and boundary-scan at board level can greatly improve the accuracy in diagnosing complex, high-density boards. The potential improvement comes from two board-level testability features realized through BIST and boundary-scan:

- The self-test capability of all or most of the components on a board allows for easy isolation of complex components with internal faults.
- The TAP serves as the common medium through which the results of chip-level self-tests are polled or scanned out for diagnosis.

Compact, effective interconnect test patterns can be easily applied to the board under test without suffering from many constraints commonly encountered by conventional in-circuit testers. Specifically, boundary-scan helps eliminate most problems associated with backdriving and test access limitation.

To fully exploit these advantages, however, the complete test response must be captured for examination. Specifically, one has to determine the locations of the erroneous bits contained in the test results to isolate those components that have failed the self-tests. As for board-level faults such as short-circuits and opens, the positions of the failed bits are also vital in pin-pointing the physical locations of these faults. For instance, a simple algorithm for identifying interconnect failures requires that all responses from the board-under-test to the $\lceil 2\log_2(n+2) \rceil$ test patterns (n = number of nets on the board) be examined [6]. The entire test set consists of the well-known $\lceil \log_2(n+2) \rceil$ test patterns (the counting sequence), that detect all interconnect failures, and of their complements. It is interesting to point out that for GO/NO-GO testing, only the responses, or (for added throughput) the compressed signature, corresponding to the first $\lceil \log_2(n+2) \rceil$ patterns need to be perused. The signature analysis approach can be particularly attractive when the first-pass yield of a board is relatively high, and when sufficient empirical repair data are available to support diagnostic techniques based on statistical pattern recognition [10]. Otherwise, the *uncompressed* responses to both the $\lceil \log_2(n+2) \rceil$ patterns and their complements need to be examined to achieve more precise fault isolation.

Other sophisticated diagnostic algorithms have been published that can achieve higher diagnostic resolution than the algorithm mentioned above. Please refer to the literature for an in-depth discussion of the various boundary-scan diagnostic techniques [7,8].

8.8: Conclusion

We have presented the outline of a complete program for testing boards equipped with BIST and boundary-scan. The test program is designed to deal with the initialization of the board under test, the verification of the boundary-scan test circuitry, the application of interconnect test patterns through the TAP, and the verification of components with BIST. Issues related to testing components without boundary-scan have been briefly addressed. Finally, with reference to some of the published diagnostic techniques, we have offered some comments on issues concerning the diagnosis of a board equipped with BIST and boundary-scan.

8.9: Acknowledgment

We want to thank Colin Maunder for reviewing an earlier version of this paper and for providing us with many valuable comments and suggestions.

8.10: References

[1] IEEE Std 1149.1, *Standard Test Access Port and Boundary-Scan Architecture*, IEEE, New York, 1990.

[2] Joint Test Action Group, *Boundary-Scan Architecture Proposal – Version 2.0,* March 1988.

[3] C. L. Hudson, "Integrating BIST and Boundary-Scan," *National Communications Forum*, 1988, pp. 1796-1800.

[4] D. R. Resnick and A. G. Bell, "Real World Built-In Test for VLSI," *IEEE Compcon*, IEEE Computer Society Press, Los Alamitos, Calif., 1986, p. 436-440.

[5] D. R. Resnick, "Testability and Maintainability With a New 6K Gate Array," *VLSI Design*, Vol. 4, No. 2, March-April 1983, pp. 34-38.

[6] P. T. Wagner, "Interconnect Testing with Boundary-Scan," *IEEE International Test Conference, 1987 Proceedings*, IEEE Computer Society Press, Los Alamitos, Calif., pp. 52-57.

[7] C. W. Yau and N. Jarwala, "A New Framework for Analyzing Test Generation and Diagnosis Algorithms for Wiring Interconnects," *IEEE International Test Conference Proceedings,* IEEE Computer Society Press, Los Alamitos, Calif., 1989, pp. 63–70.

[8] N. Jarwala and C. W. Yau, "A Unified Theory for Designing Optimal Test Generation and Diagnosis Algorithms for Board Interconnects," *IEEE International Test Conference Proceedings,* IEEE Computer Society Press, Los Alamitos, Calif., 1989, pp. 71–77.

[9] Texas Instruments, Inc., *Testability: Test and Emulation Primer*, Texas Instruments Inc., Austin, Tex., 1989.

[10] C. W. Yau, "ILIAD: A Computer–Aided Diagnosis and Repair System," *IEEE International Test Conference Proceedings,* IEEE Computer Society Press, Los Alamitos, Calif., 1987, pp. 890–898.

9. Diagnosing Faults in the Serial Test Data Path

Rod Tulloss and Chi Yau
AT&T Bell Labs
Engineering Research Center
Princeton, NJ 08540, U.S.A.

Lee Whetsel
Texas Instruments
6500 Chase Oaks Boulevard
Plano, TX 75086, U.S.A.

9.1: Objective

Before the serial test data path can be used to test the chips on a board and, through the boundary-scan registers, their normal functional interconnections, it must itself be tested for common production defects -- for example, solder shorts and opens. The design of the instruction register within the IEEE Std 1149.1 architecture includes facilities to assist in this task.

9.2: A Basic Path Test

The first step is to initialize the test access port (TAP) and instruction register. This can be achieved by holding the test mode select (TMS) signal (which is broadcast to all devices) high and applying five rising edges to the test clock (TCK). Where provided, the optional test logic reset (TRST*) inputs can also be used for this task. At the end of this process the TAP controller in each chip will be in the *Test–Logic–Reset* controller state which will cause other circuitry in the IEEE Std 1149.1 architecture to be initialized. For example, the instruction register's latched parallel output will be set to either the *IDCODE* or the *BYPASS* instruction, depending on the availability of a device identification register within the chip.

The second step is to move through the *Run–Test/Idle*, *Select–DR–Scan*, and *Select–IR–Scan* controller states and enter the instruction register scanning sequence. The initial *Capture–IR* controller state will cause the instruction register to be loaded with the {X...X01} pattern as specified by the standard.† Note that, so far, the only connections that have been involved are TMS and TCK. No shifting is needed during initialization.

By entering and remaining in the *Shift–IR* controller state for a number of TCK pulses equal to 2 plus the number of bits in *all* instruction registers of *all* chips, the constant patterns loaded into the least significant bits of the instruction registers in each chip will be observed at the serial output of a board with an error-free serial path. An additional 2-bit sequence {01} is shifted into the serial input of the first chip to check that connection and the part of the serial data path in the first chip that is between the fixed bits loaded into its instruction register and its test data input (TDI) pin. For example, the pattern at the serial output of a fault-free board containing three chips each having a 4-bit instruction register would be:

† As in IEEE Std 1149.1, we will use the convention that the least significant bit of a register is that nearest TDO. The convention that bit streams are shown with the least significant bit on the right is also adopted, so the 1 in the pattern shown will be shifted out first, the 0 second, etc.

EH0321-0/90/0000/0115$01.00 © 1990 IEEE

01XX01XX01XX01

where the bit at the right is shifted out first.

In the event of faults, the patterns from several chips will be observed followed by erroneous data -- allowing the nearest fault to the board's serial output to be localized. This allows faults to be diagnosed and removed from the board one at a time. Consider, for example, the circuit shown in Figure 9-1, which contains an open-circuit fault between chips IC2 and IC3.

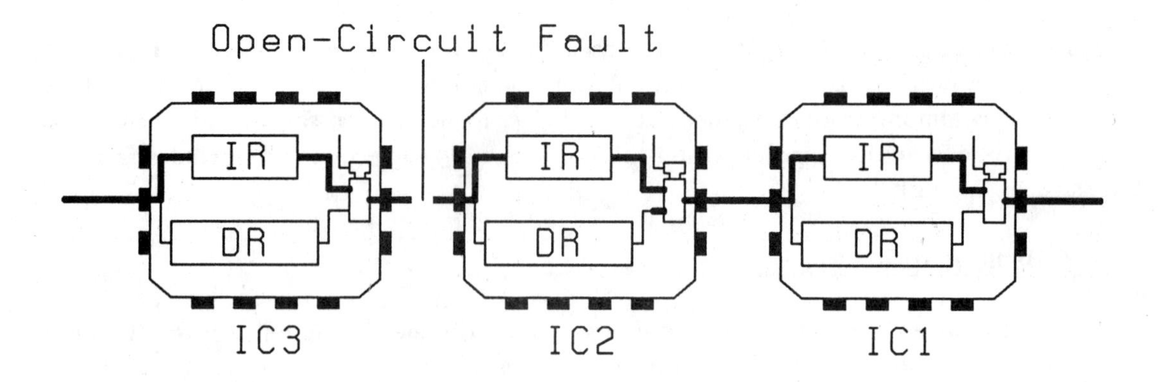

Figure 9-1: Testing for an open-circuit fault in the serial path.

Again, assume that the instruction register in each chip contains four shift-register stages and that a 2-bit {01} shift-terminating sequence is shifted into the board's serial input from the automatic test equipment (ATE) or bus master chip. In this case, the output observed at the board's serial output would be:

111111XX01XX01

where, again, the bit at the right is shifted out first. The open-circuit fault can be detected because the pull-up on the TDI input of IC2 causes a constant 1 to be shifted into that chip instead of the expected pattern, which starts {01} (read from left to right).

9.3: Use of the Device Identification Register

If a device identification register is present in the design, it is possible to combine a check of the assembly process with a test of the integrity of the board-level serial test data path by shifting all device identification codes out of the board in one pass. This approach is adopted in some proprietary boundary-scan implementations -- for example, as discussed in [1].

As described in Chapter 5, all available identification codes from chips on a board-level serial path will be shifted out for examination if the sequence {1111101000...} is applied at

TMS, one bit for each cycle of TCK. This sequence causes the TAP controllers in the driven chips to move first to the *Test–Logic–Reset* state and then through *Run–Test/Idle, Select–DR–Scan,* and *Capture–DR* to *Shift–DR*. In the *Test–Logic–Reset* controller state, the output of the instruction register is set to the *IDCODE* instruction in all chips that have the device identification register and to *BYPASS* in all other chips. As a result, the available identification codes will be shifted out, interspersed with strings of 0s output from chips that do not contain a device identification register.

A complete description of this process, and of a method for decoding the output data stream, is contained in Chapter 5.

9.4: More Complex Methods

The above diagnostic method adequately deals with many of the faults in the serial test data path (e.g., internal faults in the instruction registers and external faults resulting from solder opens and shorts). However, it does not guarantee the diagnosability of internal faults in the scannable test data registers. Next, we will describe two diagnostic methods that can alleviate this problem.

9.4.1: Method 1

This method requires the test data registers to be designed so that, in each of the registers except the single–bit bypass register, the two bits nearest to the serial test data output (TDO) can be initialized to a binary {01} pattern upon *Test–Logic–Reset*. With this added hardware provision, the method of testing the integrity of a serial test data path consisting of cascaded test data registers can be easily derived from the preceding paragraphs. Because the bypass register consists of only a single bit, we cannot locate a fault in the bypass bit of an arbitrary chip with only one scan pass. (Method 2 can be adapted here for locating a fault in the bypass register within two scan passes.)

9.4.2: Method 2

This method saves the hardware overhead required by Method 1 at the expense of decreased diagnostic throughput. This method applies after the instruction registers have checked out as good by using the method described in Section 9.2. In this method, a board is set up so that alternating integrated circuits (ICs) on the serial test data path between bus master TDO and bus master TDI receive TMS signals from two different, but coordinated, sources. This allows one set of chips (for example, those in the even–numbered spots along the serial path) to be set up for instruction register scanning while the other set (those in the odd–numbered spots) are set up for test data register scanning. The fixed bits of the instruction registers in the even–numbered ICs can then be used to diagnose the integrity of the selected test data registers in the odd–numbered ICs.

As an example, we will apply this method to the diagnosis of a fault in the bypass register of an IC. Any fault equivalent to a stuck–at fault on the input or output of an IC's bypass register can be located in no more than two scan passes. Because the bypass registers and the instruction registers are scanned in two different controller states --

Shift–DR and *Shift–IR* respectively -- this approach requires that two separate TMS wires (designated TMS1 and TMS2) be distributed. TMS1 controls the odd-numbered ICs, while TMS2 controls the even-numbered ICs as shown in Figure 9–2.

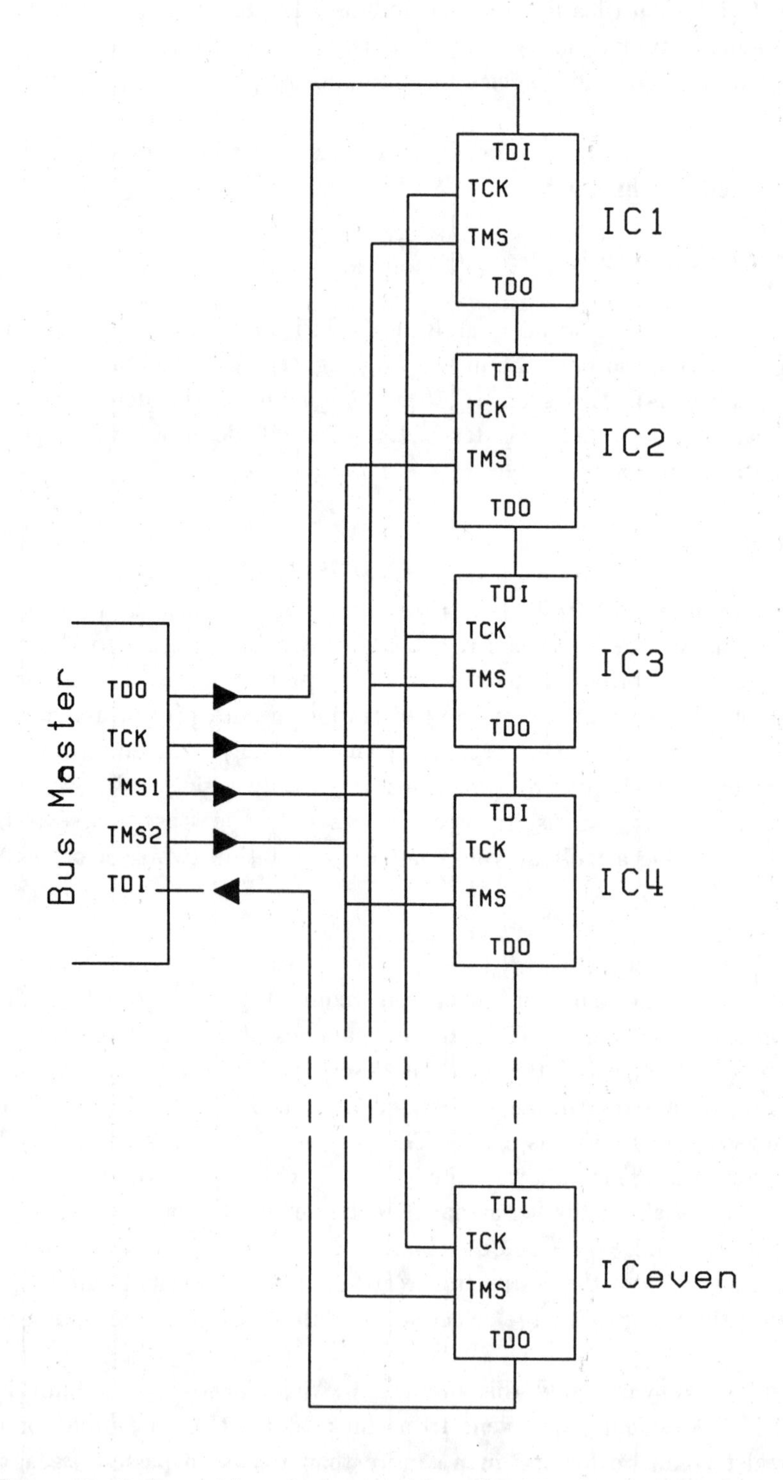

Figure 9–2: Board level connection of TAP pins for Method 2.

In Figure 9–2, it is assumed arbitrarily that there is an even number of ICs on the board–level serial path, and this assumption is carried forward through the example in this section. Note that TMS1 and TMS2 can come directly from the TAP bus master or they can come from a board–level controller.

To illustrate this approach, let us assume that all chips on a board's test data path have been initialized to the *Test–Logic–Reset* controller state. Then we can apply the following control sequences on TMS1 and TMS2 so that, on each of two separate scan passes, half of the chips at a time will be bypassed and the other half will have their instruction registers scanned.

At the start of the test, all instruction registers are set to the *BYPASS* instruction by holding the board's serial input at 1 and completing an instruction scan sequence. This must contain sufficient clocks in the *Shift–IR* controller state for *all* instruction register stages in *all* chips to be set to 1. This requires that the same control sequences are applied to the two TMS lines -- TMS1 and TMS2. At the end of the instruction scan sequence, the TMS lines are controlled such that all chips enter the *Run–Test/Idle* controller state. At this point, the bypass registers in all chips have been selected for a following data register scan sequence. From this starting state, testing proceeds as shown in Table 9–1.

Table 9–1: Coordinated TMS values for method 2.

Clock Cycle	TMS1	Odd Chip Controller State	TMS2	Even Chip Controller State
1	0	Run-Test/Idle	0	Run-Test/Idle
2	0	Run-Test/Idle	1	Select-DR-Scan
3	1	Select-DR-Scan	0	Select-IR-Scan
4	0	Capture-DR	0	Capture-IR
5	0	Shift-DR	0	Shift-IR
6	0	Shift-DR	0	Shift-IR
.	.	...	.	...
M-3	0	Shift-DR	0	Shift-IR
M 2	1	Exit1 DR	1	Exit1-IR
M-1	1	Update-DR	1	Update-IR
M	0	Run-Test/Idle	0	Run-Test/Idle
.	.	...	.	...
N	0	Run-Test/Idle	0	Run-Test/Idle
N+1	1	Select-DR-Scan	0	Run-Test/Idle
N+2	1	Select-IR-Scan	1	Select-DR-Scan
N+3	0	Capture-IR	0	Capture-DR
N+4	0	Shift-IR	0	Shift-DR
N+5	0	Shift-IR	0	Shift-DR
.	.	...	.	...

Note that, in the first scan pass, TMS1 holds the odd-numbered components in the *Run–Test/Idle* state for one more cycle than the even-numbered components, so that shifting of the data/instruction registers of all components is synchronized. In the second scan pass (starting at clock cycle N), TMS2 holds the even-numbered components in the *Run–Test/Idle* state for one more cycle than that of the odd-numbered components. Also, note that a fault in the bypass register of an odd-numbered component will be located in the first scan pass, and a similar fault in an even-numbered component will be located in the second scan pass.

In Table 9–2, control sequences similar to those in Table 9–1 are applied to diagnose a stuck-at-1 fault in the bypass register of IC_4. As in the earlier example, all chips are depicted as having instruction registers that are four bits long.†

Table 9–2: Detection of a stuck-at fault.

Chip	Test 1		Test 2	
	Expected	Observed	Expected	Observed
Bus master TDO	n/a	n/a	1	1
IClast	XX01	XX01	[0]	[1]
...	...	...	...	...
IC6	XX01	XX01	[0]	[1]
IC5	[0]	[0]	XX01	1111 └C
IC4	XX01	XX01	[0]	[1] └B
IC3	[0]	[0]	XX01	XX01 └A
IC2	XX01	XX01	[0]	[0]
IC1	[0]	[0]	XX01	XX01

The first incorrect value of observed output 2 is the 1 from IC_4. However, this error only implies that a bad bit occurred between points A and B (marked on Table 9–2). The

† Brackets are used to delimit the single-bit serial outputs from those chips that are bypassed.

bad bit must occur first at B because the instruction register of IC_4 has previously been tested and found fault-free when using the method described in Section 9.2. Similarly, the bit in IC_4 stuck-at-0 can be detected by finding a 0 at position C in the analog of observed output 2. Note that a stuck-at-0 fault in a bypass register can be detected, but *not* located, by scan operations using only one TMS line. This is because, in the single-TMS-line case, there is no way to place 1s in registers between the fault and the board-level serial output.

Note: Because a test must be performed for the stuck-at-0 fault in the bypass register of IC_{last}, it is necessary to shift an additional 1 into the TDI input of that component in the second test. Stuck-at faults on TDI itself will have been tested by the test described in Section 9.2.

9.5: Reference

[1] R. Lake, "A Fast 20K Gate Array with On-Chip Test System," *VLSI Systems Design*, Vol. 7, No. 6, June 86, pp. 46-65.

10. In–Circuit Testing

Bob Russell
Bull HN Information Systems
38 Life Street
Brighton, MA 02135, U.S.A.

In the immediate future, occasions will frequently arise in which not all the integrated circuits used to construct a loaded printed wiring board contain the features defined by IEEE Std 1149.1. For such boards, there may be a continued need to use in–circuit test techniques as a part of the overall test process. This chapter discusses how integrated circuits (ICs) compatible with IEEE Std 1149.1 may be designed so that such testing of non–conformant chips can be reliably performed.

10.1: Mixed In–Circuit and Boundary–Scan Testing

During in–circuit testing of chips on a board, it is necessary for the tester to be able to determine the signals fed into the chip under test. On occasions, tester signals will be applied to the outputs of chips that conform to IEEE Std 1149.1. Where these outputs can be placed in an inactive drive state (e.g., high–impedance) or can be set to a logic level that can be safely and effectively backdriven, this is readily achieved without risk of damage. In other cases, backdriving must be carefully controlled to ensure that no damage is caused to the chips that normally determine the signals to be supplied to the chip under test. Such controls place limits on the length of test that can be applied and may therefore adversely impact test quality.

It is advisable, therefore, to provide a means for setting the outputs of all chips -- including those compatible with IEEE Std 1149.1 -- to a state that can be safely backdriven during in–circuit test.

Figure 10–1 consists of six examples of system output (F1 through F6) and shows how each can be placed in a state that can be safely backdriven under control of a signal ICT* (ICT* = 0 for in–circuit test):

- F1 and F2 are set to high–impedance during the in–circuit test mode by using a single added AND gate (or, if the high–impedance control was previously driven from an AND gate, by adding an additional gate input).

- F3 represents one or more 3–state drivers requiring an extra AND gate for independent enabling in the system mode.

- F4 and F5 represent, respectively, outputs capable of being backdriven from zero (but not from one) and from one (but not from zero).

- F6 represents an output that can be backdriven when at either logic level and, therefore, requires no modification.

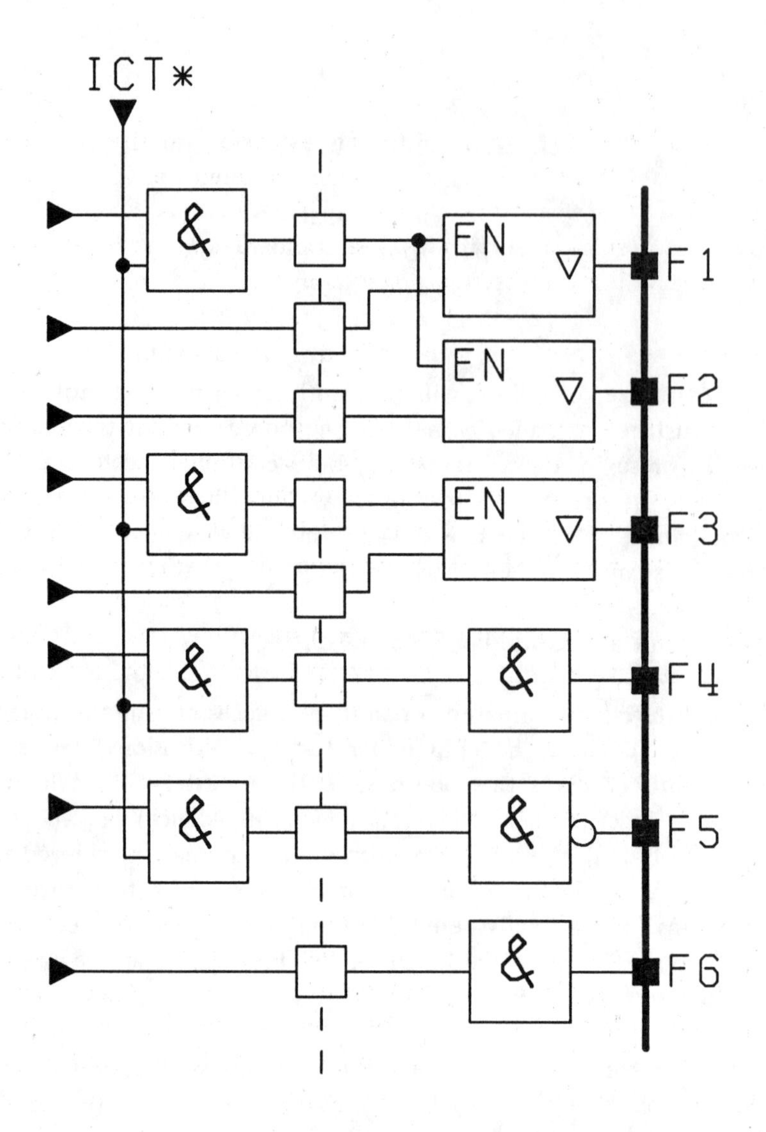

Figure 10–1: Control of outputs into an overdrivable state.

NOTE: The open squares in this figure indicate boundary–scan cells; the dotted line is the serial "TDI–to–TDO" connection between them.

The following sections discuss methods for allowing chips compatible with the standard to be configured such that they can be safely backdriven.

10.2: Method 1

The first method described is based on an extension to the test logic functionality defined by IEEE Std 1149.1. It allows signals supplied to the loaded board by the in-circuit test system to place the chip in an "in-circuit-safe" state. This is achieved through use of the test pins defined by the standard and, as will be discussed, has a minimum impact on the test and system logic in an IC.

The method is based on two properties defined by the standard:

1. The driver for the test data output (TDO) pin is a 3-state device that is active only when data or instructions are being shifted through the chip. As a result, the connection from the TDO output of one chip to the test data input (TDI) of the next on the board-level serial path will be floating (i.e., not driven) while the chips are set for normal (i.e., non test) operation of the on-chip system logic.

2. The TDI input to a chip must be designed such that, when not externally driven, it behaves as though a logic 1 was being applied. Typically, but not universally, the latter requirement will be met through inclusion of an internal pull-up resistor. The method described here assumes that a pull-up resistor is present at the TDI input of the next chip on the serial board-level path. Where this is not the case, a pull-up resistor must be added externally to the connected chips to ensure correct operation of the method described.

Typically, then, the TDO-TDI connection between a pair of chips will be pulled to logic 1 when the components are in their normal (i.e., non-test) mode of operation -- that is, when their test access port (TAP) controllers are in the *Test-Logic-Reset* state. Thus, if a bed-of-nails probe were to be connected to this signal, its state could be changed to 0 without the need for backdriving provided that the chips had previously been configured for normal operation (e.g., through application of a 0 at the test logic reset (TRST*) input).

As shown in Figure 10-2, the addition of a small amount of logic at the TDO output allows the condition where the TDO driver is inactive, but the driven connection is at logic 0, to be detected and used to control entry into an "in-circuit-safe" test mode (i.e., to generate the ICT* signal required in Figure 10-1).

Where no TRST* input is available, the "in-circuit-safe" test mode can be entered by holding TMS high and applying five or more rising edges at the test clock input (TCK) so that the *Test-Logic-Reset* controller state is reached. The TDO-TDI connections can then be pulled low as required.

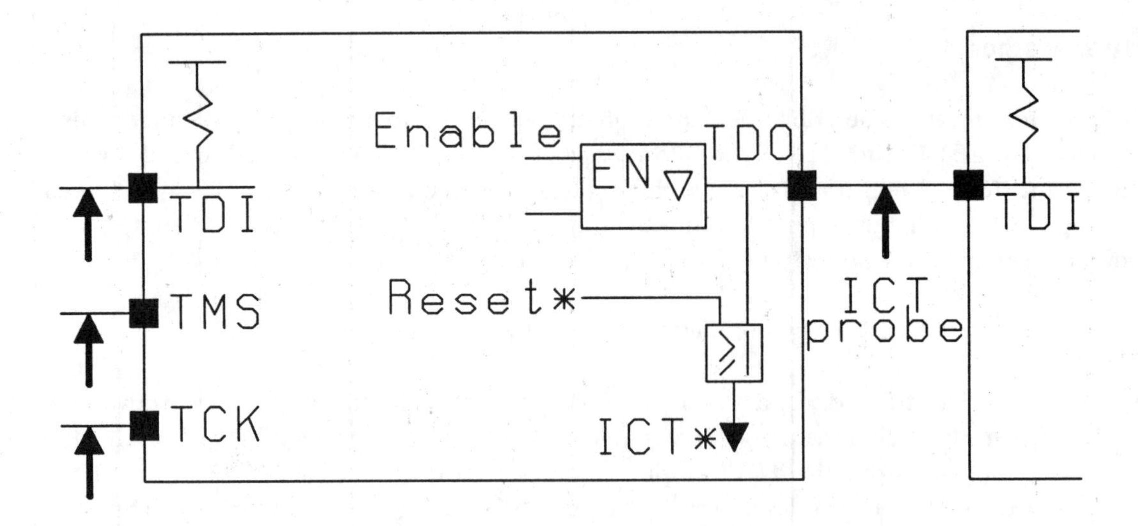

Figure 10–2: In–circuit test application.

10.3: Method 2

As an alternative to method 1, the "in–circuit–safe" test mode may be provided through inclusion of a dedicated instruction for the purpose. Such an instruction could be serially entered into the chip prior to the application of in–circuit tests to other chips on the board. In this case, the ICT* signal would be generated by the instruction decoder.

10.4: Method 3

In IEEE Std 1149.1 it is recommended that the "in–circuit safe" test mode be attainable by means of data loaded into the boundary–scan path while the *EXTEST* instruction is selected. This requires that the user knows which state at each pin can be safely backdriven and also that the automatic test equipment (ATE) is able to control the chip's TAP interface.

10.5: Conclusions

Several means of placing a chip compatible with IEEE Std 1149.1 in a state where its outputs can be safely backdriven during in–circuit testing have been discussed. These methods vary in the complexity of their use. Method 1 allows the chip to be set into the "in–circuit–safe" state simply by applying a voltage level to the board through a bed–of–nails probe. In contrast, method 2 requires that the ATE first causes an instruction to be entered into the TAP controller of each chip that is to be backdriven. Method 3 additionally requires that the ATE enters data into the boundary–scan path that will set each chip output such that it can be safely backdriven and that it will then select the *EXTEST* instruction.

The advantage of simplicity in attaining pre–test set–up in practical in–circuit testing should not be ignored.

Part IV: Implementation Examples and Further Applications

The chapters in Part IV discuss the implementation of IEEE Std 1149.1 and show how it can be applied to tasks other than loaded–board testing. Our aim in bringing together this material is to illustrate as wide a range of potential applications as possible and to provide a balanced view of the costs and benefits of using the standard.

Chapter 11. Applications of IEEE Std 1149.1: An Overview†

Peter Fleming
Texas Instruments
6500 Chase Oaks Boulevard
Plano, TX 75086, U.S.A.

The original motivation for the development of IEEE Std 1149.1 was the increasing difficulty of testing newly-assembled or field-returned printed wiring boards (PWBs). Among the causes of this difficulty are increases in the complexity of integrated circuits (ICs) and use of highly-miniaturized interconnection and assembly technologies such as surface-mount.

However, loaded-board tests are by no means the only test tasks during a product's life that can be more effectively or more economically performed if IEEE Std 1149.1 is adopted at the integrated circuit level. In fact, the range of applications is very broad -- ranging from wafer to system test and from prototype debugging to maintenance and repair.

This chapter provides an overview of these applications and gives an introduction to the more detailed discussions contained both in subsequent chapters and in the reprinted papers contained in Part V.

11.1: Test Cost Reductions: Chip-to-System, Womb-to-Tomb

IEEE Std 1149.1 provides the foundation of a hierarchical approach to testing in which tests developed for use at one level in a product assembly hierarchy (for example, for an IC) can be reused at the various higher levels of assembly (for example, for testing the loaded PWB). The idea is to obtain the maximum return for each investment in design-for-test features or test data and thus reduce the overall cost of testing.

Not only does the standard provide the basis for an hierarchical approach from chip to system that allows efficient and economic testing at one stage during the product's life (for example, at production testing), it also supports testing throughout the total life cycle of a product -- from womb to tomb. This gives the opportunity for further test cost savings.

Like any other design-for-test technique, of course, these savings cannot be achieved without incurring costs. The objective is that, overall, the savings should significantly outweigh the costs.

Unfortunately, both costs and savings are highly dependent on such features as the type of product and the type of company; therefore, it is difficult to provide a detailed analysis in this book. However, Chapter 12 discusses the various issues in more detail and provides a basis for an in-depth economic analysis that might be performed. This chapter outlines

† This chapter is an updated extract from an article first published in the *Texas Instruments Technical Journal*, Vol. 5, No. 4, July-Aug. 1988.

the costs that will be incurred in implementing IEEE Std 1149.1 and shows how these might be reduced by careful design.

In the remainder of this introductory chapter it is assumed that IEEE Std 1149.1 is fully exploited in a product -- not only through implementation of the mandatory features (e.g., the boundary-scan path), but also through provision of test access to key internal registers in ICs. The aim is to describe what could ultimately be achieved when the standard is widely adopted across the electronics industry.

11.2: Applications During Design and Development

11.2.1: Integrated Circuit Debug

Traditionally, the greatest resistance to design-for-testability has come from IC designers, who feel they pay the most significant penalty for its inclusion while reaping the least benefit. However, today these designers are paying far greater attention to testability than ever before. They are also doing this voluntarily, with little pressure from the test community.

The reason is that, while IC designs are approaching the complexity of boards, they do not provide the probeability necessary to debug the design. Industry reports indicate that half of the application-specific IC (ASIC) designs produced do not work on the first pass because of inadequate testing and simulation. To successfully debug a complex design, designers have begun to explore scan paths as a technique for improving observation and control.

Unfortunately this has often been accomplished by multiplexing the scan path input and output connections onto functional pins, precluding the use of the paths during later testing (e.g., when the chip is mounted on a board). In some complex chips, however, four to six pins are fully dedicated for testability to allow access to the serial scan paths for test and debugging, for both stand-alone IC testing and chip-on-board testing.

The emergence of an industry-standard serial test data interface (the test access port (TAP)) will allow for the development of robust debugging environments based on personal computers or engineering workstations (Figure 11-1). Software tools provided on these machines will allow designers to conduct register level transactions interactively and to view the results on a personal computer or workstation in a graphic waveform format. States of internal registers need no longer be hidden, since inclusion of optional test data registers within the IEEE Std 1149.1 architecture will provide for test access and allow faster debugging and confirmation of designs.

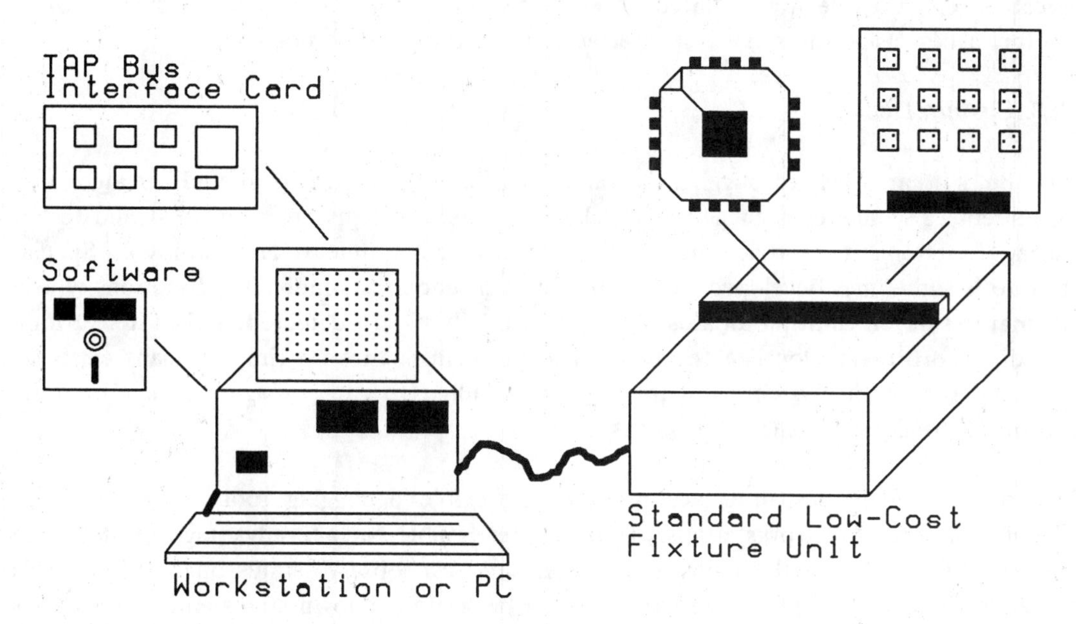

Figure 11–1: Low–cost debug/test station.

11.2.2: Loaded PWB Debug

Even with the greater accessibility afforded by loaded PWBs, design verification can often be a long and tedious process. Special software may need to be written and use of test equipment such as logic analyzers, oscilloscopes, and multimeters may be required.

For example, a major limitation of board debugging today is the difficulty of setting the design into the state the designer needs. Many instruments are available to observe those nodes that can be physically probed, but driving nodes to desired states is usually far more difficult. The outputs of chips that normally drive the nodes must be inhibited so that the nodes can be driven by the tester without risk of damage. An added complication is that the critical points to be controlled and observed frequently do not exist as probeable chip pins -- consider the key internal registers of a microprocessor, for example.

The provision of a standard test interface eliminates the necessity for physically probing the loaded board to control it. The designer can set up specific conditions (for example, in internal registers that can be accessed through a chip's TAP) and can observe how the design responds under software control via the serial test interface. Moving the points of observation is achieved simply by typing in commands, as opposed to reconnecting the logic analyzer to change the nodes being monitored. The designer can visualize the board from a register level and can use diagnostic tools to query the state of the hardware. All of the visibility provided during IC debugging remains available through the dedicated test interface.

For the future, where designs may be based on advanced surface mount technologies, physical access may be very limited, precluding the use of logic analyzers altogether. However, access using the serial test interface will continue to be possible.

11.2.3: System Debug

At the system level, debugging is rarely performed by using a sophisticated test environment. Techniques such as hot mock-ups, where systems are assembled and tested by using functional tests that emulate the end-user environment, are employed. Special test code is sometimes developed, but, typically, this code tests individual functions whose logic may reside on multiple boards. Thus, when failures are detected, it is often a long and tedious process to localize the fault. The problem is that no simple means exists to access the core of the system to help identify the failing board -- diagnosis must be done based on externally-observable symptoms.

Use of a standard test interface provides a flexible debugging tool, again based on personal computers or workstations. The designer is able to take advantage of the same debug routines used for the chips and boards, and can observe states on multiple cards simultaneously on the display. The hardware can be set into known states and its responses to these tests can be observed. Boards do not have to be placed on card extenders for probing, nor do special instruments have to be connected directly, avoiding the risk of affecting the parameter being measured.

11.2.4: Hardware/Software Integration

For many complex systems, the hardware design effort is dwarfed by the magnitude of the software development task. The most complex aspect of this task is the successful integration of the hardware and software. In cases where the system does not perform as anticipated, it is extremely difficult to resolve the failure between the hardware and software because of the poor visibility and controllability of the integrated system.

A consistent platform for debugging ICs, loaded boards, and systems, that also supports software testing tools, can significantly reduce the effort required to debug hardware/software systems. Robust software running on cost-effective hardware provides a single platform for downloading, uploading, and executing application software on target designs.

Significant debugging capabilities exist that can be windowed-in to provide improved knowledge of how the system performs, with the ability to access internal nodes that cannot possibly be viewed by using current instrumentation. Registers, program counters, arithmetic logic units, address and data busses, and other key areas become both controllable and viewable at the terminal. Instruction op-codes can be traced and converted by the debugging tool into mnemonics the programmer can more easily follow and understand. In some cases, code patches can be interactively generated and checked without the need to recompile.

For complex problems, hardware states can be captured and dumped to disc for off-line analysis.

11.2.5: Environmental Testing

Often the final step in qualifying a design is to verify that it operates correctly under a wide range of thermal, vibrational, and other environmental stresses.

During these tests, the instrumentation relied upon for design validation (e.g., oscilloscopes, logic analyzers) cannot be used because the system is enclosed by its case and is housed in a "hostile" environment. Therefore, the instruments cannot easily be connected to the desired points.

By making the standard test interface accessible on a connector of the fully assembled product, internal nodes continue to be both controllable and observable under software control. As at other test stages, the dependency on physical access is broken. Use of the interface can allow downloading and execution of special test software, and for monitoring of the system under external control. If failures occur, engineers may troubleshoot the problem while the subsystem remains in the environmental chamber under the conditions that caused the fault to occur. The added visibility may also provide useful data to system engineers in learning how the hardware responds to environmental stresses. An example would be in setting false alarm filter values for built-in test.

11.3: Applications During the Production Cycle

11.3.1: IC Testing

For relatively simple IC designs, testability features are sometimes included for design validation that are accessed through special wafer-probe pads. These features are usually not available for later test stages because the necessary connections are not bonded to pins on the packaged chip. A new test must therefore be developed for the packaged chip that accomplishes high detection without use of the test features.

This has grown to be unacceptable in complex designs. Often, the test features are preserved only by creating access through multiplexed use of functional pins. Unfortunately, the result is that the test developed for production testing of the IC is of no further use for assembled board and system test. It cannot be used once the chip is installed in a loaded board.

With the IEEE Std 1149.1 TAP, four (or five) pins are dedicated to ensure permanent access to the test features. The design of the test for the packaged chip becomes more straightforward because no multiplexing need be involved. If the IC has boundary-scan, the static vectors used for packaged-chip tests can be reused when the chip has been assembled onto the board.

The availability of the boundary-scan path also offers a simplified technique for achieving a reasonable confidence level during wafer test with greatly simplified fixturing.

By using the four (or five) pin test interface instead of providing connections to every I/O pin of the part, a significant portion of the logic can be exercised prior to packaging. This assumes that the dropout rate caused by faults in the I/O region is acceptably small enough to defer detection until the part is tested on a packaged part tester.

11.3.2: Parametric Testing of ICs

The boundary-scan register can be used to simplify the creation and application of parametric tests for ICs.

To perform a parametric test by using the boundary-scan register, a test program loop is entered. First, the boundary-scan register is set to test board interconnections by shifting the *EXTEST* instruction into the instruction register. A data register scan cycle is then entered, which causes the data applied at the system input pins to be captured in the boundary-scan register. The logic signal value perceived at each input pin can be examined by shifting the latched values through the test data output (TDO) pin. This load-shift cycle is repeated for different input voltages until all required voltage levels have been applied, The test program loop then ends.

Similarly, the boundary-scan register can be used to facilitate measurements on output drive capability, slew rates, etc. Further, the inclusion of cells in the boundary-scan register that allow each 3-state output pin or bidirectional pin to be forced to high-impedance allows this aspect of chip performance to be tested easily.

A parametric test constructed by using the boundary-scan register may be significantly shorter than that of a conventional equivalent. For example, a test of the input switching thresholds of an IC would normally require paths to be set up through the circuit so that each input can be observed by monitoring chip outputs. The resulting test sequence could be extremely long -- perhaps up to 50,000 vectors. When the boundary-scan register is used, each test cycle contains roughly as many patterns as there are pins on the chip -- typically, many fewer test patterns than would be required to propagate signals through the chip from input to output.

11.3.3: Incoming Goods Testing

Companies continue to test integrated circuits prior to introducing them into stock. This usually requires a multi-million dollar capital investment in automatic test equipment (ATE) and leads to demands for design data to be supplied by the chip vendor. Not surprisingly, these vendors are not keen to support the multiple types of ATE used by their customers or to part with their design data.

In some cases, the user has enough confidence in the foundry's quality levels to allow "ship-to-stock" without incoming inspection. It can be very expensive, however, to isolate even the limited number of faults that slip through when each echelon of test adds a tenfold increase in cost.

Depending on the fault spectrum of chips after once successfully passing the packaged–chip test, potential exists for a cost–effective static tester that could be used for many chip types. If such a test were based on features accessible when using IEEE Std 1149.1, the personal–computer– or workstation–based test environment would be capable of simultaneously directing the test of multiple ICs, reusing a subset of the vectors developed for the packaged part test.

Assuming that the most severe faults would be detected in this relatively low–cost environment, companies could feel more comfortable about "ship–to–stock." This highly portable test environment would also support retest of chips thought to have subsequently failed during the manufacturing process or field use.

11.3.4: In–Circuit Loaded–Board Test

The continued need for low–cost manufacturing defect test environments gave birth to the Joint Test Action Group and the sudden momentum in the electronics industry toward provision of standard test busses and boundary–scan.

Today, in–circuit test is used to detect the large majority of faults introduced during the manufacturing process -- damaged parts, wrong parts, misoriented parts, opens, and solder shorts. However, many major companies shared concerns over the ability to continue performing bed–of–nails testing when confronted with dual–sided surface mount boards populated with complex ASICs packaged with 25 mil lead spacing. While this suggested a need to give up this type of testing it was felt that the alternative of using functional (edge–connector) testers was unattractive, because the equipment is generally more expensive and can be far slower and more expensive for isolating typical production faults.

Current technology for in–circuit testing has many undesirable shortcomings that make it unattractive as a long–term solution. ATE vendors have attempted to overcome 50 mil spacing with clamshell fixtures whose reliability over thousands of actuations in a high volume environment remains unproven. Progression to 25 mil spacing will, in all likelihood, exceed the mechanical capabilities of these fixtures.

The alternative is to provide special staggered probe pads that cause the board to become larger, defeating the purpose of using surface mount technology. The effects are far more drastic with pin grid arrays and where the connections that link the layers of the PWB (the vias) are buried. Further, the parts themselves may be damaged from the lengthy tests that are applied by backdriving.

Boundary–scan allows a "virtual probe" to access the node between the I/O buffer and the core logic (Figure 11–2). Testing proceeds in two stages:

- *Pins–in testing:* A subset of the vectors developed for packaged–chip test are applied via the boundary–scan path to exercise the core logic in each component. These vectors are usually developed to validate the design but, hitherto, they have not been able to be used in PWB/system test.

- *Pins–out testing:* Simple vectors are propagated from scannable device output to input to detect and isolate manufacturing and other connectivity–related problems. This provides a capability superior to that of an in–circuit test, and without the need for probing.

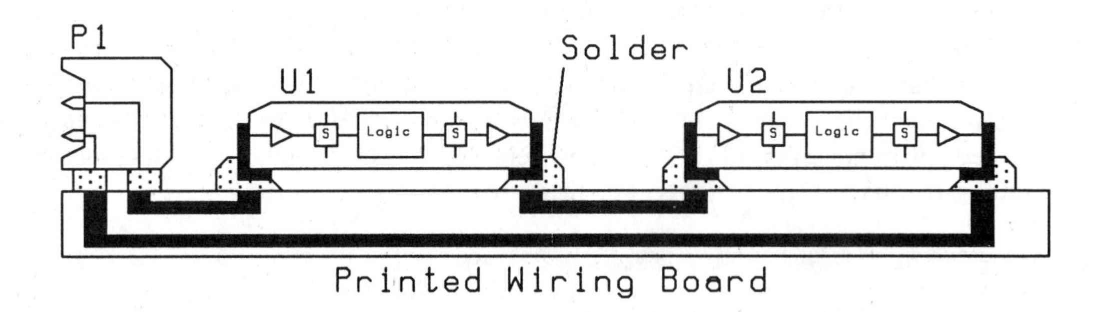

Figure 11–2: Boundary–scan approach.

No physical contact is required and backdriving is not necessary. Ones and zeroes are easily generated and applied out of the devices (pins–out testing) to confirm the integrity of the input or output (I/O) buffer, package lead, solder, and etch for the printed wiring network. Subsets of the static packaged part test are applied from the inputs across the core logic (pins–in testing) to confirm the functionality and integrity of the part.

Manufacturing faults typically detected by in–circuit testers can be localized when using personal computers and a desktop fixture, in sharp contrast to the $250K+ test systems of today. When the board design changes, the test is modified in software by linking in new device test files and by describing the new configuration of the board. This flexibility is a large improvement compared with generating a new fixture.

An important advantage of the test structures on the loaded boards is their ability to partition the design into segments small enough for computer automated test pattern generation. With proper levels of testability, the future may hold virtual turnkey test generation for patterns that detect stuck–at–one and stuck–at–zero conditions.

This new, low–cost manufacturing defect test capability can be applied cost–effectively to development efforts at far lower volumes than those required for in–circuit test investments. The low capital investment and high portability allow manufacturing screening to be introduced during design validation and leveraged across all phases of the product life cycle. The manufacturing defect test, once developed, can be used for subsequent board level tests in the factory, field, and depot.

11.3.5: Functional Loaded–Board Test

Functional testing today still relies on physical probing by the operator to isolate detected faults. Sequencing of the probing may be fixed, left to the operator, or determined on the fly by guided probe algorithms. Probing of very fine pitched pads without glitching will be a challenge for even the most skilled of operators, potentially necessitating robotic probing. Testing itself will be more complex if the additional observation is limited because only chip I/O pins can be probed.

The boundary-scan path can provide a capability referred to as "virtual probing," where the condition of nodes is retrieved by software without requiring instrumentation. A straightforward software layer can intercept the normal directions to the user to probe a node and can determine whether it is accessible via the scan path. If it is, the path can be accessed and the value can be returned to the test program without the operator having to participate. More importantly, the number of accessible nodes expands to include internal points located along the scan paths.

The improved visibility and control, combined with at-speed test capabilities incorporated into the design, can greatly improve the fault detection and isolation of functional testing. Integrated circuit built-in self-test (BIST) and board/system level built-in test (BIT) capabilities can enhance the performance of functional board testers.

11.3.6: Subsystem and System Test

The benefits provided to subsystem and system debugging may be leveraged against the production test problems with great impact. More importantly, manufacturing tests developed for the loaded boards may be reapplied and augmented to form the subsystem test at reduced cost. Eliminating the need for physical contact allows these tests to be reused in the testing of the fully assembled product, even when its enclosure prevents physical contact with internal connections.

11.4: Completing the Leverage into Field Test

11.4.1: Built-In Test

Built-in test (BIT) features have become increasingly more complex as systems have absorbed more and more functionality within a given constant volume. Software has become more complex and hardware speeds have escalated. The ability to monitor and detect system faults and to successfully isolate them has become a tremendous challenge.

The ability for firmware-based BIT to adequately exercise the full functionality of a system through instruction execution is rapidly being reducing. It may soon become unfeasible. Techniques that provide a more thorough test to smaller functional segments are therefore required. A transition into pattern-oriented BIT offers opportunities to improve BIT performance but it must be carefully measured against the impacts in terms of timelines and test data storage.

BIT techniques using pseudo-random pattern generation (PRPG) and parallel signature analysis (PSA) offer the ability to exercise hardware at full operating speed with minimal throughput and storage impact. Connectivity tests of similar or better performance to those generated for the in-circuit testers today can be algorithmically generated and rapidly executed. Pattern-oriented testing is better suited for fault simulation, offering an alternative for test grading to the physical fault insertion often conducted today. The emergence of an industry standard serial test interface provides an opportunity to provide additional control and to obtain better data within this environment.

11.4.2: Run Time Diagnostics

The boundary–scan path has the ability to capture data and make them available for examination without having any effect on the functional logic during its normal operation. This provides an avenue to the establishment of test processes in a background mode, that executes during operational time windows. One can take "snapshots" of the system and scan them out for external review. In this manner, useful information can be obtained to support run time diagnostic requirements.

11.4.3: Reconfiguration and Graceful Degradation

The operation of many systems is critical; consequently they cannot be allowed to fail catastrophically. Typically these systems feature redundancy, allowing tasks to be redistributed to fault–free resources when a failure occurs.

As with other cases discussed previously, the structured test access based on IEEE Std 1149.1 can provide greatly improved localization and monitoring of failing hardware. Upon detection of a failure, a system manager function can reallocate the task or function of the failed hardware to a backup node or it can reconfigure the hardware to allow the least critical function to be dropped temporarily. Thus graceful degradation occurs while the system manager executes diagnostics on the failed function in an attempt to confirm and isolate the fault.

The system manager is able to maintain a near real–time assessment of the system's capabilities and to rededicate resources as required. Additionally, the failed function can be continuously retested in the background to determine if the failure was intermittent or transient. Having determined a function to be "restored," the system manager can gracefully recover and bring the system back up to full performance.

11.4.4: Off–line Diagnostics

The structured test and debugging capabilities provided allow sophisticated highly portable tools to re–execute BIT and factory tests in the operational environment of the system. Manufacturing tests for digital boards can be rerun on boards still in the chassis through a system–level maintenance bus (e.g., the VHSIC TM–bus [1]†). Compact computers equipped with relatively simple interfaces can isolate failures to single boards with minimal activity on the part of the maintainer.

In cases where off–line test procedures are required, possibly augmented by portable maintenance aids, the structured test architecture acts like a built–in instrument and

---

† The VHSIC TM–bus has been accepted as the basis of a companion project to IEEE Std 1149.1 –– the P1149.5 Module Test and Maintenance Bus.

provides a path to the failure data collected during on-line operation. A smart controller within the product can interface to a host computer via phone or radio link to remotely execute diagnostics maintained at a base repair facility.

11.4.5: Test and Repair of Field Returns

Testing to detect and repair failures of boards returned after field repair of systems is an expensive and often capital-intensive area for many companies. In the military arena, and in some areas of the commercial sector, in-circuit test techniques cannot be used because the conformal coating used to protect loaded boards from damage cannot be easily removed or penetrated by probes. Depots and repair facilities have to rely on multi-million-dollar functional testers, that are good for detection, but often poor on diagnosis. The functional tests differ significantly from the on-line and off-line diagnostics used in the operational environment, causing fault repeatability problems.

Because standardized test interfaces reduce the need for physical contact, depots can use the same low-cost manufacturing tests run in the production facility. The inherent modularity of the tests provides good isolation, and when replicating field test sequences reduces the chance of "cannot-duplicate" problems. Warranty repair facilities for commercial products are small operations that cannot justify large capital investments for troubleshooting or repair. Again, the possibility of being able to use test environments similar to those used in the factory can offer greater repair efficiency at costs lower than those achievable today.

11.5: Conclusion

Implementation of a structured chip-through-system test architecture requires an investment at the IC level toward the solution of system level problems that are becoming major barriers to profitability and performance. This investment can pay for itself many times over in reduced costs throughout the product life cycle. The process of test generation and verification for digital logic can be constrained into the region capable of being handled by current-day computer-aided tools. The tests can be reutilized throughout all test phases of the product's life.

By removing the dependence on complex fixturing, the potential exists for simpler personal-computer-based systems for testing and debugging. These systems are cost-effective enough to be introduced during product debug and test of initial prototypes when volumes are still too low to justify current approaches. Basic hardware/software building blocks, combined with application software, provide highly functional yet portable debugging and testing environments. Production tests and production ATE can be utilized cost-effectively in the field to reduce the costs of field and warranty support.

Having reduced the test interface to a four-wire port, the majority of the test capability lies within the linking and execution of previously developed tests for ICs combined with automatically-generated connectivity tests. This environment is far more flexible and robust, allowing test program sets to quickly adapt to design changes. Set-up and take-down time for tests is minimal.

Additionally, tests can be extended into numerous environments previously unavailable because of constraints on connecting instrumentation. Tests can be applied in closed boxes within environmental chambers, equipment bays, or difficult-to-access places using portable, reusable test programs.

The robustness, flexibility, and performance of such a test architecture will allow many companies to meet their obligations to their customers while containing test costs and achieving greater profitability. While an investment is needed during IC design, this will be leveraged against a broad range of problems spanning the entire product cycle.

11.6: Reference

[1] IBM, Honeywell, and TRW, *VHSIC Phase 2 Interoperability Standards: TM–Bus Specification -- Version 3.0,* Nov. 9 1987. (Copies can be obtained from J.P. Letellier, Naval Research Lab, Code 5305, Washington, D.C. 20375, U.S.A.)

Chapter 12. Benefits and Penalties of Boundary–Scan

Richard Sedmak
Self–Test Services
6 Lindenwold Terrace
Ambler, PA 19002, U.S.A.

Colin Maunder
British Telecom Research Labs
Martlesham Heath
Ipswich IP5 7RE, U.K.

An analysis of the economics of boundary–scan begins with consideration of the benefits and penalties associated with the technique. In some cases, the penalties may appear to outweigh the benefits if considered only at the integrated circuit (IC) level. However, the benefits usually far outweigh the penalties when we consider a more comprehensive analysis spanning all levels of assembly from chip to system and consider all test phases during the life cycle of a system.

12.1: Benefits

12.1.1: Lower Test Generation Costs

Costs of test generation can be lowered. At the board level, boundary–scan testing provides the equivalent of in–circuit testing without the cost and need for a bed–of–nails fixture. This is true even when assembly techniques that impede in–circuit testing are used -- for example, conformal coating, surface–mount technology, and double–sided boards. By being able to use boundary–scan based testing as the primary means of testing loaded boards, perhaps supplemented by a reduced functional test, a company can avoid the enormous costs of test generation associated with pure functional or edge–connector test. In addition, the presence of boundary–scan permits some reuse of test patterns up through the hierarchy of packaging levels. For example, at the board level, a portion or full set of chip level test vectors can be reused as a nucleus for the board level test.

When boundary–scan is the a basis for built–in self–test (BIST) at the chip, loaded board, or system level, the cost of test generation can be substantially reduced because the test stimuli (such as pseudo–random patterns) are generated automatically and algorithmically within the product.

12.1.2: Reduced Test Time

Another benefit of boundary–scan is the possibility of reducing test time, particularly in the diagnostic area. In regard to GO/NO–GO testing, the use of boundary–scan may, at first glance, seem to lead to increased test time because of the serialization of test stimuli and circuit responses. This may be particularly true when comparing boundary–scan testing with in–circuit testing. However, as described earlier, the limitations imposed during the latter type of testing, as well as the complications caused by board packaging methods, may make it difficult (if not impossible) to use the in–circuit test approach. A true comparison of boundary–scan test times with functional test times requires more careful scrutiny of the assumptions. If we assume a very complex board and a required level of single stuck–at fault coverage -- for example, in the high 90 percentages -- some segments of the electronics industry feel that boundary-scan test times may actually be less than the equivalent times for functional testing because of the divide-and-conquer approach used. Achievement of high-fault coverage without adequate design-for-test provision can be a lengthy and expensive task.

Few people take exception to the claim that, for a given level of fault coverage, lower test times result when boundary-scan based BIST is used in conjunction with, or in lieu of, externally-applied functional or in-circuit tests.

12.1.3: Reduced Time to Market

Use of scan design techniques at the chip level can have a significant positive impact on time to market, because less time needs to be spent on test generation [1].

Similar benefits result at the board level through use of boundary-scan. For example, where a considerable amount of engineering effort would previously have been required to develop an in-circuit test module for a new state-of-the-art IC, this task can now be completed in a matter of hours because it is no longer necessary for the board test engineer to understand the detailed operation of the new chip.

In highly competitive markets, the saving in test development time for a new product, even where only a small percentage of the chips on a board include boundary-scan, can help ensure its commercial success.

12.1.4: Additional Benefits

Three additional benefits result:

- simpler and less costly testers;
- commonality of interface with the tester; and
- the ability to accommodate high-density and poor-access packaging approaches.

Boundary-scan based testing can be performed regardless of any constraints imposed by new packaging methodologies -- therefore allowing a reduction and, in some cases, the elimination of the need for expensive bed-of-nails test fixturing. Furthermore, since IEEE Std 1149.1 establishes a common four- (or five-) pin interface and protocol with the tester, such commonality across all board types will save even more in fixturing or interface adapter costs, particularly if one considers the cost impact of engineering changes.

12.2: Penalties: Additional Circuitry

The first and most obvious penalty is the cost of the additional circuitry.

The effect on circuit size of adding boundary-scan capability is, as for other design-for-test changes, difficult to predict because much depends on the detail of the implementation -- for example: Are "holding" registers or latches provided in all boundary-scan cells? What is the geometry and positioning of the cells? Etc.

The following examples provide estimates for the overall size of the circuitry required to give conformance to IEEE Std 1149.1, but without extensions to the facilities defined in the standard. An important consideration when it comes to an analysis of the penalties of boundary-scan, including the amount of added circuitry, is that their impact can be reduced by early planning in the development cycle, by good design practices, by the use of automated tools, and by exploiting boundary-scan in all life-cycle phases of testing as discussed in Chapter 11.

It should also be emphasized that a good many ICs are pin limited -- that is, the size of the chip is determined by the space required along the chip sides to provide sufficient bonding pads for all inputs and outputs, and not by the number of gates or transistors required to implement the function of the chip. Therefore, there may be "spare" gates or silicon area within the chip that can be used to construct the boundary-scan test logic. Under these circumstances, the *real* cost of implementing boundary-scan is at least considerably reduced and may, in some cases, be zero.

12.2.1: Example 1

The first cost example is for a full-custom 6 mm. x 6 mm. IC built in a 2.0 micron single-layer metal complimentary metal-oxide semiconductor (CMOS) process.

- *Test access port (TAP) controller:* Implementation of the TAP controller requires on the order of 80 NAND gates. A more efficient implementation could, however, be achieved using a transistor level state machine design. An initial implementation in the stated technology requires a silicon area of approximately 0.3 sq. mm.

- *Instruction and bypass registers:* An instruction register containing two bits (the minimum configuration) would occupy on the order of 0.02 sq. mm. The bypass register is approximately one half of the size of the minimum instruction register, or 0.01 sq. mm.

- *Boundary-scan register:* An estimate of the total size of the boundary-scan register can be obtained by looking at the size of the boundary-scan cell for an output pin illustrated in Figure 12-1. A circuit that implements this design requires around 0.015 sq. mm. It can be expected that boundary-scan cells for input and other pin types would be of similar size. Therefore, for an IC with 40 system pins (input or output), implementation of the boundary-scan register would require some 0.6 sq. mm. silicon area.

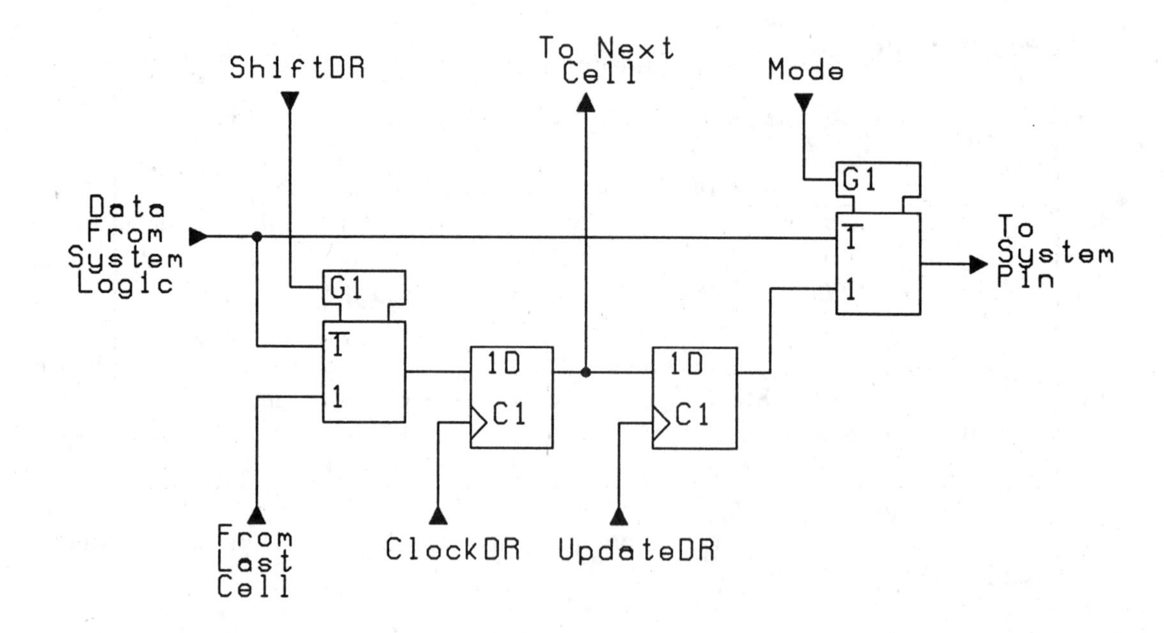

Figure 12–1: A boundary–scan cell for an output pin.

The combined silicon area for a minimal implementation comprising a TAP controller, a two–bit instruction register, a bypass register, and a 40–bit boundary–scan register would be approximately 1 sq. mm. from the above figures (including a small allowance for the multiplexers, etc., required to complete the minimum implementation of IEEE Std 1149.1). This represents an increase in size of 3 percent for the 36 sq. mm. chip. Clearly, this figure is significantly affected by changes in circuit size, component geometries, and other changes (such as the use of two metal layers). It also does not take any account of any increase in the size of the on–chip system logic, for example because of increased separation of cells caused by increased pitch between input/output pads.

12.2.2: Example 2

This second example details the cost of the additional circuitry in a library–based application–specific IC (ASIC) design environment. The assumption is made that the ASIC cell library does not include custom–designed ("hard") boundary–scan cells or other cells to support IEEE Std 1149.1. Therefore, all the required features are provided by using "soft" macros (i.e., cells constructed as interconnections of "hard" cells) in the vendor–supplied library or are constructed by the user from "hard" cells.

The gate counts given are based on those shown in [2] for basic logic gates and flip–flops and relate to a 10,000 gate design with 40 system pins. They relate to implementations of the example circuits shown in IEEE Std 1149.1.

Table 12–1: Gate requirement for a semi–custom implementation.

Item	Gate Equivalent
TAP controller	131
Instruction register (2 bits)	28
Bypass register	9
Boundary-scan register (40 cells)	680 approx.
Miscellaneous logic	20 approx.
TOTAL	**868 approx.**

In total, construction of the various building blocks required by IEEE Std 1149.1 from the available macrocells requires an equivalent of 868 gates, broken down as shown in Table 12–1. The reduction in usable capacity from 10000 to 9200 gates gives an estimated overhead of 8 percent.

Two comments must be made on this cost estimate:

1. It has been assumed that the chip has only input and 2–state output pins. Because IEEE Std 1149.1 requires additional circuitry in the boundary–scan cells placed at 3–state and bidirectional pins, the cost could rise if the design included any such pins.

2. The cost is based on the use of macrocells from a version of the cell library [2] created prior to publication of IEEE Std 1149.1. It is therefore assumed that all the required circuitry is constructed in the area available for the user's circuit design. If specific cell designs were available to support IEEE Std 1149.1 or if the vendor were to place the boundary–scan circuitry in areas of the chip not available for the user's design, then the cost could be considerably reduced. (Some methods of reducing the cost are discussed in Section 12.2.4 and Chapter 13.)

Further examples of costs using the same ASIC product are contained in [3]

12.2.3: Example 3

Reference [4] discusses the costs of implementing a built–in self–test architecture based on the principles of cellular automata in a circuit that includes a boundary–scan path. The architecture is based on the Joint Test Action Group (JTAG) version 2.0 definition, a precursor to IEEE Std 1149.1 (see Chapter 3).

The paper estimates that a boundary–scan cell with BIST facilities would occupy approximately 0.065 sq.mm. in a 3 micron CMOS process. Estimates are given for the overall cost (measured as a reduction in usable silicon area) for a range of chip sizes. These estimates vary from 17 percent down to 6.7 percent as the size of the chip increases from 11.8 sq. mm. to 64.3 sq. mm. (with pin counts varying from 28 to 84 pins, respectively).

12.2.4: Reducing the Cost of Added Circuitry

The amount of circuitry required to implement IEEE Std 1149.1 can be reduced in several ways, dependent on the circuit design. The following list gives some examples:

1. Boundary-scan register cells can be integrated with the input or output buffer stages in the circuit design [5].

2. The TAP and the boundary-scan register cells can be implemented in "dead" area around the periphery of the circuit. In the implementation discussed in [6], for example, the cells are located beneath power distribution busses. Others have discussed the possibility of locating the cells between the input and output bonding pads on the IC.

3. Circuitry can be shared between the various shift-register-based features of the test logic (e.g., the instruction, bypass, and boundary-scan registers). One way of achieving this is described in Chapter 13.

12.3: Other Penalties

12.3.1: Added Pins

The second most apparent penalty is the need to add dedicated test pins to the chip. IEEE Std 1149.1 calls for a minimum of four pins. While the provision of the fifth test logic reset (TRST*) pin is optional, feedback from some IC manufacturers indicates that they may also provide this pin.

As illustrated in the following chapters, the TAP can allow access to many testability features within a design that might otherwise require package pins for additional data or control access. The four or five pins required by the TAP may therefore frequently provide for all test purposes. Viewed in this way, the requirement for a number pins dedicated to test is not unusual -- many ICs today use several dedicated test pins to allow them to be tested economically.

12.3.2: Design Effort

Since there is additional circuitry associated with boundary-scan, it can be safely assumed that some form of additional design effort will be required. The exact impact will depend on the degree of automation of the design process and on other factors:

- Some companies are already working on computer-aids that will automatically add the boundary-scan path and associated test logic to a design, for example.

- In others, application-specific ICs (ASICs) are being developed that have the boundary-scan path built into the periphery of the base logic array. It will be there whether the designer chooses to use it or not.

In either case, the amount of additional effort required to produce an IC design that conforms with the standard will be low.

12.3.3: Performance

Performance is another consideration. The multiplexer that feeds the system pin in Figure 12-1, for example, could add two gate delays that, together with the additional delay due to the input loading of the boundary-scan register, would increase the propagation delay of signals leaving the chip. Similarly, the delays experienced by signals entering the chip would be increased if boundary-scan cells were used that included multiplexers in the pin-to-logic data path (such multiplexers are required only where the *INTEST* instruction is supported).

The importance of these additional delays clearly depends on the application for which the chip is intended. However, the impact of the additional circuitry can be minimized by careful design or by combining input buffers with the boundary-scan register cells [5], etc.

In many cases, the skew between signal changes at two or more output pins of a component resulting from a common cause is more important than the absolute delay, for example, between a clock edge and a signal change at one output. Since identical cells can be introduced at each output, the pin-to-pin skew can be kept under tight control.

It must be pointed out, however, that the use of multiplexers at output pins to permit observation of test data from the core of the design is already commonplace. Many ASIC vendors require that complex macrocells are connected in this way to ensure that library test waveforms can be applied. Given this situation, there is no *additional* delay introduced by the inclusion of a boundary-scan path -- the multiplexer at the output needs only to be widened to allow for the input from the boundary-scan shift-register stage.

12.3.4: Power Consumption

Because circuitry is added to the basic design to provide the boundary-scan path, an increase in the power consumption of the component must be expected. For CMOS IC designs in which operation is controlled by gated clock signals, the increase in consumption during normal operation will be small because the boundary-scan path and much of the other test logic will be inactive. Only the TAP controller will remain active since, in the absence of a TRST* input, it must continue to be clocked with the test mode select (TMS) input driven to logic 1 to ensure that the controller can return to the *Test-Logic-Reset* state following any upset.

12.3.5: Reliability and Yield

Also resulting from the additional circuitry are the potential penalties of reduced reliability and reduced yield. While sufficient data have not yet been collected in this regard, one can say at least that any reduced "raw yield" of the integrated circuit resulting from a slightly larger die size, for example, will be off-set by improved yield measured after test at subsequent packaging levels. This yield improvement will result from the high test performance achievable by using boundary-scan.

Note also that the periphery of an IC contains the circuitry and connections that are most likely to fail during operation of the component [7]. These are the areas that are most directly addressed by the boundary-scan test technique. An improvement to system mean-time-to-repair can therefore be expected through simplified testing and diagnosis of faults in input and output buffers, bond wires, etc.

12.4: Conclusion

The principal benefits and penalties of boundary-scan have been presented and discussed. By careful design and by provision of appropriate design-support tools, the cost of implementing IEEE Std 1149.1 can be minimized. As discussed in Chapter 11, the benefits of using boundary-scan accrue at many test stages and can be significant where field support and maintenance of systems are key requirements [8]. When viewed against the escalating cost of using traditional functional or in-circuit test techniques for loaded boards, boundary-scan quickly becomes an attractive proposition.

12.5: References

[1] M.E. Levitt and J.A. Abraham, "The Economics of Scan Design," *IEEE International Test Conference Proceedings,* IEEE Computer Society Press, Los Alamitos, Calif., 1989, pp. 869-874.

[2] LSI Logic Corp., *CMOS Macrocell Manual,* LSI Logic Corp., Milpitas, Calif., September 1984.

[3] LSI Logic Corp., *IEEE P1149.1/JTAG Testability Bus,* LSI Logic Corporation, Milpitas, Calif., November 1989.

[4] C.S. Gloster and F. Brglez, "Boundary-Scan with Built-In Self-Test," *IEEE Design and Test of Computers,* Vol. 6, No. 1, pp. 36-44.

[5] S. Das Gupta et al., "Chip Partitioning Aid: A Design Technique for Partitionability and Testability in VLSI," *ACM/IEEE Design Automation Conference Proceedings,* IEEE Computer Society Press, Los Alamitos, Calif., 1984, pp. 203-208.

[6] D. Laurent, "An Example of Test Strategy for Computer Implemented with VLSI Circuits," *IEEE International Conference on Computer Design: VLSI in Computers*

and Processors, IEEE Computer Society Press, Los Alamitos, Calif., 1985, pp. 679–682.

[7] D.H. Merlino and J. Hadjilogiou, "Built-In Test Strategy for Next Generation Military Avionic Hardware," *IEEE International Test Conference Proceedings*, IEEE Computer Society Press, Los Alamitos, Calif., 1988, pp. 969–975.

[8] C. Dislis, I.D. Dear, J.R. Miles, S.C. Lau, and A.P. Ambler, "Cost Analysis of Test Method Environments," *IEEE International Test Conference Proceedings*, IEEE Computer Society Press, Los Alamitos, Calif., 1989, pp. 875–883.

Chapter 13. Single Transport Chain

Wim Sauerwald, Frans de Jong, and Math Muris
Philips Centre for Manufacturing Technology
Eindhoven, The Netherlands

The example implementations included in IEEE Std 1149.1 show instruction and test data registers implemented as a bank of parallel shift–register paths connected between the test data input (TDI) and test data output (TDO) pins. This chapter describes a more efficient, implementation called the single transport chain (STC) architecture.

13.1: Introduction

The test logic defined by IEEE Std 1149.1 can be implemented as a bank of parallel shift–register paths, for example as illustrated by Figure 13–1. The registers will, in general, contain different numbers of shift–register stages. Each stage can be visualized as being constructed from three basic elements:

1. a capture element that allows data to be loaded into the register stage;

2. a shift–register (or transport) element that allows data to be moved serially through the register stage; and

3. an update element that holds a data value at the register's output while a new value is shifted in.

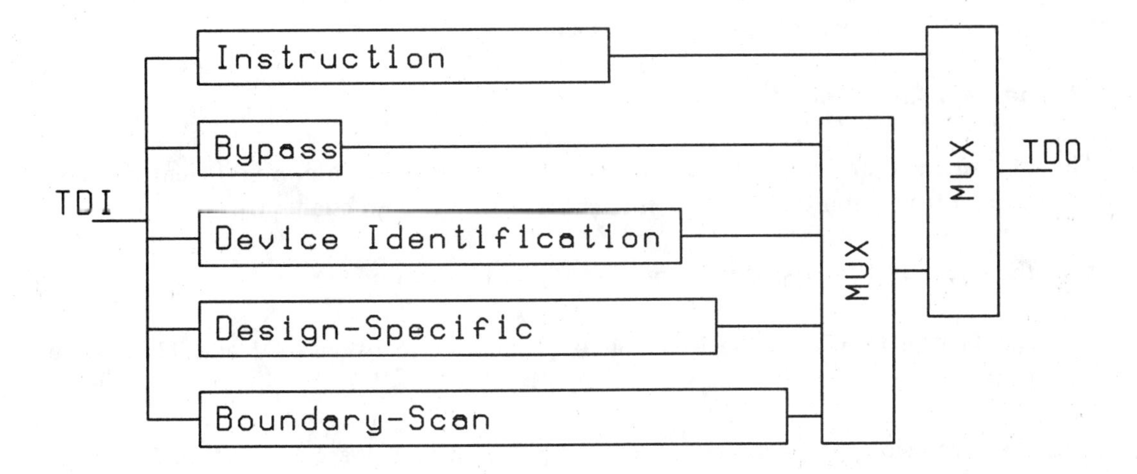

Figure 13–1: IEEE Std 1149.1 architecture.

These elements work together to perform the following functions:

1. data can be captured and transported through TDO for examination (Figure 13–2)

EH0321-0/90/0000/0151$01.00 © 1990 IEEE

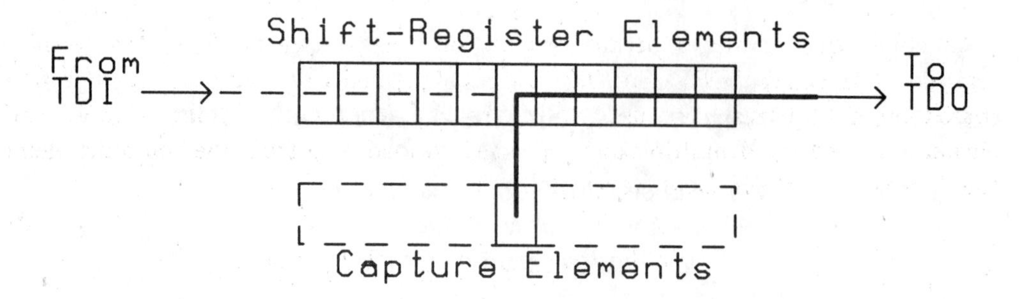

Figure 13–2: Capture then transport.

2. data can be shifted in through TDI and, when shifting is completed, made available through update elements (Figure 13–3).

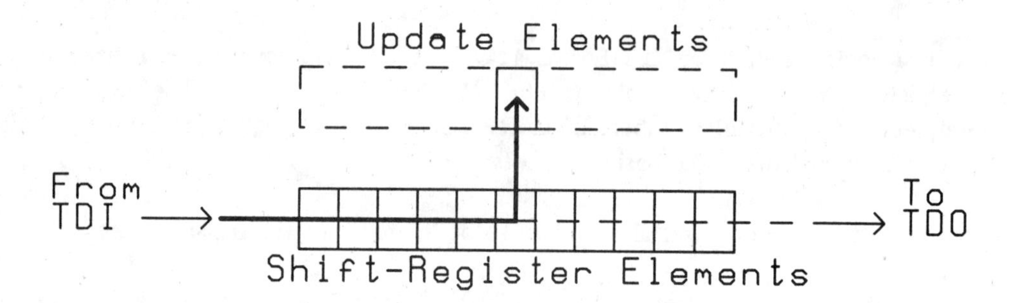

Figure 13–3: Transport then update.

Not all the registers defined by IEEE Std 1149.1 are constructed from all three kinds of element (for example, the bypass register has only capture and transport elements). However, where a register has an element of a given kind then its operation will be the same as that of any other register with that element type.

13.2: The STC Architecture

The STC architecture exploits the following features to allow a more efficient (in terms of gate count) realization that is more efficient in terms of gate count:

1. the commonality of structure and operation just described;

2. the fact that only one register can be connected between TDI and TDO at any time;

3. the permission that registers are required only when they are selected; and

4. the fact that every scan operation (instruction or test data) starts with a "capture" that overrides old data in the shift–register (transport) elements.

Together, these features permit one set of transport elements to service all the basic registers defined in IEEE Std 1149.1. As illustrated in Figure 13–4, capture elements (based on multiplexers) are used to select data from a set of data sources and update elements (based on demultiplexers) are used to load data from the transport element onto the appropriate update element output.

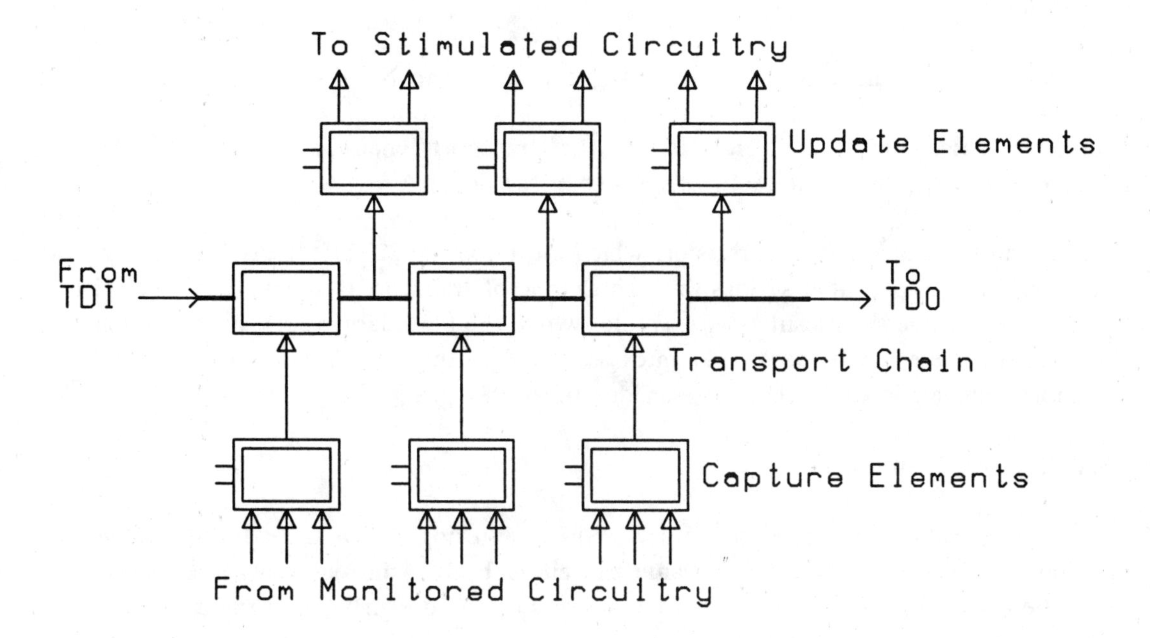

Figure 13–4: The single transport chain architecture.

13.3: The Transport Chain

The length of the core transport chain is determined by the longest of the registers to be implemented. For example, Figure 13–5 shows how segments of the transport chain that are not required for the selected register can be bypassed.

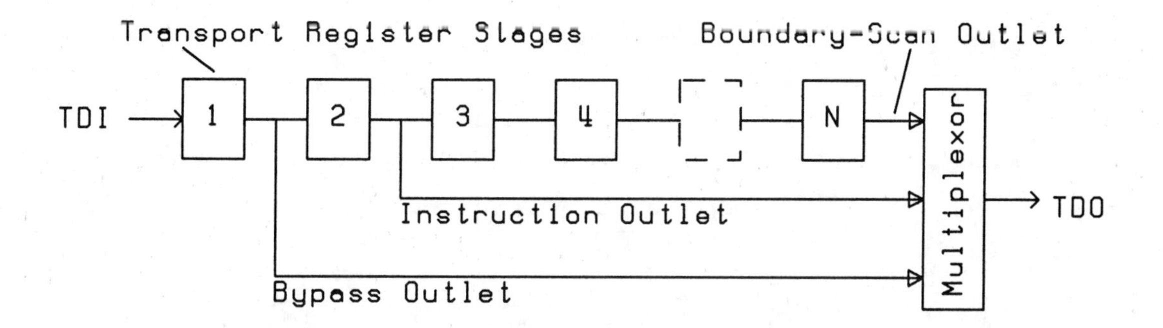

Figure 13–5: Single transport chain with various outputs.

Figure 13–5 shows a design with:

- a bypass register (transport register stage 1);
- a minimum instruction register implementation (transport register stages 1 and 2); and
- a boundary–scan register (transport register stages 1 to N).

In this case, the overall length of the transport chain is determined by the boundary–scan register -- the longest register in the design.

Where a device identification register, which must contain 32 shift–register stages, is also implemented, the potential savings through sharing of transport elements between registers as described can be significant. For example, where the boundary–scan register contains 60 shift–register–based cells, some 180 gates are saved through the reduction from 95 to 60 transport elements in the design with an identification register.

13.4: Capture Element Design

The identification code of a chip is accessed by capturing a stored, read–only, value into the transport elements. The stored value can either be held in read–only memory cells or it can be built into the design by use of two types of cell design in the register:

- one that loads a 0 in the *Capture–DR* controller state; and
- one that loads a 1 in the *Capture–DR* controller state.

In the STC architecture, the identification code can be built into the design of the capture elements for each relevant shift–register stage. As shown by Figure 13–6, the capture elements then become multiplexers that are able to select between a variety of external data sources (X1, X2, or X3) and the hard–wired identity–code bit (IDENT) -- 0 or 1.

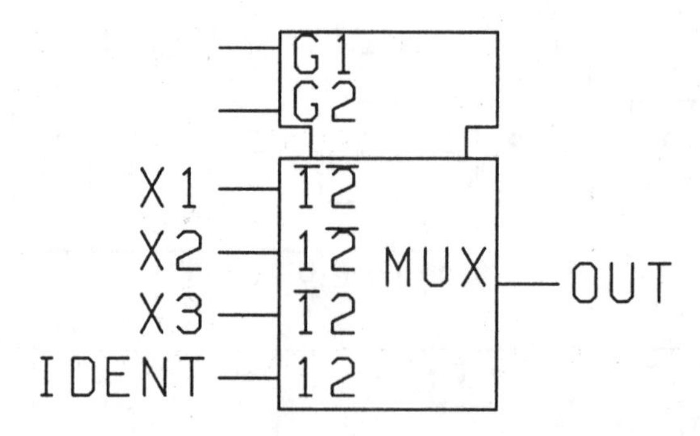

Figure 13–6: Multiplexor with built–in IDENT value.

13.5: Update Element Design

Together, the update elements operate as a bank of addressable latches -- one for the instruction register, another for the boundary-scan register, and so on. Data are loaded into one of these latches at the end of shifting, dependent on the type of scan operation being executed (instruction or test data) and the register selected by the current instruction.

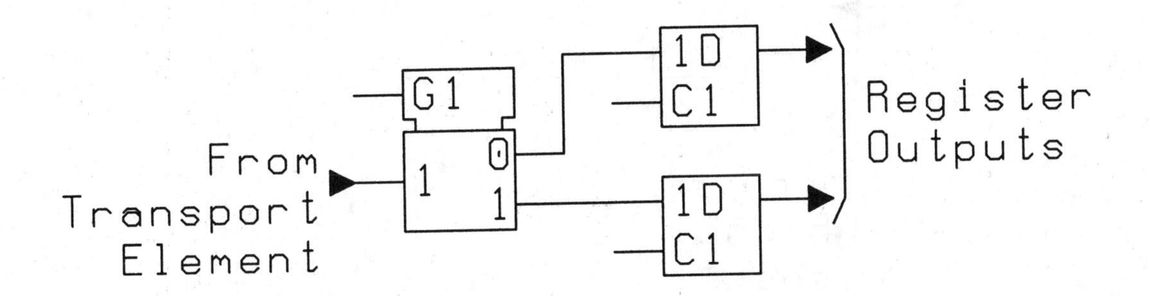

Figure 13–7: Update element design.

Note that, as shown in Figure 13–7 the latches in the update elements need only be level-operated devices (they do not have to be master-slave flip-flops). Note also that for registers that do not require a latched parallel output, the update element behaves, in effect, as a data demultiplexer.

13.6: Transport Element Design

The transport element completes the design for each register stage. Figure 13–8 shows a design with a short feedback path to allow the state of the register to be held and therefore does not require clock gating. The selection between the three modes of operation:

1. holds the present data value;

2. loads the data presented from the capture element; and

3. shifts in data from the previous transport element (or TDI)

is controlled by signals fed to the multiplexer from the test access port (TAP) controller.

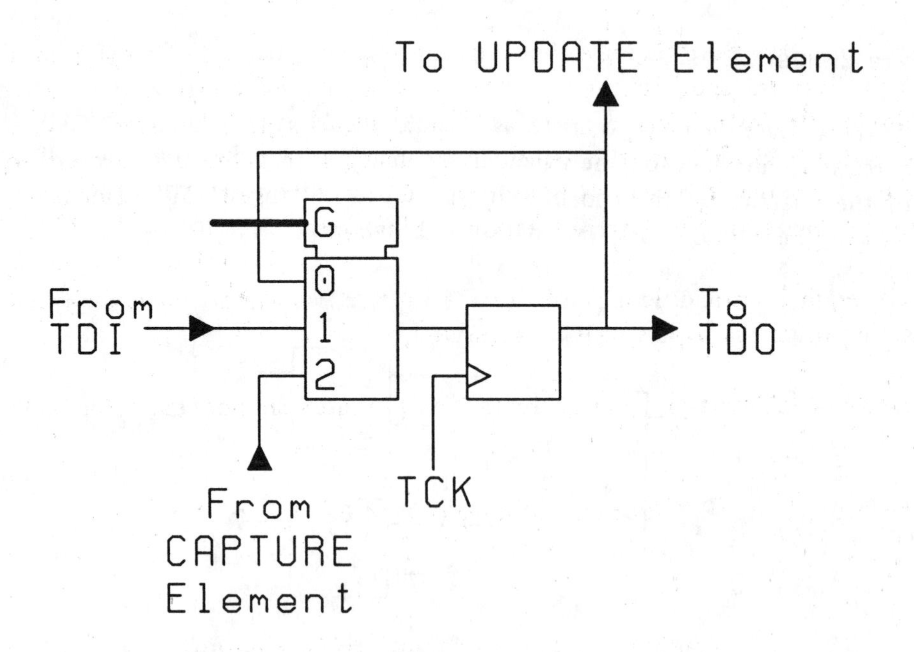

Figure 13–8: Transport element design.

13.7: A Complete STC Register Cell Design

Figure 13–9 shows a complete cell design that implements the functions of:

1. the instruction register with data input (Status) and instruction output (Instruction);
2. the device identification register with data input IDENT; and
3. the boundary-scan register with data input Data In and data output Data Out.

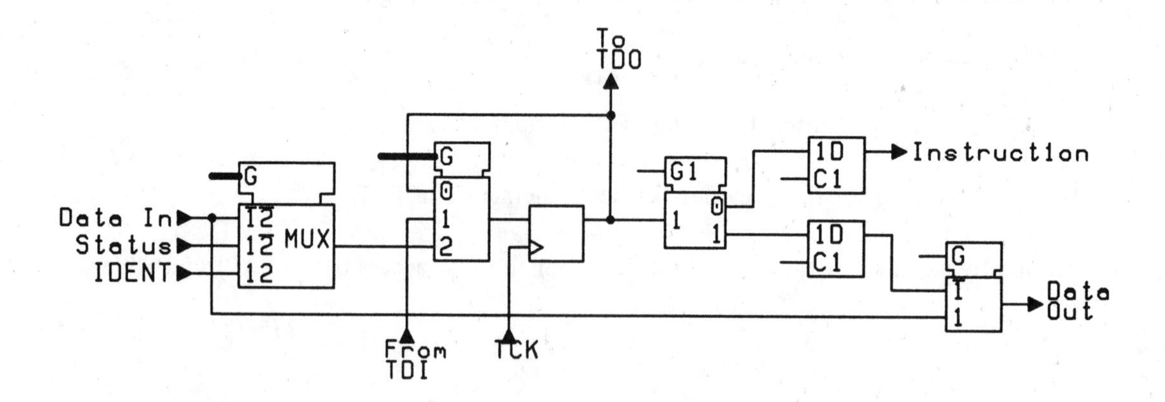

Figure 13–9: A complete cell design.

Table 13–1 shows the functions that must be implemented by each cell according to its position in the transport chain in a design example used earlier (which included a minimum–length instruction register, a device–identification register, and a 60–cell boundary–scan register). In the table, it is assumed that the cell nearest to TDI is numbered 1 and that the boundary–scan register is the longest. The complexity required for cell implementation varies from location to location:

- cell requires an additional input to the capture element to allow a constant 0 to be loaded when the bypass register is selected;
- the Data In input and the instruction output latch are not required in cells 3 to 60; and
- the IDENT input is not required in cells 33 to 60.

Table 13–1: Cell function versus position.

Register	Cell Locations Used								
	1	2	3	4	5	...	32	...	60
Bypass	x								
Instruction	x	x	x						
Boundary-Scan	x	x	x	x	x	...	x	...	x
Identification	x	x	x	x	x	...	x		

13.8: Conclusions

In this chapter, we have described a different implementation of IEEE Std 1149.1 to that given as an example in the standard. This implementation, which we call the STC architecture, exploits the potential for sharing circuitry between the registers defined by the standard and, therefore, allows a lower cost implementation.

Chapter 14. Boundary–Scan Cell Provision: Some Dos and Don'ts

Colin Maunder
British Telecom Research Labs
Martlesham Heath
Ipswich IP5 7RE, U.K.

Kenneth P. Parker
Hewlett Packard Company
P.O. Box 301A, M/S AU100
Loveland, CO 80537, U.S.A.

This chapter provides examples to illustrate the correct provision of boundary–scan cells within an integrated circuit that seeks to conform to IEEE Std 1149.1. It must be emphasized that the opinions expressed here are those of the authors, and not necessarily those of the IEEE.

14.1: Clock Pins

For system clock input pins, performance issues are often important, for example, the time taken for clock signals to reach stored–state devices within the integrated circuit. The inclusion of a boundary–scan cell at the clock pin could, therefore, have an adverse effect on the capability of the complete design to meet its performance targets.

For this reason, IEEE Std 1149.1 permits the use of cells that can monitor, but not control, the signals that arrive at clock pins. Figure 14–1 shows an example of such a cell.

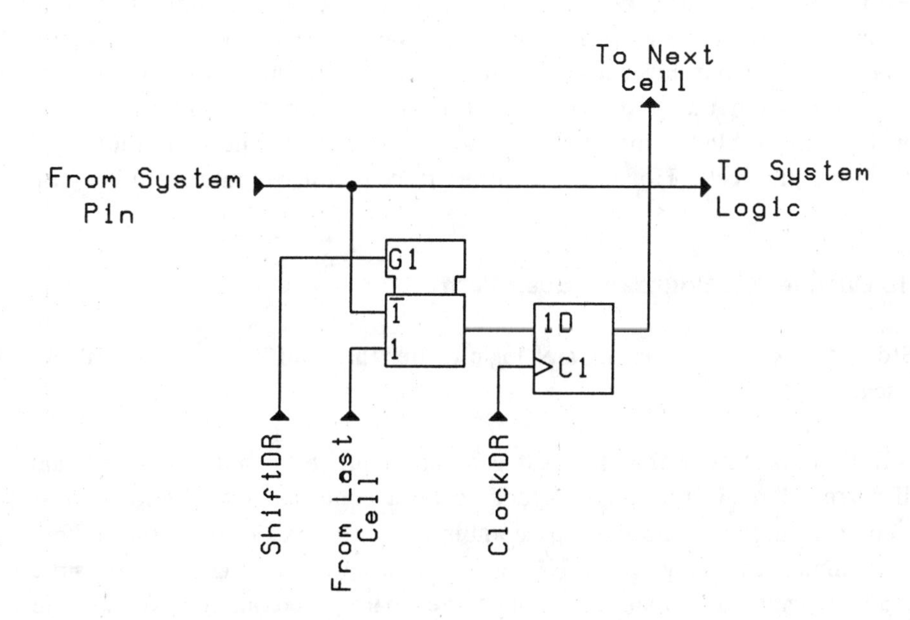

Figure 14–1: A boundary–scan cell for a clock input pin.

Further, the standard permits the data input to the boundary–scan cell to be taken from any point in the clock distribution tree, provided that there is no logic (other than buffers or inverters) between the clock pin and the monitored point. Figure 14–2 shows several points in a clock distribution tree that could be used to supply the data input of a boundary–scan cell. These points are labeled A and B. Figure 14–2 also shows a point that cannot be used as the input to the boundary–scan cell, because two signals are combined onto the monitored point.

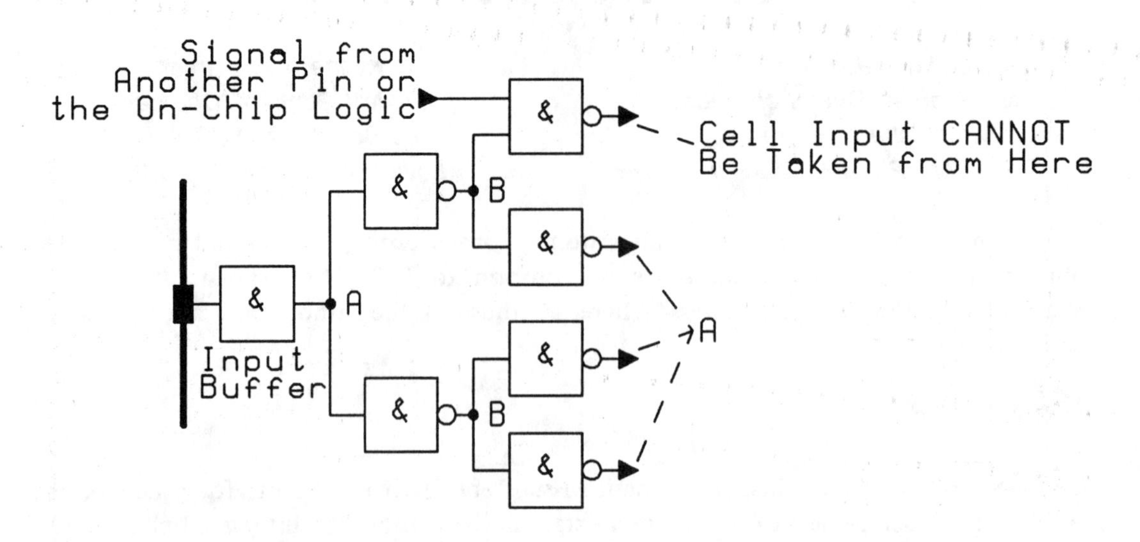

Figure 14–2: Boundary–scan cells for clock inputs.

Note that the standard requires that the value seen by loading and then scanning the boundary–scan register must be that applied at the input pin. Given that the boundary–scan cell shown in Figure 14–1 loads the value present at the data input into the shift–register stage without inversion, the monitored point must be an even number of inversions removed from the input pin (i.e., it must be driven from a point marked A). It would not be permissible to monitor the output of one of the first rank of inverters (marked B) using this cell. However, a different cell design could be used (see Section 14.4).

14.2: Logic Outside the Boundary–Scan Path

IEEE Std 1149.1 does not permit any logic (other than buffers or inverters) outside the boundary–scan path.

The motivation for this is that the test generation process would be significantly more complex if "external" logic functions needed to be accommodated. While there may be a savings in circuitry in the component by combining two signals outside the boundary–scan path (for example, by using a NAND gate) it would no longer be possible to use algorithmically generated test patterns to test the board interconnect. To be able to take into account the effect of the external logic on the interconnect test, a test generator would need to be expanded every time a chip became available that included a new circuit type external to the boundary–scan register.

Figure 14-3 shows a number of situations where logic is placed between the boundary-scan path and the input/output pins. All circuitry between the boundary-scan path and the package pins in this figure violates the rules of IEEE Std 1149.1, because it allows interaction between data received at two or more input pins or between data from two or more outputs of the on-chip system logic.

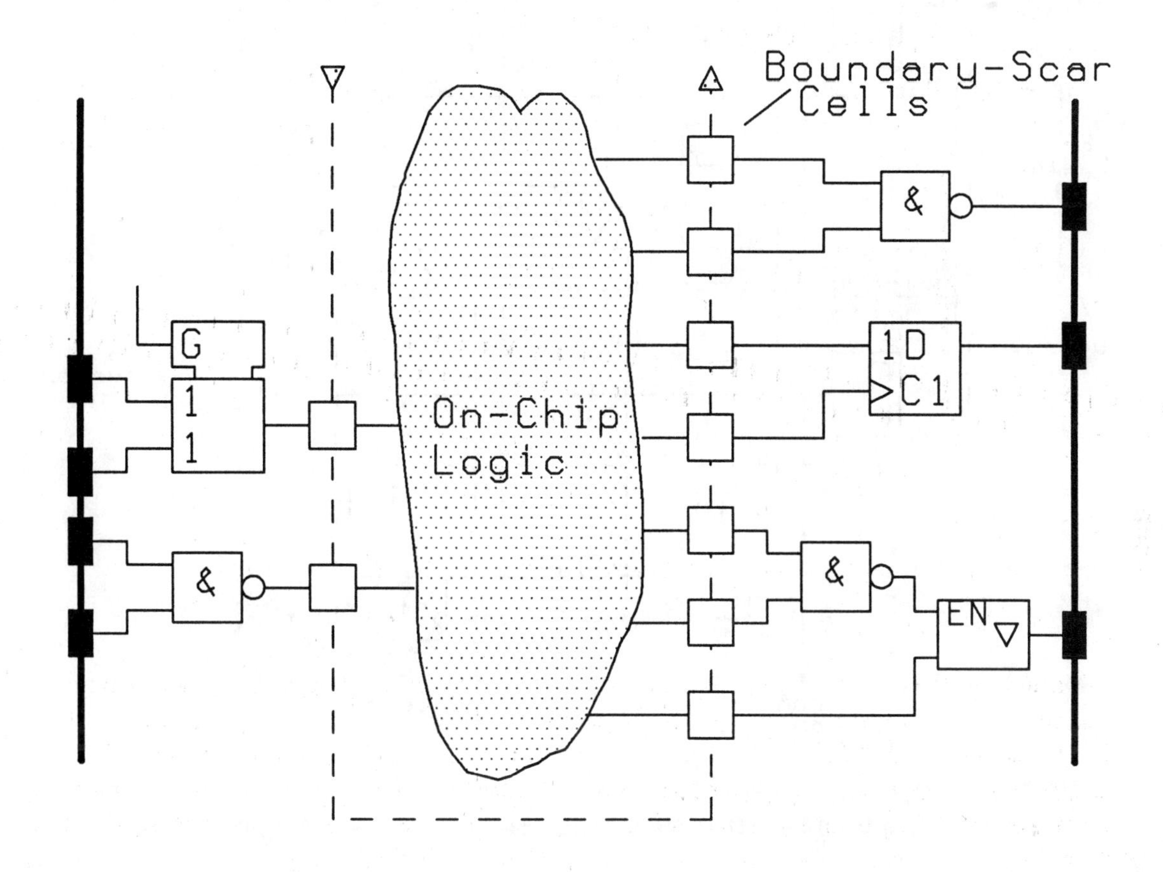

Figure 14-3: "Illegal" logic outside the boundary-scan path.

Note that, in IEEE Std 1149.1, cell designs are shown for 3-state and bidirectional pins that include a logic gate between the boundary-scan cell and the control input of the output buffer. An example is shown in Figure 14-4, where the added gate is controlled by signal CHIP_TEST*. While the provision of this gate may seem to be "illegal" according to the earlier discussion, note that the CHIP_TEST* input is generated by decoding the instruction that has been entered (CHIP_TEST* is 0 when either *INTEST* or *RUNBIST* is selected). The signal does not come from the on-chip system logic or a system pin. Therefore, when an instruction is selected that requires the system pin to be controlled from the boundary-scan cells (e.g., *EXTEST*), the added gate is transparent (CHIP_TEST* = 1) and can be ignored by software that determines how to control the system pin to the desired state (0, 1, or Z). In contrast, for the earlier examples, the test generation software would require a knowledge of the logic function provided *outside* the boundary-scan path.

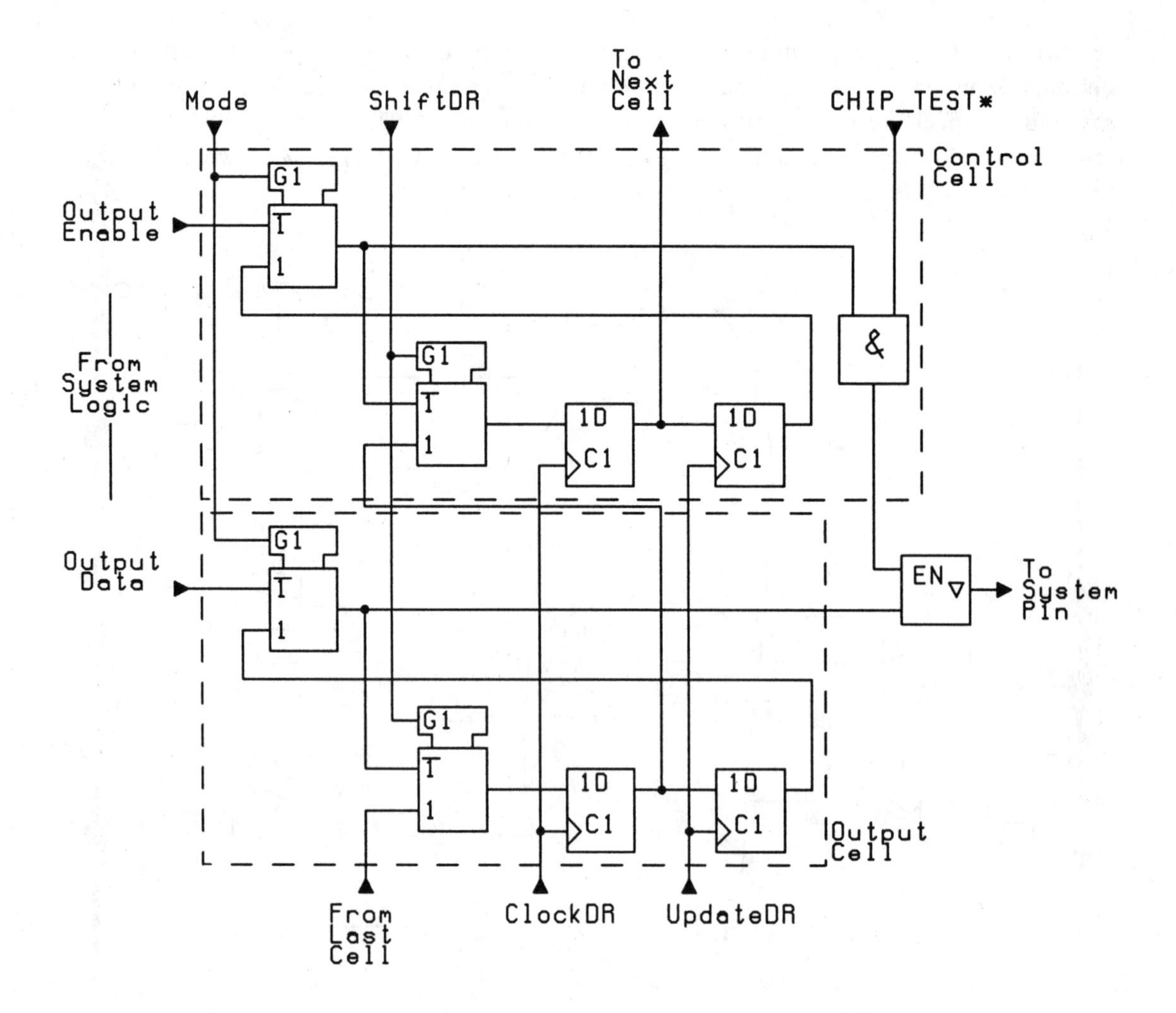

Figure 14–4: Boundary–scan cells for a 3–state pin.

14.3: Special Cases

There are a number of special cases where there is no circuitry, other than inverters and buffers, located between two system pins of a component. A common example is when a component has an output–enable input pin that serves only to control the activity of a set of output drivers, for example, as shown in Figure 14–5. In cases such as this, it is possible to use a single boundary–scan cell to meet the requirements for the input pin and output pin or driver.

Note, however, that it is only permissible to use a single boundary–scan cell in cases where the input signal is not used to feed the on–chip system logic as well as the output driver(s). Figures 14–6 and 14–7 show two "illegal" circuits where the input data are also fed to the on–chip system logic. A correct implementation, using two separate boundary–scan cells, is shown in Figure 14–8.

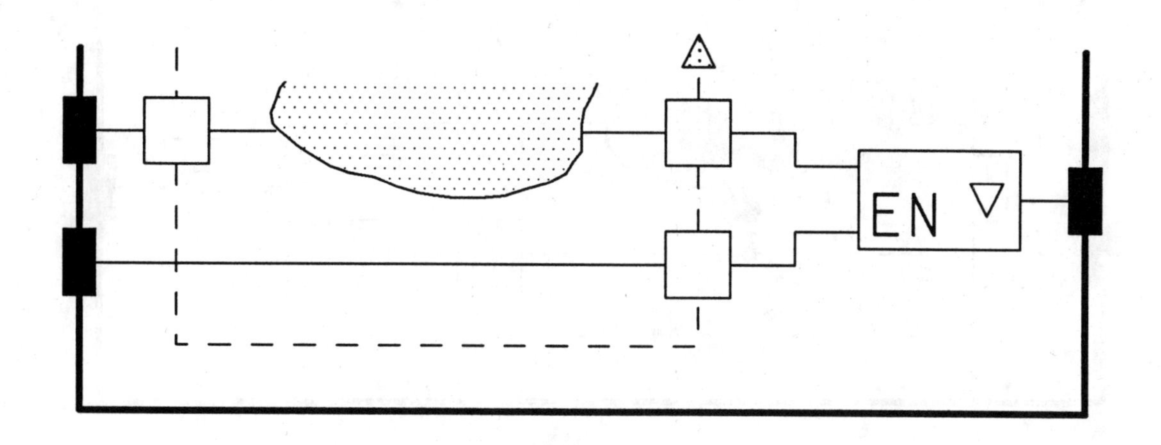

Figure 14–5: Input used only to control an output enable.

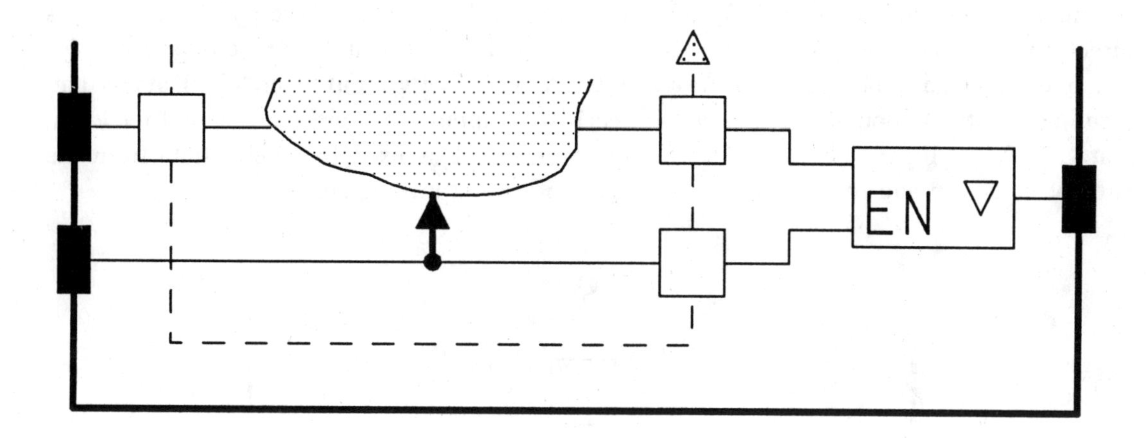

Figure 14–6: "Illegal" design: Example 1.

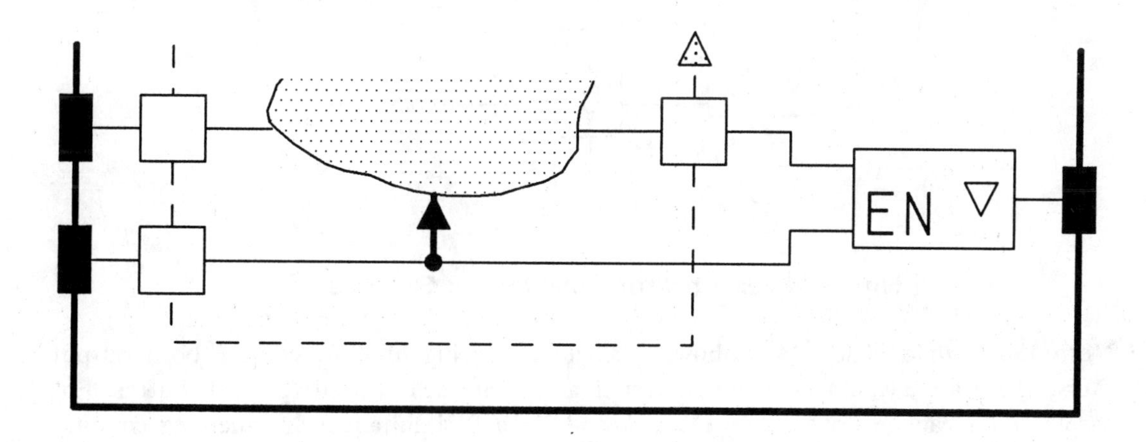

Figure 14–7: "Illegal" design: Example 2.

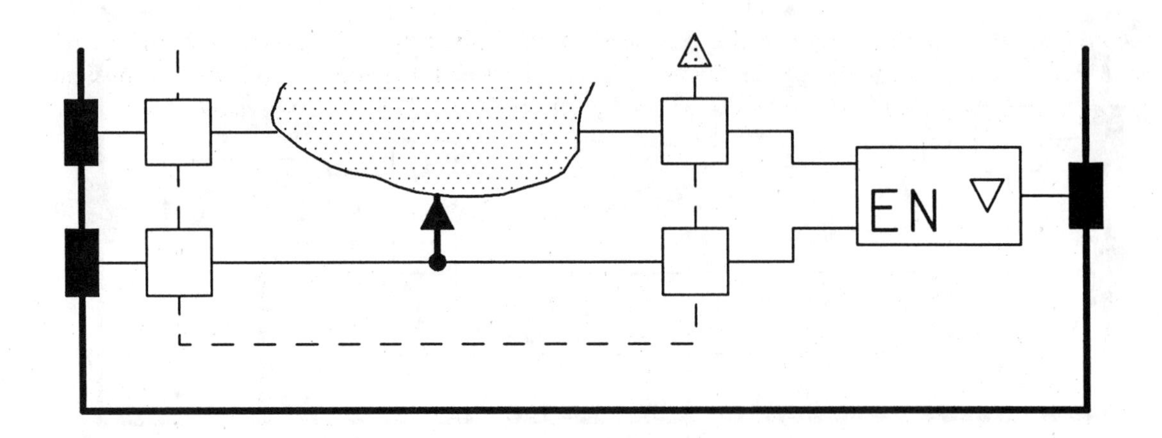

Figure 14–8: A correct design.

Similar shared use of a boundary-scan cell is possible in cases where a data input feeds directly to a data output. Figure 14–9 shows a bidirectional buffer component, for example. In Figure 14–9, boundary-scan cell A receives its input from pin Data_A and feeds pin Data_B. Boundary-scan cell B receives its input from pin Data_B and feeds pin Data_A. As in Figure 14–5, a single boundary-scan cell is used to receive data from the control input pin and to supply control signals to the output buffers.

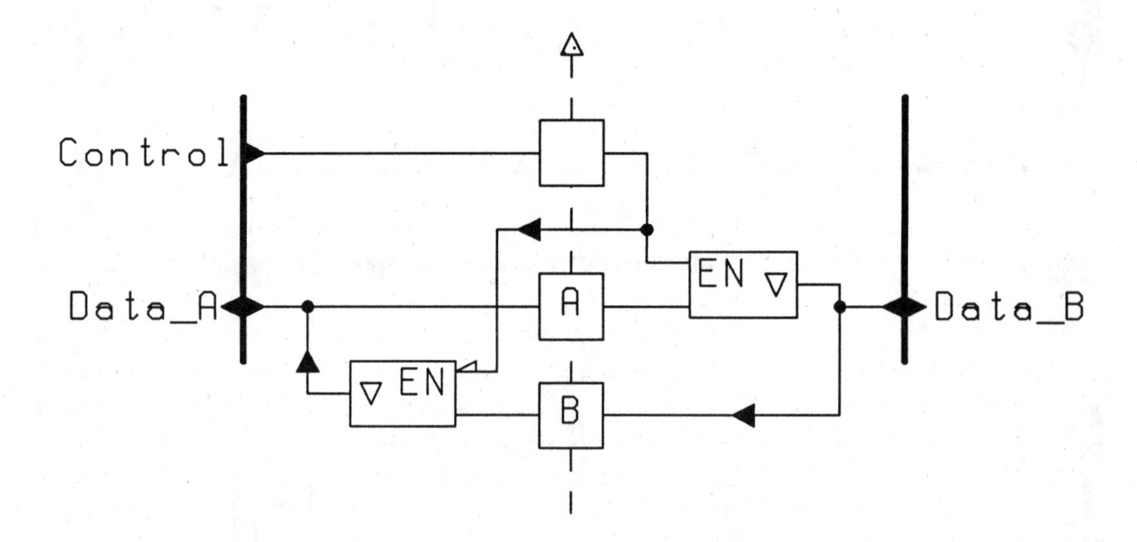

Figure 14–9: A bidirectional buffer: Example 1.

Note that, while Figure 14–9 shows a single cell being used to control both output buffers, there are advantages to board test if a separate cell is used for each buffer. For example, data can be driven onto Data_A and Data_B simultaneously, allowing circuitry on both sides of the component to be stimulated. Also, the data received at both inputs would be captured simultaneously. Together, these features would permit independent testing of the connections and/or logic on each side of the component. It is therefore recommended that the design of Figure 14–10 is used where possible. However, designers should use the design shown in Figure 14–9 where simultaneous activation of busses Data_A and Data_B would cause the power supply to the chip to be overloaded.

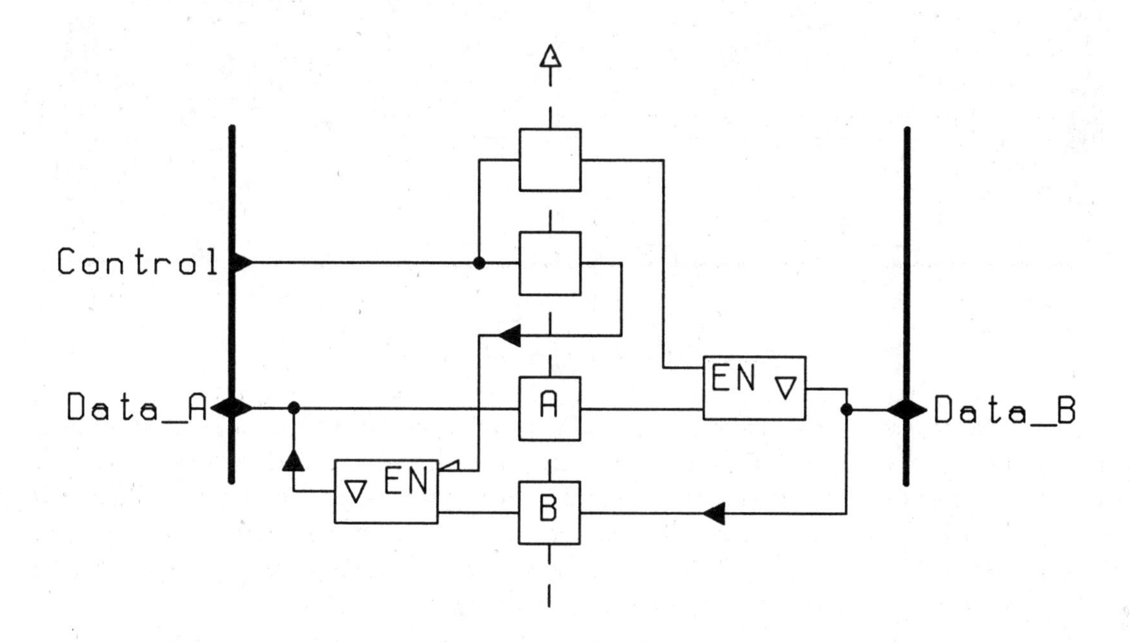

Figure 14–10: A bidirectional buffer: Example 2.

14.4: Components with Inverting Input and Output Buffers

It was briefly mentioned in the last section that IEEE Std 1149.1 requires that the data shifted into or from a boundary–scan cell must be identical to that driven from or applied at the corresponding package pin, respectively. For example, when the *EXTEST* instruction is selected, a logic 1 applied to an input pin of a component should result in a logic 1 being captured into the corresponding shift–register stage. Equally, a logic 1 shifted into a shift–register stage should result in a logic 1 being driven through a connected output pin. The aim of these requirements is to ensure that the data that are shifted into or out of the component's boundary–scan path is exactly that which would be seen by connecting probes at the system pins.

The example boundary–scan cell designs included earlier in this book and in the standard assume that non–inverting input and output buffers are used at the component's system pins. All paths in these cell designs are non–inverting, so the requirements are met. However, if inverting input and/or output buffers are used, then a number of inversions must take place in the boundary–scan cells to compensate. Figures 14–11 and 14–12 give examples for input and output pins, respectively. Note, for example, the inversions at the inputs of the multiplexers controlled by ShiftDR and at the outputs of the flip–flips controlled by UpdateDR.

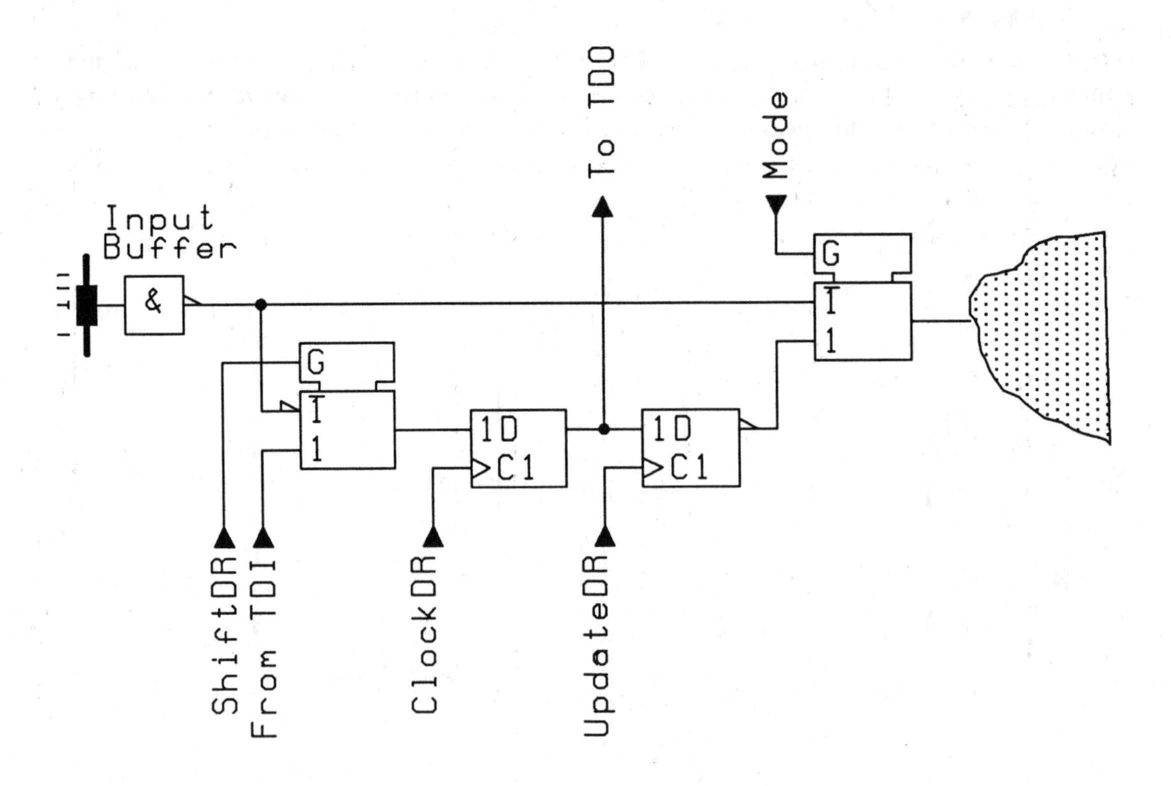

Figure 14–11: A boundary–scan cell for an input pin with an inverting input buffer.

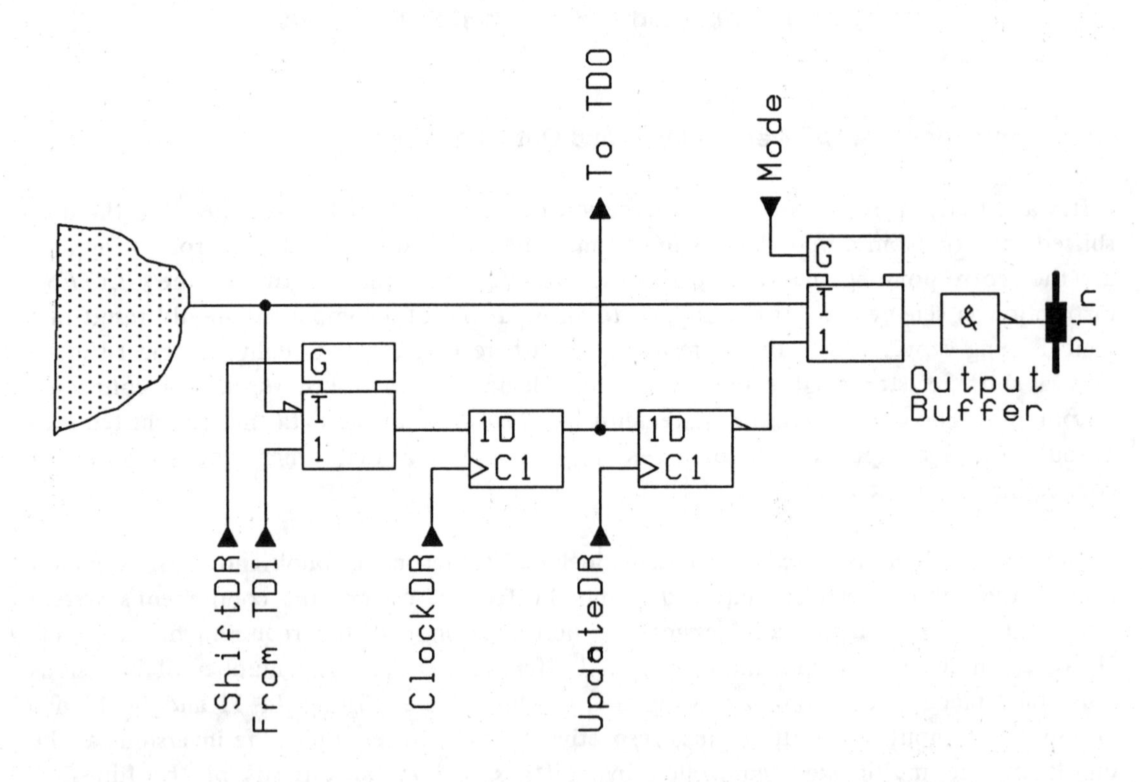

Figure 14–12: A boundary–scan cell for an output pin with an inverting output buffer.

In Figure 14–12, for example, a logic 1 output from the on–chip system logic would result in a logic 0 being driven from the component pin during normal chip operation. When the *SAMPLE/PRELOAD* instruction is used (Mode = 0), the logic 1 output from the on–chip system logic will (due to the inverting $\overline{1}$ input to the input multiplexer) result in a logic 0 being loaded into the shift–register stage. That is, the value loaded into the shift–register stage will be the same as that driven through the component pin.

When the *EXTEST* instruction is selected (Mode = 1), a logic 1 shifted into the shift–register stage will result in a logic 1 being driven through the component pin, this time due to the inversion at the output of the parallel output flip–flop.

A further example is given in Figure 14–13. This figure shows a design for an inverter chip which, in effect, is a component with a non–inverting input buffer, an inverting output buffer and no on–chip system logic.

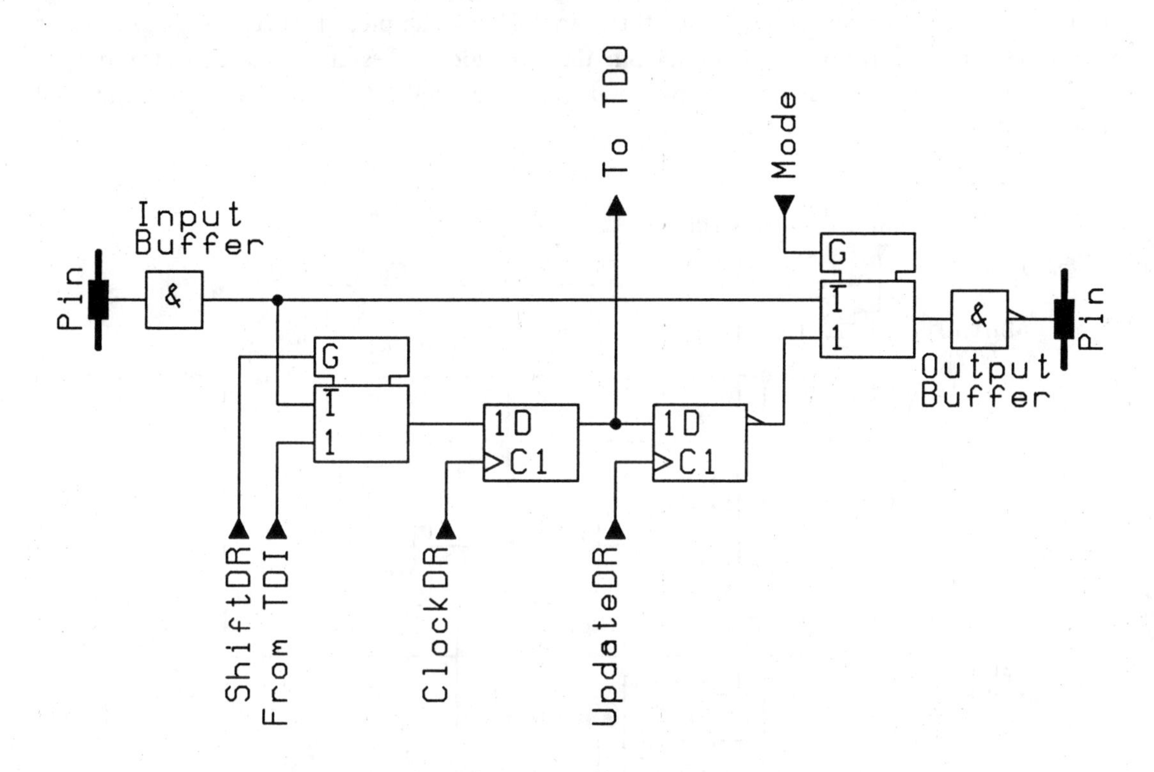

Figure 14–13: An inverter chip with boundary–scan.

As discussed in Section 14.3, one boundary–scan cell can meet the requirements for both the input and output pin in this case. In this case, the requirements for the data in the shift–register stage to match that at the pins has the following impact: there can be no inversion between the input pin and the shift–register stage, therefore the inversion between the shift–register stage and the inverting output buffer is required.

14.5: Complex Boundary-Scan Cells

IEEE Std 1149.1 addresses the four most common types of pin on an integrated circuit:

- input pins;
- 2-state (including open-collector) output pins;
- 3-state output pins; and
- bidirectional pins with 3-state output capability.

Other types of pin are occasionally found on integrated circuits. In these cases, an appropriate combination of the "basic" input and output pin cells must be constructed to meet the special requirements of the pin. For example, Figure 14-14 shows a boundary-scan cell for a bidirectional pin that includes a 2-state open-collector output buffer. This combines a 2-state output boundary-scan cell (at the top) with an input cell (at the bottom).

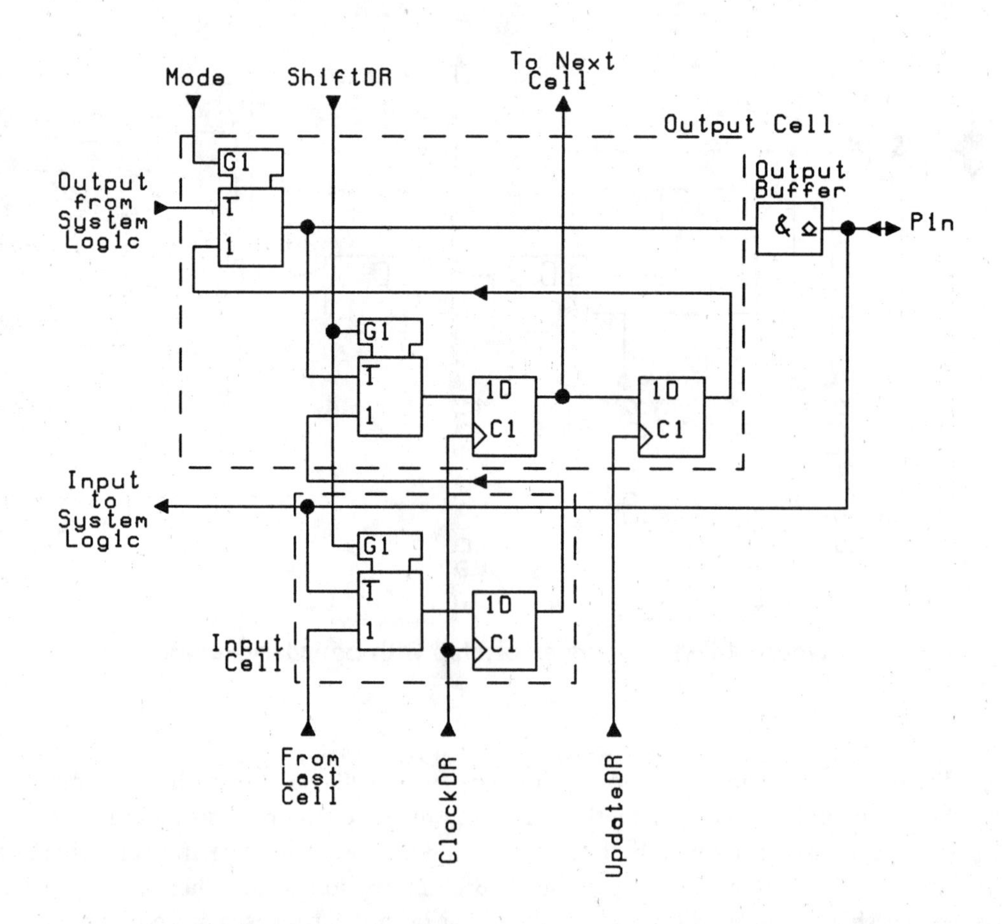

Figure 14-14: Boundary-scan cells for 2-state open-collector bidirectional pins.

Figure 14–15 shows two components that use a more complex type of pin for chip–to–chip communication. In effect, a 3–state bus flows into each component. Two 3–state drivers and one input are included in each chip, connected to the bus.

In a case such as this, each component would need to have five boundary–scan cells connected to the pin:

- one input cell;
- two data output cells; and
- two output enable cells.

This combination of cells will permit the ability of each output buffer in the component to drive the pin to be tested, as well as the ability of the component to receive data from the pin.

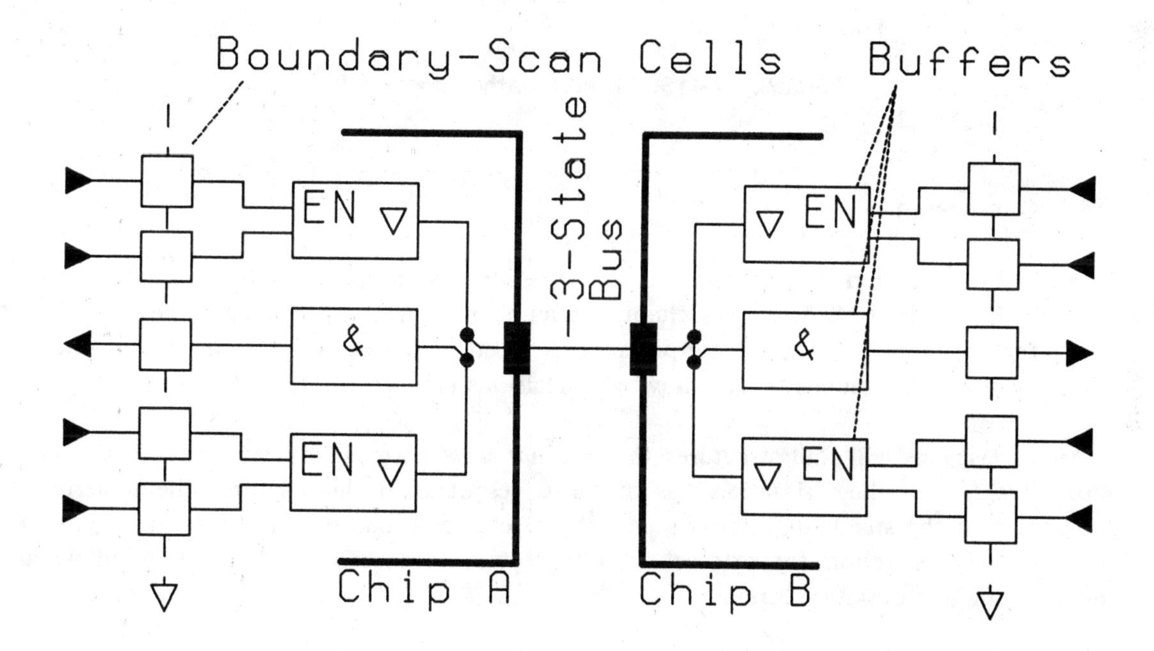

Figure 14–15: A complex chip–to–chip connection.

An alternative arrangement would be to redesign the circuitry as shown in Figure 14–16.

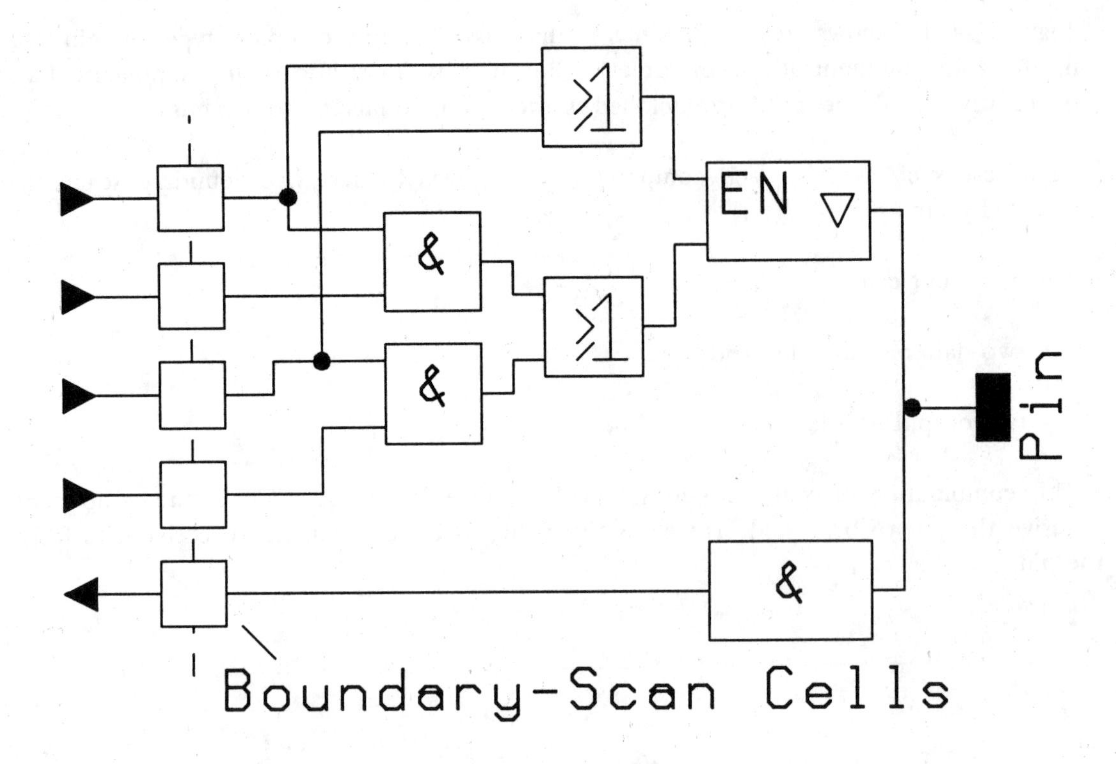

Figure 14–16: An alternative approach.

14.6: Conclusion

Inevitably, situations will arise from time–to–time that are not covered explicitly by the rules in IEEE Std 1149.1. In this chapter, a number of such situations has been discussed and the authors' personal interpretations of the standard have been presented, accompanied with comments that show why decisions have been made.

More designers will find structures that are not covered explicitly by the standard as its use increases. In these situations, as in those described in this chapter, the underlying objectives of the standard -- clear separation of chip and loaded–board test, simplicity of test pattern generation for external circuitry, etc. -- should be borne in mind when determining appropriate solutions.

Chapter 15. Providing Boundary–Scan on Chips with Power or Output–Switching Limitations

Lee Whetsel
Texas Instruments
6500 Chase Oaks Boulevard
Plano, TX 75086, U.S.A.

This chapter will identify problems that may arise during boundary–scan testing of inter–component connections in cases where a chip is not designed to support simultaneous enabling or switching at all its output pins. Some solutions are given to these problems.

15.1: Problem Statement

A principal motivation for including boundary–scan in a chip is to be able to efficiently verify the wiring interconnects between multiple chips on a loaded board. Initially these test patterns will be developed manually by a test engineer familiar with the circuit board design. In the future, this effort will become automated as test pattern generation tools are developed. In either case the test patterns will be designed to verify that all possible wiring paths between chips in the circuit can be set to both logic zero and logic one and that no short– or open–circuit faults exist.

Ideally, only an understanding of the board interconnects and of the boundary–scan configuration or each chip should be required to allow the development of test patterns for board–level interconnect. However, if chips are not designed to fully support the operation of the boundary–scan circuitry, other factors may need to be included to insure proper operation of such a test. The result is that test generation, and the tools that support it, will be more complex.

In Figure 15–1, an example circuit is shown to illustrate problems that could occur during boundary–scan testing of board–level interconnections, but that can be avoided by correct design of the test logic (as will be described later).

IC1 has three output busses. In normal operation, these busses would be controlled either so that they change state at different times (Figure 15–2) or so that only one of the three output busses is active at any given time (Figure 15–3). Controls such as these may be required to minimize the power consumption requirements for the chip, allowing the number of power and ground pins needed by the chip to be reduced. For example, more power is consumed when outputs change state so sequencing of output changes as shown in Figure 15–2 will result in lower power consumption.

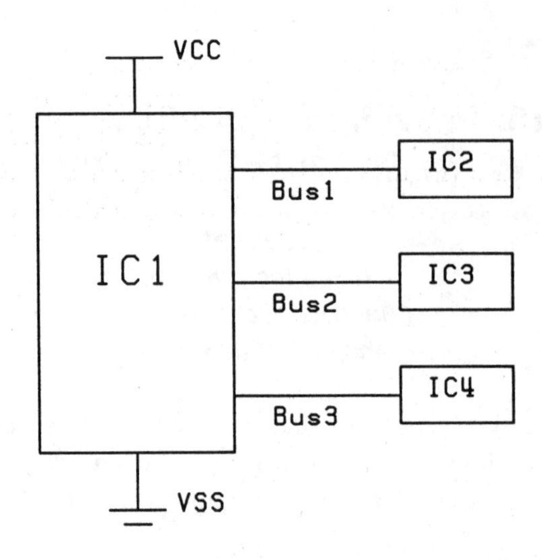

Figure 15–1: Example component with three output busses.

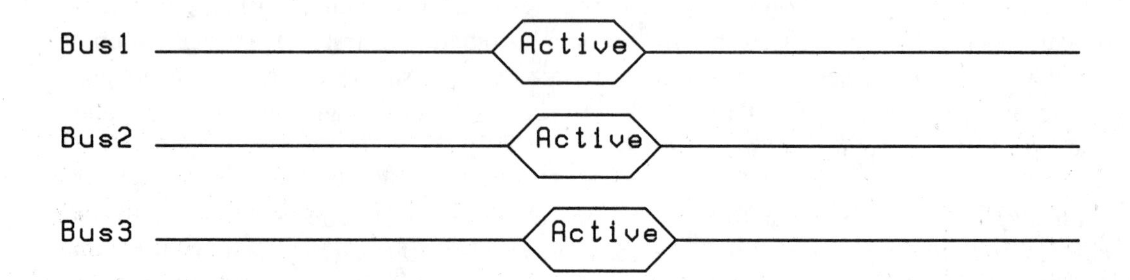

Figure 15–2: Normal component operation: Case 1.

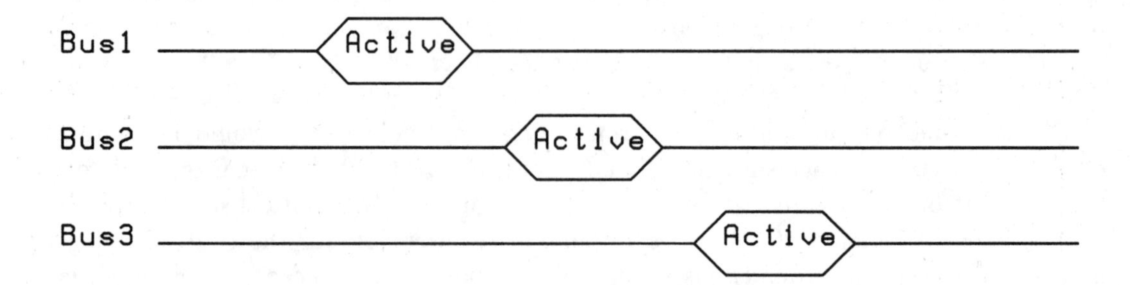

Figure 15–3: Normal component operation: Case 2.

The problems arise when the pins of a chip that does not normally support simultaneous switching or enabling of all outputs are controlled from the boundary–scan path, rather than from the on–chip system logic. For example, this situation arises when the *EXTEST* instruction is selected. As shown in Figure 15–4, it is probable that the tests supplied through the boundary–scan path will cause all outputs to change state and/or be active simultaneously. In the former case (Figure 15–2), the sum of the switching currents of all pins may produce VCC and VSS glitches that might very well exceed the tolerance level of IC1, and the core logic as well as the boundary cells and TAP would then be subject to interference. In the latter case (Figure 15–3), the power consumption could be increased beyond the capacity of the power pins for a prolonged period, with the probable result that incorrect operation of the chip will occur.

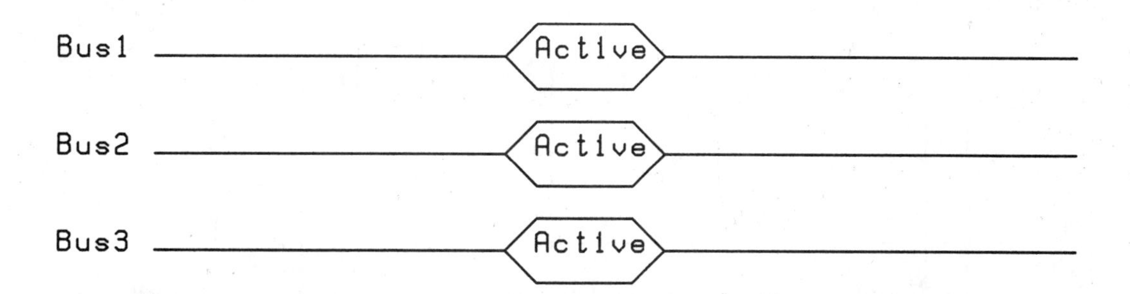

Figure 15–4: Possible component operation during a boundary–scan interconnect test.

Both problems can be solved if the boundary–scan test patterns are constrained to conform to the normal requirements of each IC on the board design, for example, by ensuring that only one output bus changes state between adjacent test patterns. Unfortunately, however, this complicates the test generation task and requires that additional information on chip operation is made available to, *and used by*, the test engineer or test pattern generation tool. Alternative solutions, where features built into the chip ensure that it cannot overload its power pins, are preferred because they do not require provision, storage, and use of this additional information. Some example solutions are presented in the following sections.

15.2: Provide More Power Pins

The first (and least practical) solution is for the chip manufacturer to provide the additional power and ground pins required to ensure tolerance of simultaneous switching or enabling of all output pins. This solution may, however, not be practical because it may force an increase in the size of package required for the chip, which in turn may affect the cost to the customer.

15.3: Preventing Simultaneous Switching of Output Pins

Figure 15–5 shows how delays can be added to prevent simultaneous switching of outputs when pins are driven from the boundary–scan register [1,2]. The added delays should be small in comparison with the minimum period of TCK, but should be sufficient to ensure that the power–current demand arising from the change of state at one pin does not overlap with that from another.

Note that the added delays impact only changes at the pin due to:

- a change of instruction (the delays in the Mode distribution network); or
- a change in test pattern (the delays in the UpdateDR distribution network).

Signals received from the on-chip logic propagate through the boundary-scan cells without added delay.

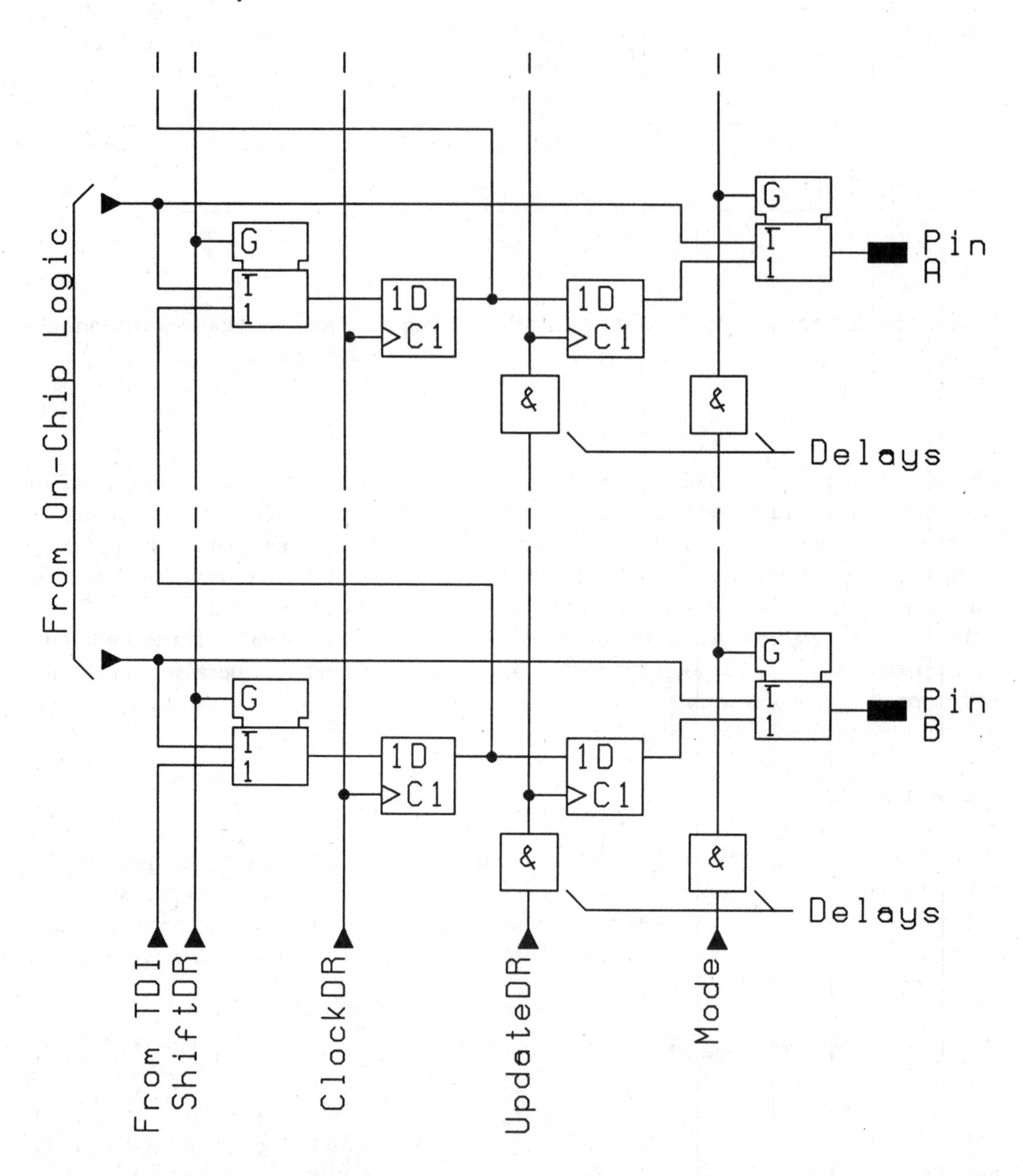

Figure 15–5: Adding delays to prevent simultaneous switching of outputs.

15.4: Do Not Allow Pins to be Enabled Simultaneously

Where output pins or busses cannot be enabled simultaneously, this limitation should be met as a result of features built into the chip, rather than through constraints imposed on the test pattern generation process. For example, because two boundary–scan cells (numbered 1 and 3) are provided in Figure 15–6, pins A and B can be enabled independently. The test engineer or, more realistically, the test generation software can enable both pins, disable both, or enable just one pin as required. In this case, the chip should be provided with sufficient power and ground pins to support simultaneous enabling of the two pins –– avoiding the need to constrain the test generation process.

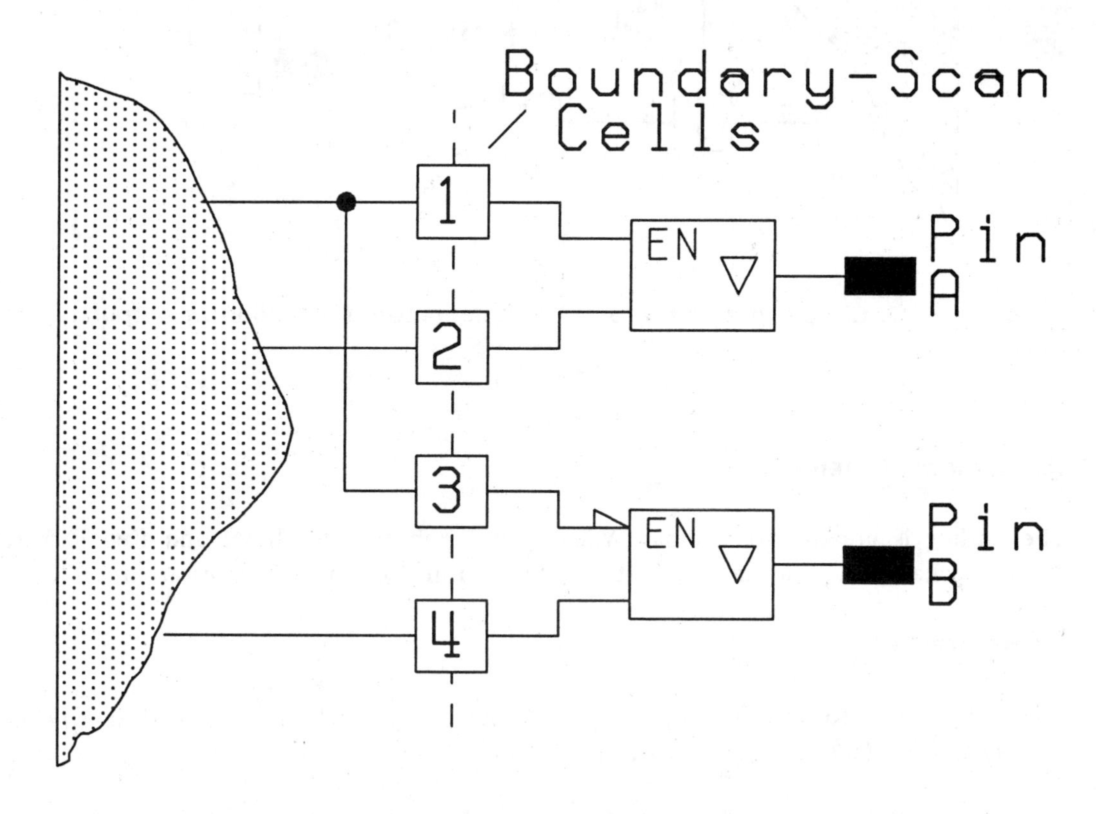

Figure 15–6: Circuit that allows simultaneous enabling of outputs.

In contrast, only three boundary–scan cells are provided in Figure 15–7, although the circuit performs the same normal function. In this circuit, both output buffers are controlled from a single boundary–scan cell (numbered 1) –– a logic 1 in this cell enables pin A and disables pin B. Since there is only one boundary–scan cell, it is not possible for both pins to be enabled simultaneously during testing –– the restriction is inherent in the chip's design and is not only one that needs to be imposed on the test generation process. The test engineer, or test generation software, cannot inadvertently cause the power supply to be overloaded.

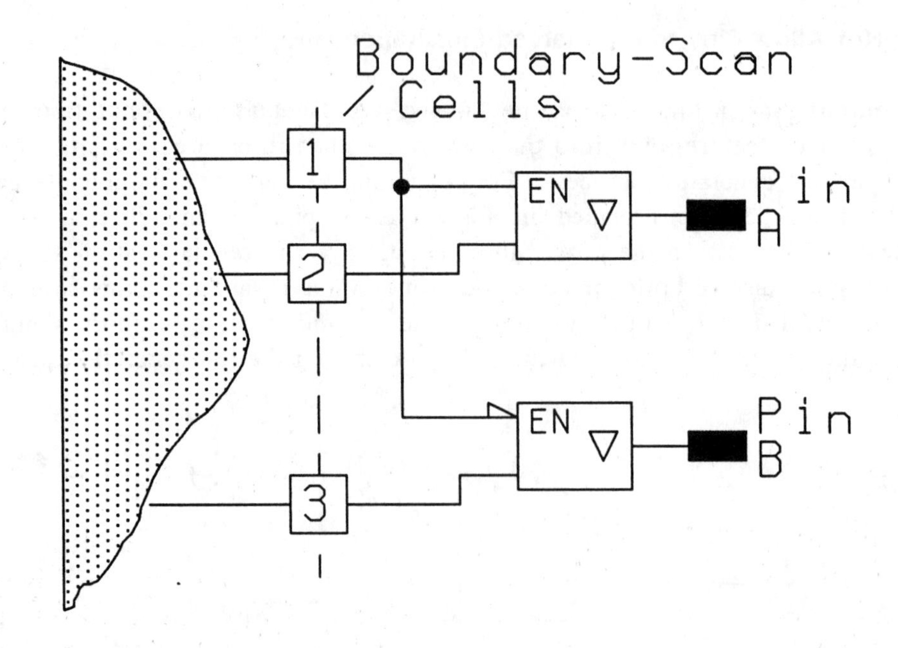

Figure 15–7: Circuit that does not allow simultaneous enabling of outputs.

15.5: Acknowledgments

The author is greatful to Thomas Williams and Bob Bassett, IBM, and Ken Parker, Hewlett–Packard, for bringing the solution presented in Section 15.3 to his attention.

15.6: References

[1] Anon., "Improved Off–Chip–Driver Sequencer for LSSD Testing," *IBM Technical Disclosure Bulletin,* Sept. 1989, pp. 422–423.

[2] Anon. "Inhibit Sequencing Delay Circuit," *IBM Technical Disclosure Bulletin,* June 1986, pp. 251–252.

Chapter 16. Tapping into ECL Chips

Lee Whetsel
Texas Instruments
6500 Chase Oaks Boulevard
Plano, TX 75086, U.S.A.

This chapter will illustrate some problems and suggested solutions for the addition of the test access port (TAP) interface to emitter-coupled logic (ECL) chips.

16.1: The Problem

Several semiconductor technologies -- for example, transistor-transistor logic (TTL) and complimentary metal-oxide semiconductor (CMOS) -- use compatible input and output voltage levels for the two logic states (0 and 1). Chips constructed using these technologies can therefore be easily connected together, for example to form the serial test data path at the board level. However, to construct a scan path through chips that use different voltage levels for each logic state -- for example, through chips built using TTL and ECL technologies -- it is necessary to convert the various test signals from one set of logic voltage levels to another at appropriate points along the serial path. This level translation can be performed by either an external level shifting circuit residing between the device boundaries or in the input and output buffer regions of chips incorporating the TAP.

The focus of this chapter is the problem created by the way that ECL technology reacts to open-circuit conditions at the TAP inputs, which is significantly different from the reaction of a TTL chip. The problem stems from the fact that to conform to IEEE Std 1149.1 the output state of a non-driven TAP input buffer must be set to a logic 1 level.

In TTL technology a non-driven input buffer is usually pulled up to logic 1 by an internal resistance incorporated in the buffer, because this condition draws minimal current. In CMOS technology, a non-driven input buffer can be pulled up or down, with neither logic state having an apparent advantage over the other. Both of these technologies adapt easily to the rules of IEEE Std 1149.1 regarding non-terminated TAP input buffers.

For performance and biasing reasons, a non-driven ECL input buffer is usually pulled down to logic 0 by an internal resistance incorporated in the buffer. At first, it would appear that substituting a pull-up resistor for the pull-down resistor on the input buffers for the TAP signals would solve the problem. However, the circuit shown in Figure 16-1 (in which the test data input (TDO) output of one ECL chip is connected to the test data input (TDI) input of another) identifies three problems that prevent this from being the desired simple solution:

1. An open circuit at wiring point 1 causes the receiving chip's input to be pulled down to −V (typically −2 or −5 volts). Since R1 << R2, the condition results in a logic 0 output from the chip's input buffer.

EH0321-0/90/0000/0177$01.00 © 1990 IEEE

2. An open circuit at wiring point 2 causes the receiving chip's input to be pulled up to GND (typically -0.7 volts), resulting in a logic 1 output from the chip's input buffer.

3. An open circuit at wiring point 3 results in improper termination and loss of operation between the ECL chips since the output buffer in the driving chip has no resistive path to -V that would enable it to source the required output biasing current.

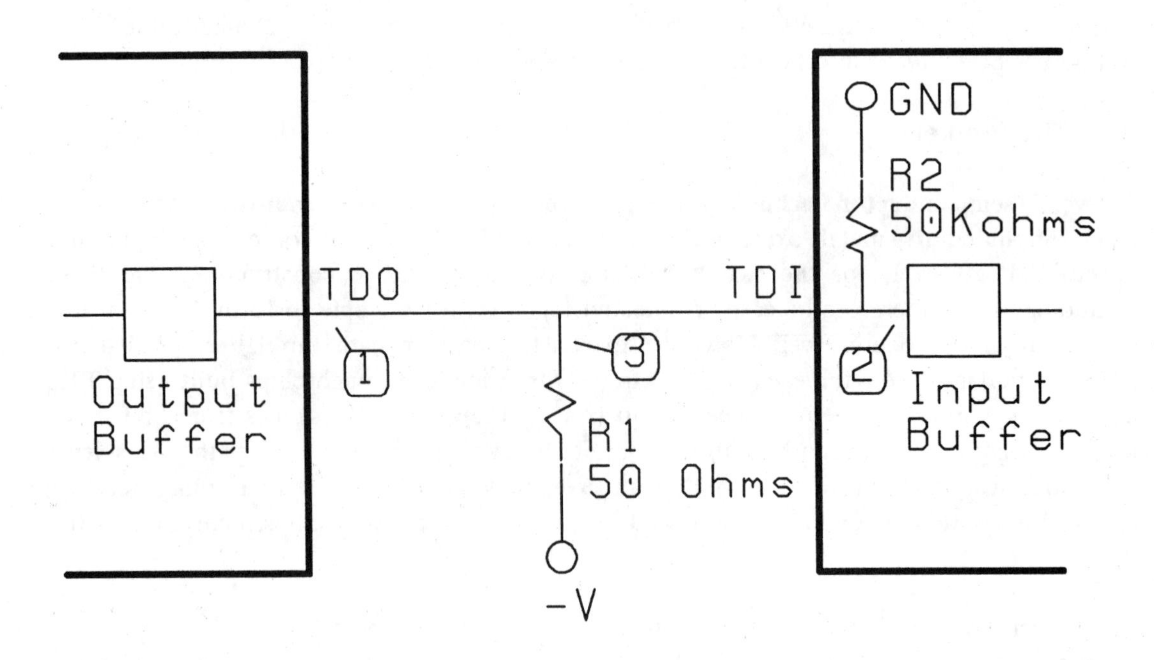

Figure 16–1: A TDO–to–TDI connection between ECL components.

It is clear, therefore, that placing a pull-up resistor on ECL inputs is not a suitable solution. The result of an open-circuit in the serial path between ECL chips would be dependent on the location of the fault in the wiring path and could disable data transmission.

The following are suggested solutions that can be implemented in ECL chip designs to allow compliance with IEEE Std 1149.1 regarding open-circuit, non-driven TAP inputs.

16.2: Incorporating TTL/CMOS TAP Connections on ECL Chips

The first option is to incorporate TTL/CMOS compatible TAP inputs and outputs into ECL designs. Using this approach requires the use of an additional 5 volt power supply pin on the package for the TAP input and output level shifting buffers. If the additional 5 volt supply pin is not a problem, this is probably the preferred approach.

16.3: Using a Special ECL Input Buffer for TDI, TMS, and TRST*

The second option is to design a special ECL input buffer that can differentiate between an open-circuit input and a normally-biased logic 0 input. These special ECL input buffers would be provided at the TAP input pins: TDI, the test mode select (TMS) input, and the optional test reset (TRST*) input.

In Figure 16-2, a typical interconnection between an ECL device output and ECL device input is shown. During normal operation, the ECL output buffer provides the output biasing current (Iout) required to develop the correct ECL logic voltage levels across load resistor R1. Typical ECL output voltages are -0.9V for a logic 1 and -1.75V for a logic 0.

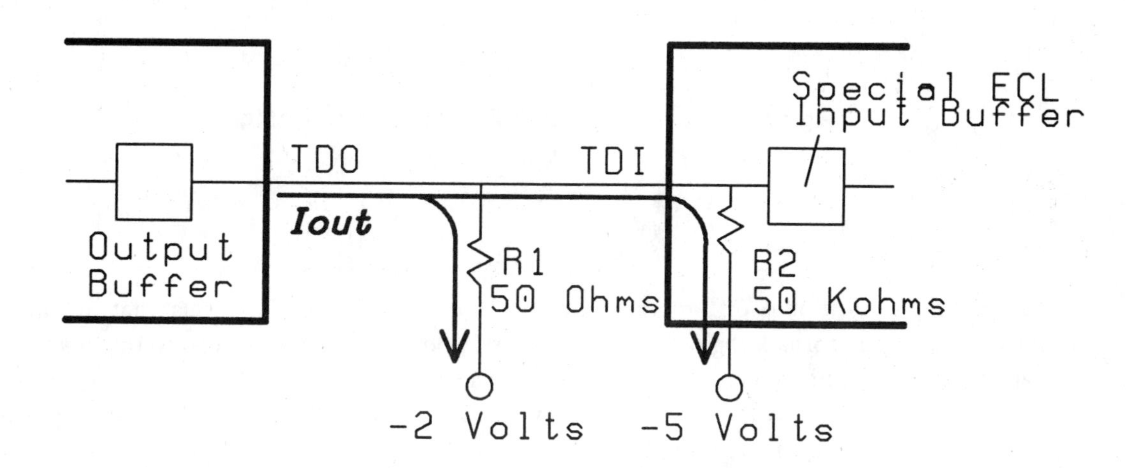

Figure 16-2: Current flow for an ECL TDO-to-TDI connection.

As long as the interconnection between the ECL output buffer and load resistor R1 remains intact, the input voltage to the ECL input buffer will remain in the normal ECL logic level switching range of either a logic 1 or logic 0.

In Figure 16-3, a detailed view of the special ECL input buffer is illustrated. The input buffer contains two single-ended differential amplifiers, Damp1 and Damp2. Damp1 is the normal differential amplifier seen in ECL input buffers and is used to determine the input logic state by comparing the input voltage (Vin) against a reference switching threshold voltage VR1.

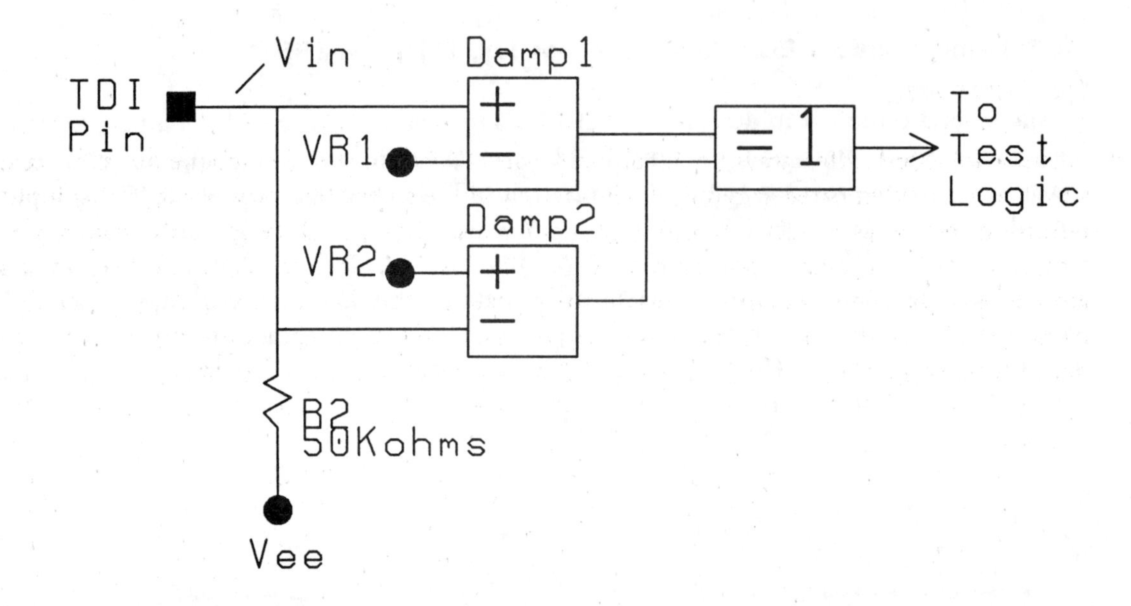

Figure 16–3: A special ECL input buffer design.

Damp2 is an additional differential amplifier used to detect Vin levels that are more negative than the normal logic 0 voltage levels produced by ECL output buffers. The voltage reference input (VR2) to Damp2 is set to detect Vin levels falling below the typical ECL logic 0 level. VR2 should be set to allow Damp2 to detect the following two types of open–circuit fault (see Figure 16–2):

- *A fault between the ECL output buffer and the load resistor R1*. If this fault occurs, the ECL input buffer will be driven to a static voltage level determined by the voltage divider effect of R1 and R2. Since R1 << R2, the open–circuit input voltage to the ECL input buffer and Damp2 is slightly less than –2V.

- *A fault between load resistor R1 and the ECL input buffer*. If this fault occurs, the voltage applied to the ECL input buffer will be the –5V level attached to pull–down resistor R2. Since VR2 is set to detect voltages below –2V (to detect the first stated open–circuit condition) this type of open–circuit is detected by Damp2.

When either of the above faults occurs, the non–inverting output of Damp2 is set to a logic 1. This logic 1 output is routed to the exclusive–OR gate in Figure 16–3 and causes the logic level output to the test logic to be inverted. By using this type of ECL input buffer, it is possible to differentiate between an ECL logic 0 input level and an open–circuit input level. Therefore, conformance to the IEEE Std 1149.1 specification for undriven TAP inputs is achieved.

16.4: Summary

These solutions offer ways to incorporate the TAP into ECL components. Option 2 needs to be implemented carefully to insure that noise spikes that may occur during input transition between a logic 1 and 0 do not cause Damp2 to temporarily switch on. However, this is probably not a problem for the TAP because the TDI and TMS inputs are basically data inputs and will be in a stable state by the time the rising edge of the test clock (TCK) arrives. If a 5 volt power supply pin is already implemented in a device of mixed technologies, option 1 is probably the most logical choice to implement.

Chapter 17. Cell Designs that Help Test Interconnect Shorts

Dilip K. Bhavsar
Digital Equipment Corporation
Semiconductor Design and Engineering
Hudson, MA, U.S.A.

This chapter describes a problem that may, under some circumstances, arise when using a boundary-scan register to test interconnection shorts on loaded boards. A remedy to overcome the problem is proposed.

17.1: Introduction

One of the major test problems addressed by IEEE Std 1149.1 is that of testing interconnection faults on densely populated printed wiring boards. With the advent of surface-mount components and the use of buried and blind vias for mounting chips on these boards, the access available to in-circuit-testers is rapidly disappearing

Interconnection defects are introduced during the printed wiring board manufacture and assembly processes. In general, these defects fall into two categories:

1. *Opens*: These include defective solder joints and open-circuits in the interconnection tracks of the board.

2. *Shorts*: These include shorts between adjacent pins on the same chip and shorts between adjacent interconnecting tracks on a board. In either case, the defect manifests itself as a short between two signals or nets.

A simple and straight-forward test for both fault categories can be achieved by using the boundary-scan register.

This chapter will focus on the testing of shorts and will point out a problem that may occur in certain semiconductor technologies, such as complimentary metal-oxide semiconductor (CMOS) and transistor-transistor logic (TTL). The case to be examined is that in which output buffers are not designed to withstand a short to another output driving the opposite logic state. We must point out that the severity of the problem highlighted in this chapter depends significantly on the implementation details of a chip's output buffers and, in many cases, will be minimal.

17.2: The Problem

Consider that we are testing the interconnection between two chips (chip A and chip B) on a board. Assume that the chips are implemented in CMOS technology, that both implement the IEEE Std 1149.1 boundary-scan architecture, and that both are connected in the same boundary-scan ring on the board. Now consider the simple interconnection as shown in Figure 17–1 where a short occurs between the two adjacent output pins P1 and P2 fed from ordinary drivers D1 and D2. The tests that detect this fault consist of driving a differential pattern. One of the patterns, "10" or "01", must be applied to the shorted pins. Applying one of these tests requires several steps.

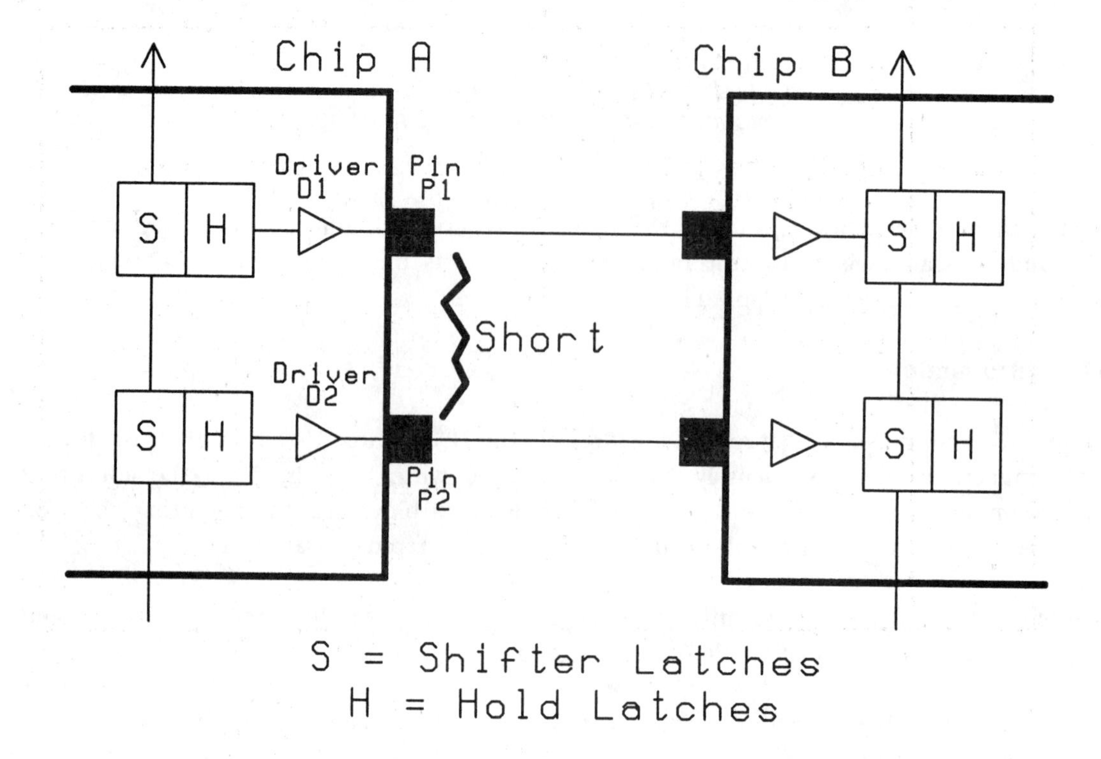

Figure 17–1: Testing a pin short with IEEE Std 1149.1.

Assuming that the appropriate instruction has been loaded in the instruction register and that the test access port (TAP) controller has been appropriately initialized, then a portion of the steps involved and the corresponding TAP controller state transitions are shown in Table 17–1.

During Step 3, if the pins have no short, the receiving boundary scan cells in chip B observe the response pattern "01." However, if the pins are shorted, then the receiving boundary scan cells will sense a "00" pattern and the fault will be detected. Notice that to successfully detect and diagnose the presence of shorts via the boundary scan it is essential that the receivers in chip B must have their switch–over voltages away from the voltage expected to be reached by the shorted drivers. In our analysis, we arbitrarily chose that the receivers should sense "0." All arguments hold true if "1" were to win and the observed pattern were to be "11."

Table 17-1: Steps for testing for shorts.

Step	Action	Remarks
1	Shift in the [...01...] pattern	
2	Apply the [01] pattern to the shorted pins.	Say 0 is applied to P1 and 1 to P2.
3	Capture the response	Boundary scan cells at input pins of chip B capture the response.
4	Shift the response out	The duration of this operation depends on the total length of the boundary scan registers on the board.

The problem is that, whereas the above test procedure succeeds in detecting the short, it may cause permanent damage to the drivers D1 and D2. This is because the differential pattern "01" enables a power-to-ground path via the turned-on P and N transistors and the shorted P1 and P2 pins. This is shown in Figure 17-2.

Notice that this short will persist for the duration of the entire shift operation and until a safe pattern ("00" or "11") can be shifted in and applied to the pins. This period can be arbitrarily long because it depends on the total length of the boundary-scan registers in the board-level path containing chips A and B, the frequency of the test clock (TCK) and on any interruptions (e.g., pause cycles) to the shift operation. Whereas a short on pins is generally considered a repairable fault, in some cases the above test procedure may destroy the high-price chip, making the repair meaningless.

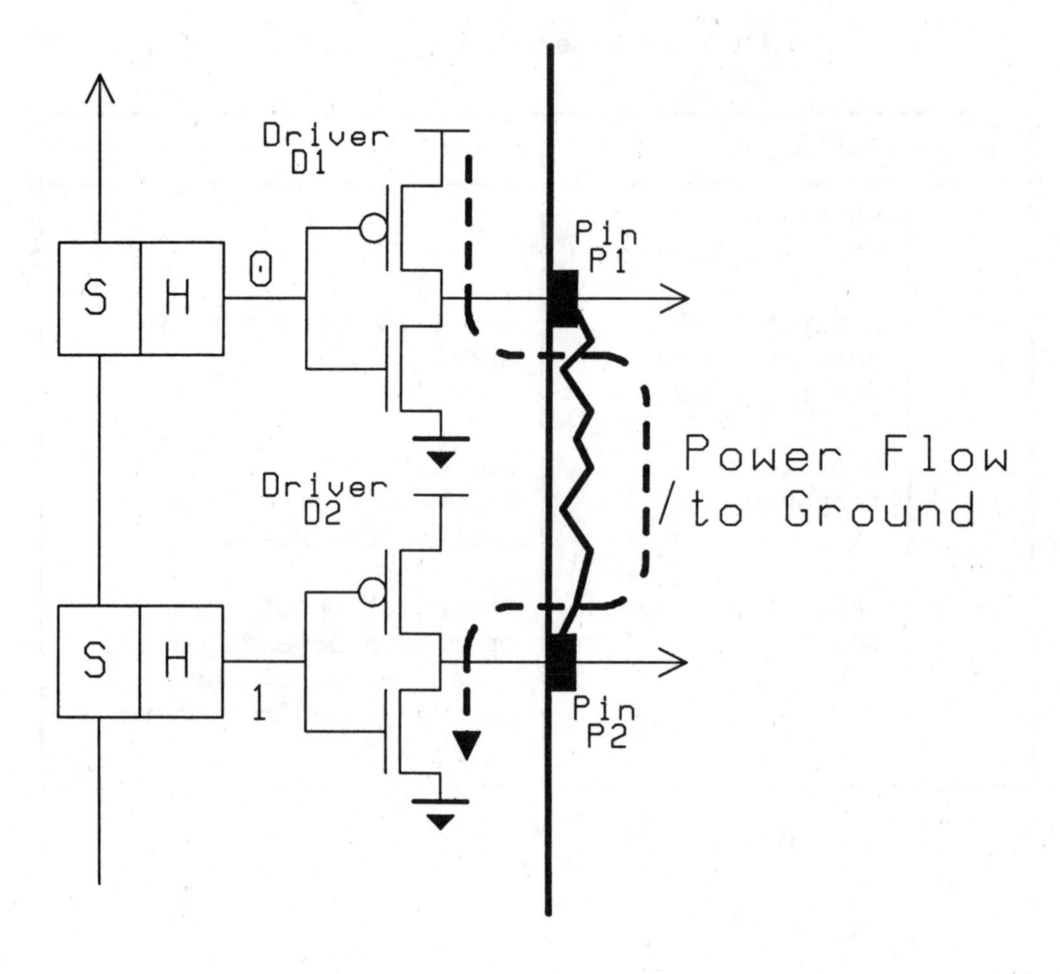

Figure 17–2: Power–to–ground path enabled during shorts test.

17.3: A Proposed Solution

The above problem can be overcome by using a special boundary–scan cell design at the output pins and by slightly modifying the test steps used. A proposed cell structure that is compliant with IEEE Std 1149.1 is shown in Figure 17–3. Notice that the cell has a second observation tap taken from the output of the driver via a dedicated receiver. During application of a board interconnect test using the *EXTEST* instruction, the instruction register will set the multiplexer controls on the shift latch such that the data are captured from this special observation tap. The rest of the controls and operations remain as usual.

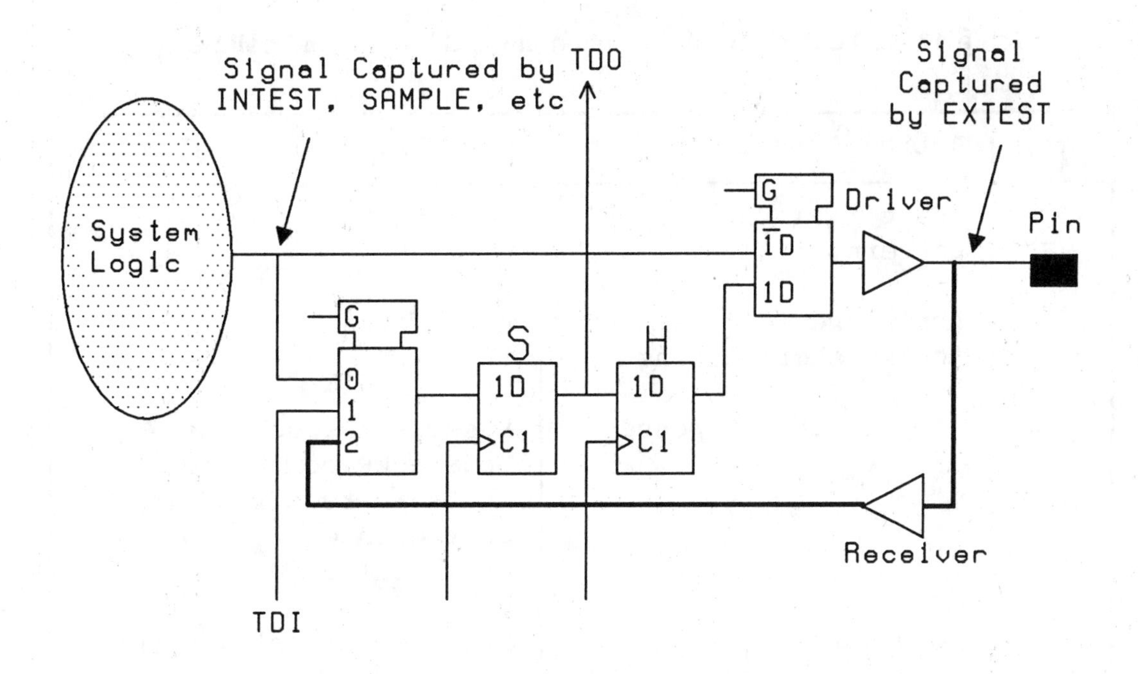

Figure 17–3: Proposed boundary–scan cell for output pins.

With this cell structure in place at the drivers D1 and D2, the test operation will use slightly different test steps as shown in Table 17–2.

By virtue of the boundary–scan cell design, during Step 3a the response pattern at the output of the drivers is captured and loaded into the shifter latches of the cell. If the pins P1 and P2 are not shorted, then the pattern captured will be the same as the pattern applied, namely, "01." If the pins are shorted, the captured pattern will be "00" (per the previous assumptions).

The additional Steps 3b, 3c, and 3d re–apply the captured pattern to pins P1 and P2. Thus, if the pins P1 and P2 were indeed shorted, the differential pattern "01" is removed and the safe pattern "00" will be applied. The power–to–ground path will be disabled. With this test sequence, the power–to–ground short is enabled only for a fixed duration of 4 clock cycles of TCK, independent of factors such as the duration of the shift operation and the total length of all the boundary scan registers on the board–level path. The test operation for detecting shorts is thus made very safe for the drivers, assuming that the clock applied at TCK is sufficiently fast.

Table 17–2: Testing for shorts when using the proposed cell design.

Step	Action	Remarks
1	Shift in a [...01...] pattern	
2	Apply the [01] pattern to the shorted pins	As before
3a	Capture the response	Passage through the Update-DR controller state is required. In consequence, P1 and P2 will be fed [00].
3b	Move through the Capture-DR controller state to Exit1-DR	Exit immediately, without shifting.
3c	Exit the scan cycle	P1-P2 fed [00] if shorted or [01] if healthy.
3d	Re-capture the response.	Response pattern is re-captured without any modifications.
4	Shift the response out	As before.

17.4: Conclusion

In this chapter, we have indicated a potential danger in using boundary-scan for testing interconnection shorts among components in certain technologies, such as CMOS and TTL. We have proposed a solution that uses a special design for boundary-scan cells used at output pins and a slightly modified test operation. We used simple output pins to illustrate the application. However, the same solution is applicable to 3-state output and bi-directional pins and can handle shorts to power or ground.

Besides overcoming the potential danger to the drivers, the proposed boundary-scan cell also offers the following advantages:

1. Because the cell uses a dedicated receiver, this receiver can be especially designed to guarantee the success of the interconnection short test without any adverse impact on system operation.

2. Shorts on output pins (or nets) can be tested on chips whose outputs do not feed any chip or whose outputs feed chips that do not have boundary scan implemented in them. This also means that the ability of the proposed scheme to test pin shorts on a board is unaffected by the presence of chips that implement boundary-scan cells differently, although the test procedure becomes more complex.

3. During testing of an integrated circuit (e.g., in various stages of chip manufacturing), the cell structure provides convenient observability of output drivers for detecting defects in the drivers.

4. This cell, when used on 3-state output and bi-directional pins, provides additional visibility into busses to facilitate isolation of bus problems.

In the end, it must be emphasized that the severity of the problem caused by shorted output drivers is not clear. At best, it is highly dependent on processing technology and may vary considerably from one chip manufacturer to another.

If the threat is serious, a key question is still left unanswered by the solution presented. Could the powering up of a board leave drivers of shorted pins (especially, 2-state output pins) at opposite logic states? If so, how likely are the drivers to survive the short before testing begins? If damage is possible, it can only be avoided by careful design of the power-up initialization routines for a board -- the solution presented in this chapter is limited to test-operation-induced contention between shorted output drivers.†

Finally, note that if the method described in this chapter is used in a catalog IC, the specification must clearly indicate the cell design and the test strategy required to make best use of it. Further, if the cell can be damaged by inappropriate test sequencing, chip specifications should include a clear and conspicuous warning of this fact.

† Note that such problems can occur on any board, with or without boundary-scan., They will be encountered equally frequently in cases where functional or power-up in-circuit tests are applied without prior screening of a board for short-circuit faults (e.g., by using a manufacturing defects analyzer).

Chapter 18. Integrating Internal Scan Paths

Colin Maunder
British Telecom Research Labs
Martlesham Heath, Ipswich IP5 7RE, U.K.

At first sight, integration of the scan path for the internal (system) logic of a component with the IEEE Std 1149.1 test logic appears straight-forward. The internal scan path could, for example, be connected into the test logic as a user-defined test data register so that it could be accessed through the test data input (TDI) and test data output (TDO) pins when an appropriate instruction was present. Further examination, however, shows a number of technical, commercial, and logistic problems. The objective of this chapter is to discuss these problems and to show how they can be resolved.

18.1: Problems at the Chip Level

There are two key technical problems at the chip level.

18.1.1: Dead States

The state diagram for the test access port (TAP) controller includes at least three dead states (states in which no activity occurs) between completion of the inward shifting of a test pattern and the time when the results of the test are captured into the shift-register path from its parallel inputs. In the optimum case, the controller must sequence through *Exit1–DR*, *Update–DR*, and *Select–DR–Scan* as shown by the highlighted path in Figure 18–1.

The need to cycle through these dead states has several effects:

1. There is a marginal impact on test length, because three additional clocks need to be applied for every scan cycle.

2. The scannable registers used to build the internal scan path must have the ability to enter a "hold" state, in addition to "shift" and "load." This will increase the size of the registers in cases where the "hold" operation is not needed for normal system operation, as shown in Figure 18–2.

3. Faults that cause increased propagation delays through the combinational logic between the scannable registers cannot be detected. While the larger delays may cause failure when the capture clock occurs immediately following the last shift clock (as would be the case for conventional scan testing), it is extremely unlikely that the increased delay will span three clock cycles.

EH0321-0/90/0000/0191$01.00 © 1990 IEEE

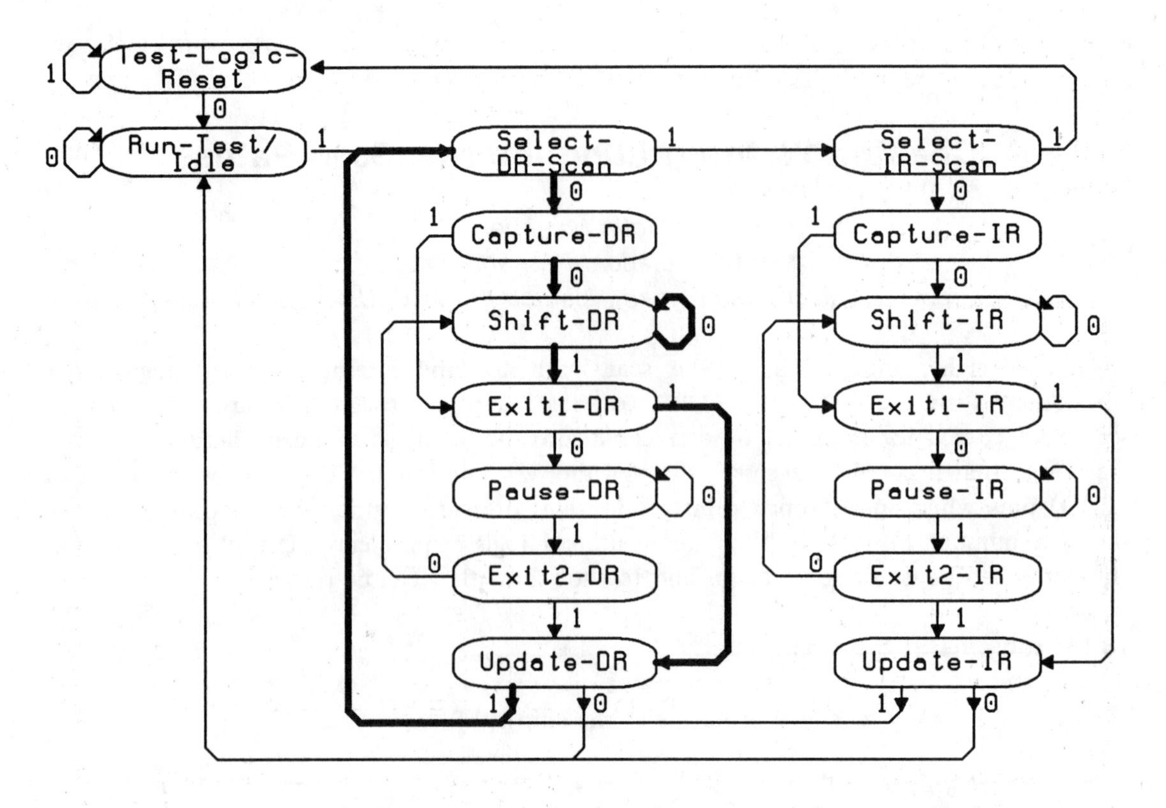

Figure 18–1: TAP controller state diagram.

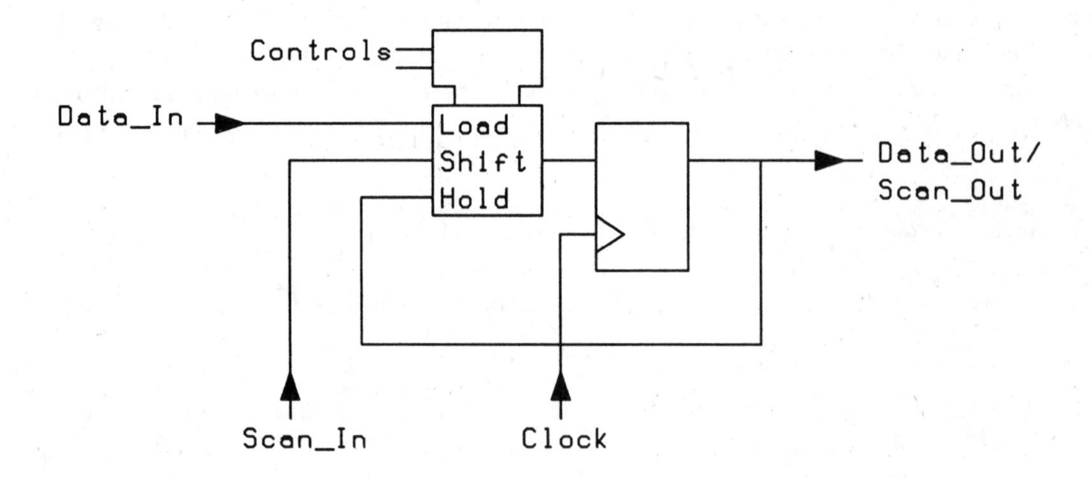

Figure 18–2: Provision of a hold mode on a scannable register.

18.1.2: Multiple Scan Paths

At the integrated circuit level, it is common to connect scannable registers within a component into several independent scan paths, each of which has its own serial input and output connections. Such use of multiple scan paths in a chip allows test times to be reduced, since the number of clocks–per–scan–cycle is reduced. It also allows a scan implementation to be achieved at lower cost in cases where several different clocks are used within the chip design. For example, all registers controlled by clock CK1 may be connected into the first scan path, those controlled by clock CK2 into the second, and so on (Figure 18–3). This avoids the need for clock signals to be switched between the chip's normal and test modes of operation.

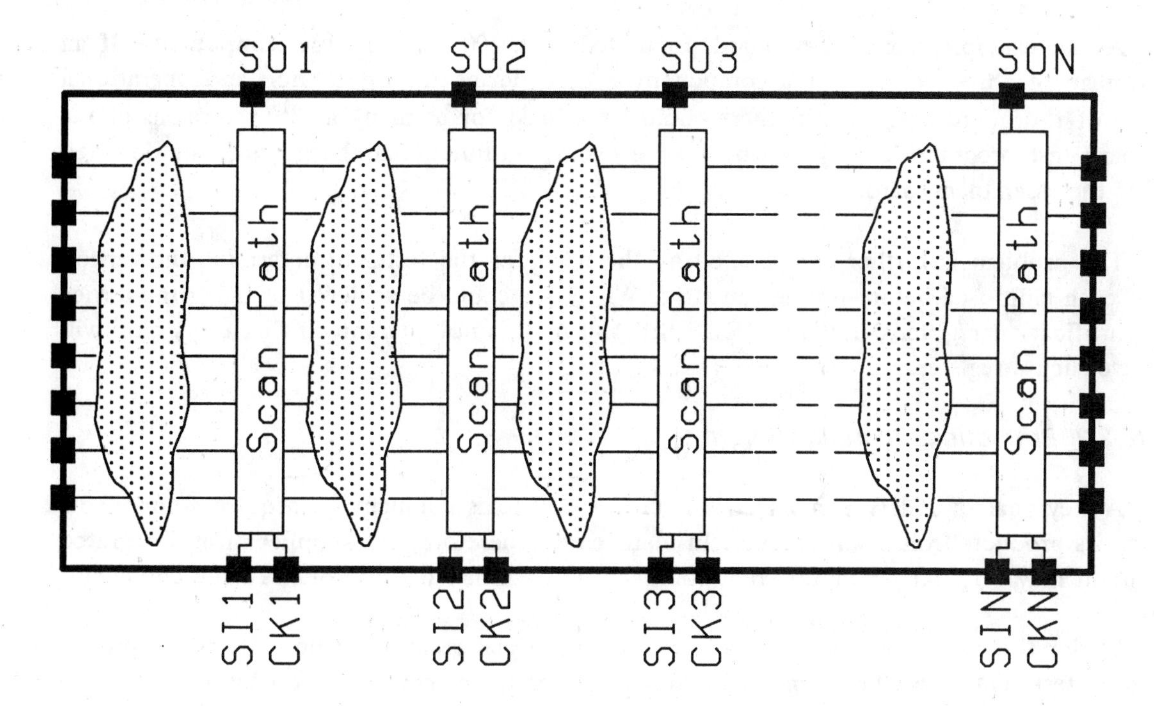

Figure 18–3: Use of multiple scan paths.

In contrast, IEEE Std 1149.1 dictates a single serial path between TDI and TDO. If the internal scan paths were to be accessed through the test access port, they would need to be connected in series and controlled by a common clock during testing.

18.2: Problems at the Board Level

Logistic and commercial problems come to light at the board level, as discussed below.

18.2.1: Volume of Test Data

An internal scan path may contain many hundreds of shift–register stages. To permit use of the scan test once the chip has been assembled onto a board, this would need to be extended by the chip's boundary–scan path and (at a minimum) the bypass registers for all other chips on the board–level serial path.

The volume of test data to be stored on an automated test equipment (ATE) system for a board populated entirely with scannable chips would therefore be very large.

18.2.2: Version Dependence

The scan test data for a component will change from one version of the design to the next due to the impact of design modifications. Because of this, severe logistic problems could result through the use of internal scan test data at the board level.

As an example, consider a board populated with 100 scan testable components. If an average of three versions of a component is used during the production and operational life of the board design, then there could be a need for as many as 3^{100} versions of the board test program. This need would arise even if no functional change could be detected for the assembled board.

The problem is further complicated by the need for the tester to determine the version of each chip used before testing can start. While this could be achieved by using the device identification register defined by IEEE Std 1149.1, it is not guaranteed that every chip will have such a register.

18.2.3: Protecting Proprietary Information

A key goal of IEEE Std 1149.1 is to allow a systems company to acquire components for its products from many sources. In many cases, therefore, the supplier of an integrated circuit may very likely be a different company (not another division of the same company).

Under these circumstances, there may be commercial issues that limit the availability of scan test data to the component purchaser and, in consequence, limit the board assembler's ability to test the loaded board. For example, it may be possible (with effort) to create a copy of the original design by examining the scan test data (assuming that fault coverage is high).

18.3: A Solution

A solution to these problems is to combine scan testing (for use by the chip manufacturer) and self-test (for use by the purchaser).

18.3.1: The Chip Level

The objective of IEEE Std 1149.1 is to ensure that integrated circuits from multiple vendors can cooperate during the process of testing a loaded board. As long as the standard can be met with regard to the operation of the defined test features (e.g., the boundary-scan path, the instruction register, and the bypass register) and with regard to any other test feature that is to be offered for "public" use, there is no reason why additional "private" test features should be designed while fully complying with the standard.

Bearing this in mind, a solution to the use of scan test techniques for stand-alone integrated circuit testing can be obtained. First, the instruction register can be used as a means of selecting scan test operation of the integrated circuit. Second, the manufacturer can provide a **private** *SCANTEST* instruction for this purpose†. Entry of this instruction would be done in accordance with the operation of IEEE Std 1149.1.

When the *SCANTEST* instruction is present, certain states of the TAP controller can be redefined as shown in Figures 18–4 and 18–5:

1. *Exit1–DR* and *Exit2–DR* cause data to be captured into the scan paths in the same way as would normally occur in the *Capture–DR* controller state.

2. *Pause–DR* causes data to be shifted in the same way as *Shift–DR*.

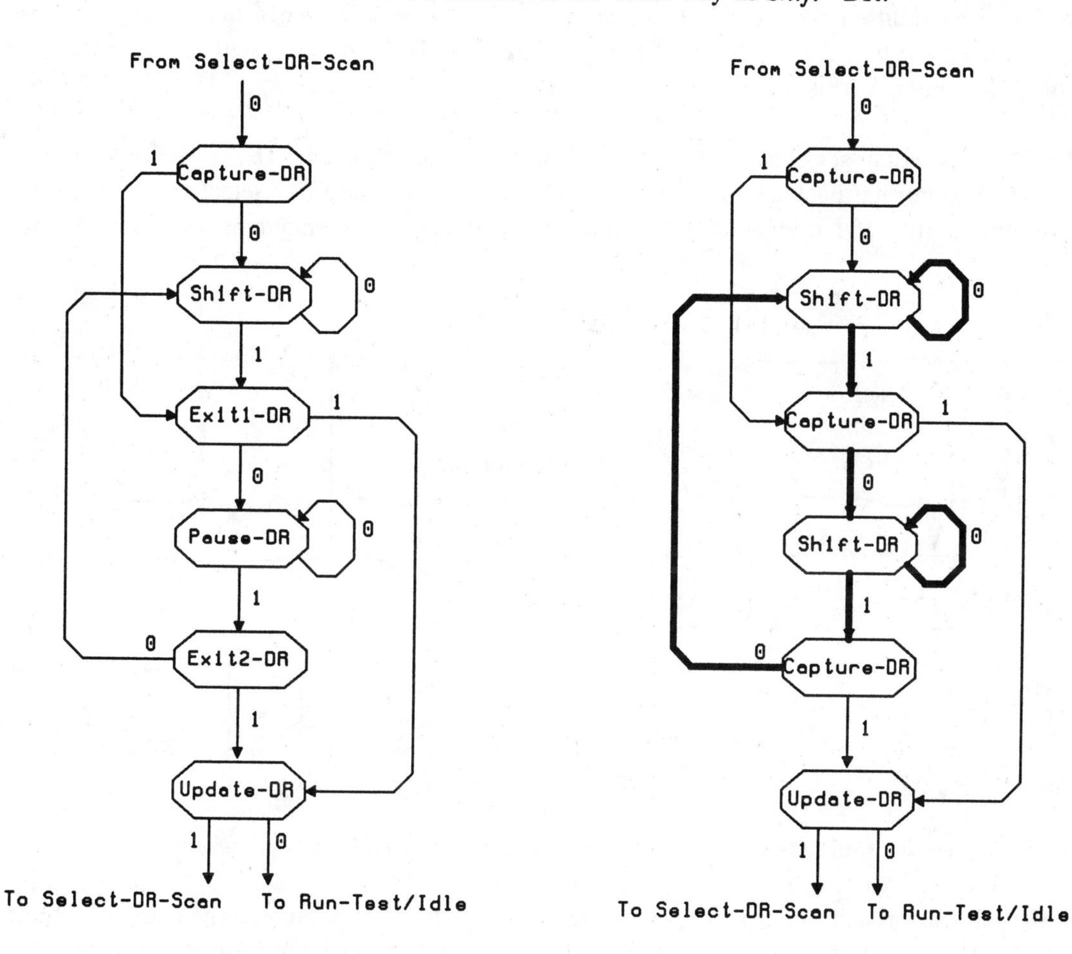

Figure 18–4: Basic scan state diagram.

Figure 18–5: Scan state diagram when SCANTEST is selected.

Note that the effect of this redefinition is to make test mode select (TMS) (which controls movement between controller states) almost equivalent to the test mode control

† The *SCANTEST* instruction described could not be offered for use by the component purchaser, because it requires dynamic alteration of the TAP controller state machine and, thus, does not conform to IEEE Std 1149.1.

for a conventional scan circuit (which causes movement between "shift" and "load") while the *SCANTEST* instruction is selected. When TMS = 1 (which lasts for only one clock cycle), data are loaded into the scan path and while TMS = 0 data are shifted. The bold paths in Figure 18–5 show the cycle that would be followed during scan testing. In contrast to the conventional operation of IEEE Std 1149.1, the transition from "shift" to "load" (i.e., capture) can be effected without leaving the data register scan states of the TAP controller. Therefore, there are no dead states and the need to provide "hold" operation on the scannable registers is avoided.

In components that have a single internal scan–path, the TDI and TDO pins could be used for "scan–in" and "scan–out." Note that this would require the control of the TDO driver to be modified to allow it to be active in the *Pause–DR* controller state (redefined to behave as *Shift–DR*) whenever the *SCANTEST* instruction is present -- normally, it would be inactive in this state.

Multiple internal scan paths can be provided by multiplexing the serial inputs and outputs onto normal package pins when TMS = 0 and *SCANTEST* is selected. At outputs, this requires that the design of the boundary–scan cell is extended as shown in Figure 18–6.

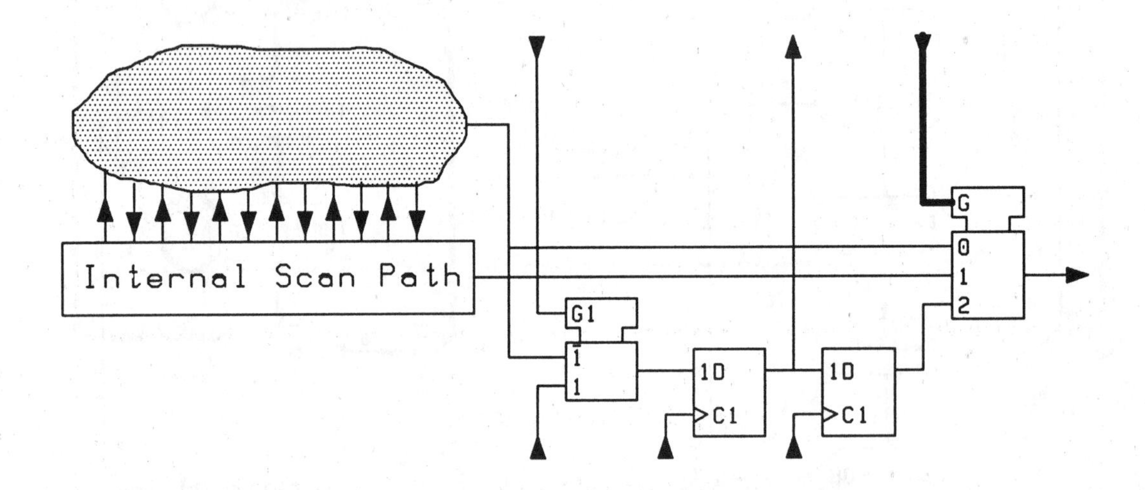

Figure 18–6: Multiplexing of an internal scan path onto a system output pin.

Using this approach, internal scan testing can be achieved in almost the same way as it would without IEEE Std 1149.1. **Again, it must be emphasized that the *SCANTEST* instruction is intended only for the private use of the integrated circuit manufacturer because it does not comply with IEEE Std 1149.1.**

18.3.2: The Board Level

It is clear from the problems highlighted earlier that internal scan testing is not the ideal basis for an hierarchical chip–through–system test approach. Further, if the solution just proposed is adopted to allow stand–alone scan testing of the integrated circuit, then there may be problems in reusing the scan test data at the board level. For example, the multiple scan paths would need to be connected into a single path to allow access through the TDI and TDO pins.

These problems can, however, be overcome by combining the internal scan design with self-test facilities. IBM's LSSD† on-chip self-test (LOCST) approach [2], for example, shows how linear feedback shift-registers and signature analyzers can be used to convert a scan/boundary-scan design into a self-testing circuit at moderate cost (Figure 18-7). The paper also shows how a self-testing circuit can be created when multiple internal scan paths are used.

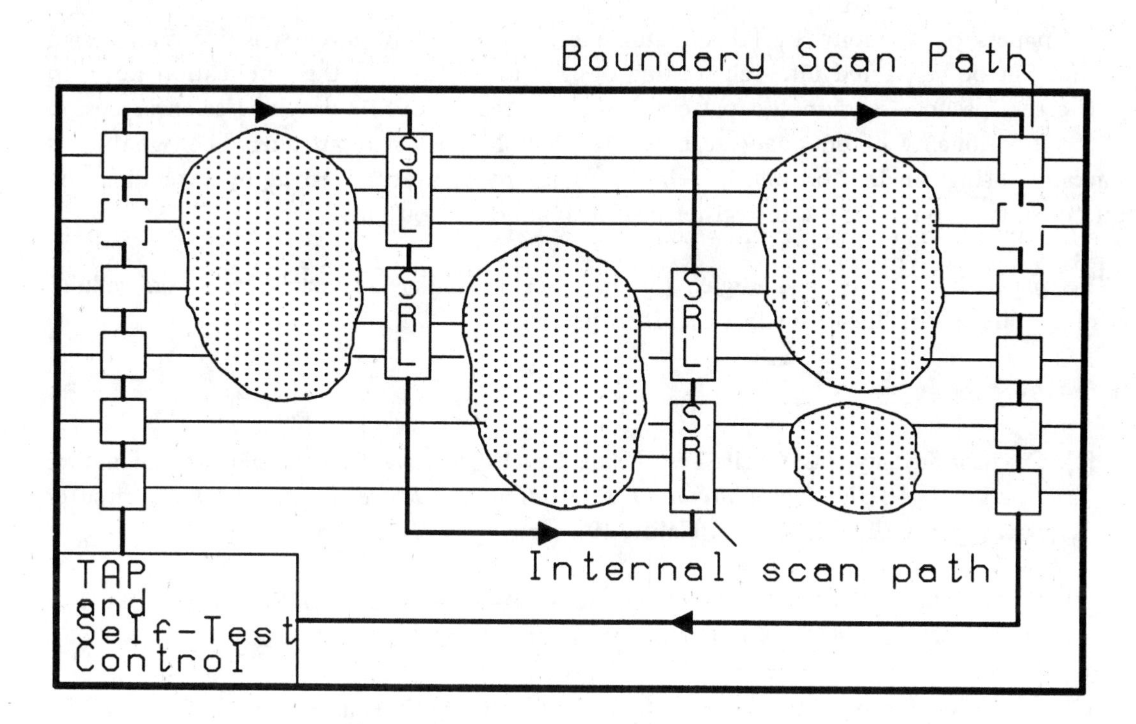

Figure 18-7: The LOCST scan and self-test approach.

The combination of scan and self-test allows the requirements of both the integrated circuit manufacturer and the component purchaser to be met. The manufacturer can use the scan test facilities through the **private** *SCANTEST* instruction as defined in the previous section; the purchaser can use the self-test operation through a **public** *RUNBIST* instruction.

From the purchaser's viewpoint, this has the following advantages:

1. The test is compact and can be run easily when the component is mounted on the board. For example, there is no need to store and shift large amounts of test data.

† Level-sensitive scan design (LSSD).

2. The manufacturer can arrange that all versions of a component will yield the same results from self-test execution. Where linear feedback shift-registers (LFSRs) are used to produce the self-test signature, this can be achieved by choosing the initial state of the linear-feedback shift-register (LFSR) such that the final state will be the required constant value.

The use of self-test to provide a manufacturer-supported test of the component will also allow the manufacturer to limit access to detailed design information.

18.4: Further Reading

The papers by Komonytsky [1] and LeBlanc [2] discuss how level-sensitive scan design circuits can be converted into self-testing designs as outlined in this application note. In both cases, pseudo-random test patterns are generated by LFSRs provided as an extension to the functionality of boundary-scan register cells located at input pins. The signature is generated using single- or multiple-input signature analyzers formed by extending the functionality of boundary-scan register cells at component outputs.

The paper by Gloster and Brglez [3] discusses a similar approach based on cellular automata instead of linear-feedback shift-registers.

18.5: References

[1] D. Komonytsky, "LSI Self-Test Using Level-Sensitive Scan Design and Signature Analysis," *IEEE International Test Conference Proceedings*, IEEE Computer Society Press, Los Alamitos, Calif., 1982, pp. 414-424.

[2] J.J. LeBlanc, "LOCST: A Built-In Self-Test Technique," *IEEE Design and Test of Computers*, Vol. 1, No. 4, Nov. 1984, pp. 45-52.

[3] C.S. Gloster and F. Brglez, "Boundary-Scan with Built-In Self-Test," *IEEE Design and Test of Computers*, Vol. 6, No. 1, Feb. 1989, pp. 36-44.

Chapter 19. Testing Mixed Analog/Digital ICs†

J. Hirzer
Siemens AG
Munich, West Germany

This chapter discusses the design and use of boundary-scan in mixed analog/digital integrated circuits.

While the prime thrust of the boundary-scan path defined by IEEE Std 1149.1 is to reduce the complexity (and hence the cost) of testing miniaturized digital circuits, there are also benefits to be gained through provision of such a path in mixed analog/digital circuits. Test costs can be high for such designs unless design-for-test features are included and, as will be described in this chapter, boundary-scan can be a valuable tool for simplifying the creation and application of parametric and functional tests.

19.1: The Location of the Boundary-Scan Path

In mixed analog/digital integrated circuits the boundary-scan path must be designed to visit each purely digital pin -- other than the test access port (TAP) pins -- and each digital signal received from, or supplied to, the analog block within the design. Figure 19-1 illustrates this in a component that contains a large digital block and an A-to-D converter.

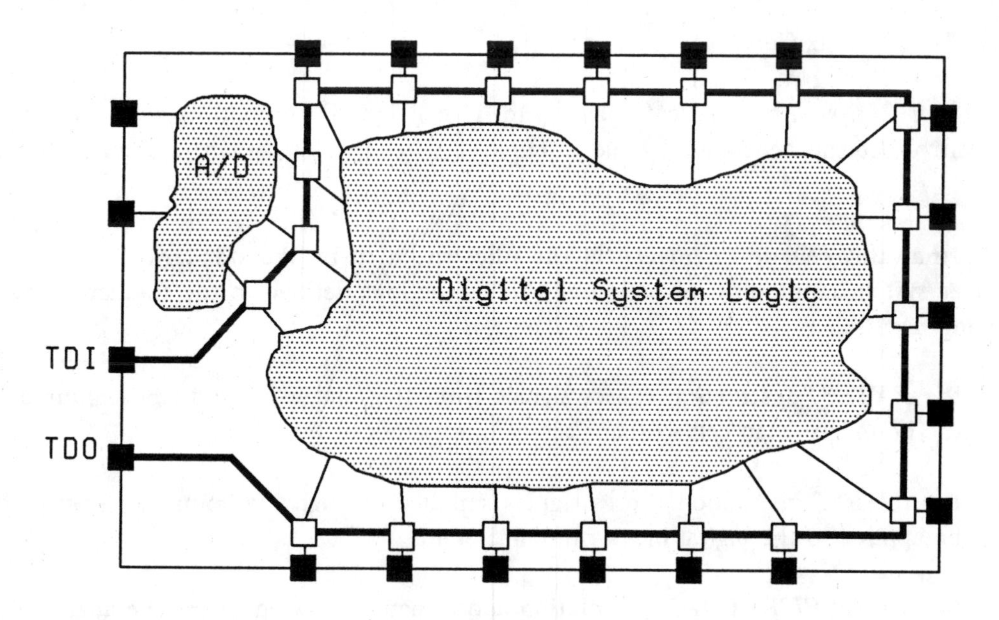

Figure 19-1: The location of a boundary-scan path at the analog/digital interface.

† The work described in this chapter was performed within the AIDA collaborative project of the ESPRIT research programme, supported by the Commission of the European Community.

The provision of access to the analog/digital interface separates the analog and digital blocks and allows them to be tested individually using the test techniques and strategies best suited to the block designs:

- tests for the digital block can be performed without having to propagate signals through the analog block

- the analog block can be tested functionally without having to propagate signals through the potentially complex digital block.

Some of the complexity in testing a complete mixed-signal integrated circuit arises due to the tolerances inherent in the A-to-D converter. Due to these tolerances any given voltage applied at the analog input can give rise to one of a range of digital codes at the converter's outputs. During testing, such uncertainty in the pattern applied to the digital circuit block is difficult to accommodate (digital testing requires precise knowledge of the pattern being applied at any test step).

19.2: Boundary-Scan Cell Design

Two types of boundary-scan cell are required at the analog/digital interface: an A-to-D type and a D-to-A type. The A-to-D cell is placed at the analog/digital boundary on any unidirectional digital signal that feeds from an analog block into a digital block, while the D-to-A cell is placed on any signal from a digital block that feeds into an analog block. The design of each cell type, and the reasons for differences between them, are discussed below.

19.2.1: The A-to-D Cell

Figure 19-2 gives a schematic for an A-to-D cell that is compatible with the clocking and control scheme generated by the example TAP controller shown in Figures 4-8 and 4-9.

Note that this cell design meets all the rules specified by the standard for cells to be placed at system input pins. Selection of the instructions defined in the standard gives the following results:

1. the *EXTEST* instruction causes signals from the analog block to be captured into the boundary-scan cell so that they can be examined by shifting;

2. the *INTEST* instruction causes signals supplied through the boundary-scan path to be applied to the digital block on-chip; and

3. the *SAMPLE*/PRELOAD instruction allows signals flowing across the analog/digital boundary to be examined without interfering with normal circuit operation.

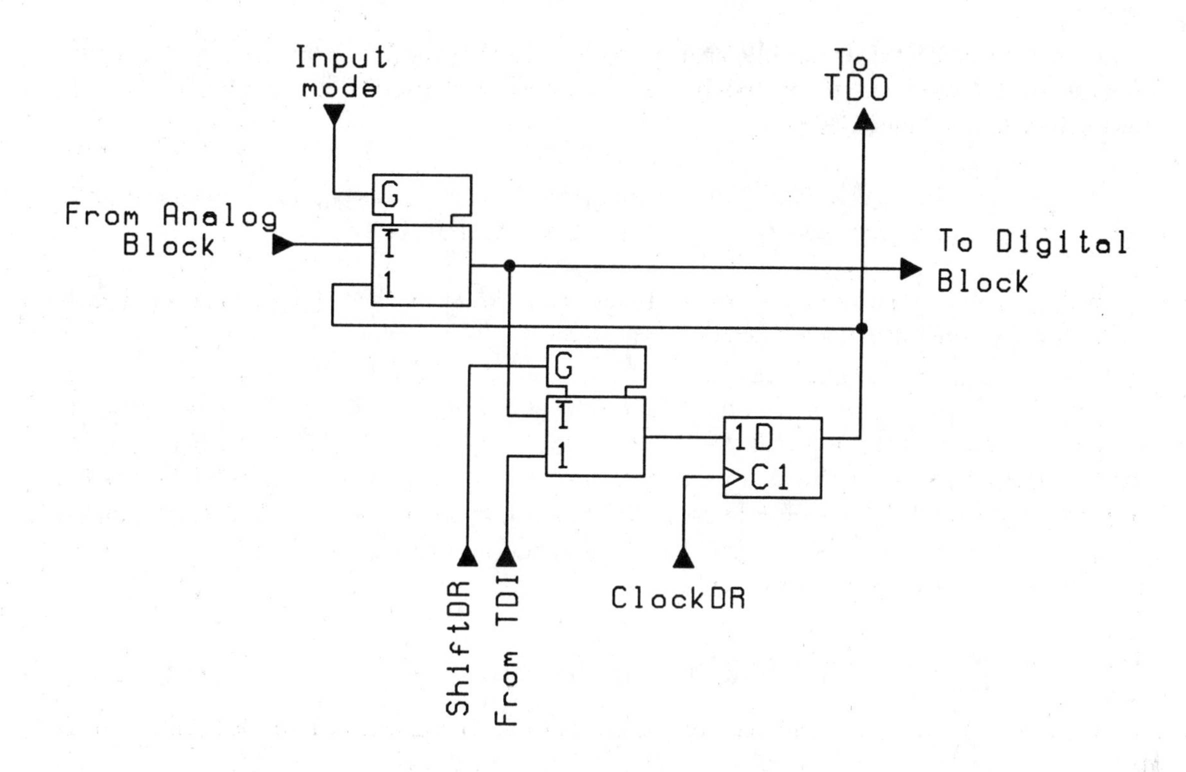

Figure 19–2: An A–to–D cell.

Note that no provision is made in the example cell design to prevent the signals (applied to the digital block when the *INTEST* instruction is selected) from rippling as data are shifted into or out of the boundary–scan path. If such rippling signal values were likely to cause unwanted operation of the digital block (e.g., because they were fed to asynchronous or clock inputs), then additional holding latches or registers would need to be provided at the parallel output from the shift–register stage.

19.2.2: The D–to–A Cell

In general, it is impossible to preserve the state of an analog circuit if the signals at its inputs are allowed to change. This situation also occurs for some digital circuits (for example, asynchronous state machines), but not for others (for example, synchronous sequential circuits change state only when clocked; changes at data inputs between clocks have no effect).

Consequently, it is necessary to ensure that no interruption occurs to a test sequence and that input signals applied to an analog block change only from one valid value to another. Therefore, during shifting of the boundary–scan register, it is vital that the signals driven to the analog/digital block from D–to–A cells do not ripple. The value at the cell's data output must be held until the shift operation is complete.

The requirements for the D–to–A cell design are identical to those of boundary–scan cells to be placed at 2–state digital pins of the component. A suitable design is shown in Figure 19–3.

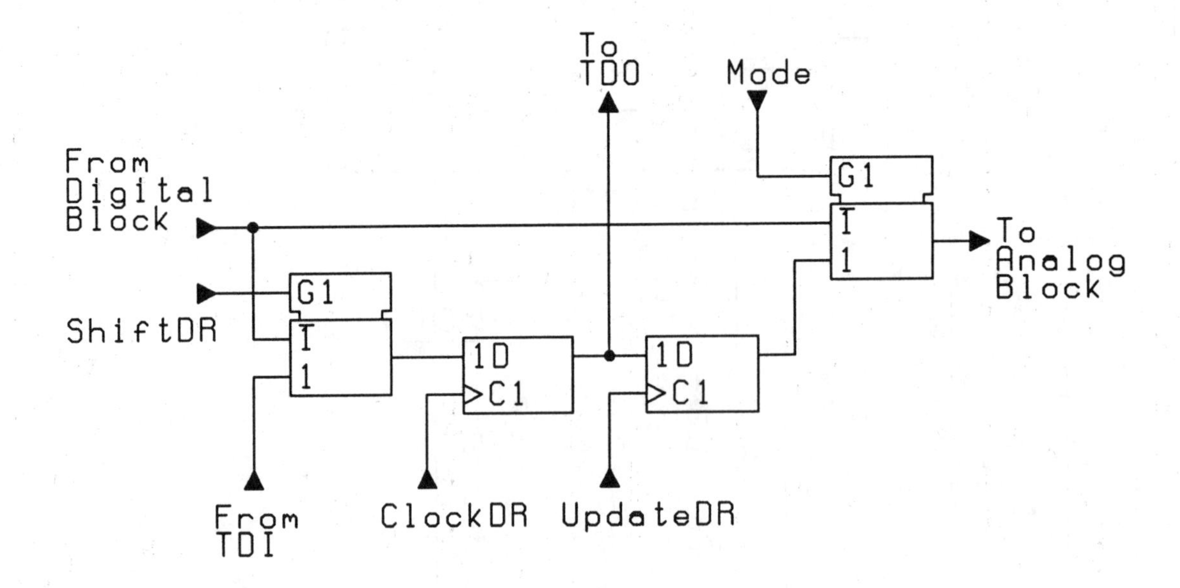

Figure 19–3: A D–to–A cell.

Selection of the instructions defined in the standard gives the following results:

1. the *INTEST* instruction causes signals from the digital block to be captured into the boundary–scan cell so that they can be examined by shifting;

2. the *EXTEST* instruction causes signals supplied through the boundary–scan path to be applied to the analog block; and

3. the *SAMPLE*/PRELOAD instruction allows signals flowing across the analog/digital boundary to be examined without interfering with normal circuit operation.

19.3: Testing Analog Blocks Using Boundary–Scan

This section gives an example of how a test on an analog block in a mixed analog/digital integrated circuit design can be achieved by using a mixture of direct connections through the chip pins and indirect connections through a boundary–scan path placed at the analog/digital interface.

The example is based on the use of digital signal processing (DSP) test techniques [1] in which analog signals applied to the circuit are generated by digital programmable function generators, and those received from the circuit are analyzed by using DSP techniques.

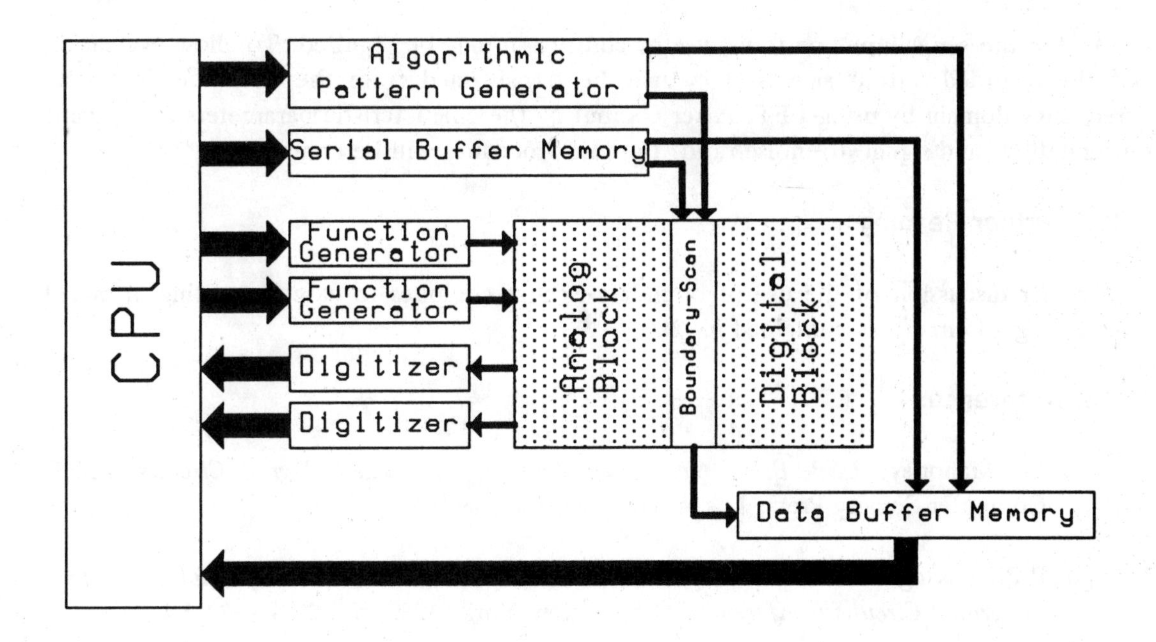

Figure 19–4: Test system configuration for analog test.

Figure 19–4 shows the design of a test system that couples DSP test techniques with support for the boundary-scan path. With this test system, dynamic analog measurements are performed in the time domain by application of continuous waveforms or digital pattern sequences. In some cases (e.g., filters), it is useful to describe the analog function in the frequency domain, in which case the inverse fast Fourier Transform (IFFT) technique can be used to create the test signals. The function generators can be loaded with the array of real numbers representing the amplitude values of the required waveform at discrete points in time.

The serial buffer memory is used to scan digital vectors onto the digital/analog interface. Each of these vectors is related to a single point of time and to one distinct input signal. Clearly, the boundary scan path at the analog/digital interface may be split up logically into one or more input and output vectors. Also, layout optimization and other requirements may lead to scrambling of the boundary-scan register cells, so test vector conversion may be required to map the test vectors onto the actual structure of the boundary scan path. For example, the order bits at a digital-to-analog converter's input lines may be reversed such that it is not consistent with the binary number representation of the fast Fourier transform (FFT) and inverse FFT (IFFT) operations of the array processor.

Analog measurements are performed in a similar manner. Analog output signals are digitized and recorded in the digitizer's random-access memory (RAM) and the scan-out signal (which is again a serial digital data stream) is selectively analyzed. Only one analog output signal is evaluated at a time, with the binary vector representing this signal being strobed bitwise to the data buffer memory. The bit strobe signal originates from the serial buffer memory that generates the scan-in signal to achieve synchronization.

As for the serial input patterns, vector conversion may be required. To allow evaluation of the recorded output signals, they may be transformed from the time domain to the frequency domain by using FFT. After computing the characteristic parameters (e.g., gain, attenuation, and signal–to–noise ratio) the test procedure will be completed.

19.4: Further Reading

Further discussion of the use of boundary–scan techniques to ease the testing of mixed analog/digital circuits is contained in [2].

19.5: References

[1] M. Mahoney, *DSP–Based Testing of Analog and Mixed–Signal Circuits*, IEEE Computer Society Press, Los Alamitos, Calif., 1987.

[2] P.P. Fasang, "ASIC Testing in a Board/System Environment," *IEEE Custom Integrated Circuits Conference*, IEEE, New York, 1989, pp. 22.4.1–22.4.4

Chapter 20. Adding Parity and Interrupts to IEEE Std 1149.1

Patrick F. McHugh
Electronics Technology and Devices Lab.
US Army LABCOM
Fort Monmouth, NJ 07703, U.S.A.

Lee Whetsel
Defense Systems and Electronics Group
Texas Instruments Inc.
Plano, TX 75086, U.S.A.

Trends within the U.S. Department of Defense (DoD) are forcing system integrators to use industry standard interfaces to fulfill DoD system requirements. The test access port and boundary scan architecture defined by IEEE Std 1149.1 meets many of the requirements set forth by the DoD for the testability of very high-speed integrated circuit (VHSIC) components, with the exception that it does not support parity checking of instructions and test data or the flagging of interrupts to the device controlling test operations (the bus master). These capabilities are considered important in both military and commercial systems where high operational reliability is required.

In this chapter, a method is proposed whereby parity checking and interrupt capability can be provided within the framework of IEEE Std 1149.1 that could be implemented to meet the DoD's requirements.

20.1: Introduction

The principal application of the proposed parity checking method is verification of instruction data input to a component's test logic. The parity checking scheme can be extended to cover test data input and output data from the test logic. Both these applications will be discussed in this chapter.

There must be a means of flagging parity errors to the data source (i.e., the bus master or automated test equipment (ATE)). To achieve this, an additional signal must be added to the test access port defined by the standard. The relationship between parity coding of instruction data and the additional interrupt signal, the test interrupt (TINT*), is also described. A method for using the TINT* signal to flag other types of error to the master device is also proposed.

The cost of implementing the proposed extensions to the standard is modest. A small amount of logic must be added to the instruction register and instruction codes must be extended to include a parity check bit. Finally, one signal, TINT*, must be added to the test access port.

20.2: Why Use Parity?

Where highly reliable operation is required, the coding of instructions and data (for example, by adding a parity bit) is a valuable means of detecting any data corruption involving an odd number of bits that might arise during transmission.

For example, in Figure 20–1 a circuit is shown that consists of two slave devices (each compatible with IEEE Std 1149.1) interfaced to a test bus master controller. When the master transmits information (e.g., an instruction) to the slave devices, it assumes this is correctly received. However, in actual practice there could be corruptions to the serially–transmitted information caused by external electrical or mechanical interference. When the corrupted information is acted on by the receiving slave, an incorrect test operation may be performed, possibly causing malfunction of the complete system. For example, a transmitted *BYPASS* instruction could be corrupted into an instruction that would cause the slave to enter a self–test mode of operation.

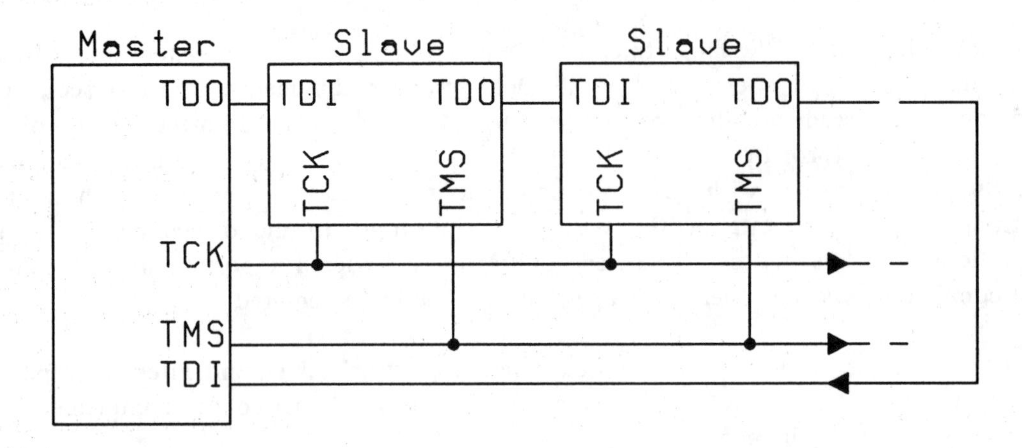

Figure 20–1: A basic master–slave system.

By adding a parity check bit to each item of information transmitted, the slave devices can check the received data for corruptions that effect odd numbers of bits (e.g., single bit errors). Figure 20–2 shows how the master–slave system of Figure 20–1 could be enhanced to allow parity encoding and checking.

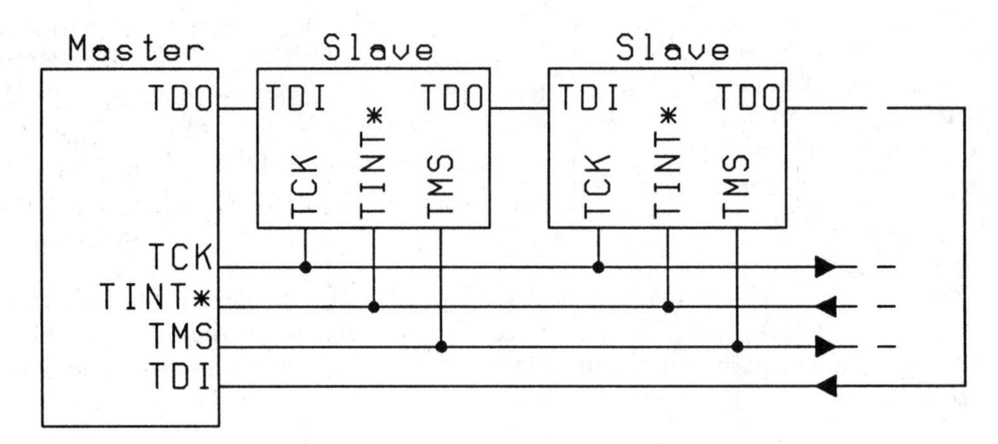

Figure 20–2: A master–slave system with parity coding and interrupts.

In Figure 20–2, each slave device has been enhanced to include parity checking of received instructions. Detected errors are flagged back to the bus master by using an additional interrupt signal, TINT*. In the example, the TINT* signals from the two slave devices are capable of wire–AND operation, so the master receives an interrupt when an error is detected by either slave. Other methods of combining the interrupts generated by the various slave devices controlled by a single master could also be used (e.g., a priority encoder).

By the means illustrated, the master receives confirmation that valid information has been received by the slave(s). When an error is detected, the information transmission can be repeated until the information is transmitted and received correctly.

20.3: Adding Parity to Instructions

The most critical information items transmitted to a component compatible with IEEE Std 1149.1 are the instructions which control the operation of the test logic and, in particular, the way that the test logic can alter or impede the operation of the on–chip system logic. Transmission errors can convert a public instruction into a public one or a *BYPASS* instruction into *RUNBIST*. In some cases, such a change can have a significant impact the functional integrity of the complete board.

This section proposes changes to the test logic defined by the standard which will permit parity coding of instructions. Enhancements to this basic scheme will be introduced in later sections.

*20.3.1: The Test Interrupt Signal, TINT**

As mentioned earlier, a signal must be added to the test access port to allow a slave component to bring parity errors to the attention of the master.

TINT* should be an active–low output from the test logic capable of wire–AND connection to the TINT* outputs of other components that offer parity checking. TINT* would normally be held at 1, and should be set to 0 when a parity error is detected. When the TINT* outputs of several components are wired together, the resultant signal should be 0 whenever any of the connected components sets its TINT* output to 0. To allow different technologies to be used, the output characteristics of TINT* must be defined to allow an output buffer to be connected to buffers constructed in different logic technologies (e.g., open collector or emitter–coupled outputs).

20.3.2: The Instruction Register and Instruction Coding

IEEE Std 1149.1 requires that:

- the *BYPASS* instruction must have the all–1s value; and

- the *EXTEST* instruction must have the all–0s value.

Because the parity bit must be decoded as part of the instruction, it must be a 0 in the encoded *EXTEST* instruction. This implies that even parity is required. Again, because it will be decoded as part of the instruction, the parity bit must be a 1 in the encoded *BYPASS* instruction. If we add to this requirement the necessity of even parity, we find that an instruction register *with parity* must have a total length that is an even number of bits. The unencoded instruction (without the parity bit) must be an odd number of bits in length (Figure 20–3).

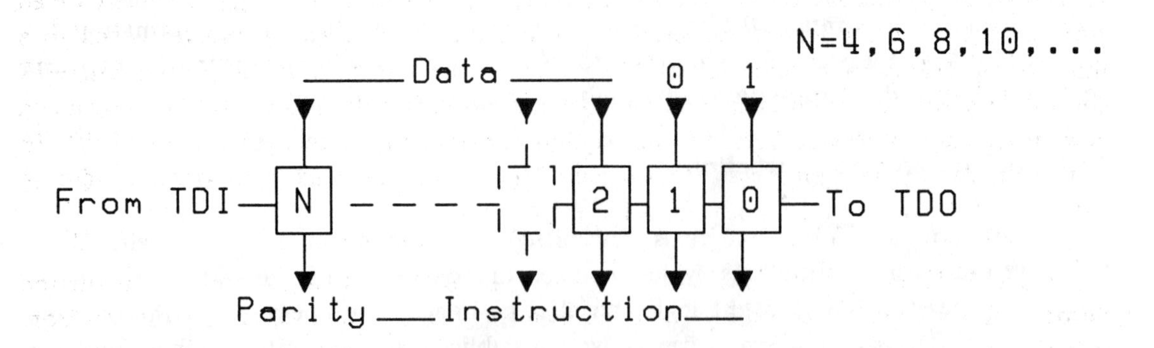

Figure 20–3: Instruction register with parity bit.

Note that it is proposed that the parity bit is the most significant bit of the complete instruction (i.e., that it is the last bit shifted into the chip). Note also that, where parity coding is used, the minimum length of the instruction register is four stages. This is because a component must support at least three instructions if it is to conform to the standard: *BYPASS, EXTEST,* and *SAMPLE/PRELOAD.*

Figure 20–4 gives an example implementation for an instruction register that includes even parity detection logic and associated interconnections. This figure does not show the connection of the clock and control signals to the shift–register stages.

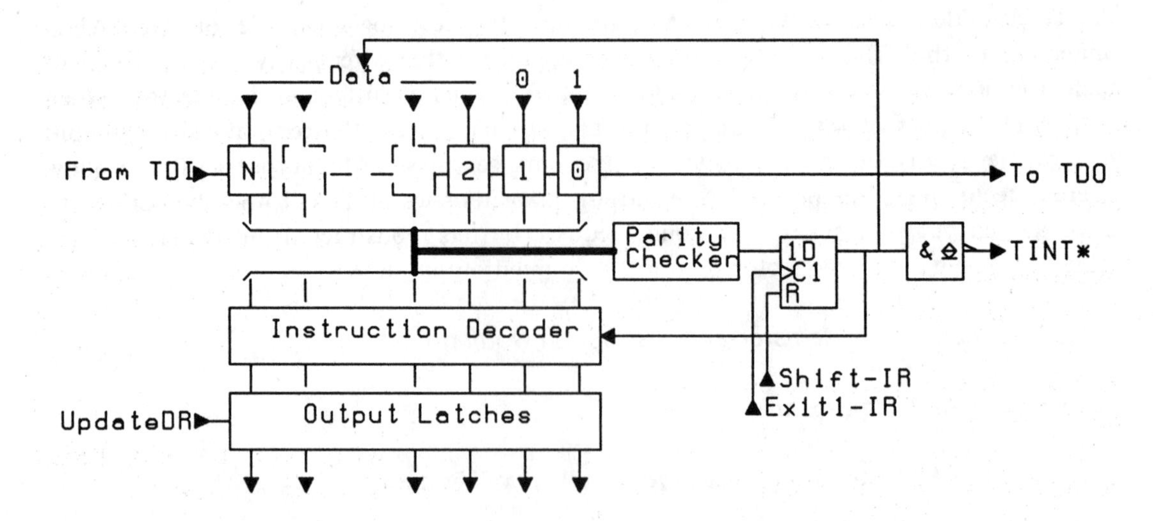

Figure 20–4: An enhanced instruction register design.

An instruction register that complies with the standard consists of a shift-register section which is connected between the test data input (TDI) and test data output (TDO) pins, logic to decode the received instruction, and an output holding latch that retains the previous instruction until a new instruction has been latched in the *Update–IR* controller state.

To achieve even parity checking, a parity checker must be connected between the shift-register stages and the instruction decoding logic. This checker might consist of an array of exclusive-OR gates, for example. The output of the parity checker is fed to a flip-flop that is clocked only on the rising edge of the test clock (TCK) in the *Exit1–IR* controller state. This latch is included so that the resulting output does not change while a new instruction is shifted into the instruction register. It is reset when in the *Shift–IR* controller state. The design should also cause TINT* to be released to its inactive state on entry into the *Test–Logic–Reset* controller state.

The parity latch output must feed both to the TINT* output and to the instruction decoding logic. This latter connection is necessary to prevent the received instruction from being applied to the test data registers in the event that a parity error is detected. It is recommended that, when a parity error is detected, the instruction decoder's output should be forced to the state that would normally result when the *BYPASS* instruction is received. In this way, a non-damaging instruction is presented to the test data registers if the user should cause the test logic to pass through the *Update–IR* controller state.

Additionally, the output of the parity checker may be latched and fed back to a data input of the instruction register, thus allowing the bus master to interrogate slave components to determine which had received the corrupted instruction. Note that, to allow the parity flag to be examined, it is necessary to move through the *Update–IR* controller state to the *Capture–IR* state where the data will be loaded into the instruction register.

If desired, the scheme shown in Figure 20–4 can be extended to allow masking of interrupts. In this case, instructions for enabling and disabling interrupts must be provided. The addition of these two instructions will not increase the length of the instruction register beyond the size required by the three mandatory instructions. Since the minimum length for the instruction register with parity is four stages, there are unused opcodes when the minimum instruction set is implemented. Note that, if the capability to mask interrupts is provided, it must be possible to read the state of the interrupt mask register within the component. This could be achieved in a number of ways, for example, by making it a user-defined register in the standard architecture.

20.3.3: A Typical Operating Sequence

The sequence of operations for a IEEE Std 1149.1 interface with parity coded instructions and a non-maskable interrupt for a parity errors is described below.

The bus master would first control the test mode select (TMS) and TCK signals to initiate scanning of the instruction register. When the *Capture–IR* controller state is entered, the required "10" pattern is loaded into the least significant bits of the instruction register. When a design allows the state of the parity checker output latch and the status of the interrupt mask to be loaded, then this will also occur in the *Capture−IR* controller state.

The captured data are shifted out through TDO during the *Shift−IR* state while an instruction with even parity is shifted in through TDI. Note that the data being shifted out through TDO (captured during *Capture−IR*) need not be parity coded -- in fact, they may have odd parity if they happen to be data from a failed attempt at loading an instruction. When shifting of the captured data and new instruction is completed, parity of the received instruction will be checked. If the received data does not have even parity, TINT* will be asserted (i.e., set to 0).

The bus master would typically cause the slave components to enter the *Pause−IR* controller state to allow time to sample its TINT* input and determine its next action. If a parity error is detected, the master can return the slaves to the *Shift−IR* controller state, whereupon the TINT* signal will be released. The instruction sequence can be transmitted to the slaves again.

If a parity error has been detected and the slaves are moved into the *Update−IR* controller state, then the instruction decoder will be forced to operate as if the *BYPASS* instruction had been received. The *BYPASS* instruction is a safe default instruction in the event that a parity error cannot be corrected by repeated transmission. The recovery action by the master, in this case, would require initiating the instruction scanning sequence again. During this second instruction scan operation, the test bus master would be able to identify which slaves were flagging the parity error.

20.4: Extending Parity to Received Test Data

The parity scheme can, dependent on the application, be extended to include test data received by a component as well as instructions. As for the instruction register, the most significant bit of the test data register would be required to be the parity bit. In contrast to the instruction register, no matter what the implementation details there is no requirement for the length of the test data registers to be even for the restriction to even parity coding. It is, however, suggested that even parity coding is used since this maintains consistency with the instruction register.

The operation of the component when shifting in parity coded test data would be analogous to that described above for encoded instruction loading. If a parity error were detected, TINT* would be asserted in the *Exit1−DR* controller state and released in the *Shift−DR* or *Test−Logic−Reset* controller states. Note that, in contrast to the instruction register case, the updating of any latched parallel output should be inhibited on entry into the *Update−DR* controller state when incorrectly encoded data are present in the selected test data register. Note that, in general, there will be no guaranteed safe state for the output latches of a data register.

20.5: Parity Coding of Output Data

The previous sections have addressed only the application of parity coding to instructions and test data received by a component. The data output by a component could also be parity coded such that the receiving device could determine its validity. However, there are more limitations to the usefulness of this application than in the previous cases (see below). This would require the most significant bit of the data shifted out to be the parity bit.

An encoder would need to be provided to generate the parity code from the data presented to the other inputs to the instruction register or selected test data register. The encoder's output would then be loaded into the register in the *Capture–IR* or *Capture–DR* controller state, respectively.

A limitation of this scheme is that, unless specific provision is made within a component or in the operation of the master device, it will not be possible to request retransmission of the information should it be found to be corrupt on receipt. This is because it is necessary to terminate the register scanning operation (i.e., enter the *Update–IR* controller state) to revisit the *Capture–DR* or *Capture–IR* controller state.

In the case of the boundary-scan register where the *EXTEST* instruction is used solely to test interconnections to adjacent components compatible with the standard, it is possible to make a repeat attempt to read the test results by feeding the original test pattern in again before exiting from the scan operation through the *Update–DR* controller state. Note, however, that this cannot be done where stored-state circuitry outside the component is being tested, since it is not possible to "undo" previous tests.

20.6: Other Uses of TINT*

In addition to its use to flag errors in received instructions or test data, the TINT* output could be used to indicate other error conditions within the component to a master device. Examples of such "error" conditions might include

- completion of a test task; or
- an abnormal event in the system operation of the component (e.g., memory overflow).

If the TINT* is to be used to flag such interrupts in addition to its use to flag parity errors, then it is important that the master device is able to distinguish between the different types of interrupt being transmitted.

To allow parity errors in received information to be distinguished, it is suggested that specific test access port (TAP) controller states be reserved for this application. While scanning of instructions or test data is in progress (i.e., between *Capture–IR* and *Update–IR*, or between *Capture–DR* and *Update–DR*), TINT* should indicate parity errors in the received data as discussed in the previous sections. In the other controller states (*Test–Logic–Reset, Run–Test/Idle, Select–DR–Scan,* and *Select–IR–Scan*) TINT* may be used to indicate other error conditions. Figure 20–5 shows how this could be achieved by inclusion of a multiplexer.

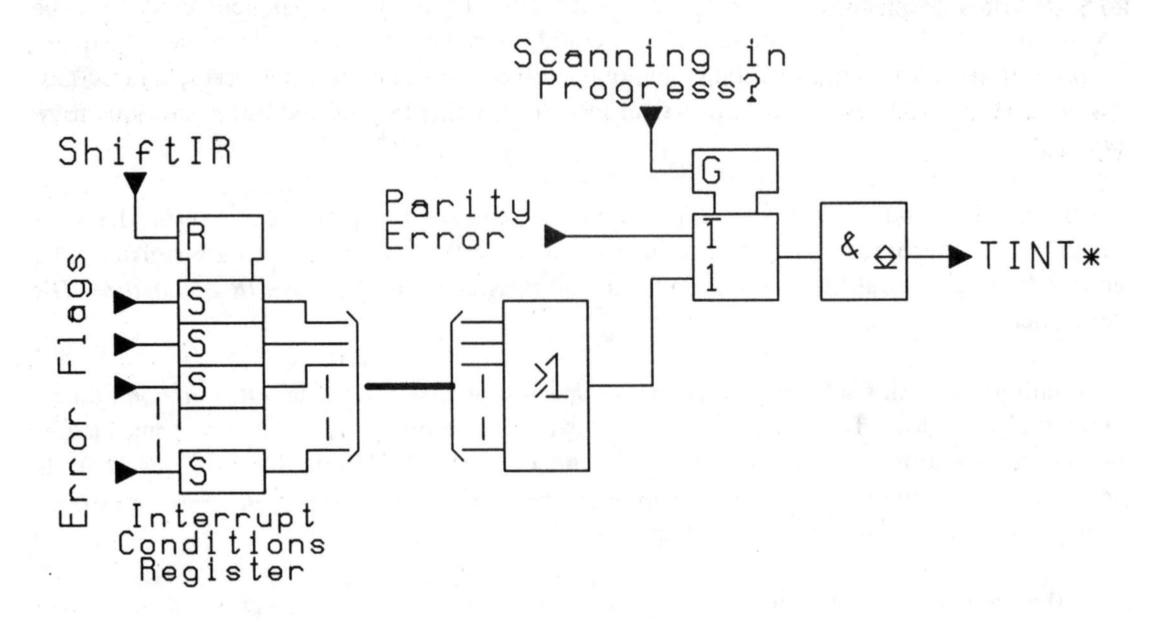

Figure 20–5: Multiplexing of interrupts onto TINT*.

It is also necessary to allow the master to determine which component has generated an error condition interrupt and, where a component is able to flag several error conditions, which specific condition exists. It is recommended that this is achieved by inclusion of an interrupt conditions register whose outputs feed the data inputs of the instruction register. Appropriate data bits in this register would be set when the interrupt was generated and be reset either following the rising edge of TCK in the *Capture–IR* state or when the test logic reset (TRST*) pin is asserted. (If TRST* is not provided, then the interrupt conditions register must reset on power–up).

20.7: Conclusion

The addition of parity checking capability to IEEE Std 1149.1 in the manner described allows a component to quickly check received instructions and, where appropriate, to test data for transmission errors. The TINT* output is added to the TAP to allow the component to notify the bus master that data corruption has occurred. Both these features are added in a manner that ensures that the component design remains fully compatible with the standard.

The TINT* signal can also be used to indicate that other error conditions have occurred, provided that diagnostic information is available to explain why TINT* has been asserted. Interrupts can also be masked if instructions to enable and/or disable interrupts are provided and if it is possible to externally determine the interrupt enable status.

Together, these capabilities will allow DoD systems integrators to use a commercial interface to fulfill DoD requirements for a chip level test interface.

20.8: Acknowledgments

The authors wish to thank Colin Maunder, British Telecom Research Labs, and Chuck Hudson, Honeywell Inc., for reviewing drafts of this chapter and for the suggestions they supplied.

Part V: Further reading

Part V provides an annotated bibliography and contains reprints of papers that either discuss the general topic of boundary-scan or provide specific examples of applications and developments based on IEEE Std 1149.1.

Readers should be aware that several of the reprinted papers discuss boundary-scan testing in general, or applications of the various versions of the Joint Test Action Group (JTAG) proposals that preceded the development of IEEE Std 1149.1. Some papers may therefore provide implementation examples that are not compliant with the standard.

Chapter 21. Bibliography

The following is an annotated bibliography of papers covering the development and application of boundary-scan test techniques and, in particular, IEEE Std 1149.1. Reprints of papers marked **REPRINT** are included following this bibliography.

[1] P. Goel and M.T. McMahon, "Electronic Chip in Place Test," *IEEE International Test Conference Proceedings,* IEEE Computer Society Press, Los Alamitos, Calif., 1982, pp. 83–90.

In ECIPT, a chip designed according to Level-Sensitive Scan Design (LSSD) principles is enhanced such that a shift-register latch (SRL) is connected directly to each package pin. Where an SRL is available at the input or output as a part of the normal functional design, no additional SRL is added. However, where a pin feeds or is fed by combinational logic, an SRL is added purely for test purposes. Together, these SRLs can be used to perform boundary-scan-like tests of chip-to-chip interconnections.

[2] D. Komonytsky, "LSI Self-Test Using Level Sensitive Scan Design and Signature Analysis," *IEEE International Test Conference Proceedings,* IEEE Computer Society Press, Los Alamitos, Calif., 1982, pp. 414–424.

[3] J.J. Zasio, "Shifting Away from Probes for Wafer Test," IEEE COMPCON, IEEE Computer Society Press, Los Alamitos, Calif., 1983, pp. 395–398.

This paper discusses the use of boundary-scan as a means of reducing the number of probe contacts that must be made during wafer testing. In the particular implementation described, a small number of probes is sufficient to give access to power pins, the boundary-scan path, and the internal scan path of each chip. Note that the boundary-scan cells are placed outside the bonding pads for the integrated circuit (IC) so that the integrity of the signal paths from the bonding pads to the circuitry is checked.

[4] S. das Gupta, M.C. Graf, R.A. Rasmussen, R.G. Walther and T.W. Williams, "Chip Partitioning Aid: A Design Technique for Partitionability and Testability in VLSI," *ACM/IEEE Design Automation Conference Proceedings,* IEEE Computer Society Press, Los Alamitos, Calif., 1984, pp. 203–208.
REPRINT

This paper describes IBM's proprietary implementation of boundary-scan that is based on Level-Sensitive Scan Design (LSSD).

[5] J.J. LeBlanc, "LOCST: A Built-In Self-Test Technique," *IEEE Design and Test or Computers,* Vol. 1, No. 4, Nov. 1984, pp. 45-52.
REPRINT

This paper describes an IC architecture that combines boundary-scan, on-chip test control, and self-test.

[6] R. Lake, "A Fast 20K Gate Array with On-Chip Test System," *VLSI Systems Design,* Vol. 7, No. 6, June 86, pp. 46-65.
REPRINT

This paper describes the architecture and operation of a gate-array family that includes boundary-scan and self-test.

[7] F.P.M. Beenker, "Systematic and Structured Methods of Digital Board Testing," *IEEE International Test Conference Proceedings,* IEEE Computer Society Press, Los Alamitos, Calif., 1985, pp. 380-385.

It was this paper that started the drive toward a standard boundary-scan architecture. The paper reviews the problems faced by companies attempting to use advanced surface-mount interconnection technologies and considers the value of boundary-scan as a solution to the problems identified.

[8] D. Laurent, "An Example of Test Strategy for Computer Implemented with VLSI Circuits," *IEEE International Conference on Computer Design: VLSI in Computers and Processors,* IEEE Computer Society Press, Los Alamitos, Calif., 1985, pp. 679-682.

This paper discusses Bull's implementation of boundary-scan. Note two key features. First, the boundary-scan cells at output pins are not able to drive signals through the pins -- they can only monitor the signal values driven through the pins by the system logic. Second, the cost of implementing boundary-scan is reduced by placing the boundary-scan cells beneath power distribution busses around the periphery of the chip.

[9] P.T. Wagner, "Interconnect Testing with Boundary-Scan," *IEEE International Test Conference Proceedings,* IEEE Computer Society Press, Los Alamitos, Calif., 1987, pp. 52-57.
REPRINT

This paper discusses how tests may be generated for a board populated with boundary-scannable chips and provides formulae that can be used to compute the overall test length.

[10] C.M. Maunder and F.P.M. Beenker, "Boundary-Scan -- A Framework for Structured Design for Test," *IEEE International Test Conference Proceedings,* IEEE Computer Society Press, Los Alamitos, Calif., 1987, pp. 714-723.

This paper discusses the JTAG version 1.0 architecture.

[11] L. Avra, "A VHSIC ETM–Bus Compatible Test and Maintenance Interface," *IEEE International Test Conference Proceedings,* IEEE Computer Society Press, Los Alamitos, Calif., 1987, pp. 964–971.

This paper describes the U.S. DoD's VHSIC ETM and TM busses. Many ideas from the ETM–bus were incorporated in the design of the IEEE Std 1149.1 test logic. Other papers included among these reprints show how board–level test busses formed by interconnecting chips compatible with IEEE Std 1149.1 can be interfaced to a system–level test and maintenance bus (the TM–bus).

[12] M.M. Pradhan, R.E. Tulloss, H. Bleeker, and F.P.M. Beenker, "Developing a Standard for Boundary–Scan Implementation," *IEEE International Conference on Computer Design: VLSI in Computers and Processors,* IEEE Computer Society Press, Los Alamitos, Calif., 1987, pp. 462–466.

A review of the development of the JTAG standard proposal up to the middle of 1987.

[13] IBM, TRW, and Honeywell, *VHSIC Phase 2 Interoperability Standards: TM–Bus Specification –– Version 3.0,* November 9 1987 (available from J.P. Letellier, Naval Research Lab., Code 5305, Washington DC 20375, U.S.A).

The TM–bus is a test and maintenance bus intended for use at the system level. For example, the bus may be used to convey test data to and from a printed wiring board (PWB) in a rack of equipment. The TM–bus is the basis of the IEEE P1149.5 project.

[14] P. Barton and C. Dolan, "ASICs and Testability Devices Revolutionize Testability Design," *Texas Instruments Technical Journal,* Vol. 5, No. 4, July/August 1988, pp. 86–97.

This paper shows how, by selectively replacing key chips on a PWB with compatible devices that include a boundary–scan register, the overall testability of the loaded board can be increased considerably.

[15] A. Hassan, J. Rajski, and V.K. Agrawal, "Testing and Diagnosis of Interconnects Using Boundary–Scan Architecture," *IEEE International Test Conference Proceedings,* IEEE Computer Society Press, Los Alamitos, Calif., 1988, pp. 126–137.
REPRINT

The application of self–test techniques, based on an extension to the functionality of boundary–scan cells, to the testing of chip–to–chip interconnections is described. The technique offers easy diagnosis of detected faults.

[16] M.A. Breuer and J.-C. Lien, "A Test and Maintenance Controller for a Module Containing Testable Chips," *IEEE International Test Conference Proceedings,* IEEE Computer Society Press, Los Alamitos, Calif., 1988, pp. 502-513.

This paper describes how a controller could be designed that would interface between a system-level TM-bus and a board-level test bus based in IEEE Std 1149.1.

[17] B.I. Dervisoglu, "Using Scan Technology for Debug and Diagnostics in a Workstation Environment," *IEEE International Test Conference Proceedings,* IEEE Computer Society Press, Los Alamitos, Calif., 1988, pp. 976-986.

While the architecture described is not based on the use of IEEE Std 1149.1, the paper gives a good example of how a chip-through-system test and maintenance architecture can be achieved and of the resulting benefits.

[18] C.L. Hudson, "Integrating BIST and Boundary-Scan on a Board," *National Communications Conference,* 1988, pp. 1796-1800.

This paper considers several options for the integration of boundary-scan and self-test.

[19] J. Sweeney, "JTAG Boundary-Scan: Diagnosing Module Level Functional Failures," *National Communications Conference,* 1988, pp. 1801-1804.

This paper discusses how the *SAMPLE/PRELOAD* instruction defined by IEEE Std 1149.1 can be used during fault location during functional (edge-connector) loaded-board testing or during system-level built-in self-test.

[20] P.A. Uszynski and A.C. Erdal, "Hybrid Global Test Strategy," *High Performance Systems,* Vol. 10, No. 1, Jan. 1989, pp. 68-74.

A test approach that combines boundary-scan and self-test is described. This approach employs linear-feedback shift-registers and signature analysis, coupled to a scan path through the on-chip system logic.

[21] C.S. Gloster and F. Brglez, "Boundary-Scan with Built-In Self-Test," *IEEE Design and Test of Computers,* Vol. 6, No. 1, Feb. 1989, pp. 36-44.
REPRINT

This paper describes how boundary-scan cells can be designed to act as cellular automata as a part of a self-test scheme for on-chip logic.

[22] R.P. van Riessen, H.G. Kerkhoff, and A. Kloppenburg, "Design and Implementation of a Hierarchical Testable Architecture Using the Boundary-Scan Standard," *European Test Conference Proceedings,* IEEE Computer Society Press, Los Alamitos, Calif., 1989, pp. 112-118.

This paper shows how a macro-oriented design-for-test strategy might be integrated with the IEEE Std 1149.1 test logic.

[23] L.-T. Wang, M. Marhoefer, and E.J. McCluskey, "A Self-Test and Self-Diagnosis Architecture for Boards Using Boundary-Scan," *European Test Conference Proceedings,* IEEE Computer Society Press, Los Alamitos, Calif., 1989, pp. 119-126.

The authors present a low-cost self-test and self-diagnosis architecture, using boundary-scannable chips, that allows location of defective chips and chip-to-chip interconnections.

[24] D. van de Lagemaat, "Testing Multiple Power Connections with Boundary-Scan," *European Test Conference Proceedings,* IEEE Computer Society Press, Los Alamitos, Calif., 1989, pp. 127-130.

Where multiple power and ground connections are provided on an IC, it is essential that all are correctly connected to the PWB if the assembled product is to operate reliably. For example, if one power connection is open-circuit, failures will only occur when sufficient current is drawn by the integrated circuit to cause the power voltage within the chip to fall. This paper shows how faults in board-to-chip connections at power pins can be detected using extensions to the boundary-scan path.

[25] P.P. Fasang, "ASIC Testing in a Board/System Environment," *IEEE Custom Integrated Circuits Conference,* IEEE, New York, 1989, pp. 22.4.1-22.4.4.
REPRINT

This paper discusses the use of boundary-scan in mixed analog/digital ICs.

[26] J-C. Lien and M.A. Breuer, "A Universal Test and Maintenance Controller for Modules and Boards," *IEEE Transactions on Industrial Electronics,* Vol. 36, No. 2, May 1989, pp. 231-240.
REPRINT

This paper looks at architecture for the design of maintainable systems and at the design of a board-level maintenance controller compatible with IEEE Std 1149.1.

[27] P. Hansen and T. Borroz, "Tough Board Test Problems Solved with Boundary-Scan," *Electronics Test,* Vol. 12, No. 6, June 1989, pp. 34-40.

The authors review the application of boundary-scan to the process of testing loaded boards.

[28] C.H. Hao et al, "Computer Aided Structured Design for Testability of ASICs," *8th Australian Microelectronics Conference*, Brisbane, July 1989, pp. 116-121.

This paper outlines how a range of design-for-test approaches, including boundary-scan and self-test, might be employed in the design of an application-specific IC (ASIC).

[29] S. Evanczuk, "IEEE 1149.1: A Designer's Reference," *High Performance Systems,* Vol. 10, No. 8, Aug. 1989, pp. 52-60.

This article provides a brief tutorial introduction to IEEE Std 1149.1.

[30] D. McLean, S. Banerji and L. Whetsel, "Bringing 1149.1 Into the Real World," *High Performance Systems,* Vol. 10, No. 8, Aug. 1989, pp. 61-70.

A view is given of the way that one company has chosen to support IEEE Std 1149.1 in its product range.

[31] P. Hansen and C. Rosenblatt, "Handling the Transition to Boundary-Scan for Boards," *High Performance Systems,* Vol. 10, No. 8, Aug. 1989, pp. 74-81.

A test equipment vendor gives a view of the impact of the availability of boundary-scan chips on the overall loaded-board test process.

[32] K.P. Parker, "The Impact of Boundary-Scan on Board Test," *IEEE Design and Test of Computers,* Vol. 6, No. 4, pp. 18-30.
REPRINT

This paper assesses the impact that boundary-scan may have on the board test task. It includes an analysis of the impact of boundary-scan in reducing the need for bed-of-nails probes on in-circuit test fixtures, for example in the case where a board contains a mixture of chips with and without IEEE Std 1149.1 features.

[33] S. Vining, "Prototype JTAG Controller Trade-Off Decisions," *IEEE International Test Conference Proceedings,* IEEE Computer Society Press, Los Alamitos, Calif., 1989, pp. 47-54.

This paper looks at the issues involved in the design of an interface between a low-cost test processor (say, a personal computer) and the serial IEEE Std 1149.1 test data path.

[34] A. Dahbura, M.U. Uyar, and C.W. Yau, "An Optimal Test Sequence for the JTAG Boundary-Scan Controller," *IEEE International Test Conference Proceedings,* IEEE Computer Society Press, Los Alamitos, Calif., 1989, pp. 55-62.
REPRINT

Systems companies will wish to be assured that the chips they buy are truly conformant to IEEE Std 1149.1. This paper looks at how a conformance test could be constructed that, from the package pins, will verify that the requirements of the standard have been met. The resulting test also gives good fault coverage of the test access port (TAP) controller and other mandatory features of the standard.

[35] C.W. Yau and N. Jarwala, "A New Framework for Analyzing Test Generation and Diagnosis Algorithms for Wiring Interconnects," *IEEE International Test Conference Proceedings*, IEEE Computer Society Press, Los Alamitos, Calif., 1989, pp. 63–70.
REPRINT

This paper develops a method of analyzing interconnect test patterns for use in boundary-scan environments -- for example, in terms of their length, fault coverage, and diagnostic capability. Based on a comparison of previously-published algorithms for generating test patterns, a new adaptive algorithm is presented.

[36] N. Jarwala and C.W. Yau, "A Unified Theory for Designing Optimal Test Generation and Diagnosis Algorithms for Board Interconnects," *IEEE International Test Conference Proceedings*, IEEE Computer Society Press, Los Alamitos, Calif., 1989, pp. 71–77.
REPRINT

This is a companion to the previous paper. It shows how tradeoffs can be made between test length and diagnostic resolution given information on the size of defects typically introduced by the assembly process.

[37] P. Hansen, "Testing Conventional Logic and Memory Clusters Using Boundary Scan Devices as Virtual ATE Channels," *IEEE International Test Conference Proceedings*, IEEE Computer Society Press, Los Alamitos, Calif., 1989, pp. 166–173.

For some time, board designs will continue to include a number of chips that do not offer boundary-scan. These components may continue to be tested via bed-of-nails techniques if sufficient access for probing is available. Where access is limited, the boundary-scan paths in chips surrounding such chips may be used as "virtual" ATE channels. This paper looks at the way that an ATE may be designed to support this type of testing.

[38] A. Halliday, G. Young, and A. Crouch, "Prototype Testing Simplified by Scannable Buffers and Latches," *IEEE International Test Conference Proceedings*, IEEE Computer Society Press, Los Alamitos, Calif., 1989, pp. 174–181.

The authors look at the impact of IEEE Std 1149.1 on the testing of prototype boards.

[39] W.D. Ballew and L. Streb, "Board-Level Boundary-Scan: Regaining Observability with an Additional IC," *IEEE International Test Conference Proceedings*, IEEE Computer Society Press, Los Alamitos, Calif., 1989, pp. 182–189.

This paper describes the design of a chip that can be added to a board to provide controllability and observability to sections of the design that do not offer boundary-scan. The chip is provided with a TAP compatible with IEEE Std 1149.1.

[40] F. Brglez, C. Gloster, and G. Kedem, "Hardware-Based Weighted Random Pattern Generation for Boundary-Scan," *IEEE International Test Conference Proceedings,* IEEE Computer Society Press, Los Alamitos, Calif., 1989, pp. 264–274.

This is a continuation of the work presented at the 1988 International Test Conference and in IEEE Design and Test of Computers magazine (see reference 21).

[41] D.L. Landis, "A Self-Test System Architecture for Reconfigurable WSI," *IEEE International Test Conference Proceedings,* IEEE Computer Society Press, Los Alamitos, Calif., 1989, pp. 275–282.
REPRINT

This paper looks at the application of IEEE Std 1149.1 in the test and configuration of wafer-scale ICs. Each chip on the wafer is tested using an internal test mode of the boundary-scan path. Chips that pass this test are then interconnected via laser-programmable interconnections. As interconnections are programmed, they are tested using the external test mode of the relevant chip boundary-scan paths.

[42] A. Hassan, V.K. Agarwal, J. Rajski, and B.N. Dostie, "Testing Glue Logic Interconnects Using Boundary-Scan Architecture," *IEEE International Test Conference Proceedings,* IEEE Computer Society Press, Los Alamitos, Calif., 1989, pp. 700–711.

This is a continuation of the work presented at the 1988 International Test Conference (see reference 15).

[43] Y. Zorian and N. Jarwala, "Designing Fault-Tolerant, Testable VLSI Processors Using the IEEE P1149.1 Boundary-Scan Architecture," *International Conference on Computer Design: VLSI in Computers and Processors*, IEEE Computer Society Press, Los Alamitos, Calif., 1989.

The production of defect and fault tolerant VLSI processors poses many problems in debugging and production testing. This paper proposes IEEE Std 1149.1 as a solution to these problems.

[44] P. Hansen, "Strategies for Testing VLSI Boards Using Boundary-Scan," *Electronic Engineering,* Vol. 61, No. 755, Nov. 1989, pp. 103–111.

This paper discusses programming, tester, and diagnostic requirements for boundary-scan-based loaded-board testing.

[45] R.P. van Riessen, H.G. Kerkhoff, and A. Kloppenburg, "Designing and Implementing an Architecture with Boundary-Scan," *IEEE Design and Test of Computers,* Vol. 7, No. 2, Feb. 1990, pp. 9-19.
REPRINT

The authors describe a standardized structured test methodology based on IEEE Std 1149.1. The architecture ensures testability of the hardware from the printed wiring board level down to the chip level. In addition, the architecture has built-in self-test at the IC level and is implemented in a silicon compiler.

[46] R.W. Bassett, M.E. Turner, J. H. Panner, P.S. Gills, S.F. Oakland, and D.W. Stout, "Boundary-Scan Design Principles for Efficient LSSD ASIC Testing," *IBM Journal of Research and Development,* Vol. 34, No. 2/3, Mar./May 1990, pp. 339-354.

A boundary-scan design method based on LSSD design principles is presented and compared to IEEE Std 1149.1.

[47] D. Landis and P. Singh, "Optimal Placement of IEEE 1149.1 Test Port and Boundary-Scan Resources for Wafer-Scale Integration," *IEEE International Test Conference Proceedings,* IEEE Computer Society Press, Los Alamitos, Calif., 1990, Paper 5.1.

This paper identifies the tradeoffs of using a standardized serial test interface for wafer-scale integration. The test circuitry area overhead and yield loss are compared to the benefits of reduced input/output connections and improved wafer testability.

[48] J. Maierhofer, "Hierarchical Self-Test Concept Based on the JTAG Standard," *IEEE International Test Conference Proceedings,* IEEE Computer Society Press, Los Alamitos, Calif., 1990, Paper 5.2.

A hierarchical self-test architecture from system down to macrocell level is presented. At the board level, the IEEE Std 1149.1 is used to activate both chip-level self-test and board-level interconnect testing. A special-purpose BIST bus is used to provide test communications within the chip.

[49] P. Dziel, "Integrating DFT into the ASIC Design Process", *Semicustom Design Guide,* 1990, pp. 37-43.

A complete design-for-test methodology is discussed, based on IEEE Std 1149.1. An internal partial-scan path is used that is integrated with the test access port.

[50] L. Whetsel, "Event Qualification: A Gateway to At-Speed Functional Testing," *IEEE International Test Conference Proceedings,* IEEE Computer Society Press, Los Alamitos, Calif., 1990, Paper 5.4.

This paper describes an event-qualification architecture and interface that can be designed into integrated circuits to facilitate functional testing at the board level.

[51] K. Parker and S. Oresjo, "A Language for Describing Boundary-Scan Devices," *IEEE International Test Conference Proceedings,* IEEE Computer Society Press, Los Alamitos, Calif., 1990, Paper 9.1.
REPRINT

A simple, machine-readable, language for describing the chip-specific characteristics of an IEEE Std 1149.1-compatible integrated circuit is presented. The language, which is a subset of VHDL, is proposed as a standard for communication of data on chips compatible with IEEE Std 1149.1.

[52] F. de Jong, "Boundary-Scan Test Used at Board Level: Moving Towards Reality," *IEEE International Test Conference Proceedings,* IEEE Computer Society Press, Los Alamitos, Calif., 1990, Paper 9.2.

To be able to use boundary-scan facilities effectively at the board level, a method is required for communicating test patterns between test pattern generation tools and boundary-scan test systems. This paper describes a test-specification language, called BITL, that is intended for this purpose.

[53] D. Sterba, A. Halliday, and D. McClean, "ATPG Issues for Board Designs Implementing Boundary-Scan," *IEEE International Test Conference Proceedings,* IEEE Computer Society Press, Los Alamitos, Calif., 1990, Paper 9.3.

This paper describes experiments conducted to study the use of automatic test pattern generation tools and scan-based fault dictionaries on loaded boards that use IEEE Std 1149.1.

[54] M. Lefebvre, "Functional Test and Diagnosis: A Proposed JTAG Sample Mode Scan Tester," *IEEE International Test Conference Proceedings,* IEEE Computer Society Press, Los Alamitos, Calif., 1990, Paper 16.1.
REPRINT

The *SAMPLE* instruction is used to allow logic analysis and diagnostic probing of a loaded board while it is performing its system function or executing a functional test.

[55] M. Fichtenbaum and G. Robinson, "Scan Test Architectures for Digital Board Testers," *IEEE International Test Conference Proceedings,* IEEE Computer Society Press, Los Alamitos, Calif., 1990, Paper 16.2.

This paper looks at how features can be added to a digital board tester to allow effective use of boundary-scan facilities in the unit under test.

[56] C. Yau and N. Jarwala, "The Boundary-Scan Master: Target Applications and Functional Requirements," *IEEE International Test Conference Proceedings,* IEEE Computer Society Press, Los Alamitos, Calif., 1990, Paper 16.3.

A bus-master chip is described that can control the standardized test features a board populated with chips compatible with IEEE Std 1149.1. The paper describes both the chip architecture and some possible applications.

[57] G. Robinson and J. Deshayes, "Interconnect Testing of Boards with Partial Boundary-Scan," *IEEE International Test Conference Proceedings,* IEEE Computer Society Press, Los Alamitos, Calif., 1990, Paper 27.3.

Test generation for and diagnosis of short- and open-circuit faults on loaded boards that have partial boundary-scan can be achieved by combining boundary-scan and conventional in-circuit interconnect testing.

CHIP PARTITIONING AID: A DESIGN TECHNIQUE FOR PARTITIONABILITY AND TESTABILITY IN VLSI

S. DasGupta
IBM Corporation
P.O. Box 390
Poughkeepsie, NY 12602

M. C. Graf
IBM Corporation
Route 52
Hopewell Junction, NY 12533

R. A. Rasmussen
IBM Corporation
Route 52
Hopewell Junction, NY 12533

R. G. Walther
11400 F.M.R.D. 1325
IBM Corporation
Austin, Texas 78759

T. W. Williams
IBM Corporation
P.O. Box 1900
Boulder, CO 80314

ABSTRACT

This paper presents a structured partitioning technique which can be integrated into the design of a chip. It breaks the pattern of exponential growth in test pattern generation cost as a function of the number of chips in a package. In one of its forms, it also holds the promise of parallel chip testing, as well as migration of chip-level tests for testing at higher package levels.

INTRODUCTION

Level Sensitive Scan Design (LSSD) [1, 2] is one method to solve controllability and observeability problems in sequential networks and hence, ease the problem of test pattern generation. This is achieved by incorporating all memory elements in a sequential network in shift register latches (SRLs) and then connecting all SRLs into one or more shift registers so that the internal state of the network can be controlled or observed at any time through the shift register path. LSSD also permits software-based partitioning techniques [3] to divide a large network into manageable, independent networks, each of which is separately addressed by test pattern generators. This LSSD-based approach to partitioning will be discussed in the next section prior to the main topic of this paper. While this partitioning approach was adequate for large scale integration (LSI), it is inadequate for networks in very large scale integration (VLSI) [4] due to the rapid increase in test generation complexity. Several solutions have been suggested to solve this partitioning problem. Hsu, et al [5] , and Tsui [6] have recommended ad hoc techniques for controlling and observing the outputs of chips, and, hence, inputs of other chips fed by the former on a common package. Goel and McMahon [7] have proposed another method where extra circuitry in system latches and multiplexors on chip outputs are required to control and observe chip boundaries.

This paper presents a structured, logical partitioning technique called Chip Partitioning Aid (CPA) that can be designed into a VLSI chip technology. In its simplest form, called Half-CPA (HCPA), it is a structured technique that partitions a network into nearly disjoint, physical segments that are approximately a chip's worth of logic in size. Thus, test generation cost, at higher package levels, drops from the normal exponential cost function to a straight multiple of the number of chips in the package. In the complete version of CPA, called Full-CPA (FCPA), the logic network is partitioned into subnetworks virtually along chip boundaries with built-in latch isolation around chip inputs and outputs (I/O's) that allows the potential reapplication of chip level test data and, more importantly, the potential for simultaneous testing of the internal logic of all chips. This latter version is, of course, the ultimate in the "divide and conquer" approach to test generation. The only addition at each package level is the incremental set of tests for interconnection faults at that package level which can be derived from a considerably simpler model of the network.

Next, we will define the rules associated with the two versions of CPA, and, finally, there will be a discussion on the effect of CPA on sytem design and how it can be mitigated by proper implementation of CPA.

HALF-CPA (HCPA)

Figure 1 shows a conceptual diagram of HCPA. It shows that in this version of CPA, all logic outputs of a chip are buffered by shift register latches (SRLs), called CPA-SRLs here, before being driven off-chip. CPA-SRLs are similar to standard SRLs, an example of which is shown in Figure 2. However, control outputs, such as clocks, are treated differently. They feed off-chip drivers unimpeded as required by system function. However, to ensure that this control function can be tested properly, it is required to also feed a CPA-SRL which is left to the side, out of the system path.

The HCPA structure at the chip boundary described above does not play any significant role in test pattern generation at the chip level. The only difference from a chip without CPA-SRLs is that measurements at chip outputs can be made only after the data has been clocked into the CPA-SRLs. The HCPA structure, however, has a considerble influence in test pattern generation at higher package levels, forcing creation of partitions of approximately a chip's worth of logic.

Reprinted from *IEEE Proceedings 21st Design Automation Conference,* 1984, pages 203-208.

EH0321-0/90/0000/0228$01.00 © 1984 IEEE

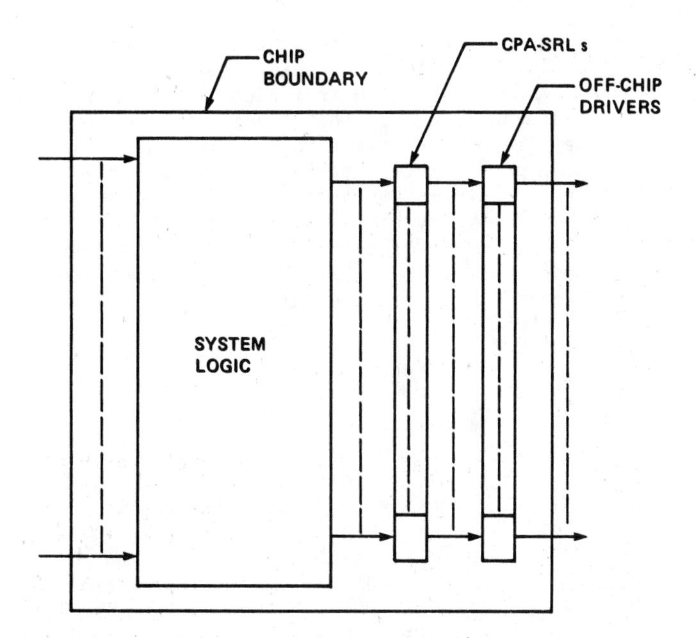

Figure 1. Half-CPA (HCPA) Structure

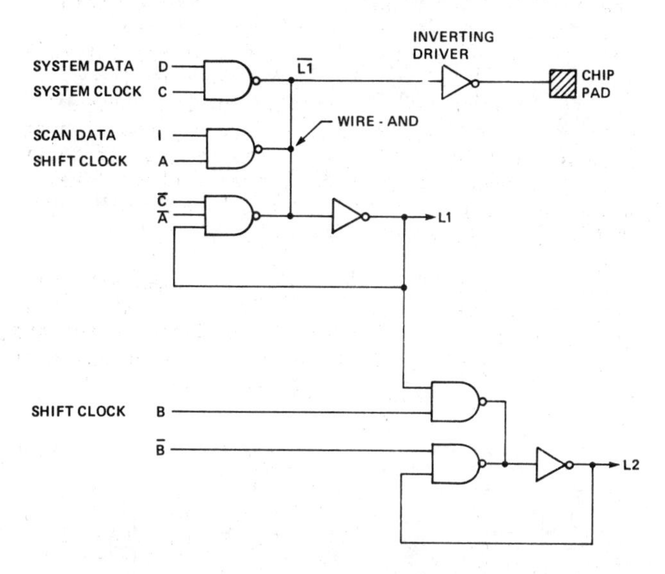

Figure 2. Example of CPA-SRL

An understanding of some partitioning concepts [3] is necessary here to appreciate the effect of HCPA. In an LSSD environment, SRLs, like package outputs, are considered observable nodes since the values in SRLs can be shifted out and observed. Therefore, to divide a network into smaller, independent subnetworks, a back-trace is performed from each observable node, stopping only at package inputs or SRLs, since the latter can be considered as a controllable input. All logic encountered in this back-trace constitutes an independent partition since it contains all the logic that can ever affect this SRL or primary output (PO). An example of this partitioning approach is shown in Figure 3. The unfortunate problem with this approach is that once the design is done, there is no way to bound or change the sizes of these partitions without a redesign; in fact, experience has shown that in many cases a significant segment of the entire network may be accounted for in a single partition.

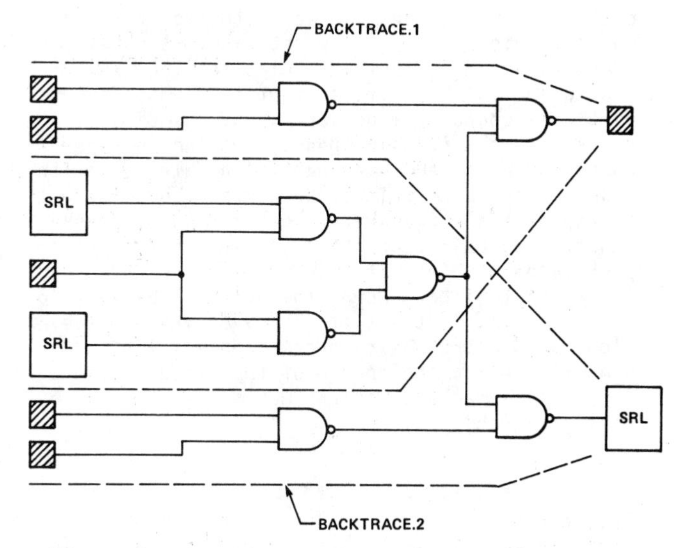

Figure 3. Examples of Partitioning Backtraces

A particularly good example of this is a bus-architected design where backing up from the bus, one can pack up just about the entire network in a single partition. A second problem is partition overlap in which a gate appears in more than one partition back-trace. This gate is considered at least for signal propagation during test generation and fault simulation, thus, effectively increasing the total number of gates that are evaluated by test generation/fault simulation programs.

The HCPA structure of Figure 1 changes the above situation. Figure 4 shows a module with several chips with HCPA structure. Since the CPA-SRLs also satisfy the property that they are controllable/observeable points, they serve both as "start points" and "stop points" of partitioning back-traces. Thus, starting from any HCPA-SRL, a

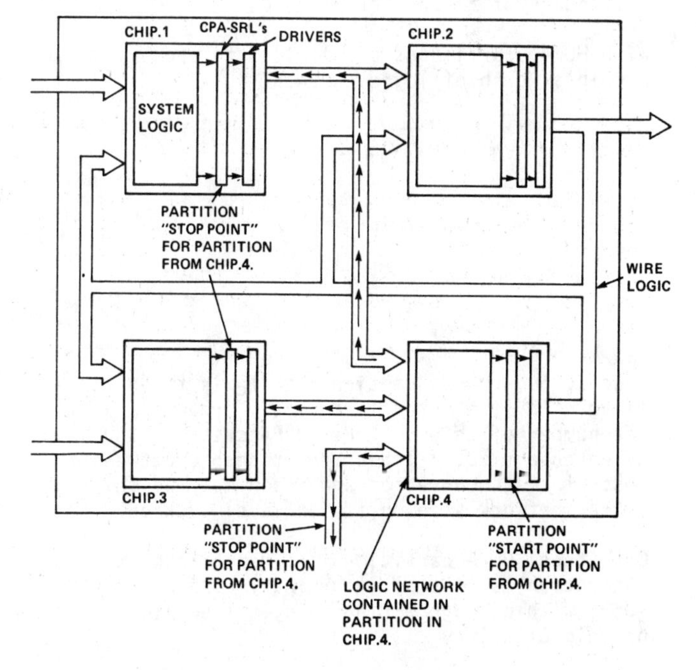

Figure 4. Partitioning with HCPA

back-trace propagates backwards through the logic on that chip and, in the worst case, stops at CPA-SRLs on the outputs of chips feeding the chip from where the back-trace started. Thus, each partition contains a network about the size of a chips's worth of logic, hence, putting an upper bound on the size of the partition. The question now is: how does this concept break the trend of an exponential rise in test generation cost as a function of chip count? To answer that, consider a package of n chips with m circuits on each chip. Assume that without HCPA, the worst case partition is approximately the size of the entire package. Also, assume that test generation cost is proportional to the square of the circuit count. Then, for a package without HCPA, test generation cost for a chip is:

$$T_c = km^2$$

i.e., $$m^2 = T_c/k$$

Test generation cost at package level, T_p, is given by,

$$T_p = k(nm)^2$$

$$= kn^2m^2$$

or, $$T_p = kn^2T_c/k$$

$$= T_cn^2$$

For a fixed chip size, T_c can be considered to be fixed. Hence, T_p varies as the square of n, i.e., the square of the number of chips.

With HCPA, partitions are limited to approximately a chip's worth of logic. Hence,

$$T_p = (km^2)n$$

$$= (kT_c/k)n$$

$$= T_cn$$

Once again, the assumption that T_c is fixed for a fixed chip size makes Tp directly proportional to the number of chips. Thus, HCPA creates, for a given chip size, a linear relationship between the cost of test generation and the number of chips and hence, network size on a high level package.

Once these HCPA partitions are determined, test generation is done as in ordinary LSSD networks [8,9] with package wiring being tested along with on-chip circuitry.

The only exception to what has been said about partitioning in HCPA relates to control outputs of chips. It is possible to back-trace through multiple chips, starting from a control output, but experience has shown us that these paths, while they may traverse multiple chips, are sparse in logic content. These outputs, therefore, are not expected to have large partitions, even in dense VLSI networks. The CPA-SRLs, that are fed by control outputs and sit on the side, aid in testing the logic since these SRLs act like intermediate observation points for the logic.

FULL-CPA (FCPA)

This is the complete version of CPA and is built on the benefits of HCPA. Unlike in HCPA where only system logic outputs are buffered by CPA-SRLs, in FCPA, both system logic inputs and outputs are buffered by CPA-SRLs, as shown in Figure 5. The only exceptions are control inputs and outputs. In the case of control inputs, they are required to feed CPA-SRLs on the side along with the system logic they are designed for, while control outputs are treated the same way they are treated in HCPA. The FCPA structure has two benefits over HCPA:

1. Though a FCPA chip needs at least two test clocks, they can be shared with all chips at higher package levels.

2. Latches on all system logic inputs/outputs effectively isolate the internal logic of chips allowing <u>all chips with the potential to be tested simultaneously</u> (hence, saving time on the tester) along with the potential to apply tests that were generated for the individual chip.

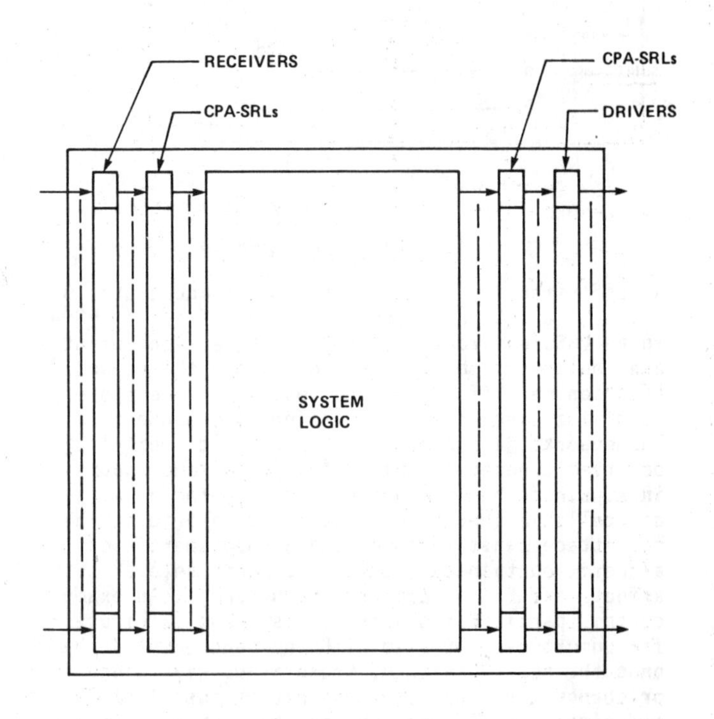

Figure 5. Full-CPA (FCPA) Structure

Test generation in the FCPA environment is now done in two stages:

1. Test generation of the internal logic of chips which is done either at the chip level and migrated up through the packaging levels or are generated again at the package level.

2. Test generation for stuck-at-faults in the package wiring and drivers/receivers on chips for which a simple model is created (see Figure 6) since they are bounded by SRLs. If any logic is performed at this package level, with wire-ORs or wire-ANDs, the number of test required is (r+1) where r is the maximum number of wires that is tied together to perform the largest AND or OR. If no such functions are performed, the package wiring can be tested with two tests.

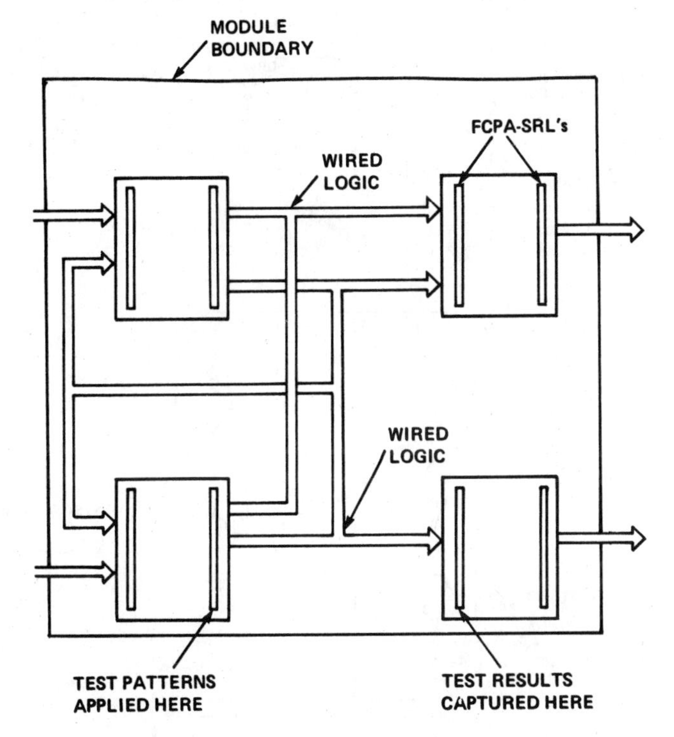

Figure 6. Simplified Model for Wiring Test on FCPA Module

Note that in FCPA, as in HCPA, test generation cost can be shown to be a linear function of the number of chips.

The real advantage offered by FCPA over HCPA is in parallel testing of the internal logic of all chips in a package. While in an idealized environment, all chips can be tested simultaneously, in a more realistic environment, parallel chip testing is affected by the way clocks are shared between chips and the order in which they need to be sequenced during a test. For example, if a particular test for one chip requires a C1-C2 sequence, while another chip requires the opposite order, these patterns cannot be merged into one common pattern for the package, even if everything else in the two patterns match. This limitation, however, is not expected to be a serious problem.

CPA RULES

The rules for HCPA are as follows:

1. Each chip "data output" signal feeding a chip output driver must be fed directly from a single CPA-SRL.

2. Each chip control output (for example, RAM control, shift clocks, tri-state inhibits) must feed a CPA-SRL, as well as it's off-chip driver.

In addition to the above rules, FCPA has the following additional rules:

1. Each chip "data input" must directly feed a CPA-SRL.

2. Each chip control input must feed directly to a CPA-SRL, as well as the system logic that it normally drives.

SYSTEM CONSIDERATIONS

From a system viewpoint, the choice of HCPA and FCPA is dependent upon density, system architecture and performance. At sufficient densities, latches naturally migrate to chip boundaries. In an LSSD environment, these latches would be embodied in SRLs, thus, satisfying the CPA requirements. However, there will be situations where an SRL will be required for CPA only, that is, a test-only SRL with no system application. In this situation, the clock(s) to that SRL will be used to control or observe a chip boundary during test.

When test-only SRLs are required for CPA, several steps can be taken to mitigate the real-estate and delay penalty of CPA-SRLs. In the case of HCPA, the CPA-SRL can be merged with its output driver to minimize both real-estate and delay. Also, the output from the CPA-SRL is taken from its L1 latch to the driver, thus saving the delay of the L2 latch. Figure 7 shows an example of an integrated CPA-SRL and driver where the performance detractor is the loading of the wired-AND function in the CPA-SRL. Estimates have shown that the above techniques can be used to limit real-estate overhead to less than 10% of the chip area and the delay penalty to a fraction of the delay of a logic circuit. And, finally, the test clock that sets data into the L1 latch of the CPA-SRL can be held "on" during system operation so that data can be flushed through it. Note that this test clock would constitute an overhead and at higher level packages, in a worst case situation, each chip might require a separate test clock for race-free testing. However, in a typical multi-chip package, it is possible to have many chips share the same test clock and still have race-free testing.

In the case of FCPA, the above ideas can be applied for the CPA-SRLs on both chip inputs and outputs. Additionally, the latches at the inputs and outputs can be merged into a single SRL with the L2* latch [10] , as shows in Figure 8, so that the L1 latch

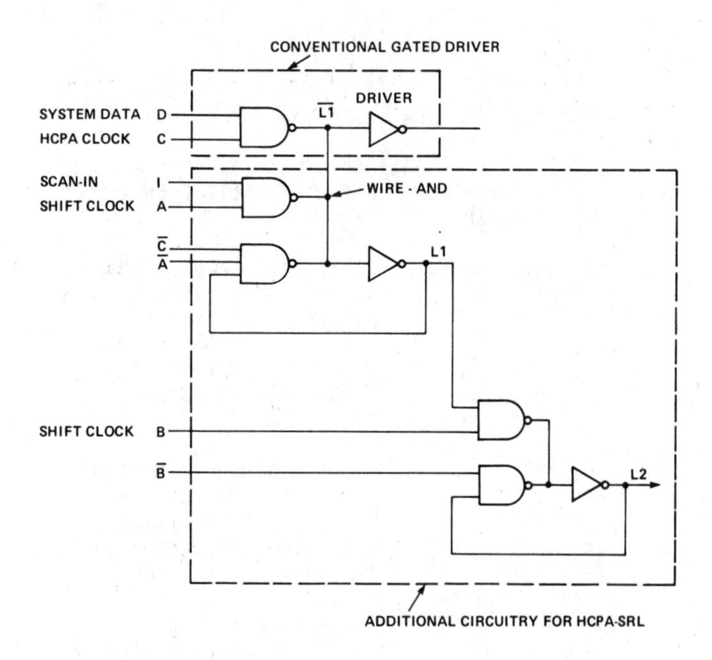

Figure 7. Example of Integrated HCPA-SRL/Driver

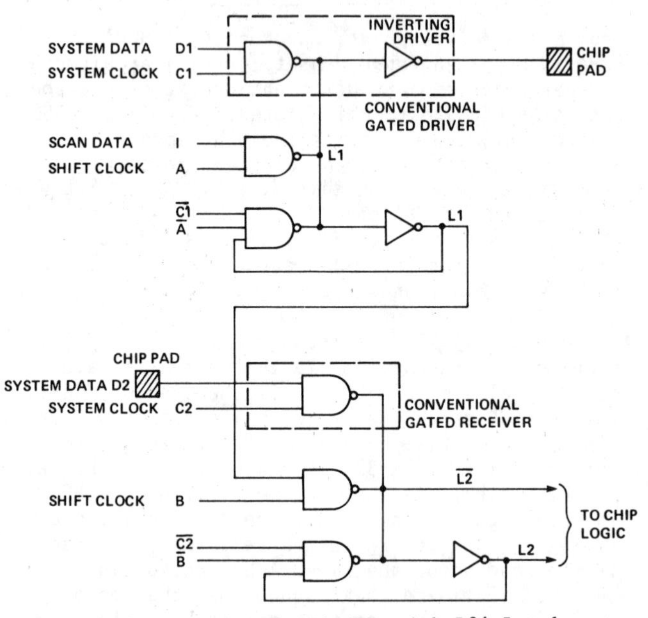

Figure 8. Example of CPA-SRL with L2* Latch Integrated with Receiver and Driver

could serve as the CPA boundary for an output and the L2* latch could serve as the CPA boundary for an input (see Figure 9). This provides a further reduction in the real-estate overhead for FCPA. The two test clocks that set data into these CPA latches are now part of the CPA overhead. However, at higher package levels, these two clocks can be shared between other chips. Figure 10 shows an example of a CPA-SRL implementation from [11] to show the delay impact on a system data path.

One final note on CPA! Whether the latches at chip boundaries are system usable or not, CPA-SRLs can be used to trap machine states when desired and in the event of an error/fault, can, in most cases, be used to pinpoint the failing chip [12] .

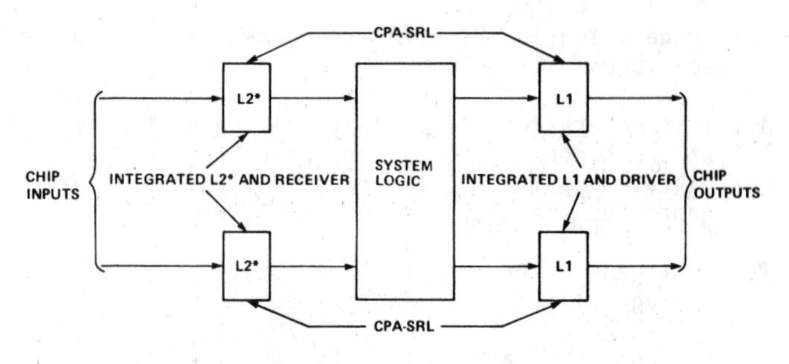

Figure 9. FCPA Chip with CPA-SRL Built with L2* Latch Integrated with Driver and Receiver

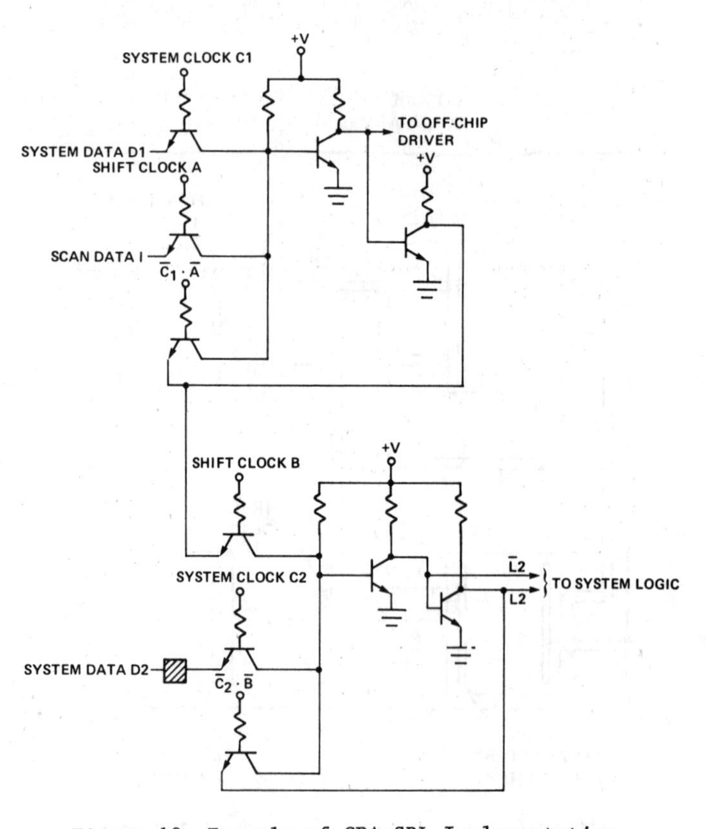

Figure 10. Example of CPA-SRL Implementation for FCPA

CONCLUSIONS

In this paper, we have described a partitioning technique that removes the uncertainty of partitioning sizes, since each partition is forced around chip boundaries and contains approximately a chip's worth of logic. Test generation, at higher package levels, now increases as a linear function of the number of chips and, in one of the versions, allows parallel chip testing which saves time on the tester during manufacturing. We have also defined the design rules and discussed system design aspects of CPA.

REFERENCES

[1] Eichelberger, E. B. and Williams, T. W., "A Logic Design Structure for LSI Testability," Proc. 14th Design Automation Conf., June 1977, pp. 462-468.

[2] DasGupta, S., Eichelberger, E. B. and Williams, T. W., "LSI Chip Design for Testability," Digest of Technical Papers, 1978 International Solid-State Circuits Conference, February 1978, pp.216-217.

[3] Bottorff, P. S., France, R. E., Garges, N. H. and Orosz, E. J., "Test Generation for Large Logic Networks," Proc. 14th Design Automation Conf., June 1977, pp. 479-485.

[4] Goel, P., "Test Generation Costs Analysis and Projection," Proc. 17th Design Automation Conf., June 1980, pp. 77-81.

[5] Hsu, F., Solecky, P., and Zobniw, L., "Selective Controllability: A Proposal for Testing and Diagnosis," Proc. 1978 Semiconductor Test Conf., October 1978, pp. 170-175.

[6] Tsui, F., "In-situ Testability Design (ISTD) - A New Approach for Testing High-Speed LSI/VLSI," Proc. IEEE, Vol. 70. No. 1, January 1982, pp. 59-78.

[7] Goel, P. and McMahon, M. T. "Electronic-Chip-In-Place Test," Proc. 19th Design Automation Conf., June 1982, pp. 482-488.

[8] Goel, P., "An Implicit Enumeration Algorithm to Generate Tests for Combinational Logic Circuits," Proc. 10th International Symposium on Fault Tolerant Computing, October 1980, pp. 145-151.

[9] Goel, P. and Rosales, B. C., "PODEM-X: An Automatic Test Generation System for VLSI Logic Structures," Proc. 18th Design Automation Conf., June 1981, pp. 260-268.

[10] DasGupta, S., Goel, P., Walther, R. G. and Williams, T. W., "A Variation of LSSD and Its Implication on Design and Test Generation," 1982 International Test Conf., November 1982, pp. 63-66.

[11] Culican, E. F., Diepenbrock and Ting, Y. M., "Shift Register Latch for Package Testing in Minimum Area and Power Dissipation," IBM Technical Disclosure Bulletin, Vol. 24, No. 11A, April 82, pp. 5598-5600.

[12] DasGupta, S., Walther, R. G., Williams, T. W. and Eichelberger, E. B., "An Enhancement to LSSD and Some Applications of LSSD in Reliability, Availability and Serviceability," Proc. 11th International Symposium on Fault Tolerant Computing, June 1981, pp.32-34.

Reprinted from *IEEE Design and Test*, November 1984, pages 45-52.

LOCST: A Built-In Self-Test Technique

With its low hardware cost, simple implementation and excellent coverage, this technique promises to meet the needs of a variety of VLSI environments.

Johnny J. LeBlanc, IBM Federal Systems Division

The advent of very large scale integration technologies has increased interest in built-in self-test as a technique for achieving effective and economical testing of VLSI components. As used in this article, the term "built-in self-test" refers to the capability of a device to generate its own test pattern set and to compress the test results into a compact pass-fail indication. Many buit-in self-test techniques have been proposed over the past 10 years, ranging from self-oscillation to functional pattern testing of microprogrammed devices to random-pattern testing (for examples, see papers by Mucha et al.,[1] Sedmak,[2] and McCluskey et al.[3]). These various techniques provide different capabilities for defect detection and self-test execution time. They also impose different requirements for implementation and control.

Summary

A built-in self-test technique utilizing on-chip pseudorandom-pattern generation, on-chip signature analysis, a "boundary scan" feature, and an on-chip monitor test controller has been implemented on three VLSI chips by the IBM Federal Systems Division. This method (designated LSSD on-chip self-test, or LOCST) uses existing level-sensitive scan design strings to serially scan random test patterns to the chip's combinational logic and to collect test results. On-chip pseudorandom-pattern generation and signature analysis compression are provided via existing latches, which are configured into linear-feedback shift registers during the self-test operation. The LOCST technique is controlled through the on-chip monitor, IBM FSD's standard VLSI test interface/controller. Boundary scan latches are provided on all primary inputs and primary outputs to maximize self-test effectiveness and to facilitate chip I/O testing.

Stuck-fault simulation using statistical fault analysis was used to evaluate test coverage effectiveness. Total test coverage values of 81.5, 85.3, and 88.6 percent were achieved for the three chips with less than 5000 random-pattern sequences. Outstanding test coverage (>97%) was achieved for the interior logic of the chips. The advantages of this technique, namely very low hardware overhead cost (<2%), design-independent implementation, and effective static testing, make LOCST an attractive and powerful technique.

Benefits to be gained from self-test, however, are common to all implementation techniques and include

- reduced test pattern storage requirements,
- reduced test time, and
- defect isolation to the chip level.

Since test patterns are generated automatically, only self-test initialization, control, and pass-fail comparison patterns need be stored, significantly reducing pattern storage requirements. Test time is reduced because one can use simple hardware devices (e.g., counters or linear-feedback shift registers) to control test execution, rather than retrieving test patterns from storage devices (e.g., disks) and applying them to the component under test. When components with built-in self-test are mounted on higher-level packages, the self-test pass-fail indication provides defect isolation to the chip level (e.g., during card repair testing).

At the IBM Federal Systems Division we have implemented a VLSI built-in self-test technique, which can be incorporated at very low hardware cost into any chip conforming to level-sensitive scan design (LSSD) rules, on three VLSI signal-processing chips. Our method (designated LSSD on-chip self-test, or LOCST) uses on-chip pseudorandom-pattern generation and on-chip signature analysis result compression. This is not a new self-test method; LOCST utilizes the serial-scan, random-pattern test technique pioneered by Eichelberger et al.[4,5] and Bardell et al.[6] of IBM. This article (1) details the adaptation of this technique to our existing chip testability architecture, (2) details the implementation of LOCST on three VLSI chips designed and fabricated by IBM FSD, and (3) discloses the results of the test coverage evaluations performed on these three chips. (For a thorough understanding of the principles of serial-scan, random-pattern testing, I strongly recommend a review of references 4, 5, and 6 and also a very comprehensive paper by Komonytsky.[7])

Standard FSD VLSI testability features

For a better understanding of the self-test architecture chosen for LOCST, a discussion of design features typical to IBM FSD's products is warranted. Figure 1 illustrates the three standard testability features incorporated in our VLSI products. They include

- level-sensitive scan design,
- "boundary scan" latches, and
- a standard maintenance interface, the on-chip monitor, or OCM.

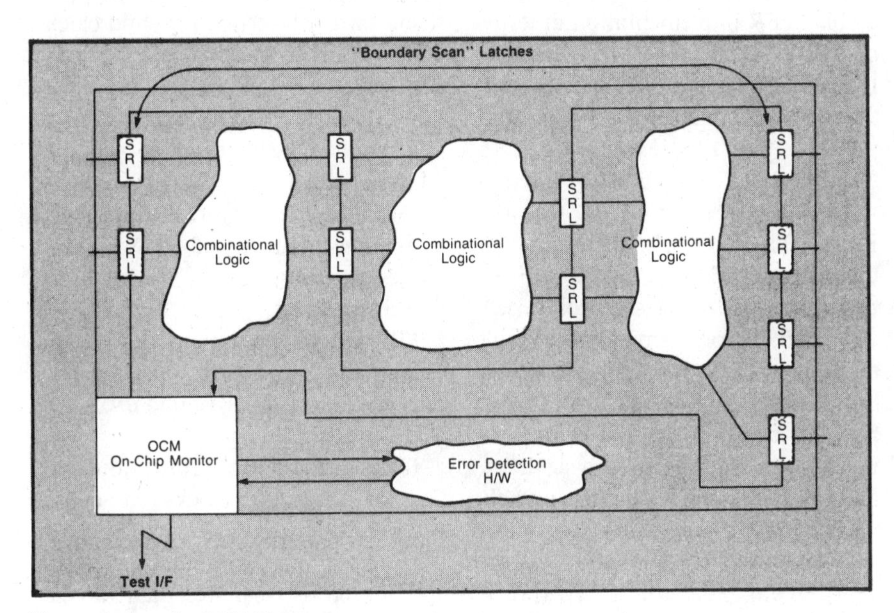

Figure 1. Standard VLSI features.

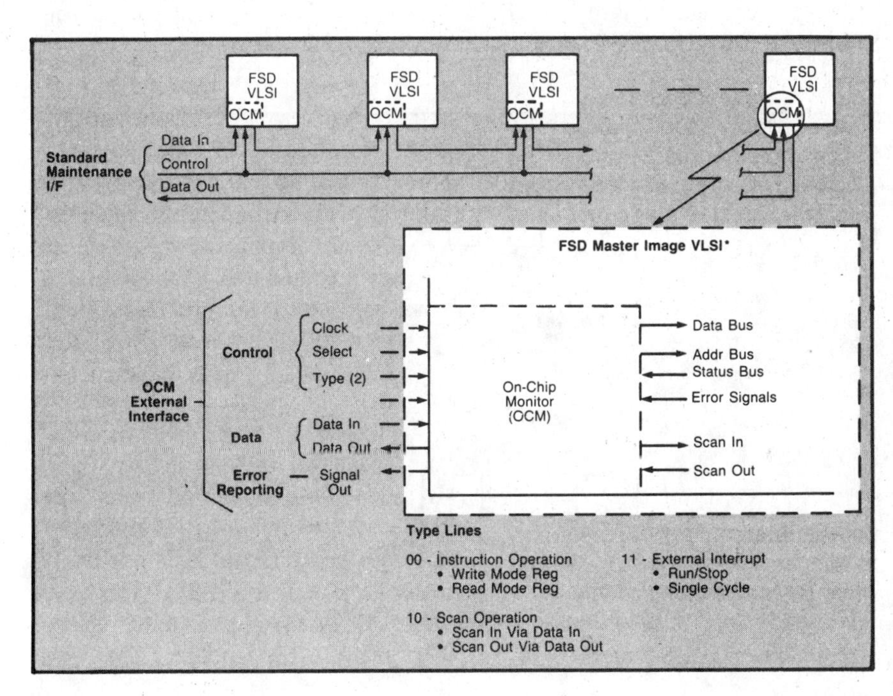

Figure 2. On-chip monitor.

All chips are designed following IBM's LSSD rules (see Eichelberger and Williams[8]) to ensure high test coverage and high diagnostic resolution during chip manufacture testing. "Boundary scan" is a requirement that all primary inputs (PIs) feed directly into shift register latches (SRLs, or LSSD latches) and all primary outputs (POs) are fed directly from SRLs. Boundary scan greatly simplifies chip-to chip interconnect testing and also provides an ideal buffer between LSSD VLSI products and non-LSSD vendor components, thereby reducing the complexity of testing "mixed-technology" cards.

The OCM is a standard maintenance interface for our VLSI chips (Figure 2). It consists of seven lines: two for data transfer, four for control, and one for error reporting. The OCM maintenance bus can be configured as either a ring, a star, or a multidrop network, depending on system maintenance requirements. The four major functions of the OCM are

- scan string control,
- error monitoring and reporting,
- chip configuration control, and
- clock event control: run/stop, single cycle, and stop on error.

During LSSD testing (chip manufacture testing), scan strings are accessed via either dedicated or shared PIs and POs. (Note: The OCM is not used as a test aid during LSSD testing; it is simply logic to be tested by LSSD test patterns.) During card and system test, however, chip scan strings are accessed via the OCM interface.

The error detection hardware depicted in Figure 1 consists of on-chip error checkers used for on-line system error detection and/or fault isolation (described by Bossen and Hsiao[9]). When these checkers are triggered by an on-chip error, an attention signal is sent to the system maintenance processor through the OCM interface. The system maintenance processor reads internal chip error registers (or writes internal chip mode control registers) via OCM "instructions."

LOCST architecture

The basic self-test methodology used in LOCST is to (1) place pseudorandom data into all chip LSSD latches via serial scan, (2) activate system clocks for a single cycle to capture the results of the random-pattern stimuli through the chip's combinational logic, and (3) compress the captured test results into a pass-fail signature. With the existing testability features (LSSD, boundary scan, OCM) on each chip, it was a simple matter to incorporate a self-test capability.

To perform the pseudorandom-pattern-generation and signature-compression operations while in LOCST self-test mode, functional SRLs are reconfigured into linear-feedback shift registers, or LFSRs. The pseudorandom-pattern generator, or PRPG, is 20 bits in length, and the signature analyzer (SA) is 16 bits in length (see Figure 3). It should be noted that the devices shown in Figure 3 operate as normal serial-scan latches *and* as linear-feedback shift registers. The transformation from normal serial-scan mode to LFSR mode is controlled by multiplexing the scan inputs with a self-test enable signal (controlled via the OCM interface). The parallel data ports of these latches are not modified in any way. During self-test the data port clocks (system clocks) are disabled to prevent outside data from disrupting the deterministic sequences of the LFSRs.

The feedback polynomial for the PRPG was chosen because it is the least expensive "maximal-length" 20-bit LFSR implementation in terms of XOR gates required. For the LOCST implementation, the characteristic polynomial of the PRPG and the SA is fixed. Differing test pattern sequences can be obtained by altering the initial value (or "seed") of the PRPG. The feedback polynomial for the SA was chosen because of its proven performance (see Frohwerk[10] and Smith,[11] for example). The result of using a 20-bit PRPG and a 16-bit SA is a self-test capability with $2^{20} - 1$ possible random-pattern sequences and a very low probability of signature analysis fault masking (approximately $1/2^{16}$ or 0.0015 percent).

A high-level block diagram of the LOCST implementation structure is shown in Figure 4. In self-test mode the initial 20 SRLs of the chip's scan strings are configured into a PRPG LFSR, and the last 16 SRLs are configured into an SA LFSR. For normal LSSD chip manufacture testing, a chip usually contains several scan strings—each accessible from chip input and output pins. During LOCST testing, however, all scan strings except the one containing the OCM latches are configured into a single scan string. (Note: Random test patterns are scanned into the single scan string under OCM control. SRLs that are part of the OCM and any chip clock generation circuitry *cannot* be included in the LOCST scan string since self-test control and clock control cannot be disrupted by random data.)

The following is a description of the LOCST sequence:

(1) Initialize all internal latches: scan known data into all SRLs; this includes scanning "seeds" into PRPG and SA registers.

(2) Activate self-test mode: enable PRPG and SA registers; disables system clocks on input boundary SRLs and LFSRs.

(3) Perform self-test operation:

(a) Apply scan clocks until entire scan string (up to the SA LFSR) if filled with pseudorandom patterns. This step also scans test data into the SA LFSR for test result compression.

(b) Activate system clocks for single-cycle operation.

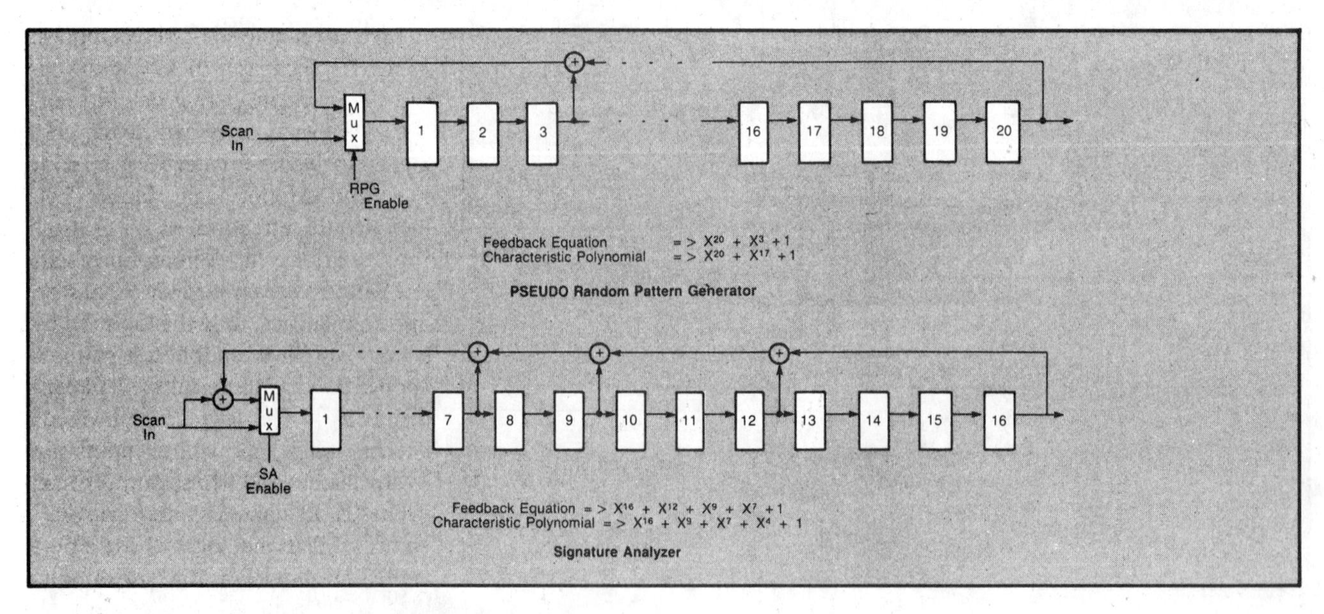

Figure 3. Linear-feedback shift register implementations.

(c) Repeat (a) and (b) until finished.

(4) Read out test result signature and compare with known "good" value.

The "good" value from step 4 can be obtained in two ways; (1) simulation of the entire self-test sequence, or (2) the "golden chip" approach (that is, determine what the "good" value is by performing the LOCST self-test operation on chips which have passed all other forms of manufacture and functional testing). Due to the high cost of the first method, the second is currently being used. If the correct "good" signature value were known (via simulation) during the chip design phase, a hardware comparator could be placed on the chip to provide an immediate pass-fail indication. Our implementations of LOCST require that the 16-bit signature be read by an external processor for comparison against the stored 16-bit "good" value.

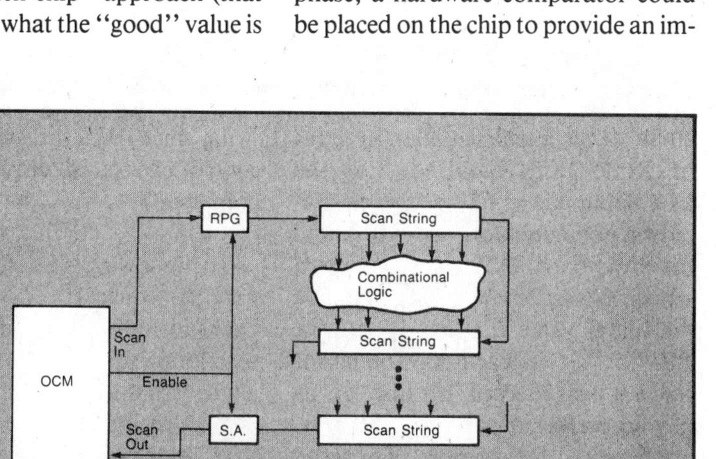

Figure 4. LOCST architectures.

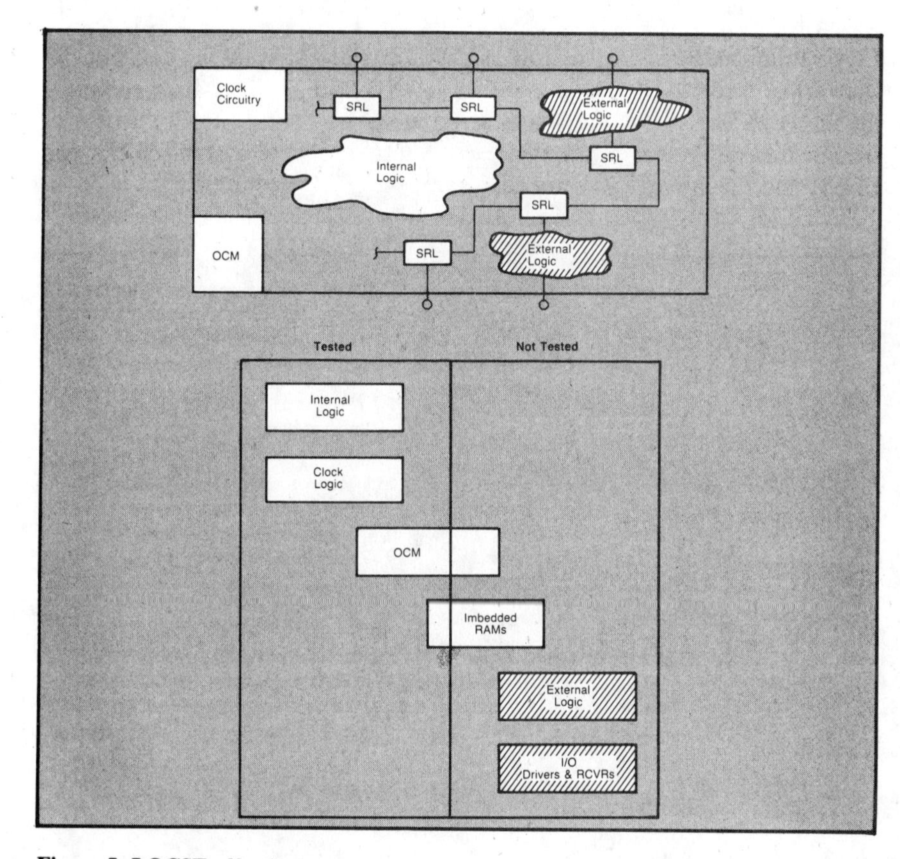

Figure 5. LOCST effectiveness.

The entire LOCST self-test operation is controlled by an external processor via the OCM interface. The external processor may be a chip or card tester or a system maintenance processor, depending on the testing environment. The OCM provides the following self-test control functions:

- PRPG and SA enable control,
- scan access to internal SRLs for random-pattern insertion and test result compression,
- chip clock control for single-cycle operation (if on-chip clock generation is used), and
- access to self-test results via direct register read or via scan.

If a chip does not have an OCM, control of these functions must be provided by some other means.

The data port clocks of input SRLs (i.e., boundary scan LSSD latches fed directly by primary inputs) are inhibited during self-test mode to prevent unknown data from corrupting the self-test sequence. If the input latch clocks are not disabled, known values *must* be ensured on chip PIs during self-test execution.

LOCST limitations

Like all on-chip self-test techniques, LOCST is incapable of testing the entire chip. In considering on-chip self-test effectiveness, we can divide chip logic into two basic categories: interior logic and exterior logic. Figure 5 illustrates the effectiveness of LOCST for the various chip regions. Since self-test patterns are applied via serial scan into chip latches, only the logic fed by latches will have random test patterns applied to it and test results will be captured only for logic which feeds latches. Chip logic whose inputs are fed by latches and whose outputs feed latches is designated "interior logic," and combinational logic whose inputs are fed by chip PIs and whose outputs feed chip POs is designated "exterior logic."

Obviously, external logic is completely untestable by the LOCST technique. The importance of boundary scan to on-chip self-test also should be obvious. The larger the percentage of exterior logic on a chip, the less effective on-chip self-test becomes. In the ideal case with 100 percent boundary scan, the only exterior logic would be I/O drivers and receivers (and with 100 percent boundary scan, I/O drivers and receivers would be very easy to test!)

Types of chip logic that do not clearly fall into the categories of interior or exterior are the OCM logic and embedded RAMs. Since the OCM controls the self-test operation, internal OCM logic is not tested by random patterns during self-test. Rather, the OCM is tested to the extent that all OCM functions needed to perform the self-test operation will have been exercised (i.e., scan control, clock control, loading self-test registers, etc.). Remaining OCM functions are tested by exercise of the OCM's remaining instruction set. RAMs embedded in a chip will not be completely tested by the LOCST self-test technique. Special RAM self-test circuitry would be needed to provide effective testing with random patterns. This topic is not addressed here.

The locations of the PPG and SA LFSRs are not illustrated because this would require a detailed scan string diagram. As mentioned previously, the PRPG and SA LFSRs utilize existing functional latches. The two other chips, B and C, when configured with a vendor multiply chip, perform digital filtering functions. Like Chip A, Chips B and C are primarily arithmetic data pipelines. All three chips are now incorporated in signal-processing systems.

To determine the testing effectiveness of the LOCST technique on these three chips, we performed fault simulation of the self-test procedure. Fault simulation provides a test coverage value upon which self-test effectiveness is based. The fault simulation was based on the classical stuck-fault model. Full fault simulation of the LOCST operation would have been too costly, so we followed this methodology:

- We used a statistical random sample of the full stuck-fault list. Test coverage results therefore have a 95 percent confidence level.
- Since no significant ($\ll 1\%$) error masking occurs due to the LFSR compression of the test results,[10-12] simulation of the serial compression activity of the SA LFSR was not performed. If the detection of a fault is observed at an SRL, it is assumed that this fault will be detected after LFSR compression.

We generated pseudorandom patterns placed in the latches during fault simulation via a PL/I program, using the same characteristic polynomial as the PRPG LFSR implemented on the chips (see Figure 3). A plot of test coverage vs. the number of self-test sequences for Chip A is presented in Figure 7. A total chip test coverage of 88.6 percent was achieved (with 95 percent confidence) with 3000 self-test sequences. Figure 8a displays the coverage evaluation results for Chip A in a different manner. Here Chip A's logic is divided into three categories (interior logic, exterior logic, and OCM logic) to highlight the LOCST testing effectiveness for each. LOCST test effectiveness for all three chips is summarized in Figure 8.

Implementions and coverage evaluation

The LOCST technique has been implemented on three VLSI chips used for signal-processing applications. The three chips—hereafter called Chip A,

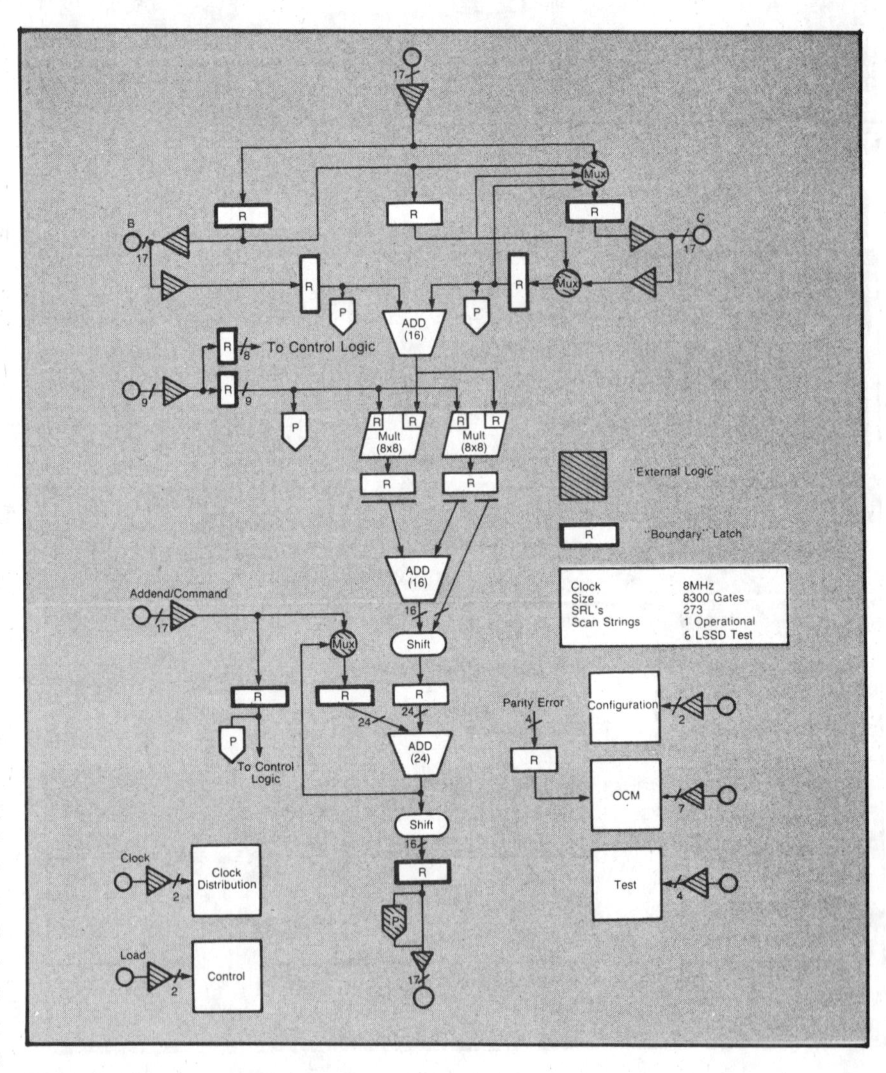

Figure 6. Signal-processing Chip A.

Chip B, and Chip C—were designed and fabricated in 1982. The addition of the LOCST capability (i.e., LFSRs for PRPG and SA functions and OCM self-test control logic) represents a hardware overhead of less than two percent. (Note: This figure does not include LSSD overhead or OCM overhead, as these features are included whether or not LOCST is implemented. Total testability overhead is in the 10-15 percent range.)

One of the three chips, Chip A, performs front-end signal-processing functions requiring high-rate, multiply-intensive algorithms such as finite-impulse response filtering, linear beam-forming, and complex band-shifting operations. Chip A performs these functions by utilizing a simple add-multiply-add pipelined data structure. A high-level diagram of Chip A is shown in Figure 6.

Overall test coverage values of 88.6, 81.5, and 85.3 percent (mean values of a 95 percent confidence interval) were obtained for the three chips respectively. Very good coverage (>97%) was obtained for the interior logic of all three chips with relatively few random-pattern loads (<5000). Test coverage obtained by deterministic LSSD test pattern generation was greater than 99 percent for all three chips. Whether or not test coverage comparable to that of LSSD testing could be obtained if more random-pattern loads were simulated (e.g., 10K, 100K, or 1M) was not evaluated because of the limited budget of this evaluation task.

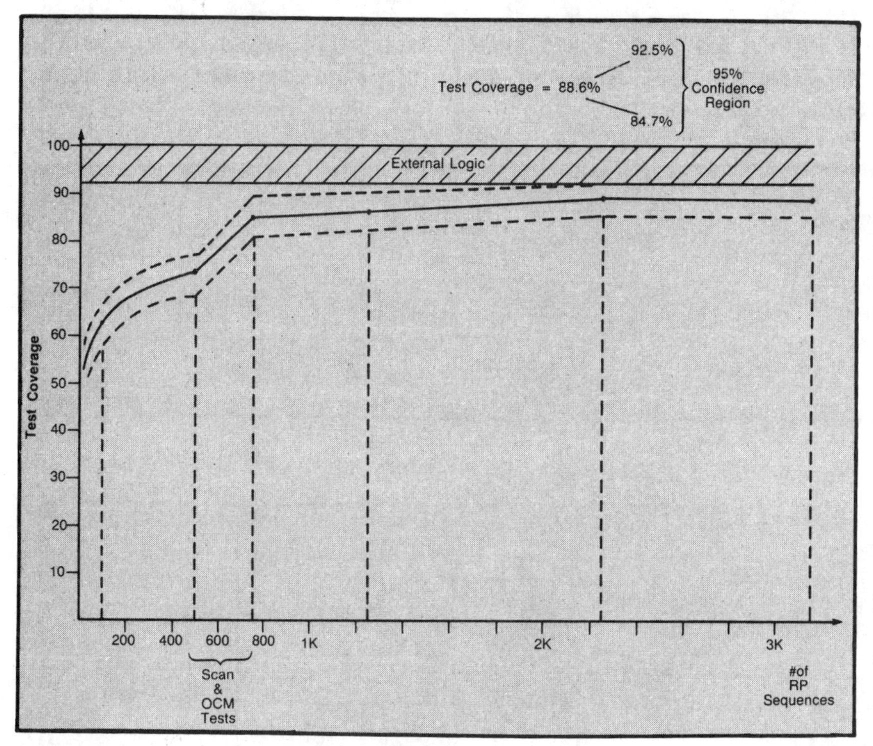

Figure 7. Test coverage results for Chip A.

Table 1.
LOCST test time.

	NO. OF RPs	NO. OF SRLs	TEST TIME
Chip A	2K	213	0.43s
Chip B	500	230	0.12s
Chip C	3K	223	0.67s

Test time = (No. of RPs/scan rate) × No. of SRLs
Scan rate = 1 MHz
RP = random-pattern sequence

LOCST execution time

In addition to providing high test coverage, a self-test technique should execute in a relatively short period of time. Table 1 presents the equation for calculating LOCST execution times and the predicted test times for the three FSD chips. For the assumed scan rate (based on existing FSD scan controllers) and the number of self-test sequences (based on the presented test coverage evaluation), subsecond execution times are achieved for all three chips.

If a large number of random-pattern loads is required to achieve adequate test coverage results, if the scan rate is slow (e.g., 1 MHz or less), or if a chip contains a large number of SRLs, LOCST self-test times may become quite large (minutes). An alternative to the basic LOCST implementation is to use many parallel scan strings feeding a multiple-input signature register, or MISR. This modification, illustrated in Figure 9, reduces the number of serial shifts required to fill all chip SRLs with random test data, thereby reducing the overall LOCST test time.

Self-test environments

One of the greatest potentials of self-test is the possibility of eliminating the need to produce a unique test pattern set for each test environment. The major test environments are

- chip manufacture test,
- card test,
- operational system test, and
- field return test (repair test).

The lack of defect diagnostic information is the key reason that self test is not considered a viable technique for chip manufacture testing. But ongoing research is investigating the use of self-test techniques for LSI devices in the chip manufacture environment. A very promising technique using random-pattern testing for diagnosing failures has been developed by F. Mo-

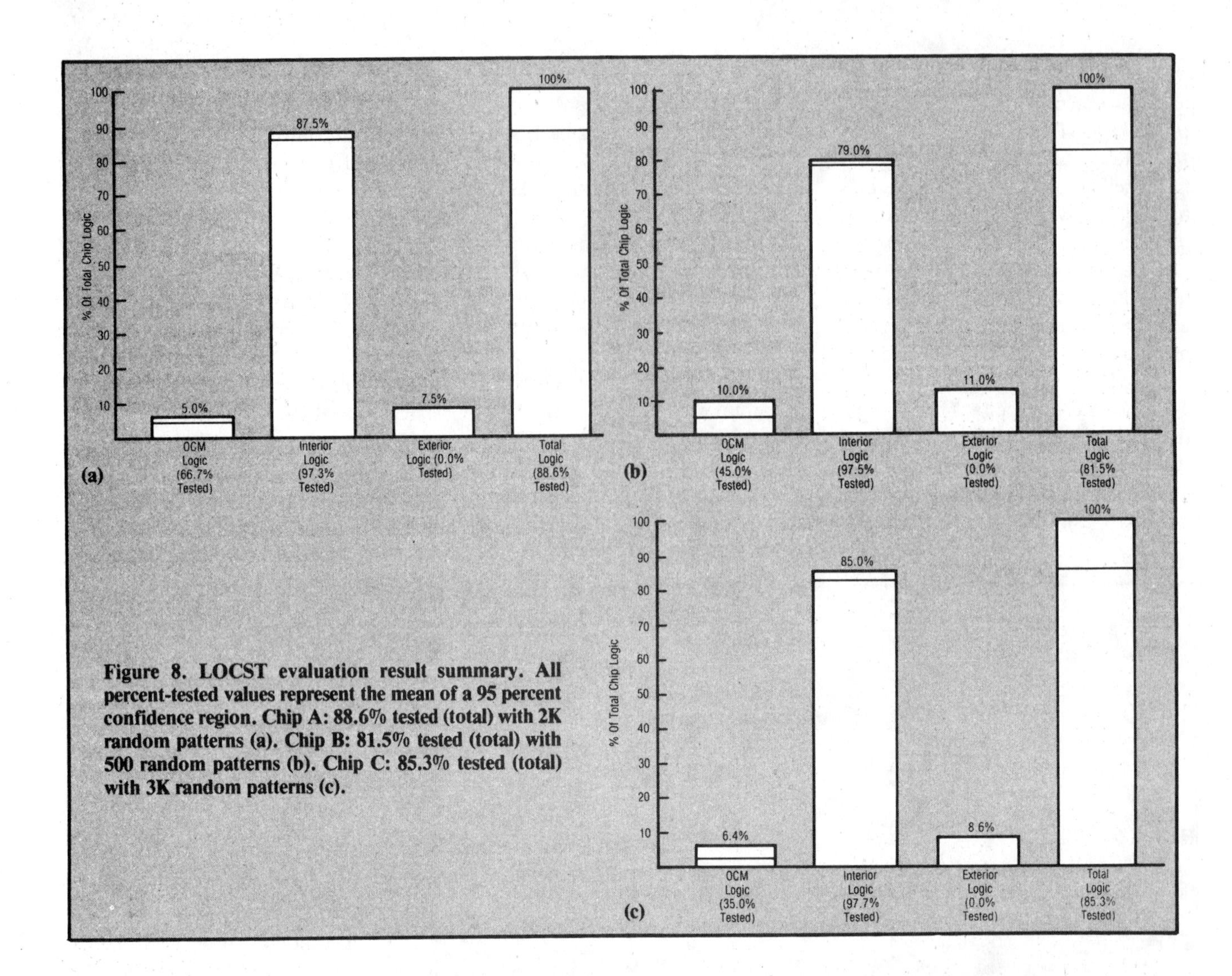

Figure 8. LOCST evaluation result summary. All percent-tested values represent the mean of a 95 percent confidence region. Chip A: 88.6% tested (total) with 2K random patterns (a). Chip B: 81.5% tested (total) with 500 random patterns (b). Chip C: 85.3% tested (total) with 3K random patterns (c).

tika et al.[13] of IBM Kingston. Presently, LOCST does not replace LSSD testing in the FSD chip manufacture test environment but is used as a supplemental chip-testing technique. As a minimum, since it provides a rapid pass-fail indication, self-testing would be useful in a production test environment to provide efficient preliminary screening of product.

The inclusion of several 10,000-gate VLSI components onto cards that have historically contained 5000 to 8000 gates of logic posed a serious problem to traditional card test methodologies. On-chip self-test offers a very effective solution. LOCST is used to verify that the FSD VLSI components on a card are defect-free. All FSD VLSI components are accessed via their OCM interface, requiring only seven card connector pins. Chip boundary scan latches (accessed via

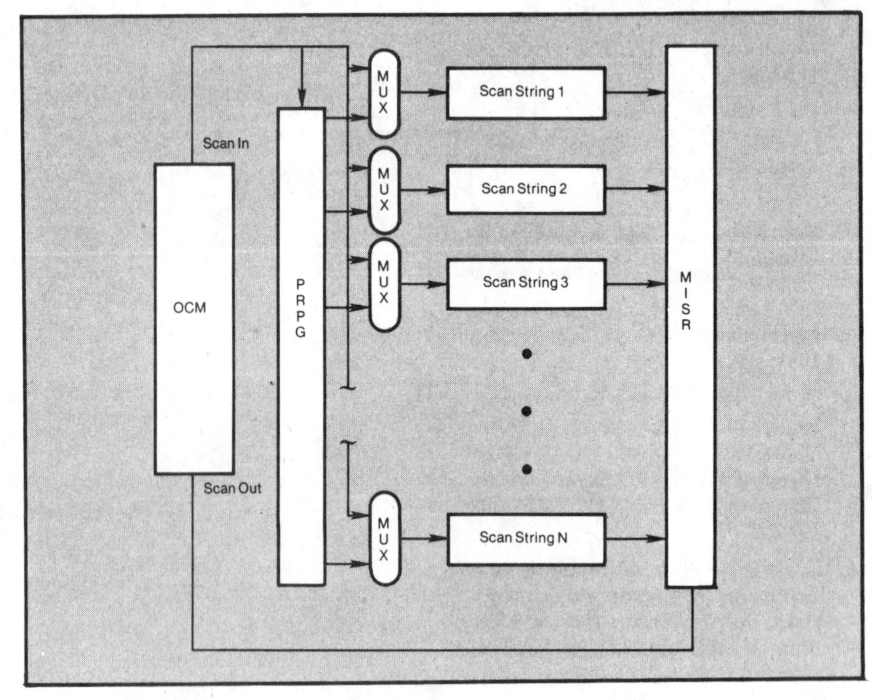

Figure 9. LOCST modification for faster execution.

the OCM) are used to apply and capture data for chip-to-chip interconnect testing. Boundary scan also effectively isolates FSD VLSI components from vendor components, enabling the use of traditional methods for testing the vendor logic on the card.

On-chip self-test supports the following types of operational system testing:

- system initialization test,
- system on-line periodic test, and
- system off-line fault localization test.

The objectives of implementing an on-chip self-test capability in our VLSI chips were to substantially reduce the plethora of unique test pattern sets for the differing test environments, reduce the volume of test vectors required to test our VLSI products, and eliminate the need for manual test pattern generation. Priorities of low hardware overhead, simple implementation, simple self-test control, high test coverage, and short self-test execution time were of prime importance. The implementation of LOCST on our VLSI products has enabled us to meet these objectives without violating any of the priorities. □

Acknowledgments

I would like to express my appreciation to E. B. Eichelberger and E. Lindbloom for their counsel and encouragement during the development and implementation of this self-test technique. I would also like to thank Tina Nguyen for her assistance with the fault simulation activity.

References

1. B. Koenemann, J. Mucha, and G. Zwiehoff, "Built-In Logic Block Observer," *Digest of Papers 1979 Test Conf. IEEE,* Oct. 1979, pp. 37-41.
2. R. M. Sedmak, "Design for Self-Verification: An Approach for Dealing with Testability Problems in VLSI-Based Designs," *Digest of Papers 1979 Test Conf. IEEE,* Oct. 1979, pp. 112-120.
3. E. J. McCluskey and S. Bozorgui-Nesbat, "Design for Autonomous Test," *IEEE Trans. Computers,* Vol. C-30, No. 11, Nov. 1981, pp. 866-875.
4. E. B. Eichelberger and E. Lindbloom, "Random-Pattern Coverage Enhancement and Diagnosis for LSSD Logic Self-Test," *IBM J. Research and Development,* Vol. 27, No. 3, May 1983, pp. 265-272.
5. T. W. Williams and E. B. Eichelberger, "Random Patterns Within a Structured Sequential Logic Design," *Digest of Papers 1977 Semiconductor Test Symp. IEEE,* Oct. 1977, pp. 19-26.
6. P. Bardell and W. McAnney, "Self-Testing of Multichip Logic Modules," *Digest of Papers 1982 Int'l Test Conf. IEEE,* Nov. 1982, pp. 200-204.
7. D. Komonytsky, "LSI Self-Test Using Level Sensitive Scan Design and Signature Analysis," *Digest of Papers 1982 Int'l Test Conf. IEEE,* Nov. 1982, pp. 414-424.
8. E. B. Eichelberger and T. W. Williams, "A Logic Design Structure for LSI Testability," *J. Design Automation and Fault Tolerant Computing,* Vol. 2, No. 2, May 1978, pp. 165-178.
9. D. C. Bossen and M. Y. Hsiao, "ED/FI: A Technique for Improving Computer System RAS," *Digest of Papers 11th Ann. Int'l Symp. Fault-Tolerant Computing,* June 1981, pp. 2-7.
10. R. A. Frohwerk, "Signature Analysis: A New Digital Field Service Method," Hewlett-Packard Application Note 222-2, pp. 9-15.
11. J. E. Smith, "Measures of the Effectiveness of Fault Signature Analysis," *IEEE Trans. Computers,* Vol. C-29, No. 6, June 1980, pp. 510-514.
12. D. K. Bhavsar and R. W. Heckelman, "Self-Testing by Polynomial Division," *Digest of Papers 1981 Int'l Test Conf. IEEE,* Oct. 1981, pp. 208-216.
13. F. Motika, et al., "An LSSD Pseudo Random Pattern Test System," *Int'l Test Conf. 1983 Proc. IEEE.,* Oct. 1983, pp. 283-288.

Johnny J. LeBlanc is a staff engineer working in the System/Subsystem Testability department of the IBM Federal Systems Division in Manassas, Virginia, where he is responsible for the development of VLSI design for testability methodologies, including VLSI test generation, built-in test, and on-chip self-test.

LeBlanc received his BS in electrical engineering from the University of Southwestern Louisiana in 1976 and an MS in electrical engineering from North Carolina State University in 1978.

Questions concerning this article may be addressed to LeBlanc at Bldg. 400/044, IBM Federal Systems Division, 9500 Godwin Drive, Manassas, Va. 22110.

A Fast 20K Gate Array with On-Chip Test System

Ron Lake, Honeywell Inc., Colorado Springs, CO

High performance VLSI system design is demanding advances in ASIC technology from IC vendors. Process development must provide dense circuits capable of efficient high-speed performance. Test techniques must verify chip functionality at all levels of integration, from wafer sort to in-system diagnostics, without requiring expensive high-pin-count testers or exhaustive test development.

In response to these demands, ETA Systems Inc. (St. Paul, MN) has designed a high-performance CMOS gate array with an on-chip self-test system known as BEST, for *B*uilt-in *E*valuation and *S*elf *T*est. Derived from Control Data Corp.'s On-Chip Maintenance System (Resnick, 1983), the BEST system provides the designer with an effective method of verifying chip functionality and ac performance using output signature analysis. The BEST system provides this test function with little input from the designer—effectively removing the time-consuming effort of test vector generation for fault coverage from the design cycle. The gate array product is licensed to Honeywell for commercial sale and is referred to as the HC20000 (HC20K).

The HC20K is a CMOS gate array with a density of 20,000 NAND gates (Figure 1). The chip is fabricated in CMOS-III, a 1.25-micron, dual n-well epitaxial process with oxide isolation and double-level metal interconnect. The array contains 12,065 internal logic cells of six transistors each, arranged in a matrix structure. This equates to 18,097 internal 2-input NAND gates; 80% utilization is recommended. Typical 2-input NAND gate delay is 450 ps at 25 °C (fan-out of one). Worst-case performance is 600 ps over the commercial temperature range and 900 ps for military temperatures. A two-tier structure of 284 I/O pads rings the periphery of the internal matrix. The BEST self-test network is incorporated into this I/O structure, and requires 2,000 gates of internal logic.

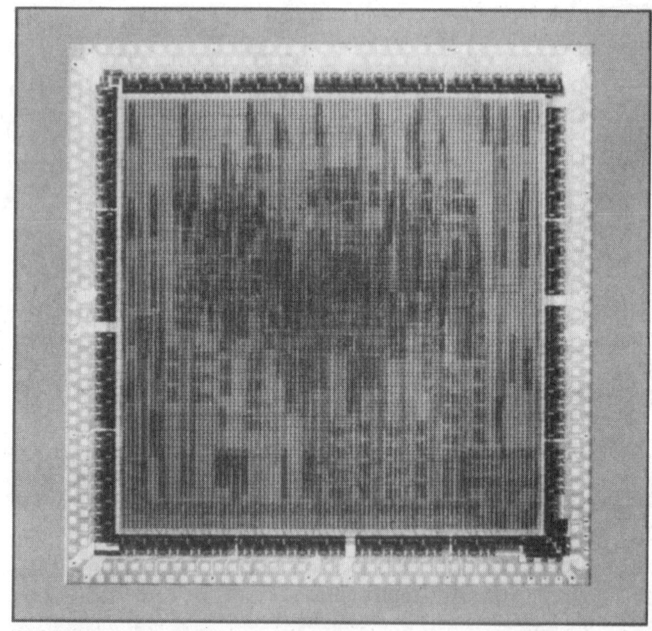

FIGURE 1. HC20K die photograph.

HC20K I/O Structure

The 284 I/O pins on the HC20K are divided into several functions: 40 pins for power and ground, four pins for the BEST system, one system clock pin, one hold-off function pin, and 238 data I/O pins. The data I/O pins are further subdivided into 140 bidirectional pins and 98 input-only pins. Each of the 238 possible input buffers may be selected for either TTL or CMOS trigger levels; all contain an input protection network. All output buffers contain dedicated force-active (FAC) and force-off (FOF) pins, which permit the BEST network to force all outputs active or tristate for parametric testing. The hold-off pin (HOF) is used to synchronize chip-to-chip data transfer within a system.

The 40 power and ground pins help minimize the effects of current spikes. Separate power and ground buses for internal array logic and I/O buffers maintain internal logic integrity even with large transient currents. Output buffers are further subdivided into three groups of 20, four groups of 18, and one group of eight, each with a separate power and ground bus. This subdivision maintains output buffer performance even when large numbers of these outputs switch simultaneously.

The system-clock pin brings the clock signal through an input buffer and distributes it to four separate sets of programmable clock drivers, which are distributed one to each side of the die. The clock drivers all drive a clock bus network, which surrounds the internal matrix and distributes the clock signal with minimum skew to each cell of the array. Each programmable driver contains transistors of three different sizes: the transistor selected for a given application depends upon the total capacitive load that must be driven by the clock bus.

Providing for data synchronization in a system environment, the hold-off pin tristates outputs from a single chip for a user-specified duration before allowing active data to pass.

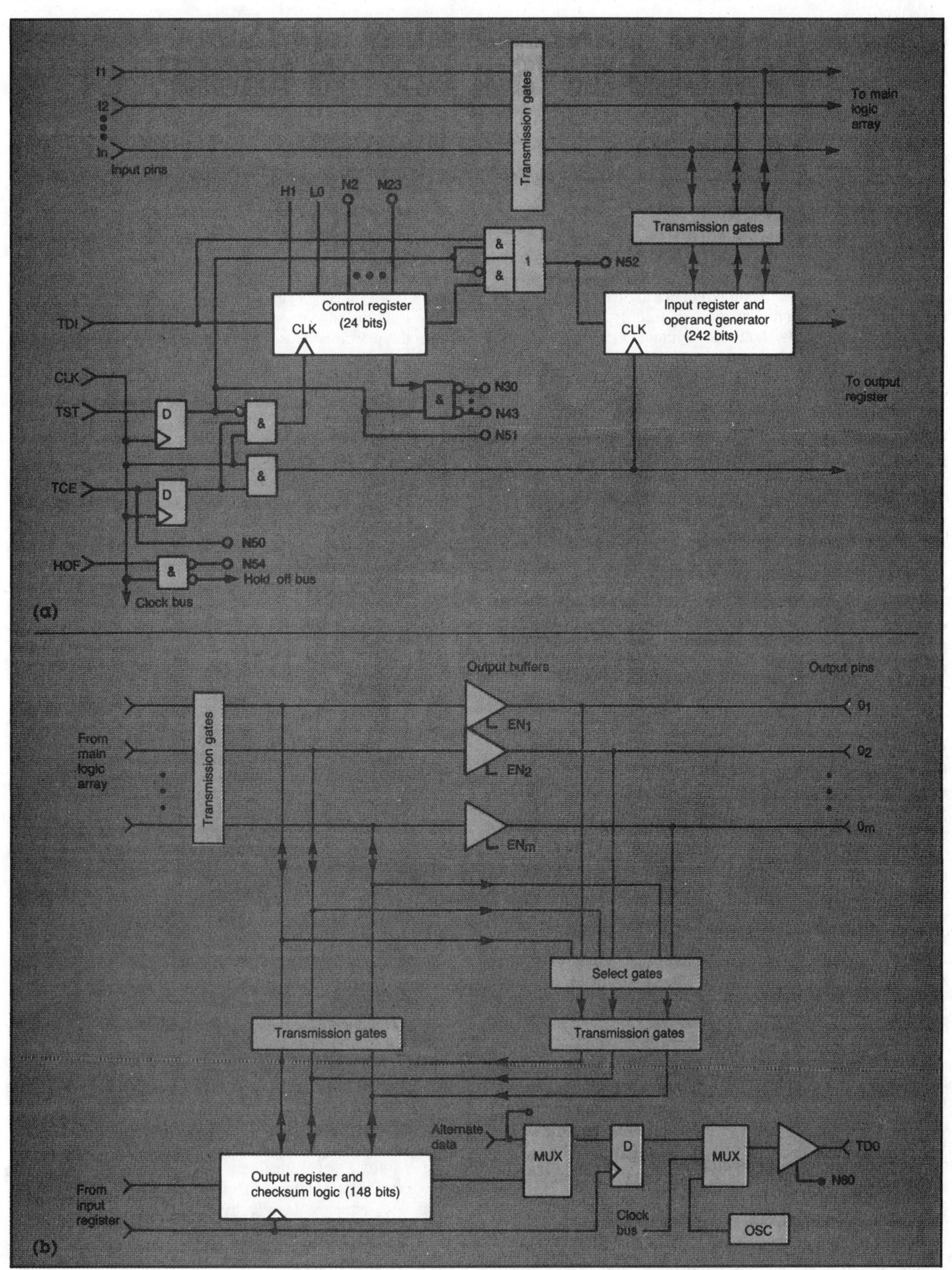

FIGURE 2. Built-in Evaluation and Self Test: input block (a) and output block (b).

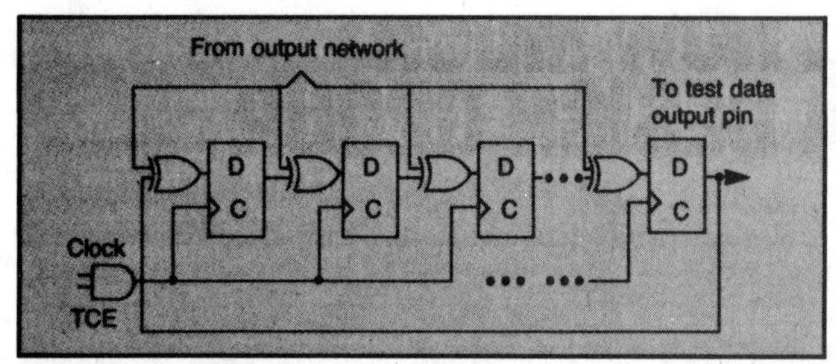

FIGURE 3. Output register in the checksum mode.

This feature is particularly useful in high-performance systems running at or near maximum operating frequency. Small skews in interchip communication caused by fast data paths can be corrected by hold-off without affecting system performance.

The BEST maintenance system requires four pins: test clock enable, test strobe, test data in, and test data out. Because the BEST network is crucial to HC20K applications, several later sections are devoted to explaning its capabilities and the interaction of these four system pins.

HC20K Macrocell Functions

A macro function library supports application design on the HC20K. All macros are optimized first for performance and then for logic density. Macros are constructed using double-level intra-macro metal interconnect, which increases array routability by freeing external channels for inter-macro connections. Speed-critical macro primitives (NAND, NOR, NOT) are constructed with different speed-power-size options to allow designers to optimize critical logic paths while minimizing system power.

Register and multiplexer macros are extensible to allow for variable-length logic functions without wasting logic gates. Stackable functions are provided by defining separate control block and element macros instead of a fixed-length combination. This separation of control provides users with the flexibility of an *n*-bit register, for example, without wasting extra storage bits or duplicating the control function.

CMOS transmission gates have been used to develop both dynamic and static flip-flop macros. Because dynamic flip-flops require no feedback, they take fewer devices to implement. Fewer devices leads to less capacitance on critical nodes, thereby giving dynamic flip-flops better performance. Dynamic flip-flops have transmission gates wired in master-slave fashion with insulator gates at each output. Insulator gates draw no dc current. Gate capacitance therefore acts as a temporary storage mechanism, holding a valid logic level until leakage currents eventually destroy it. As long as the clock runs at a minimum frequency of 10 kHz, however, leakage currents will not have time to upset operation, and the flip-flops will perform properly.

If the minimum clock rate constraint cannot be met, static flip-flops with feedback are available. With a static flip-flop, the clock may be stopped high or low and data integrity will be maintained as long as power is supplied to the chip.

Built-in Evaluation and Self Test

The BEST system provides the logic designer with special features to aid in testing a design during wafer probe, packaged IC test, system test, and in-system field maintenance. The system permits probing and testing with only 30 I/O pins connected, PC board interconnect testing, standardization of test programs for different array designs, and on-line integrity checking during normal system operation. With BEST, development of long test programs is not required: the designer must merely initialize the chip logic and then access the BEST system through the control and data pins. The BEST logic generates a final output signature by summing all logic outputs during the pseudo-random test sequence. The designer need only check this final result to verify chip functionality.

The BEST system comprises a 24-bit control register, a 242-bit input register with operand generation capability, and a 148-bit output register with checksum capability (Figure 2). These three components are arranged into a serial shift register configuration. To access the BEST logic, four I/O pins are required. These are the control pins, test clock enable (TCE) and test strobe (TS); and the data transfer pins, test data in (TDI) and test data out (TDO). TCE gates the system clock to the maintenance registers, while TS engages the maintenance function. If TS is low, the maintenance registers are separated from the array logic, so that data may be serially shifted through the maintenance registers with no effect on system operation. When TS goes high, a function code is frozen in the control register, and the contents are gated to the array control nodes. TDI is used to shift data serially to the first bit of the control or input registers in order to provide the input register with an initial seed value and to define the function in the control register. TDO serially shifts data from the last bit of the output register to the outside world, allowing the designer to examine the output register results.

The use of these four control pins along with system clock, hold-off, 12 of the V_{DD} pins, and 12 of the V_{SS} pins provides full functional testing of this array at low pin count. Functional tests may be performed at system speed up to a maximum frequency of 100 MHz (at room temperature).

BEST Registers

The input register contains one bit for every input buffer on the array, plus four extra bits to multiplex data for vector generation. It serves as either a data source or destination for nodes between the input buffers and the internal gates of the logic array. The register may be loaded either in parallel from the input buffers or serially through the TDI pin. When in test mode, the logic-array gate inputs may be isolated from the input buffers, and receive instead the contents of the input register. When this input register is subsequently clocked, pseudo-random operands are generated and applied to the array inputs at the system clock rate.

Similar in operation to a cyclic redundancy code generator, the pseudo-random generator is formed by feeding back the input register's output at selected intermediate points, and half adding this result to the previous state value at this bit. Given a user-defined seed value, the number generator will define a unique set of patterns that begins to repeat after approximately 10^{35} patterns. Thus, a designer may develop a unique test sequence merely by specifying an initial seed value and a number of clock iterations. The BEST circuitry will then generate the vectors required to provide a particular level of fault coverage.

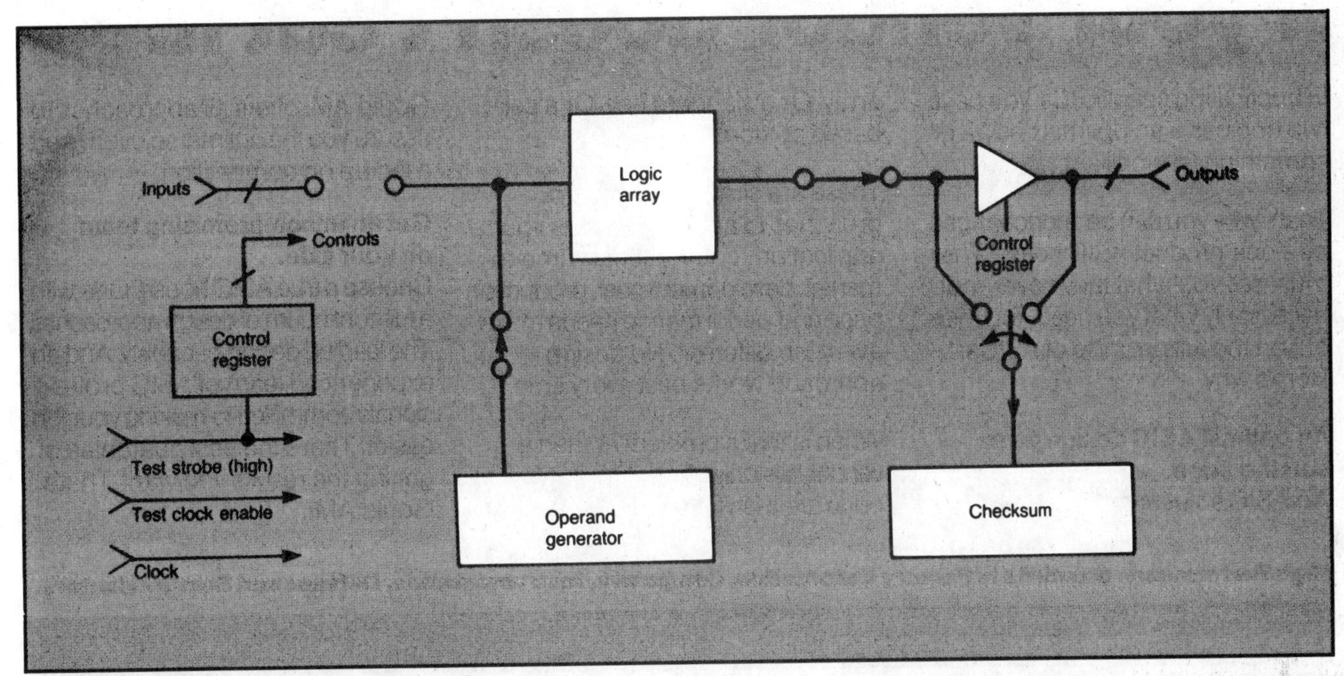

FIGURE 4. BEST in the self-test mode.

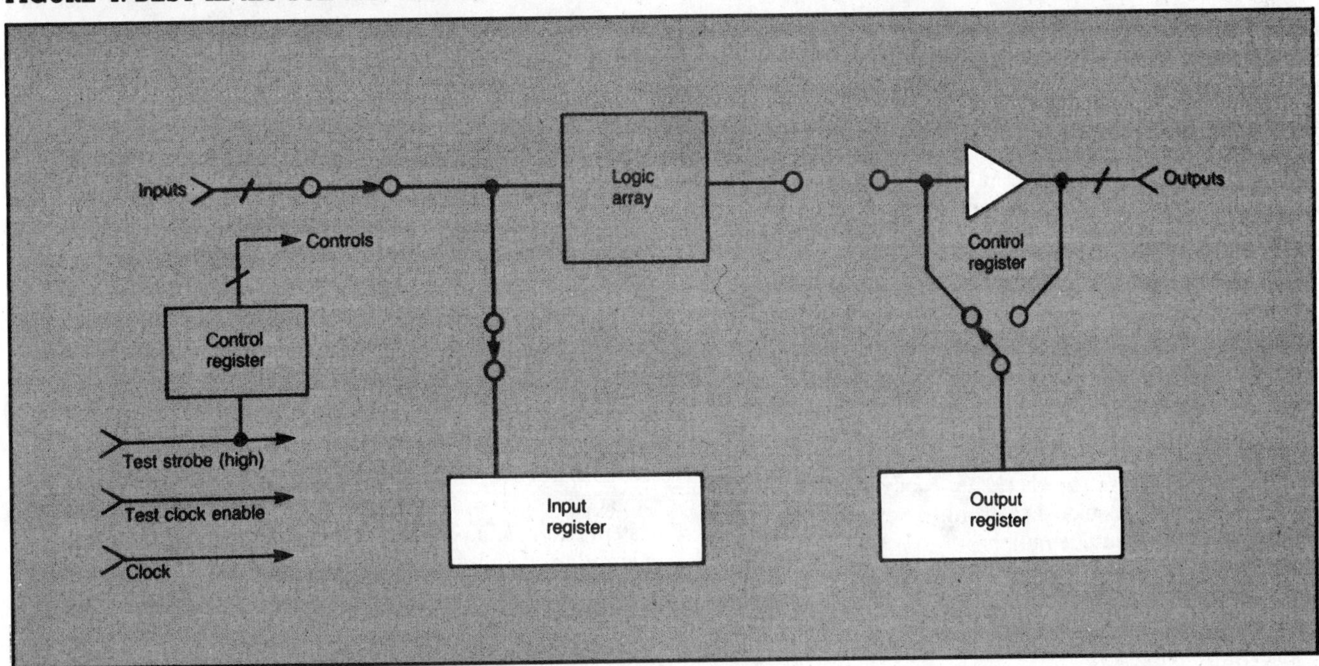

FIGURE 5. BEST in the interconnect test mode.

The output register contains one bit for every output buffer on the array plus an extra bit for data multiplexing, and may source or sink data. When sourcing, the logic-array gate outputs are separated from the output buffers, and the output register data is substituted. This allows known data to be forced through the outputs in order to verify buffer functionality. When sinking data (in checksum mode), data is loaded into the output register from either the logic-array gate outputs or the output buffers, at the user's choice. Again, this gives the user the flexibility to test separately the functionality of the logic array and the output buffers. In this mode, each bit loaded into the output register is half-added to the contents of the previous bit of the register. The result is reloaded into the output register with the data shifted by one bit. The shift is circular in that the data from the last bit is loaded into the first bit, ensuring that an error at any pin is kept in the checksum (Figure 3).

The 24-bit control register is partitioned into a 10-bit system portion and a 14-bit user portion. The individual bits in the system portion each control a distinct function in the BEST system. The first eight bits connect and disconnect data paths between the registers and internal logic gates; they also enable the input operand generation and output checksumming. Bits 9 and 10 are used to tristate or force active all output buffers. Outputs of the 14 user bits are available for definition by the designer. Several uses for these bits could be

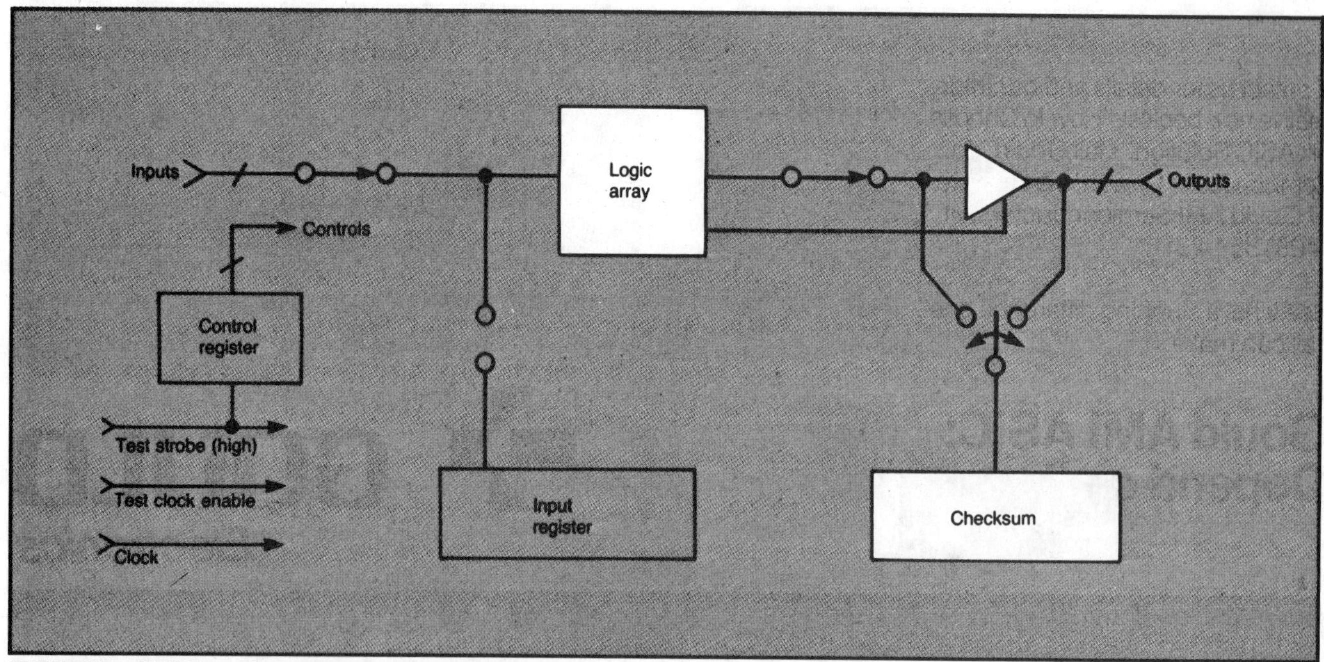

FIGURE 6. BEST in the system checksum mode.

to initialize the internal logic, set all flip-flops high or low, control a set-scan network, or multiplex internal logic nodes out the TDO pin.

BEST System Operation

The BEST maintenance system is configured to not affect system performance. Maintenance registers lie in parallel with the I/O pins instead of in series; thus, data will not pass through an input register bit when it passes from an input buffer to the logic array. The BEST system supports self test, interconnect test, ac test, and provides a logic-analyzer mode and a system checksum mode.

For self test (Figure 4), the control register must set the following conditions:

- Input register set to random generator mode;
- Output register set to checksum mode;
- Output register sinks data from either the logic-array outputs or the output buffers.

The designer must then provide:

- A chip initialization sequence, perhaps using a bit in thecontrol register;
- A seed value for the pseudo-random number generator, which is shifted in serially;
- The number of clock cycles to iterate;
- The expected checksum result.

If errors occur in the final checksum, the self test could be repeated with intermediate checksum values observed in order to isolate the test cycle that first demonstrates the error.

For the interconnect test (Figure 5), the control register must set the following conditions:

- A known operand in the output register;
- Output register sourcing data to the output buffers;
- Input register sinking data from the input buffers.

By checking input register bits in the various receiving chips in a system, a quick check can be performed to find faults such as opens, shorts, or grounded lines, which exist in the PC board configuration for the system.

For ac test, the control register must be set to select the ring oscillator in the ring periphery to be gated on, and to connect the oscillator output to the TDO pin.

In logic-analyzer mode, the control register must set the input register to sink data from the input buffers, and the output register to sink data from the output buffers. The designer must then bring TS low at predefined times during system operation. This captures I/O data in the input and output registers whenever TS toggles low. The resulting timing "diagram" for the chip may be serially shifted off chip for comparison to expected values.

For system checksum (Figure 6), the control register must set the output register to checksum mode, and connect it to either the logic-array outputs or the output buffers. After a system diagnostic program is run, the final checksum can be checked for validity. This diagnostic can be scheduled during normal operation to perform on-line checking. Due to the parallel configuration of the maintenance registers, this checksum operation has no effect on either system operation or performance. If no diagnostic program is available, normal system operation can be performed with several arrays and the resulting data compared for corroboration. The user must ensure that the test sequence in normal system operation is fully deterministic. There can be no undefined data in storage elements at the beginning of the test, and no interrupts can occur during the test. These constraints will generate a unique checksum that will remain consistent for all gate arrays of a given design. If an error is apparent from the final checksum in a gate array, diagnostic or system programs can be rerun and intermediate checksums compared to isolate the first point of error.

Control Nodes

The HC20K provides 42 internal control nodes for use at

the designer's discretion. These control nodes may be connected within the design netlist to increase a design's testability beyond that automatically achieved by the BEST maintenance circuitry. The control nodes may be divided into two separate cases: nodes that are outputs of the BEST logic, and nodes that are inputs to the BEST logic. Output nodes (N30-N43, N50-N52, and N54) provide the designer with access to the 14 user-defined bits in the control register, the BEST control signals (TCE and TS), TDI, and the hold-off bus. These nodes are intended to give the designer access to critical internal control signals, but may be ignored if the function provided is not required to increase testability.

Input nodes (N2-N23, N60, and N61) must be defined, and fall into three groups. Nodes N2-N23 are used to overwrite the contents of the control register whenever the TS toggles low. These nodes are intended to give the designer the ability to encode a chip type and revision level into the netlist. The nodes are connected to either the power or ground bus by user assignment, and their values may be observed by shifting the control register data through the maintenance registers and out the TDO. Node N60 provides hold-off control of the TDO. If hold-off is not used, this signal should be tied high. Node N61 provides a path for alternate on-chip data to be observed through the TDO pin. This node must be tied high if the alternate data function is not used.

Summary

The HC20K offers the density, efficiency, and performance demanded by VLSI system design. High pin-count I/O reduces the need for data multiplexing while incorporating discharge protection and parametric testing into the buffers. Macro functions are optimized for high performance and logic density. Power, ground, and clock bus distributions support high-performance applications while maintaining data integrity. The hold-off capability allows for fine-tuning a system after prototype delivery.

The BEST maintenance network reduces the burden of test development by providing an on-chip facility capable of verifying the performance of the array at clock rates up to 100 MHz. The system permits probing and testing with only 30 I/O pins connected. □

Reference

Resnick, D. March/April 1983. "Testability and Maintainability with a New 6K Gate Array," *VLSI Design*.

About the Author

Ron Lake holds B.S. and M.S. degrees in electrical engineering from the University of Missouri at Columbia. He is currently a system applications engineer at Honeywell's Digital Product Center. Previously, Ron held various positions in design, applications, and engineering with other semiconductor companies.

INTERCONNECT TESTING WITH BOUNDARY SCAN

Paul T. Wagner

Honeywell, Inc.
Solid State Electronics Division
12001 State Highway 55
Plymouth, Minnesota 55441

ABSTRACT

Boundary scan is a structured design technique which can be used to simplify the testing of digital circuits, boards, and systems. With boundary scan, test patterns can be generated which provide 100% stuck-at and bridging fault coverage of board interconnections. The paper describes the advantages and disadvantages of boundary scan along with the application and implementation of boundary scan circuitry. Algorithms for generating interconnect test patterns for stuck-at and bridging fault coverage are also presented.

INTRODUCTION

Advances in VLSI technology have increased the density and speed of integrated circuits. Thus, the complexity and cost of testing digital integrated circuits, boards, and systems have also increased. By providing a simple means to access the periphery of digital circuits, boundary scan can greatly simplify the task of testing and maintaining systems which use these circuits. This advantage allows boundary scan to reduce the costs of wafer-level IC testing, board and system testing, and system field maintenance.

Wafer-Level Testing

At the wafer level, boundary scan can be used to reduce the need for complex probing fixtures and high-pin-count testers[1]. By using boundary scan to access the primary chip I/O, a simple probe card consisting of power, ground, and serial test interface signals can be used to test chips with hundreds of I/O pads[2]. The decrease in fixturing complexity simplifies test setup, reduces test fixturing costs, and reduces the possibility of damaging the device-under-test during probing.

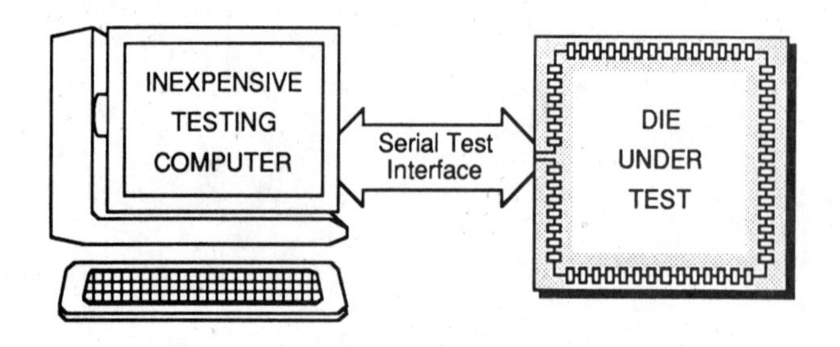

Figure 1: Wafer-Level Testing With Boundary Scan

Besides simplifying testing fixturing, boundary scan also reduces test equipment requirements. Since the boundary scan path provides access to the primary I/O, the testing process is reduced to serially shifting the test pattern into place, executing one or more clock operations, and serially shifting out the results as the next pattern is shifted in. Thus a small, inexpensive testing computer can be used to perform chip testing. This simple setup is shown in **Figure 1**.

Board-Level Testing

At the board level, boundary scan can be used to resolve testing difficulties introduced by new packaging technologies associated with surface mount devices and multi-chip packages. Traditional methods for digital board testing include through-the-hole probing to gain access to the primary component I/O with a "bed-of-nails" testing fixture. Difficulties with the "bed-of-nails" approach include degraded reliability due to over-driving connections from other board components, physical limitation of through-the-hole accessibility, difficulty of reproducing tests, and expenses involved with developing the "bed-of-nails" testing fixture[3,4]. These problems, combined with the increasing use of surface-mount technology[5] and the need for high speed and high pin count testers, have resulted in extremely expensive board-level testing costs.

Boundary scan can reduce the problems associated with board-level testing. As shown in **Figure 2**, boundary scan provides serial access to the primary component I/O and their interconnections. This allows any component to be partitioned from the rest of the board during testing and eliminates the need for a "bed-of-nails" testing fixture. Also, boundary scan reduces the time and cost associated with test pattern generation because test patterns used on the component at wafer level can be modified and applied through the boundary scan path. This can be useful when components are purchased from outside vendors and knowledge of the internal circuitry is limited. Since board interconnections are easily accessed, simple algorithms can be used to generate test patterns which provide 100% stuck-at and bridging fault coverage.

As was the case at wafer-level testing, boundary scan greatly simplifies the setup required for board testing as shown in **Figure 2**. This setup reduces testing costs because the test patterns can be applied serially with an inexpensive test computer through a simple test interface consisting of the boundary scan-in signal, boundary scan-out signal, and necessary control lines.

Reprinted from *IEEE Proceedings 1987 International Test Conference*, pages 52-57.

EH0321-0/90/0000/0248$01.00 © 1987 IEEE

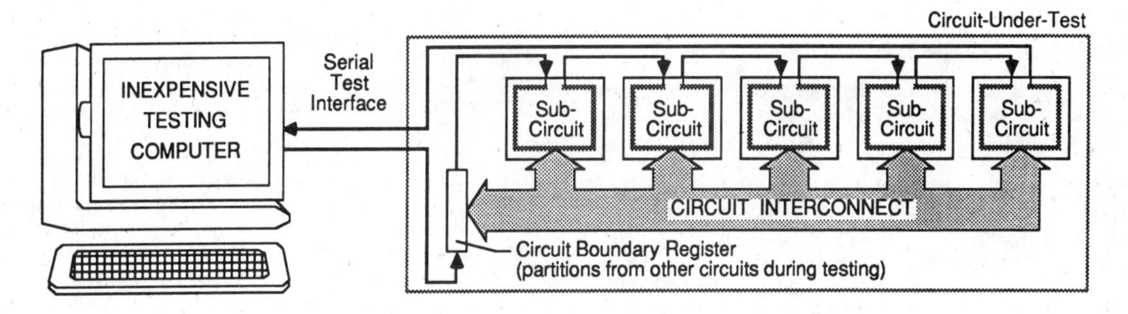

Figure 2: Board-Level Testing With Boundary Scan

Field Testing

Boundary scan can also reduce the cost of system field maintenance. Since boundary scan tests the input buffers, the output buffers, and all component interconnect, it provides excellent coverage of the most common field failures. Furthermore, the procedure for testing with boundary scan in the field is nearly identical to that described for board-level testing. Thus field testing can be performed using a simple testing computer accessing a serial test interface. Since very few interconnect test patterns are required, the testing computer can be as simple as a lap-top personal computer, which is ideally-suited for field maintenance.

Boundary scan can be used to test the system interconnections and to partition the system into separately-tested modules. In this case, testing will isolate the fault to a single module or to a faulty interconnection(s) if the individual modules can themselves be adequately tested. If boundary scan is extended to the component level, the fault can be isolated to the individual component. Thus, cost-effective repair of the module is possible since the faulty component or interconnection can be easily identified for replacement or repair.

Boundary scan can also be used as part of a system self-test strategy. By allowing a system test processor to access the boundary scan paths in the system, boundary scan can be used to test the system interconnections and to partition the system into smaller self-testable units. The easy execution of self-test and improved fault isolation provided by boundary scan reduce the mean-time-to-repair; thereby increasing system availability.

IMPLEMENTING BOUNDARY SCAN

In general, boundary scan provides a method for accessing all application inputs and outputs from an external test controller. As shown in **Figure 3**, this can be accomplished by including boundary scan registers, which are selected during the test mode, to shift in test patterns and shift out results. The boundary scan registers consist of individual flip-flops associated with each application input and output. These registers are designed to support both a parallel and a serial mode. The registers interface to both the application and its I/O during the parallel mode and can be read from and written to by means of a serial interface during the serial mode. Selecting the boundary registers can be accomplished using either MOS transmission gates[2] or the multiplexers shown in **Figure 3**.

Before actually implementing boundary scan, a number of options must be considered which affect both the design and capabilities of the boundary scan circuitry. These options include: the use of application registers as boundary scan registers, the control of output buffers, the selection of a test interface, and the implementation of the boundary registers. These options and others are addressed in the following sections on implementing the components of a boundary scan technique.

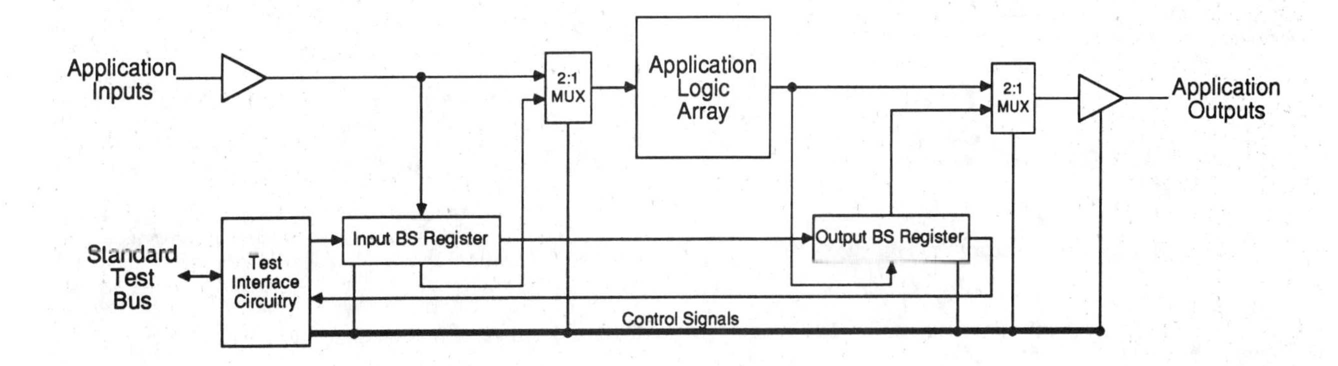

Figure 3: A Conceptual Diagram For Boundary Scan

Dedicated Boundary Scan Registers

Boundary registers can either be dedicated for boundary scan testing or they can be used in both functional and test modes. When implementing boundary scan on high-speed bipolar integrated circuits, we found that there were a number of advantages to using some functional registers for boundary scan testing. First, the high-speed of the system mandates that most of the chip inputs and outputs be registered directly at the I/O buffer. Since we already incorporate serial scan[6] in our chip designs, these registers were easily added to the boundary scan path. Dedicated boundary "shadow" registers are then added to any I/O which are not directly registered. A 2:1 multiplexer is used to make the shadow registers visible during the test mode and invisible during the functional mode. This approach of exploiting existing registers substantially reduces both the circuit and power overhead associated with boundary scan and eliminates a 2:1 multiplexer delay from the path of critical signals.

If boundary scan is to be implemented on a gate array product, associating dedicated shadow scan registers with the I/O buffers at the periphery of the array has a number of advantages. First, the user of the gate array can utilize boundary scan with little or no design effort. Furthermore, array cells are not consumed when implementing boundary scan and numerous signal routings are eliminated. Finally, implementing dedicated boundary scan registers on a CMOS gate array product[2] will not significantly increase the chip power (contrary to bipolar designs).

The Boundary Scan Bit-slice

The boundary scan registers consist of bit-slices that are attached to each application input and output. Our implementation of this bit-slice is shown in **Figure 4**. This consists of a 3:1 multiplexer which allows data to be loaded in the functional mode, serial data to be shifted in the test mode, and a reset operation to be performed. The output of the multiplexer is then fed to the scan flip-flop which in turn drives the scan out signal and the chip output.

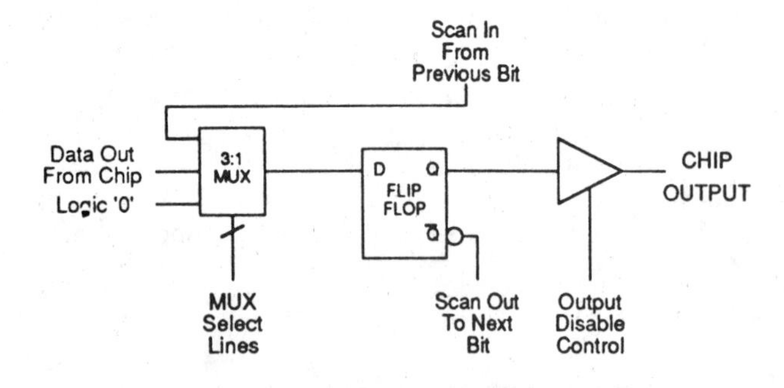

Figure 4: The Boundary Scan Bit-slice

As data is shifted through the boundary scan path, the chip outputs must be latched or disabled to prevent unwanted and possibly damaging output conditions. For example, the scan operation could damage output buffers by forcing two separate output drivers on the same net to different logic levels. Also, the shifting operation may cause a large number of output buffers to change state at the same time; resulting in excessive noise on power and ground busses. For these reasons, a global output buffer disable signal is included in our implementation of boundary scan and can be controlled by the test interface circuitry.

If the testing of asynchronous sequential logic is necessary, a latch must be added to between the flip-flop and output buffer to hold the output state during shifting operations. A similar latch would also be required at the input boundary register if the application logic array contained asynchronous sequential logic to be tested with the boundary scan circuitry. Typically, we do not include this latch because we infrequently use asynchronous sequential logic in our digital system designs.

Another implementation concern involves connecting the boundary scan bit-slices as inverting serial shift registers or as non-inverting serial shift registers. The advantages of an inverting serial shift path include the easy identification of faults in the shift path. To test the shift path, the entire path is reset to either a logic 0 or a logic 1 and the contents are shifted out. The serial output pin is then examined for an alternating pattern of ones and zeros. If the data remains at a logic 1 or 0 after k clocks, then we know that a fault exists k bits back from the output pin. With this information, we can quickly isolate the cause of the fault. Without the inverting boundary scan path, finding the fault could be tedious and difficult task. For this reason, we frequently make use of inverting serial shift paths when implementing boundary scan.

The Test Interface

Selecting an appropriate test interface is a very important part of the boundary scan implementation. A common interface will allow the boundary scan paths of multiple chips on a complex circuit board to be easily accessed. Without this common interface, many of the advantages of using boundary scan at the board level are diminished due to the difficulty in using the technique.

To resolve this problem, we are using the VHSIC standard Element Test and Maintenance Bus[7] (ETM-Bus) as our serial test interface for boundary scan and other on-chip design-for-test techniques[8]. If the serial test bus is to be connected solely to on-chip boundary scan, a simplified version of this interface logic can be used.

INTERCONNECT TEST PATTERN GENERATION

When testing interconnection nets on a digital module, both stuck-at and bridging faults must be considered. Since the boundary scan path provides direct access to these nets, test patterns can be generated which provide 100% coverage of these faults. The following sections discuss algorithms we use for the generation and application of boundary scan test patterns which detect all possible stuck-at and bridging faults.

Stuck-at Fault Test Pattern Generation

Because stuck-at faults occur on a variety of bus configurations, different test pattern generation algorithms are required for wired-AND, wired-OR, and three-state interconnect nets.

Testing wired-AND interconnection nets. As the name implies, the values forced on a wired-AND interconnection net are logically ANDed to obtain the resulting value. Thus, the wired-AND net can be treated in the same way as an AND gate where 100% of all the stuck-at faults can be detected with **k + 1** test patterns where **k** is the number of inputs. The test patterns can be divided into **k** patterns which test for stuck-at '1' faults and a single pattern which tests for all stuck-at '0' faults. **Figure 5** shows the steps we use for testing wired-AND interconnection nets.

---

1) The driver to be tested is set to a logic '0'
2) All other drivers on the net are set to a logic '1'
3) The data is clocked into the receivers
4) All receivers on the net are examined for a logic '0'
5) Repeat steps 1-4 until each driver is tested
6) Every driver is set to a logic '1'
7) The data is clocked into the receivers
8) Every receiver is examined for a logic '1'

Figure 5: S-A Faults Testing Steps for Wired-AND Nets

Testing wired-OR interconnection nets. Generating test patterns for a wired-OR interconnection net is nearly identical to the wired-AND case. For a wired-OR net with **k** drivers, 100% of all stuck-at faults can be detected with **k + 1** test patterns. In this case, the test patterns can be divided into **k** patterns which test for stuck-at '0' faults and a single pattern which tests for all stuck-at '1' faults. **Figure 6** shows the steps we use for testing wired-OR interconnection nets.

---

1) The driver to be tested is set to a logic '1'
2) All other drivers on the net are set to a logic '0'
3) The data is clocked into the receivers
4) All receivers on the net are examined for a logic '1'
5) Repeat steps 1-4 until each driver is tested
6) Every driver is set to a logic '0'
7) The data is clocked into the receivers
8) Every receiver is examined for a logic '0'

Figure 6: S-A Faults Testing Steps for Wired-OR Nets

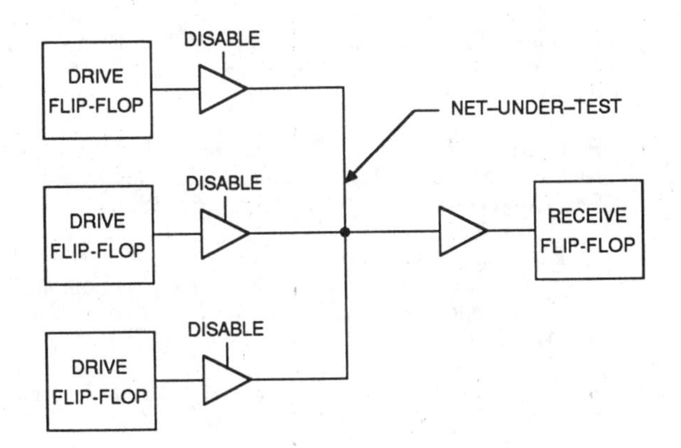

Figure 7: A Three-state Interconnection Net

Testing three-state interconnection nets. When a three-state interconnection net is used, multiple drivers control one or more receivers as shown in **Figure 7**. Since only a single driver can be enabled at any one time, a special restriction is imposed on the generation of the three-state interconnect test patterns. In order to achieve 100% stuck-at fault coverage, each driver on the net must be individually for stuck-at '1' and stuck-at '0' faults while the remaining drivers are disabled. Since this requires **2** test vectors per driver, 100% stuck-at fault coverage can be achieved using **2 • k** test vectors where **k** is the number drivers on the net. The steps we use for testing three-state interconnect nets are shown in **Figure 8.**

---

1) The driver to be tested is enabled and set to a logic '1'
2) All other drivers are set to a logic '0' and disabled
3) The data is clocked into the receivers
4) The receivers are examined for a logic '1'
6) Repeat steps 1-5 until all drivers have been tested
7) The driver to be tested is enabled and set to a logic '0'
8) All other drivers are set to a logic '1' and disabled
9) The data is clocked into the receivers
10) All receivers are examined for a logic '0'
11) Repeat steps 7-11 until all drivers have been tested

Figure 8: S-A Fault Testing Steps for Three-state Nets

Bridging Fault Test Pattern Generation

In addition to testing for stuck-at faults, we also test the interconnects for bridging faults. A bridging fault occurs when two nets are electrically connected as shown in **Figure 9**. A procedure which detects this fault is described in **Figure 10.**

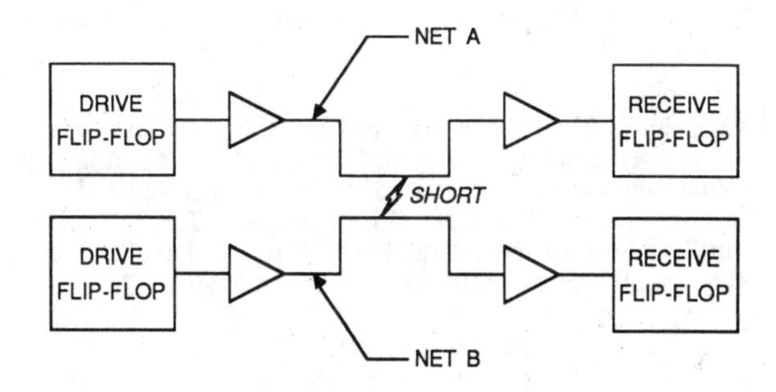

Figure 9: A Bridging Fault Between Two Nets

---

1) Enable the drivers on each net
2) Apply a logic '1' to all drivers on the first net
3) Apply a logic '0' to all drivers on the second net
4) Clock the data into the receivers
5) Examine at least one receiver on each net
6) If the data at the receiver of either net does not correspond with the data applied at the respective driver, then a bridging fault exists between the nets

---

Figure 10: The Procedure for Detecting a Bridging Fault

The procedure described in **Figure 10** operates on two nets. Since a digital module may contain hundreds of interconnection nets, this procedure must be applied to every possible pair of nets to achieve 100% bridging fault coverage. Since separate pairs of nets can be tested at the same time, 100% bridging fault coverage can be achieved with $\log_2(n + 2)$ test vectors where **n** is the number of nets on the board[9].

Bridging fault test generation example. The algorithm we use to generate the $\log_2(n + 2)$ test patterns for bridging fault detection is best illustrated through a simple example. The example given below uses a board with 8 interconnect nets.

Step 1 - Determine the total number of nets on the board. In this example, n = 8 which requires $\log_2(8 + 2)$ or 4 test vectors.

Step 2 - Assign each interconnect net a unique number. Assignments should begin with the number 1 and continue in increments of 1. In this example, the first net is given the number 1, the second net is given the number 2, and the last net is given the number 8.

Step 3 - Assign binary values to each net. Since 4 test vectors are required, assign each net the 4-bit binary equivalent of the net number assigned in the previous step as shown in **Figure 11**.

interconnect net 1 - **0001**
interconnect net 2 - **0010**
interconnect net 3 - **0011**
interconnect net 4 - **0100**
interconnect net 5 - **0101**
interconnect net 6 - **0110**
interconnect net 7 - **0111**
interconnect net 8 - **1000**

Figure 11: The Binary Numbers Assigned to the 8 Nets

Step 4 - Determine the test vectors. The first test vector is comprised of all the bits in the least significant position of the binary numbers. The second test vector is comprised of the bits in the second least significant position. This is continued until all bit positions of the binary numbers have been used. The resulting test vectors are shown in **Figure 12**.

test vector 1 - **10101010**
test vector 2 - **01100110**
test vector 3 - **00011110**
test vector 4 - **00000001**

Figure 12: The Bridging Fault Test Vectors for the 8 Nets

Isolating the faulty interconnects. The bridging fault test pattern generation scheme described in the previous section provides a quick and easy method of bridging fault detection. Although this scheme determines if any bridging faults exist, it does not isolate every interconnection net with a bridging fault. If repairing interconnection nets with bridging faults is possible, all of the faulty interconnects need to be identified. This can be accomplished using the test patterns generated by the algorithm described in the previous example along with an additional $\log_2(n + 2)$ test patterns. Thus, $2 \cdot \log_2(n + 2)$ test patterns can be used to provide complete bridging fault isolation of the interconnection nets.

test vector 5 - **01010101**
test vector 6 - **10011001**
test vector 7 - **11100001**
test vector 8 - **11111110**

Figure 13: Additional Test Vectors for Isolating Faulty Nets

The additional $\log_2(n + 2)$ test patterns are generated by simply inverting the binary values of the first $\log_2(n + 2)$ test vectors. For the previous example, the these test vectors are shown in **Figure 13**. To identify those interconnects with bridging faults, a list of the faulty nets can be maintained during testing. When a bridging fault is detected, the corresponding interconnect net can be identified and added to this list. After all the test patterns have been applied, the list will contain all of the faulty interconnection nets.

CONCLUSIONS

Boundary scan simplifies the testing of digital circuits, boards, and systems. Since boundary scan provides easy access to the periphery of digital circuits through a serial shift path, the setup needed for testing is simplified to an inexpensive computer and a simple test interface. This reduces the complexity and costs of wafer-level testing, board-level testing, and field maintenance.

Boundary scan allows easy partitioning of board components and interconnects, thus wafer-level test patterns can be modified and used to test the components on the board. Also, the simple algorithms presented generate test patterns which provide 100% stuck-at and bridging fault coverage of board interconnects. These advantages allow boundary scan to significantly reduce test and maintenance costs while maintaining a high percentage of fault coverage at the circuit, board, and system level.

REFERENCES

[1] J. J. **Zasio**, "Shifting Away From Probes For Wafer Test," *COMPCOM S'83*, San Francisco, CA, pp. 317-320.

[2] R. **Lake**, "A Fast 20k Gate Array With On-chip Test System," *VLSI Systems Design*, June 1986, pp. 47-55.

[3] F. P. M. **Beenker**, "Systematic and Structured Methods for Digital Board Testing," *IEEE International Test Conference 1985 Proceedings*, pp. 380-385.

[4] H. **Bleeker** and D. **van de Lagemaat**, "Testing A Board with Leaded and Surface Mounted Components," *IEEE International Test Conference 1986 Proceedings*, pp. 317-320.

[5] W. **Booth**, "VLSI Era Packaging," *VLSI Design*, December 1986, pp. 22-35.

[6] H. W. **Miller**, "Design for Test Via Standardized Design and Display Techniques," *Electronics Test*, October 1983, pp. 38-61.

[7] *VHSIC Phase 2 Interoperability Standards ETM-Bus Specification,* Version 2.0, December 31, 1986

[8] L. **Avra**, "A VHSIC ETM-Bus-Compatible Test and Maintenance Interface," *IEEE International Test Conference 1982 Proceedings,* pp. 83-90.

[9] P. **Goel** and M. T. **McMahon**, "Electronic Chip-in-place Test," *IEEE International Test Conference 1982 Proceedings,* pp. 83-90.

Testing and Diagnosis of Interconnects using Boundary Scan Architecture

Abu Hassan, Janusz Rajski and Vinod K.Agarwal

VLSI Design Laboratory
Department of Electrical Engineering
McGill University, 3480 University Street
Montréal, Canada H3A 2A7

Abstract

This paper proposes a new approach to built-in self-test of interconnects based on Boundary Scan Architecture. Detection and diagnosis schemes are proposed which provide minimal-size test vector set, I/O scan chain order independent test vector set and walking sequences. Properties like ease of test vector generation, structure independent detection and diagnosis, local response compaction have made the developed schemes suitable for BIST implementation. An example board interconnect test session is described using one of the proposed schemes.

Key Words : Interconnect, Boundary Scan Architecture, Detection, Diagnosis, Walking Sequence.

1. Introduction

In recent years, structured design-for-testability at the printed circuit board (PCB) level has become an activity of major interest. This is a natural evolution following a wide acceptance of the structured DFT (i.e. scan) at the IC level and the realization that the costs associated with implementing scan cannot be justified unless it can be used to simplify the testing efforts at the PCB and higher levels as well. This combined with the emergence of the very high density packaging technology at the PCB level, in particular that of surface mount interconnections, made it essential to develop the concept of boundary scan, as detailed in Boundary Scan Architecture Standard Proposal, Version 2.0, produced by JTAG [1].

The boundary scan concept allows one to access and control all the primary input and output pins on the PCB from outside. This is done by connecting all the primary inputs and outputs of an IC into a shift register which has a boundary scan input and a boundary scan output. A simple boundary scan cell is shown in Figure 1. The shift registers on all the IC's of a PCB can be connected together to form a larger shift register with a single scan in edge and a single scan out

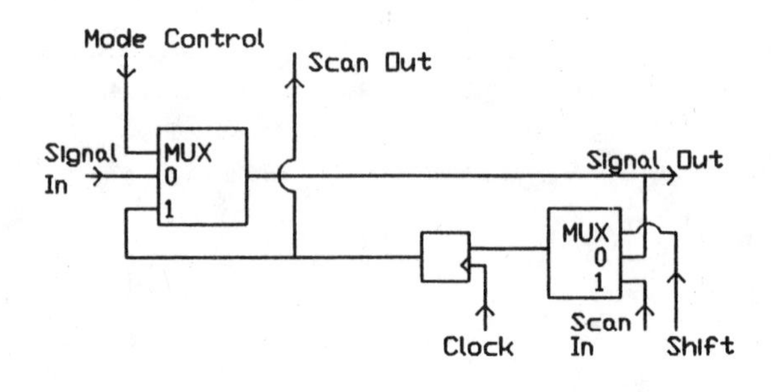

Figure 1 A Simple Boundary Scan Cell Design

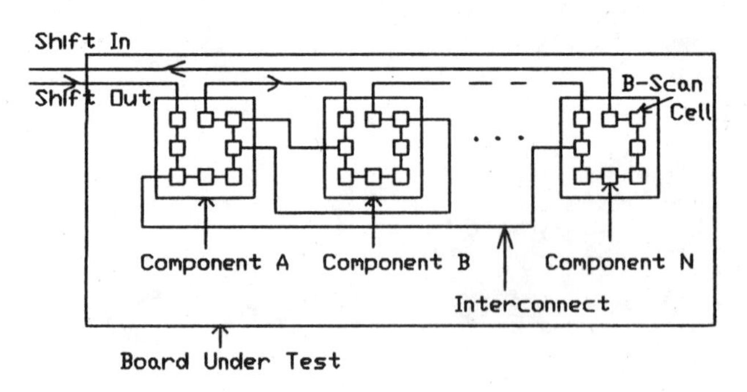

Figure 2 A Boundary Scan Board

edge, as shown in Figure 2. Thus in effect, the boundary scan concept provides a sort of electronic in-circuit testing facility.

Using this concept at the PCB level, it should be possible to confirm that each IC performs its required function, that the IC's are interconnected in the correct manner, and that the IC's interact correctly and that the complete PCB performs its intended function. The problem of interest in this paper is that of using this concept to verify that the IC's are interconnected in the correct manner.

Interconnection of IC's and other discrete components on a PCB is a complex maze of multi-layer electrical conductors which are likely to be failed by the presence of shorts, stuck-ats and stuck-open faults. In order to test such a structure in a cost-effective way structured techniques are required which can be easily automated, possibly BISTed. At the outset, it does not appear to be a simple problem when one realizes that on a single PCB there may be thousands of I/O pins

This work was supported in part by the Commonwealth Scholarship Plan of Canada and in part by the Natural Science and Engineering Research Council of Canada.

Reprinted from *IEEE Proceedings International Test Conference*, 1988, pages 126-137.

EH0321-0/90/0000/0254$01.00 © 1988 IEEE

from all different IC's which are connected to each other in many different ways (unidirectional, bidirectional, one-to-many, many-to-one, forming chains and clusters, etc.). In addition, since repair at the PCB level is a necessary activity, it is not sufficient to know if the interconnect is faulty; one also has to determine where the fault might be if indeed the board is faulty.

This paper attempts to develop a formal set of structured test generation and test diagnosis techniques for interconnect faults on PCB's with boundary scan. These techniques are easily implemented in a BIST manner. This latter requirement implies that the BIST implementations should not require any information about the actual topology of which pin is connected to which. Such techniques are the most important contribution of this paper.

The remainder of the paper is organized as follows. In section two, various basic notions related to boundary scan, different types of interconnects, and failures of interest are described. Many test generation and test diagnosis schemes are developed in section three. An example test session to test the interconnects on a boundary scan PCB using one of the proposed test schemes is described in section four. Research directions and concluding remarks are made in section five.

2. Basic Model

The test access port (TAP) concept of the boundary scan architecture facilitates standard test communication protocol between IC's on the same PCB manufactured by different vendors. In the JTAG proposal [1], the TAP consists of a test data in (TDI) pin, a test data out (TDO) pin, a test clock (TCK) pin, and a test mode select (TMS) pin. These pins are used to access (i) an instruction register in TAP; (ii) the boundary scan register; or (iii) some user defined data registers. More details can be found in [1].

As seen in Figure 1, the basic cell of the boundary scan architecture for an input pin allows one to either load data into the scan register from the input port, or drive data from the register through the output port of the cell into the core of the IC design. Boundary scan cells associated with output or bidirectional connections can be designed in a similar manner.

In a typical interconnection testing scenario, all the boundary scan cells associated with output connections of all the IC's would be first loaded with test data using the boundary scan register. In the second step, this test data would be applied and collected at the corresponding boundary scan cells associated with the input connections. In other words, interconnect tests are applied by output cells and received by input cells. In the final step, the response collected at the input cells is shifted out and verified. The example in Figure 3 illustrates all these three steps. The actual control sequence required to carry out these steps is executed with the help of the test access port and is detailed in [1].

2.1 Structure of the Interconnects

To conveniently describe various testing and diagnosis schemes, we will use the term Inet to refer to any group of

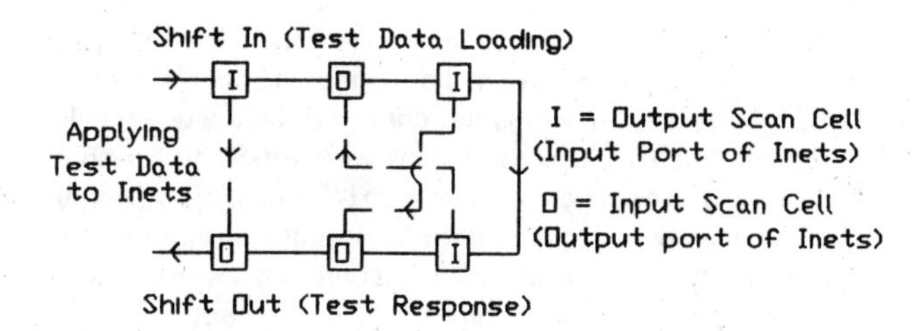

Figure 3 Inets Testing using Boundary Scan

two or more I/O boundary scan cells and the electrical conductors connecting these cells. Different Inet structures are shown in Figure 4. The simplest type of Inet is a pair of I/O cells connected by a single wire, as shown by AB in Figure 4a. When an output cell, such as G in Figure 4b, is connected to two or more input cells, fanout results. A more complex Inet is formed when multiple drivers are connected to the same bus, as shown in Figure 4c. In such a case, of course only one output cell is connected to the bus input cell at any given time. However, due to the common driving point, detection and diagnosis schemes for such Inets are slightly different for certain fault types as will be discussed later in the following. A combination of these three types of Inets can result in cluster type Inet shown in Figure 4d.

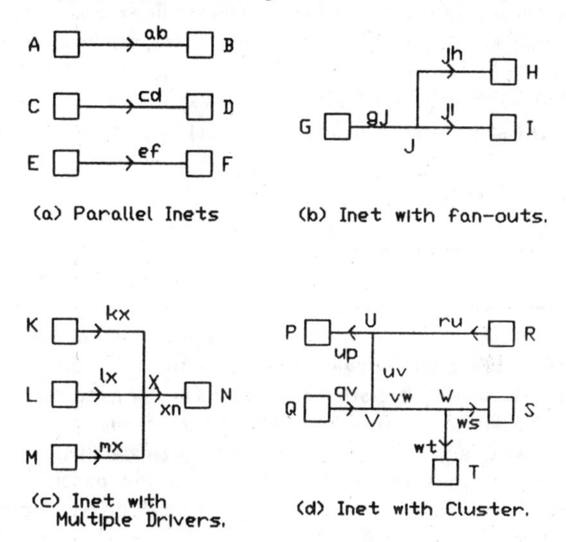

Figure 4 Different Inet Structures.

A few observations about the way the term Inet will be used in the remainder of this paper are in order here. When we refer to an Inet as a unit under test, the Inet must be such that under fault-free conditions all the I/O cells of the Inet form a single connected graph. Thus for instance, Figure 4a contains three different Inets, AB, CD, and EF; Figure 4b has one Inet, GHI; also each of Figures 4c and 4d shows a single Inet, respectively, KLMN and PQRST. The second observation is about bidirectional cells. Each of the bidirectional cells on a PCB has to be tested both as an input cell and as an output cell. This choice is of course controlled by the test access port [1]. In the following, we will assume that during testing each cell of each Inet has been controlled to be an input or an output cell, but not both simultaneously. In other words, the testing of Inets is not done with bidirectional cells floating with high impedance. Finally we will refer to a path in an Inet as any connection between two I/O cells. However, an independent path associated with an output cell in an Inet is the group of I/O scan cells and interconnection wires formed by connecting this output scan cell to all the input scan cells in that Inet. For instance, in Figure 4c, the interconnection wires 'kx' and 'xn' connect one output scan cell K to the input scan cell N. Thus, kx-xn is one independent path in the Inet KLMN. Similarly, lx-xn and mx-xn are the independent paths associated with output scan cells L and M respectively. In Inet PQRST (Figure 4d), there are two output scan cells. The two independent paths associated with these scan cells Q and R are qv-vu-up-vw-ws-wt and ru-up-uv-vw-ws-wt respectively. Thus, by definition, the number of independent paths associated with any Inet equals the number of output scan cells in that Inet.

2.2 Fault Model

The fault model of interest in Inets has to be based on the likely failures observed in interconnects on PCB's. It is well known [7,9] that the most common failure mode is shorts between any two or more Inets. These shorts can be classified as being

- AND short(where logic 0 dominates)
- OR short (where logic 1 dominates)
- weak short (where the resulting value is between logic 0 and logic 1)
- short between strong and weak drivers (where the outputs follow the strong drivers)

Of the first two types of shorts, depending upon the technology used in the individual IC component, either AND type or OR type but not both will occur. However, IC components with different technologies can be used on the same PCB. Thus, to make the testing schemes technology independent, we will, in this paper, consider the simultaneous presence of both AND and OR type of shorts on a single PCB. These two short types are treated extensively in this paper. Weak shorts and shorts between strong and weak drivers are not considered here.

Beside shorts, the following additional fault types are considered significant :

- stuck-at-one fault
- stuck-at-zero fault
- stuck-open fault
- delay fault

The schemes to be described consider single as well as multiple faults in the system. Moreover, the schemes for shorts testing allow the shorts to occur between a pair of Inets as well as among multiple Inets.

Different Inet structures and the concept of independent path have been introduced in the previous sub-section. It is interesting to note that the number of test vectors to be applied for shorts and stuck-ats testing does not depend upon the complexity of any individual Inet or the number of independent paths in any Inet. For example, if any interconnection wire 'kx', 'lx', 'mx' or 'xn' in Inet KLMN (Figure 4c) is shorted to any other Inet, then enabling only one driver, say the driver at K, will test for that short. Drivers at L and M will be kept disabled throughout the test. So, it is assumed that controls are provided for independent enable/disable of the output drivers in multiple driver Inets. The same is true for stuck-at testing.

However, this is not true for stuck-open testing. For example, in Figure 4c, if a test vector is applied, by enabling the driver at K (and disabling drivers at N and O) then any stuck-open fault in the branches 'lx' and 'mx' will remain undetected. So, by enabling the drivers at N and O, only one at a time, (and hence, enabling every independent path) all the stuck-open faults in KLMN can be detected. Thus, the testing of stuck-open faults is structure dependent. The number of vectors to be applied depends upon the number of independent paths in any Inet in the system.

3. Fault Detection and Diagnosis

This section describes some existing schemes and proposes some new schemes for testing of different types of faults in Inets. It will be seen that the existing schemes are not very efficient from implementation point of view. None of these schemes is structure independent. Thus, fault-free simulation of the Inets is required to obtain the expected response. Moreover, huge overhead is required to store this expected response for comparison with the test output. We will introduce a number of detection and diagnosis schemes to overcome the shortcomings of the existing schemes. Emphasis is given on efficient BIST implementation of these proposed schemes. Different types of deterministic vectors are used as test patterns in these schemes. Finally, some results are also presented on the detection capability of random vectors.

3.1 Detection of Shorts and Stuck-ats

3.1.1 Minimal-Size Test Set for Shorts Detection

It has been shown in [10] that a set of $\lceil log_2 n \rceil$ vectors is necessary and sufficient to detect all possible shorts in a network of 'n' unconnected terminals. The terminals are checked by physical contact using multiple probe continuity test. This set of $\lceil log_2 n \rceil$ vectors can be shown to be sufficient for testing all shorts in 'n' Inets [2,8].

The scheme is described with an example. Three vectors are required for 8 Inets as shown in Table 1. Each bit 'i' in each vector is applied to the output cell (input port) of Inet 'i', and the resulting output is collected at the corresponding input cell(s) of Inet 'i'. In the case when Inet 'i' has more than one output cell, any one output cell is arbitrarily enabled and the others are dasabled. Bit 'i' is then applied to the enabled cell.

It can be seen from the table that by applying $\lceil log_2 n \rceil$ vectors to 'n' Inets, each Inet is assigned a unique binary

V3	V2	V1	Inets
0	0	0	Inet 1
0	0	1	Inet 2
0	1	0	Inet 3
0	1	1	Inet 4
1	0	0	Inet 5
1	0	1	Inet 6
1	1	0	Inet 7
1	1	1	Inet 8

Table 1 Minimal-Size Vector Set for Shorts Detection

number. Due to this assignment each Inet input bits differ from those of all the other Inets at least by one-bit position. For example, the input assignments to Inets 1 and 2 (see Table 1) differ in V1. So, the corresponding output bits are also bound to be different in the fault free case. But in case of a short between this pair of Inets, the output bits corresponding to V1 are not different any more. Thus the short is detected at the output. This is true for every pair of Inets in the set. The same argument holds for multiple Inets shorted together.

3.1.2 Minimal-Size Test Set for Stuck-ats Detection

An Inet stuck-at-one (s-a-1) can be detected by applying a '0' as one of the input bits. Similarly a s-a-0 can be detected by applying a '1'. For example, if a bit-set '001' is applied to an Inet which is s-a-1, the faulty output is '111'. So, the s-a-1 in that Inet is detected. For this reason, stuck-at faults in most of the Inets can be detected by the set of $\lceil log_2 n \rceil$ vectors used for shorts detection in sub-section 3.1.1. However, notice that '000' and '111' are assigned to Inet 1 and 8 respectively in that example. Clearly, '000' will not detect a s-a-0 and '111' will not detect a s-a-1. So, instead of $\lceil log_2 n \rceil$ vectors, if $\lceil log_2(n+2) \rceil$ vectors are applied (thus avoiding all-zero and all-one) to 'n' Inets, all possible stuck-ats (SAs) and shorts are detected. Thus, $\lceil log_2(n+2) \rceil$ vectors are necessary and sufficient to detect all possible (single and multiple) shorts and SAs in a system of 'n' Inets.

3.1.3 Order Independent Test Set Scheme for Shorts and SAs Detection

To implement the minimal size test set scheme for shorts and SAs detection, each test vector is loaded through the scan chain, applied to the Inets and the obtained response is shifted out (Figure 3). Recall here that the I/O scan cells of different components are connected in a single scan chain. Let us assume that the total number of output scan cells in this scan chain is 'n', the total number of input scan cells is 'm' and the total number of I/O cells is (n+m) = N. In the minimal size test set scheme, $\lceil log_2(n+2) \rceil$ vectors are generated based on the number 'n'. Each of these vectors has (n+2) bits. These (n+2) bits of each vector are shifted in and applied through a scan chain which is N cells long. So, after the generation of each input vector of (n+2) bits, the vector is padded with N-(n+2) '0's, to make it compatible with the length of the scan chain. The '0's are padded in the proper order depending upon the order of the input and output scan cells in the scan chain. Thus, the generated vector is restructured or reformated before loading. This requires structural information about the scan chain as well as extra hardware and control for reformating of the input vectors.

These problems can be solved by generating $\lceil log_2(N+2) \rceil$ vectors for a scan chain N cells long. N bits from the (N+2) bits of each vector are loaded through the scan chain. The detection process works as before. But no reformating or structural information is required for test vector generation and loading. Since no information is required related to the order of the I/O cells in the scan chain, this can be termed as the order independent test set. This test set is not minimal size any more. $\lceil log_2(N+2) \rceil$ vectors are required instead of $\lceil log_2 n \rceil$ vectors. So, the time complexity becomes $O(N log_2 N)$ compared to $O(N log_2 n)$ of the minimal-size test set. But order independent test vector set is more suitable for BIST implementation due to its order-free test generation and loading property.

Test Generation Hardware :

In a BIST environment test vectors are generated on site. Thus, the test generation hardware is required to be simple and small in size to keep the BIST overhead reasonable. The test generation hardware to generate $\lceil log_2(N+2) \rceil$ vectors is shown in Figure 5. In this scheme, $\lceil log_2(N+2) \rceil$ bit counter generates the $\lceil log_2(N+2) \rceil$ test vectors. Notice that each test vector consists of (N+2) bits and is being generated serially from one of the stages of this counter. A $\lceil log_2(N+2) \rceil$: 1 MUX is then used to select which test vector should be applied during one scan cycle. The MUX is controlled by a $\lceil log_2(log_2(N+2)) \rceil$ bit counter. The state of this counter is changed after counting through (N+2) in the $\lceil log_2(N+2) \rceil$ bit counter. So the control bits are appended with the data bits in the counter. Thus, the hardware is a $\lceil log_2(N+2) \rceil + \lceil log_2(log_2(N+2)) \rceil$ bit counter with the $\lceil log_2(log_2(N+2)) \rceil$ MSBs become the control bits and the $\lceil log_2(N+2) \rceil$ LSBs become the data bits. N bits of output (excepting the first and the last bits) coming from the first LSB are chosen as the first vector by the $\lceil log_2(N+2) \rceil$:1 MUX. It can be shown that these bits form the first vector in the set of $\lceil log_2(N+2) \rceil$ vectors. In the same way the output of the second LSB register forms the second vector and so on.

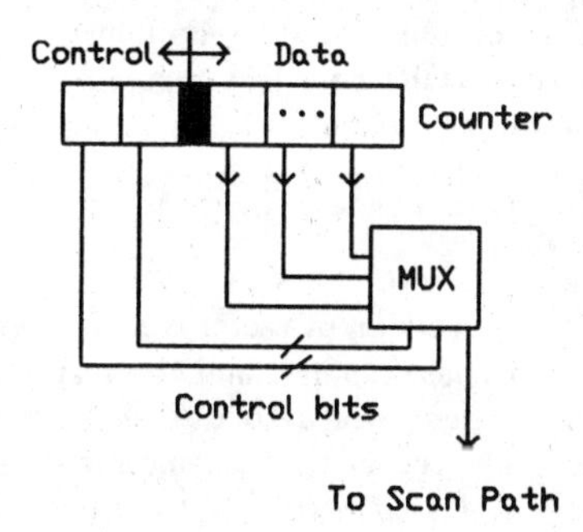

Figure 5 Generation of Order Independent Vector Set.

Response Analysis :

After loading and application of each test vector, the response is shifted out for detection of faults. The obtained response is compared with the expected response. The expected response can be determined by fault-free simulation of the system of Inets under test. This requires structural information about these Inets. Thus, fault detection is structure dependent both for minimal-size test set and order independent test set schemes. Moreover, minimal-size test set scheme requires $N\lceil log_2 n\rceil$ bits of storage for expected responses and $N\lceil log_2 N\rceil$ bits of expected response are stored for the latter scheme. Thus, although, the order independent vector set scheme makes the test generation and loading order independent, response analysis is structure dependent. Also, the storage requirement is high.

3.1.4 Walking Sequence

In order to overcome the disadvantages of minimal-size test set and order independent test set schemes, a different type of deterministic vector set is considered here. Consider a bit stream of a single '1' followed by all '0's which is shown below :

1 0 0 0 0 0 0 0 ...

This bit stream can be loaded through the scan chain as the first test vector. Then by gradually shifting this vector through the scan chain the rest of the vectors can be obtained. Since the vector set is obtained by gradually shifting the single '1' along the stream, this sequence is termed as a *walking one sequence*.

If the original bit stream (i.e., 1000...) of the walking one sequence is shifted (N-1) times, the single '1' gradually passes along all the scan cells, one at a time. Thus, in the fault-free case, the expected output on each input scan cell of each Inet is a single '1'. But in case the of a fault, number of '1's is changed (increased or decreased) at the output. Thus, by counting the total number of '1's, a faulty Inet can be detected.

A walking zero sequence (single '0' followed by all '1's, i.e., 0111... , shifted N-1 times) can also be used in the same way for detection of shorts and SAs.

Test Generation Hardware :

A walking one (or, a walking zero) sequence is very easy to generate. The outputs from the flip-flops of N-bit counter are fed to an NOR (OR) gate (see Figure 6). The resulting vector becomes 1000... (0111...).

3.1.5 Walking Sequence Scheme for Shorts and SAs Detection

In this scheme every input vector is shifted-in individually through the scan chain, applied and the response is shifted out. Walking one sequence described above is used as the input sequence. The response is taken out and fed to a compactor which is a '1's counter in this scheme. The scheme is shown in Figure 7. It can be shown that for the complete set of input vectors, expected number of ones is exactly N in the fault-free case where N is the total number of input and output scan-cells connected along the scan chain. However,

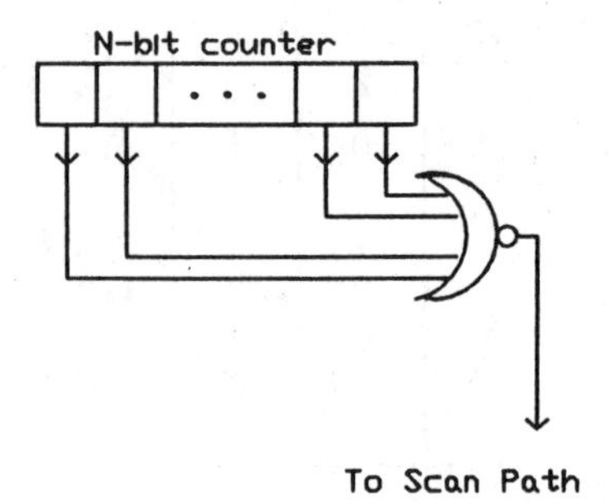

Figure 6 Generation of Walking Sequence.

in the case of a faulty Inet in the system, the total number of ones is increased or decreased depending upon the type of fault. SA-1 and OR-short increase the count whereas SA-0 and AND-short decrease the count.

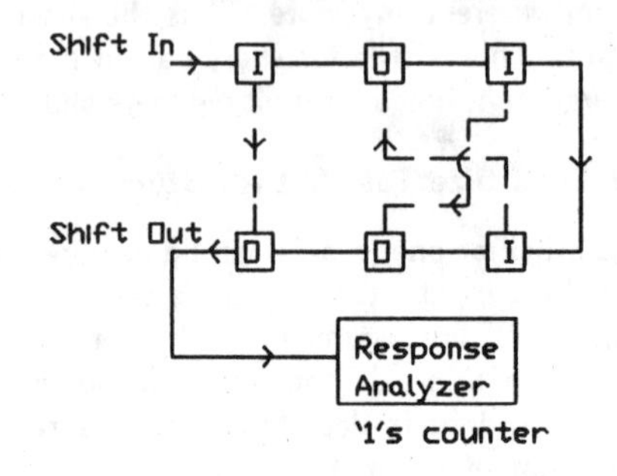

Figure 7 Walking Sequence Scheme for Shorts & SAs Detection

This scheme detects all single faults in the system. However, multiple faults can mask each other. As for example, if the count increased by an OR-short is exactly equal to the count decreased by an AND-short, then these two shorts mask each other. Similarly, a SA-1 fault can be masked by a SA-0 fault. In general, the faults can be grouped into two types. OR-short and SA-1 are the increasing count type whereas AND-short and SA-0 belong to the decreasing count type. As long as multiple faults belong to the same type, there is no masking. But multiple faults from different types can mask each other.

The scheme requires N-bits to be shifted-in and shifted-out for each vector. So, for N vectors, the time required for the complete procedure is $O(N^2)$. The response analyzer is a $\lceil log_2 N\rceil$ bit '1's counter.

This walking sequence detection scheme is independent of the order of the I/O scan cells in the scan chain for test vector generation. Moreover, structural information about the Inets and expected response storage are not required for fault detection. However, for multiple faults there are chances of masking.

3.2 Diagnosis of Shorts and Stuck-Ats

3.2.1 Existing Schemes

Goel and McMahon [8] have described a diagnosis scheme which is divided into two steps. In the first step, $\lceil log_2(n+2)\rceil$ vectors used for detection in sub-section 3.1.1 are applied to identify a subset W of the faulty Inets. The shorts must involve these Inets as well as some other Inets. In the second step, a unique test is applied to each member 'w' of W. In this test 'w' is assigned a '0' (or '1') and the remaining (n-1) Inets a '1' (or '0'). This indicates which Inets are shorted to 'w'. Thus all the shorted Inets can be identified.

Wagner [2] has proposed a diagnosis scheme which requires $2\lceil log_2(n+2)\rceil$ vectors. $\lceil log_2 n\rceil$ vectors used for shorts and SAs detection together with its complementary set forms the complete set of vectors for diagnosis. $\lceil log_2(n)\rceil$ vectors identify at least one of the Inets involved in each short. The complementary vectors then isolate the other Inets which were not identified by the first set.

None of these two schemes has addressed the implementation issues like input vector formating and loading, Inet structure dependence of response analysis, overhead required for storing the expected responses. Thus, these schemes are analytical treatment of diagnosis problem and are rather incomplete from a practical point of view.

3.2.2 Order Independent Test Set Scheme for Shorts and SAs Diagnosis

$2\lceil log_2(n+2)\rceil$ vectors used in [2] are sufficient to diagnose all possible shorts and SAs in 'n' Inets. However, for a scan chain N cells long, $2\lceil log_2(n+2)\rceil$ vectors are reformated as was described in sub-section 3.1.3. The diagnosis scheme can be made I/O scan chain order independent and the need of reformating can be avoided by using $2\lceil log_2 N\rceil$ vectors. The $\lceil log_2 N\rceil$ vectors are similar to $\lceil log_2(N+2)\rceil$ vectors described in sub-section 3.1.3. Here, all '0' and all '1' bit-sets can also be included as valid assignments. That is why $\lceil log_2 N\rceil$ vectors are used instead of $\lceil log_2(N+2)\rceil$ vectors. These $\lceil log_2 N\rceil$ vectors and their complements form the complete vector set for the proposed diagnosis scheme.

For N=6, the 6 vectors in Table 2 form the complete set.

V5	V3	V1	Scan Cell	V6	V4	V2
0	0	0	Cell 1	1	1	1
0	0	1	Cell 2	1	1	0
0	1	0	Cell 3	1	0	1
0	1	1	Cell 4	1	0	0
1	0	0	Cell 5	0	1	1
1	0	1	Cell 6	0	1	0

Table 2 $2\lceil log_2 N\rceil$ vectors for Shorts & SAs Diagnosis.

The vectors are applied in the following sequence. One vector (say V1) is applied from the set of $\lceil log_2 N\rceil$ vectors followed by its complement (say V2) from the complementary set. This is repeated until all the vectors are applied. The output bits are treated in pairs. In each pair there is a '0' and a '1' (because the components of the pair are coming from two complementary vectors). So in a non-faulty bit-pair, there is always a '0' followed by a '1' or a '1' followed by a '0'. For a s-a-1 (or, a s-a-0), the bit-pair are changed to two '1's (or, two '0's). Now let us see what happens in the case of a short. The input bit-pairs applied to two Inets can have the four different possible combinations shown in Table 3.

C1	C2	C3	C4	Inets
01	10	01	10	Inet 1
01	10	10	01	Inet 2

Table 3 Possible combinations of input bit-pairs.

In the first two cases, C1 and C2, the inputs to the two Inets are the same. So no change can be observed due to a short. For set C3, if the two Inets are shorted, the output is changed to either '00' or '11'. The same is true for set C4. Thus if any two Inets differ in input bit-pair combination (and they do differ for $\lceil 2log_2 N\rceil$ vectors at least in one bit-pair), the short can be diagnosed as a pair of '0's (or '1's) at the output.

Implementation Issues :

In this scheme, diagnosis can be done in-place or externally. In-place diagnosis is done by comparing, for each pair of input vectors, the pair of output bits obtained from each Inet within the associated input scan cell. For external diagnosis, the output bits are shifted out of the scan chain, stored in an external register and compared outside the scan chain. Notice, however, that both of these are implemented in a board level BIST environment.

For in-place diagnosis, boundary scan input cells of JTAG [1] at the output end of the Inets will have to be modified (see Figure 8). Two one-bit registers are needed to store the bit-pair at each input scan-cell. One single-bit register is already provided with the boundary scan cell. Thus, one extra one-bit register is needed per input scan cell. A two-input comparator is also added to the register pair to compare the bit-pairs. Thus 'q' input scan-cells requires 'q' extra one-bit registers and 'q' two-input comparators.

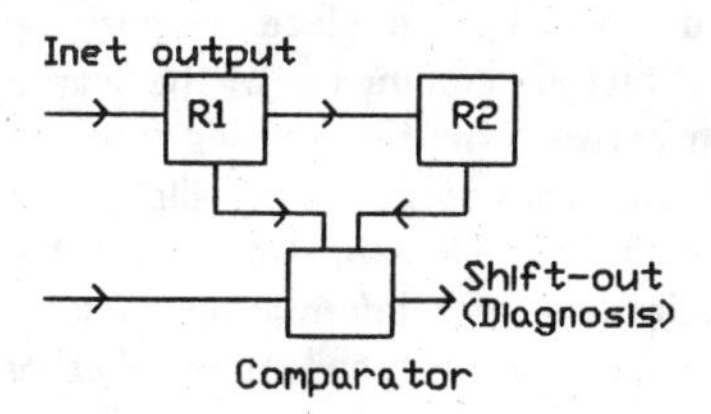

Figure 8 In-place Diagnosis using Order Independent Vector Set.

The first input vector is loaded individually through the scan chain and applied to the Inets. The output response of each Inet is stored in the one-bit register of the associated input scan cell. Then the second input vector is applied and the responses are stored in the second one-bit registers. These two bits are then compared in the comparator and the outputs of all the comparators are shifted out for diagnosis. This procedure is repeated for all the $\lceil log_2 N \rceil$ pairs of vectors in the set.

So, for each vector N bits are shifted in through the scan chain and for each pair of vectors N bits of comparator results are shifted out.
Thus, for in-place diagnosis,
Loading Time, $T_l = N.2\lceil log_2 N \rceil = O(N.log_2 N)$
Shift-Out Time, $T_o = N.\lceil log_2 N \rceil = O(N.log_2 N)$

The arrangement for external diagnosis is shown in Figure 9. The output response of the first vector is shifted out of the scan chain and loaded in an N-bit shift register. The complementary vector is then applied and while this response is shifted out it is compared with the response stored in the external register, bit by bit, through a two-input comparator. This procedure is repeated for every pair of input vectors. External diagnosis requires N-bit shift register and a two-input comparator external to the scan chain. No modification of the I/O scan cells is necessary.

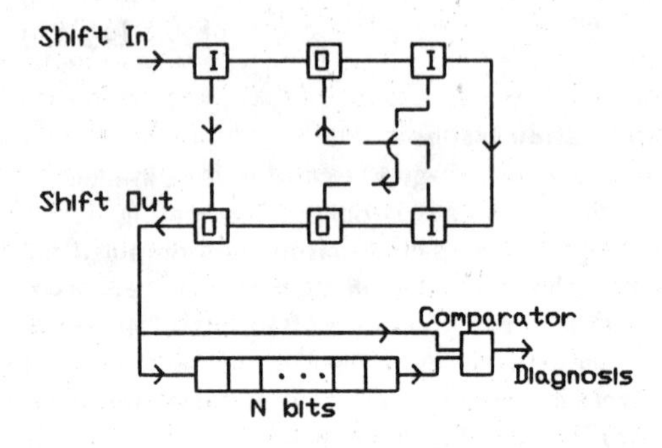

Figure 9 External Diagnosis using Order Independent Vector Set.

For each input vector, N bits are loaded and N bits of response are shifted out through the scan chain.
Thus, for external diagnosis,
Loading Time, $T_l = N.2\lceil log_2 N \rceil = O(N.log_2 N)$.
Shift-out Time, $T_o = N.2\lceil log_2 N \rceil = O(N.log_2 N)$.

So, the order of complexity remains the same for in-place and external diagnosis. For in-place diagnosis, when the comparator output bits are coming out of the scan chain, one has to distinguish between the bits coming from input scan cells and those coming from output scan cells. Thus, the order of I/O scan cells should be known. However, external diagnosis does not require any such information. The fault-free comparator output is a '0' independent of whether the bits are coming from input cells or from output cells.

Order independent vector set diagnosis scheme can diagnose all possible shorts and SAs. The scheme is independent of Inet structure and complexity and test generation is independent of the order of I/O scan cells. Diagnosis is local for each Inet which means that the diagnosis bits obtained from each Inet are sufficient to identify that Inet as fault-free or faulty. Diagnosis can be done in-place or externally. In-place diagnosis requires less time (although the order of complexity is the same) whereas external diagnosis requires no modification of the given architecture.

3.2.3 Walking Sequence Scheme for Shorts and SAs Diagnosis

As mentioned in sub-section 3.1.4, the advantage of using a walking sequence is that only a single test vector is generated and loaded through the scan chain. By gradually shifting this vector within the scan chain, the rest of the vectors can be obtained. Thus, a walking sequence is very time-efficient in terms of loading the test vectors. In the following a diagnosis scheme is described using such walking sequences. This scheme is time efficient not only in terms of loading but also from the response analysis point of view.

V4	V3	V2	V1	Inets
0	0	0	1	Inet 1
0	0	1	0	Inet 2
0	1	0	0	Inet 3
1	0	0	0	Inet 4

Table 4 Walking One Sequence for Diagnosis.

Table 4 shows the complete sequence for 4 Inets. The input scan cells have the following modifications. A single bit register and a two-input EX-OR gate is included in each input scan cell at the output or receiving end of each Inet (Figure 10). The single-bit register provided by boundary scan architecture is used only for loading and shifting of the output vectors. The second single-bit register (a shadow register) together with the EX-OR gate compacts the output response for diagnosis. No modification of the output scan cells is necessary.

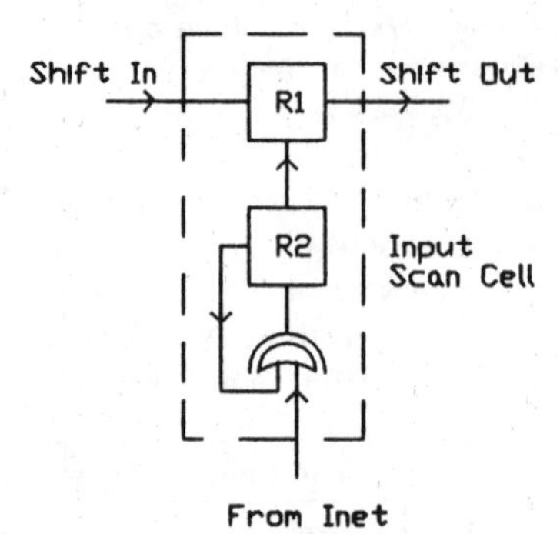

Figure 10 Diagnosis using Walking Sequence.

The first output bit coming from an Inet is stored in the shadow register R2 (Figure 10). The next bit coming from the Inet is EX-ORed with the stored bit to get a new output bit. This new bit is stored in the register and EX-ORed with the next bit coming. The procedure is repeated N times for the N vectors and finally a one-bit compacted response is obtained in the register R2. This bit is shifted out for diagnosis. For even N, following are the compacted responses :
'1': Fault-free, OR short (odd no. of Inets).

'0': SAs, OR short (even no. of Inets), AND short (odd and even).

Similar compacted responses can be obtained for odd N.

It can be observed from the above list that OR short among odd number of Inets has the same compacted response as the fault-free compacted response. Thus, this type of short cannot be diagnosed using only a walking one sequence. In order to diagnose this type of short as well as other OR and AND shorts and SAs a walking zero sequence is applied following a walking one sequence. In this scheme, N vectors of the walking one sequence are applied as before and the compacted bits are shifted out. Then the N vectors of the walking zero sequence are applied and a second set of compacted response is obtained. For even N, compacted responses for the walking zero sequence are :

'1': Fault-free, AND short (odd no. of Inets).

'0': SAs, AND short (even no. of Inets), OR short (odd and even).

By combining the two sets of compacted responses, the complete diagnosis becomes :

'11': Fault-free.

'10': OR short (odd no. of Inets).

'01': AND short (odd no. of Inets).

'00': SAs, OR short (even no. of Inets), AND short (even no. of Inets).

Time Requirement :

Complete diagnosis requires 2N vectors. For N vectors of the walking one sequence, only the first vector is loaded and shifted (N-1) times. This requires N-bits of loading and (N-1) shifts. Walking zero sequence requires the same operations. So, altogether, there are 2N bits to be loaded and (2N-2) shifts. At the output end, N bits are shifted out twice (once after every N vectors are applied). So, 2N shift-outs are done. Thus, the test time required is O(N).

Let us compare the time requirements of order independent vector set scheme and walking one/zero sequence scheme

For order independent vector set scheme (External Diagnosis):

Loading time $T_l = 2Nlog_2N$

Test application time $T_a = 2log_2N$

Shifting-out time $T_o = 2Nlog_2N$

For walking one/zero sequence scheme :

$T_l = 2N + 2N$

$T_a = 2N$

$T_o = 2N$

So, $2Nlog_2N + 2log_2N + 2Nlog_2N > 4N + 2N + 2N$

$\approx$ for, $N > 4$

Thus, although the number of vectors applied is small in the order independent vector set diagnosis scheme, walking one/zero sequence diagnosis scheme (for $N > 4$) requires less time.

Walking sequence diagnosis scheme does in-place diagnosis with time complexity of O(N). Order of I/O scan cells should be known to identify the diagnosis bits. An external diagnosis implementation is possible with the time complexity of $O(N^2)$.

3.2.4 Modifier Sequence Scheme for Shorts and SAs Diagnosis

In the walking sequence diagnosis scheme, each input scan cell has one extra single-bit register. This shadow register is used to store the compacted response so that it is not lost due to shifting of the input vectors along the scan chain. A different arrangement is possible where no shadow register is required for compaction and at the same time the compacted response obtained in each input scan cell is not affected due to the shifting operation.

In this scheme, after application of each vector and compaction of the corresponding response, the contents of all the scan cells are shifted out and modified by using a modifier vector. The objective of this modification is to generate a new vector and at the same time to not lose the compacted responses.

The arrangement is shown in Figure 11. The modifier sequence is shown in Table 5. The first input vector (V1 in Table 5) is shifted in through the scan chain and applied. Compaction is done locally in each input scan cell as was described in sub-section 3.2.3. The bit stream is then shifted out and passed through the EX-OR gate together with the next modifier vector (V2) to generate a new input vector. This procedure is repeated for all the N vectors in the modifier sequence. To explain how this works, consider the modifier sequence in Table 5. Vector V1 has a single '1'. When V1 is shifted in, this '1' goes to one of the scan cells in the scan chain. If this cell is an output scan cell then the '1' is applied to the associated Inet as the input. However, if the cell is an input scan cell, the '1' is passed through the corresponding EX-OR gate A (Figure 11) and stored back. After this, the contents of the scan chain is shifted out and passed through the EX-OR gate B together with V2. The first '1' in V2 cancels the first '1' in V1 due to the EX-OR operation. The second '1' in V2 is shifted in along the scan chain and acts in the same way as the '1' in V1 did before. This is repeated for all the modifier vectors. Thus, every output scan cell in the chain gets a single '1', one output scan cell at a time. This '1' is canceled by another '1' from the following modifier vector outside the scan chain. Similarly, the two '1's corresponding to each input scan cell cancel each other due to two EX-OR operations. Therefore, the net effect of the procedure is to apply a single '1' to each output scan cell one at time and to keep the contents of input scan cells unchanged due to shift in operations.

...	V7	V6	V5	V4	V3	V2	V1
...	0	0	0	0	0	1	1
...	0	0	0	0	1	1	0
...	0	0	0	1	1	0	0
...	0	0	1	1	0	0	0
...	0	1	1	0	0	0	0
...	1	1	0	0	0	0	0

Table 5 Modifier Sequence for Diagnosis of Shorts & SAs.

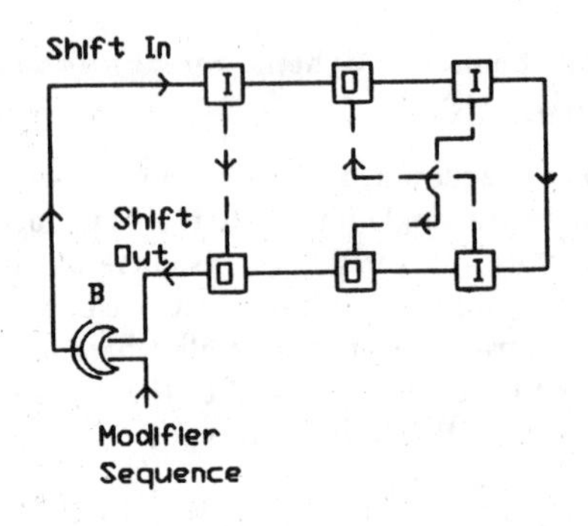

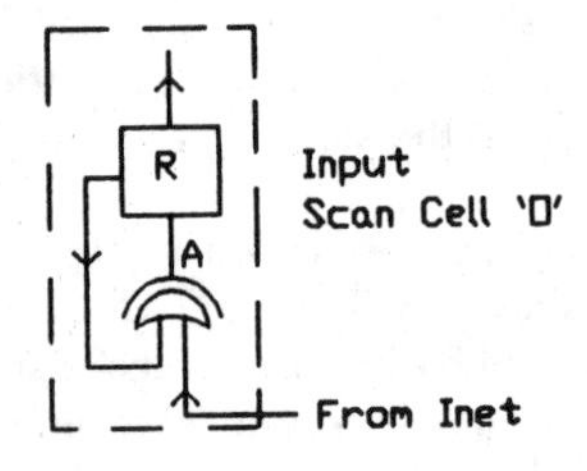

Figure 11 Diagnosis using Modifier Sequence.

Test Generation Hardware :

The modifier sequence is similar to the walking one sequence. V1 is the same as the first vector of the walking one sequence. However, starting from V2, each vector has two consecutive '1's instead of a single '1'. Thus this can be treated as a walking sequence with two consecutive '1's. So similar type of hardware can be used to generate these vectors. However, the vector V2 is to be stored somewhere and gradually shifted to get the complete sequence. Thus, the same N-bit counter (Figure 6) can be used for the generation of V1 and V2 as well as for shifting of V2 to get the rest of the sequence.

Implementation :

A two-input EX-OR gate is connected along the shift path of the scan-chain to generate the 'effective' input sequence. Moreover, each output scan-cell has a two-input EX-OR gate for local compaction and diagnosis.

Time Requirement :

In this scheme shift-in and shift-out-modification operation take place simultaneously. Thus, for N vectors of N bits each, the scheme has the time complexity of $O(N^2)$.

As mentioned in Section 3.2.3, N vectors of the modifier sequence diagnose all the shorts and SAs except odd number of Inets OR-shorted together. To take care of this type of faults, the complementary set of the modifier sequence should be applied.

This scheme can diagnose all possible shorts and SAs. Diagnosis is local and structure independent and does not require any shadow register. However, the test time is longer ($O(N^2)$ compared to O(N) or $O(N\lceil log_2 N \rceil)$) due to shift in and shift-out-modification operation.

Let us give an example of the time requirement of this scheme. For a board with 100 IC's each having 100 I/O pins, the number of I/O scan cells connected in the boundary scan chain, N = 10,000. So for a 10 MHz test clock, the time required (N^2 = 100,000,000) is 10 sec. This is quite reasonable for board-level testing.

3.3 Detection and Diagnosis of Stuck-Open Faults

Stuck-Open faults can be tested by checking for a conducting path from each output scan cell to all the input scan cells in an Inet. To do this, the input cells are initialized to a known logic value. The opposite logic value is applied from the output scan cell. In the fault-free case, the values in input cells should be changed through the conducting paths.

For detection of stuck-open faults, a single vector is shifted in through the scan chain. All the input scan cells are initialized to '0' and all the output scan cells are loaded with '1' using this vector. The vector is then applied and the response is shifted out. Since, all the output scan cells are loaded with '1's, the number of '1's in the fault-free case is exactly N where N is the number of scan cells in the scan chain. Thus, a $\lceil log_2 N \rceil$ bit '1's counter can do detection of all single and multiple stuck-open faults.

It was mentioned in section 2 that testing of stuck-open faults is structure dependent. Thus, a single vector is sufficient for detecting stuck-open faults in simple Inets without multiple drivers. For multiple driver Inets and cluster Inets, each independent path is to be tested separately. Thus, only one output driver in each Inet is enabled at one time and one test vector is applied. This vector tests one independent path in every Inet simultaneously. Thus, the number of vectors applied equals the maximum number 'p' of independent paths in any Inet in the system. The expected number of '1's is 'N' for each input vector. Thus, in the fault-free case, total expected number of '1's is 'pN' for all the 'p' vectors. However, for diagnosis each response bit coming out of the scan chain is to be checked for a fault-free value of '1'.

3.4 Summary of the Test Schemes

Table 6 is a brief summary of the proposed and existing detection and diagnosis schemes discussed in this section.

3.5 Testing with Random Vectors

Some experiments were done using the random vectors to test the Inets. In a random vector, the probability of getting a '0' or a '1' on each input bit is exactly 0.5. Thus, the probability of detecting any short is 0.5. Using this information and analyzing the complete set (2^n) of random vectors for 'n' Inets it can be shown that on the average 50% of all possible shorts are detected by a single random vector. Experiments were carried out using these average random vectors (each of which covers 50% of all possible shorts). Experimental results have shown that a very small number of random vectors (comparable to $\lceil log_2(n) \rceil$) can give close to 100% coverage of all possible shorts. However, detection and diagnosis schemes to use these random vectors are yet to be developed.

Scheme	Number of Vectors	Detection/ Diagnosis Capability	Time Requirement	BIST Hardware
(i) Minimal Size (Detection)	$\lceil log_2(n+2) \rceil$	Multiple Shorts and SAs Detection	$O(N log_2 n)$	Extremely Large
(ii) Order Independent (Detection)	$\lceil log_2(N+2) \rceil$	Multiple Shorts and SAs Detection	$O(N log_2 N)$	Extremely Large
(iii) Walking Sequence (Detection)	N	Multiple Shorts and SAs Detection (Chance of masking)	$O(N^2)$	Simple
(iv) Goel and McMahon	$\lceil log_2 n \rceil$ + W	Multiple Shorts Diagnosis	XX	Extremely Large
(v) Wagner	$2\lceil log_2(n+2) \rceil$	Multiple Shorts Diagnosis	XX	Extremely Large
(vi) Order Independent (Diagnosis, In-Place)	$2\lceil log_2 N \rceil$	Multiple Shorts & SAs Diagnosis	$O(N log_2 N)$	Simple
(vii) Order Independent (Diagnosis, External)	$2\lceil log_2 n \rceil$	Multiple Shorts and SAs Diagnosis	$O(N log_2 N)$	Simple
(viii) Walking Sequence (Diagnosis)	2N	Multiple Shorts and SAs Diagnosis	O(N)	Simple
(ix) Modifier Sequence (Diagnosis)	2N	Multiple Shorts and SAs Diagnosis	$O(N^2)$	Simple
(x) Stuck-open (Detection)	p	Multiple Stuck-open Detection	pN	Simple
(xi) Stuck-open (Diagnosis)	p	Multiple Stuck-open Diagnosis	pN	Simple

Table 6 Summary of Inet Test Schemes.

4. Example of Inets Testing using Order Independent Vector Set Diagnosis Scheme

In this section, an Inets testing session is described using the external diagnosis scheme proposed in sub-section 3.2.2. This scheme is chosen because it requires no modification of the I/O scan cells as well as the time complexity, $O(N log_2 N)$, is reasonable.

The board-under-test is chosen as an arbitrary example. There are 3 components, A, B, C, on the board each having 24 I/O pins. Seven of these 24 pins are used as VDD, GND, CLOCK and TAP (see section 1). Scan cells of the remaining 17 I/O pins form the boundary scan chain of each component. The scan cells are named as A1, B1, C1 etc., where A1 is the first cell of component A and so on. Scan chains of the three components are connected in series to form the scan path on the board. Out of the 51 I/O pins, 22 are output pins and the remaining 29 are input pins. 16 Inets are formed arbitrarily using these I/O pins. Actual connections and types of Inets are shown in Table 7. Table 8 gives the injected fault list.

Inet No.	Output Scan Cell (Input Port)	Input Scan Cell (Output Port)	Type of Inet
1	A2	B10	Simple (one output cell connected to one input cell, i.e., one-to-one)
2	A14	C15	
3	A5	B11	
4	B9	C4	
5	B12	C11	
6	B17	C5	
7	C1	A3	
8	C9	B2	
9	C10	A6, B3	Fan-out (one-to-two)
10	B15	A12, C7	
11	A4	B3, B16, C16	Fan-out (one-to-three)
12	B1	C6, A8, A9	
13	A7, B6	C13	Multi-driver (two-to-one)
14	C2, C12, B4, A16	A15	Multi-driver (four-to-one)
15	A11, C3	A10, B5, B14, C8	Cluster (two-to-four)
16	C17, C14	A1, A13, A17, B7, B8	Cluster (two-to-five)

Table 7 Inets on the Board-Under-Test.

$2\lceil log_2 N \rceil$ vectors are required to test the 16 Inets. Since, N equals 51 in this example, $2\lceil log_2 51 \rceil = 12$ vectors shown in Table 9 are applied to the Inets. Finally Table 10 shows the compacted response and diagnosis. Columns C1, C2,... are the compacted responses obtained from the comparator.

Fault Type	No. of Faults	Faulty Inet No.	I/O Scan Cells Involved	Scan Cell used for Input
S-A-1	2	3	A5, B11	A5
		13	A7, B6, C13	A7
S-A-0	1	7	C1, A3	C1
AND Short (between Inet pair)	1	5	(B12, C11)	B12
		14	(C2, C12, B4, A15, A16)	C2
OR Short (between Inet Pair)	1	1	(A2, B10)	A2
		8	(C9, B2)	C9
OR short (among 3 Inets)	1	9	(C10, A6, B13)	C10
		11	(A4, B3, B16, C16)	A4
		15	(A11, C3, A10, B5, B14, C8)	A11

Table 8 Injected Fault List.

I/O Scan Cell	V1	V2	V3	V4	V5	V6	V7	V8	...	V11	V12
A1	0	1	0	1	0	1	0	1	...	0	1
A2	0	1	0	1	0	1	0	1	...	1	0
A3	0	1	0	1	0	1	0	1	...	0	1
...	...	...	...	...	...	...	...	...	...	...	...
A17	0	1	1	0	0	1	0	1	...	1	0
B1	0	1	1	0	0	1	1	0	...	1	0
...	...	...	...	...	...	...	...	...	...	...	...
B17	1	0	0	1	0	1	0	1	...	1	0
C1	1	0	0	1	0	1	0	1	...	0	1
...	...	...	...	...	...	...	...	...	...	...	...
C17	1	0	1	0	0	1	0	1	...	0	1

Table 9 Input Vectors applied to the Inets.

The number of IC components or the number of Inets on the board is not important in this example. The objective is to show what are the various steps involved in applying the tests and diagnosing the faulty Inets. Diagnosis is done based on the comparator results without requiring any structural description of the Inets.

5. Conclusion

The various problems and complexities of interconnect testing are addressed in this paper. Schemes have been proposed for detection and diagnosis of different types of faults in the interconnects. $\lceil log_2 n \rceil$ vectors are minimal for detection of shorts in 'n' Inets. But for N I/O scan cells in the scan chain, $\lceil log_2 N \rceil$ vectors are easier to apply. External diagnosis scheme using $2\lceil log_2 N \rceil$ vectors does not require any modification of the scan cells. Walking sequence scheme is shown to be very time efficient for diagnosis of shorts and

I/O Scan Cell	V1	V2	C1	V3	V4	C2	...	V11	V12	C6	Diagnosis
A1	1	0	0	0	1	0	...	1	0	0	Fault-free
A2	0	1	0	0	1	0	...	1	0	0	Fault-free
A3	0	0	1	0	0	1	...	0	0	1	Faulty
A6	1	1	1	0	1	0	...	1	1	1	Faulty
A10	1	1	1	0	1	0	...	1	1	1	Faulty
A15	0	0	1	0	0	1	...	0	0	1	Faulty
B2	1	1	1	0	1	0	...	1	1	1	Faulty
B3	1	1	1	0	1	0	...	1	1	1	Faulty
B5	1	1	1	0	1	0	...	1	1	1	Faulty
B10	1	1	1	0	1	0	...	1	1	1	Faulty
B11	1	1	1	1	1	1	...	1	1	1	Faulty
B13	1	1	1	0	1	0	...	1	1	1	Faulty
B14	1	1	1	0	1	0	...	1	1	1	Faulty
B16	1	1	1	0	1	0	...	1	1	1	Faulty
C8	1	1	1	0	1	0	...	1	1	1	Faulty
C11	0	0	1	0	0	1	...	0	0	1	Faulty
C16	1	1	1	0	1	0	...	1	1	1	Faulty

Table 10 Diagnosis of Faulty Inets.

SAs. Modified sequence diagnosis scheme requires simple modification of the input scan cells.

One interesting feature about the schemes is that these are Inet structure independent. Based only on the number of I/O scan cells, test vector sets can be developed. Detection and diagnosis procedures are also not based on or restricted to any particular topology or structure of the Inets.

All these schemes are based on a Boundary Scan architecture on the board. The schemes are developed to be used in a BISTed environment. But these can be used in a non-BISTed DFT environment as well. Moreover, all the ideas and schemes presented here are equally applicable for testing the interconnects in a large area chip, WSI system etc. However, the modules in those systems should be isolated from each other, in the test mode, to make the design testable. Further research is being done on various unsolved problems like testing of the glue logic, testing of special I/O pins, structural testing of the Inets etc.

6. References

[1] JTAG Boundary Scan Architecture Standard Proposal, Version 2.0, Published on 30 March 1988.

[2] P.T.Wagner, *Interconnect Testing with Boundary Scan*, Proceedings of ITC 1987, pp. 52-57.

[3] F.M.Beenker et. al., *Macro Testing : Unifying IC and Board Testing*, IEEE Design and Test, December 1986, pp. 26-32.

[4] K.P.Parker, *Integrating Design & Test : Using CAE Tools for ATE Programming*, Published by Computer Society Press of IEEE, 1987.

[5] V.Ramachandran, *On Driving many long wires in a VLSI*

Layout, Journal of ACM, Vol. 33, No. 4, Oct. 1986, pp. 687-701.

[6] M.G.H.Katevenis and M.G.Blatt, *Switch Design for Soft-Reconfigurable WSI Systems*, Proceedings of Chapel Hill Conference on VLSI, 1985, pp. 197-219.

[7] J.Bateson, *In-Circuit Testing*, Published by Van Nostrand Reinhald Company, Inc. 1985.

[8] P.Goel and M.T.McMahon, *Electronic Chip-In-Place Test*, Proceedings of ITC 1982, pp. 83-90.

[9] R.G.Bennetts, *Introduction to Digital Board Testing*, Published by Crane Russard & Company, Inc., 1982.

[10] W.H.Kautz, *Testing of faults in wiring networks*, IEEE Transactions on Computers, Vol. c-23, No.4, April 1974, pp. 358-363.

Reprinted from *IEEE Design & Test of Computers*, February 1989, pages 36-44.

BOUNDARY SCAN WITH BUILT-IN SELF-TEST

CLAY S. GLOSTER
Microelectronics Center of North Carolina

FRANC BRGLEZ
Bell-Northern Research*

The authors propose a way to merge boundary scan with the built-in self-test of printed circuit boards. Their boundary-scan structure is based on Version 2.0 of the Joint Test Action Group's recommendations for boundary scan and incorporates BIST using a register based on cellular automata techniques. They examine test patterns generated from this register and the more conventional linear-feedback shift register. The advantages of the CA register, or CAR, are its modularity, which allows modification without major redesign; higher stuck-at fault coverage; and higher transition fault coverage.

Today's IC manufacturers typically use in-circuit and functional board-test systems to detect defects in their products. As designs grow more complex, however, and as we rely more on surface-mount technology, traditional testing techniques become less cost-effective. One solution to this complexity is to turn to more advanced methods, such as boundary scan. Boundary scan allows the circuit to be tested via the board-edge connector plus it introduces a shift register that is logically, and often physically, adjacent to the I/O pins of every chip on the board. Because the shift register allows test data to be shifted, applied, or captured, it can be used to test not only individual chips but also board interconnections.

There has been an industry-wide effort to standardize boundary scan techniques. The Joint Test Action Group has presented a proposal for a standard[1-3] in which boundary-scan modes are defined and guidelines are offered for implementation. The proposed standard does not explicitly address built-in self-test, but it provides for establishing a framework that would merge boundary scan and BIST. It is this type of framework that we discuss here.

BOUNDARY SCAN WITH BIST

The idea of incorporating built-in self-test with boundary scan is not new. LeBlanc,[4] Bardell and McAnney,[5] and Komonytsky[6] have introduced approaches that merge the two concepts. We also proposed a boundary-scan template at the 1988 International Test Conference,[7] which we are updating in this article to reflect the latest JTAG recommendations (Version 2.0).[3]

Figure 1 shows a block diagram of a boundary-scan template with BIST and its primary interfaces to the chip's interior. The template consists of an input register, an output register, and a controller with its own internal registers. Two additional control pins, TMS (test mode select) and TCK (test clock), are required along with two scan pins, TDI (test data in) and TDO (test data out). The registers in this template accommodate all the basic test modes proposed by JTAG along with a built-in self-test mode.

*Also with Microelectronics Center of North Carolina

The boundary-scan template has three principal tasks. It allows the circuit to function normally, it allows data to be shifted in or results to be shifted out, and it conducts several circuit tests. The template supports the following modes:

1. *External test.* This mode tests the interconnections of the printed circuit board. Data is applied to the board from the output register. The input register latches the data flowing from another chip via the board. Data can then be shifted out and verified.
2. *Internal test.* This mode tests the internal logic of the design. Data is applied from the input register to the circuit. The corresponding responses are latched in the output register. Once again, the results can be shifted out and verified.
3. *Sample test.* In this mode, the test engineer can take a snapshot of the circuit in time. Data is latched in both the input and output registers. The boundary-scan input and output registers are configured in this manner during the circuit's normal operation as well. The TCK pin must be asserted to capture the snapshot.
4. *Bypass.* This mode uses an output multiplexer to bypass the chip's lengthy boundary-scan path. Without this feature, testing a board with 100 chips, each with 100 I/O pins, would take too long. The data on the chip travels from the TDI pin, through one latch, and directly to the TDO pin.
5. *Built-in self-test.* In this mode, the input register is reconfigured to a pseudorandom pattern generator, while the output register functions as a signature analyzer. Random patterns are shifted serially into the internal scan register and are applied synchronously with patterns from the input register. The responses from these random patterns are compressed in the output register. The resulting signature can be checked to ensure proper circuit operation.

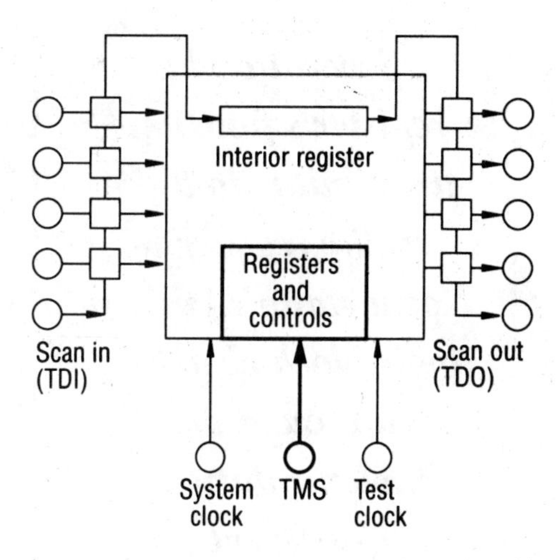

Figure 1. *Structural template for boundary scan with built-in self-test; TDI=test data in, TDO=test data out.*

HARDWARE COMPONENTS

The modes just described require several hardware components: an input register, an output register, and a controller section. Input and output registers share similar characteristics. In fact, they operate in the same way except that the input register generates patterns while the output register analyzes the signatures of multiple inputs. The controller section has its own internal registers. We have captured a complete description of the controller specified by JTAG 2.0 using a Pascal-like programming language, called Logic-III, which we compile automatically into a netlist of standard cells.[8] We discuss all the proposed hardware in more detail in an earlier report.[9]

Figure 2 shows how the input register is reconfigured during various test modes. The register must meet the following requirements:

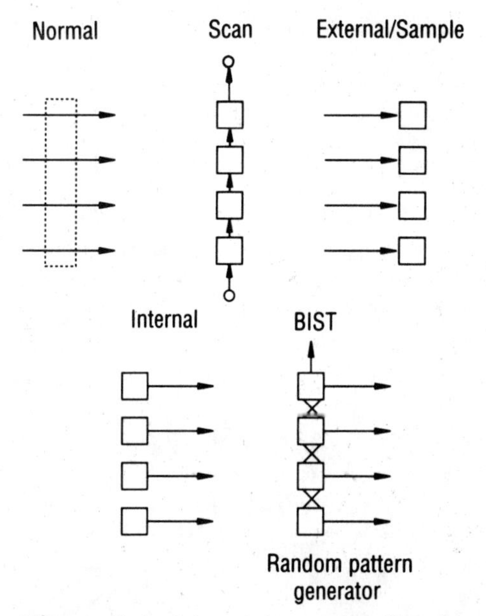

Figure 2. *Input register modes (four-bit example).*

Several of the register's functions are similar to those of the register in Koenemann's built-in logic-block observer, although our implementation is different.

- It must appear transparent in the normal mode of operation.
- It must latch the data during external and sample mode.
- It must form a scan chain during scan mode.
- It must be able to apply data in the internal mode.
- It must generate pseudorandom patterns in the BIST mode.

The register implements boundary-scan input cells as recommended by JTAG while incorporating cellular automata principles for built-in self-test.[10] Several functions are similar to those of the register in Koenemann's built-in logic block observer,[11] although our approach differs in implementation.

Approaches that generate pseudorandom patterns using cellular automata principles are relatively new. A cellular automaton, or single-cell, finite-state machine, evolves in discrete steps. The next value of each cell depends on the previous value of the cell to its left and the cell to its right. Cellular automata either are cyclically connected or have null boundary conditions. We used the null boundary condition in our research because it allowed us to remove the long feedback loop between the first and last cells. Hortensius has shown that by combining cellular automata rules 90 and 150, we can generate binary sequences of maximum length from each site.[10] Rule 90 is

$$a_i(t+1) = a_{i-1}(t) \oplus a_{i+1}(t)$$

and Rule 150 is

$$a_i(t+1) = a_{i-1}(t) \oplus a_i(t) \oplus a_{i+1}(t)$$

where i is the index of cell a. Combining these two rules gives us a sequence of maximum length, $2^s - 1$, where s is the number of cells or the length of the cellular automata. Table 1 lists the construction rules that produce this sequence.[10] We can compare the results from this table with registers that have maximum-length configurations that are based on linear-feedback shift registers.[12-14]

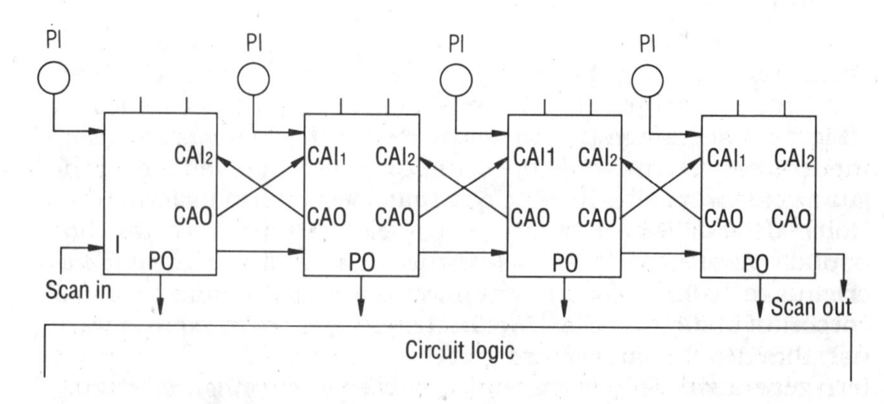

Figure 3. *Input register design (four-bit example); PI = pin input, CAI = cellular automata input, CAO = cellular automata output, PO = pin output.*

Table 1. *Construction rules for two configurations of a cellular automata register. In configuration 1, 0 represents a Rule-90 cell, 1 represents a Rule-150 cell. In configuration 2, 0 represents a Rule-150 cell, 1 represents a Rule-90 cell. The period of the sequence for either configuration is 2^s-1.*

Length (s)	Construction Rule
4	0101
5	11001
6	010101
7	1101010
8	11010101
9	110010101
10	0101010101
11	11010101010
12	010101010101
13	1100101010100
14	01111101111110
15	100100010100001
16	1101010101010101
17	01111101111110011
18	010101010101010101
19	0110100110110001001
20	11110011101101111111
21	011110011000001111011
22	0101010101010101010101
23	11010111001110100011010
24	111111010010110101010110
25	1011111010101010011110 0100
26	01011010110100010111011000
27	000011111000001100100001101
28	0101010101010101010101010101
29	00011000100011000111111100101
30	000001100010000110000100111110
31	0000110100100000110000001100101
32	00011111100100011001110110110000

A pattern generator based on the principles of cellular automata is a viable alternative to the more conventional LFSR-based generator in terms of pattern coverage as well as for transition fault testing.

Figure 3 shows an input register design. In this example, the input register is between the input pins and the circuit logic. We get a maximum-length sequence for four inputs by alternating Rule 90 (odd) and Rule 150 (even) cells. Figure 4 shows the boundary-scan cell recommended by JTAG in more detail. The original cell consists of two multiplexers and two flip-flops. To incorporate built-in self-test, we added an additional control signal, BIST, to the multiplexer and the Exclusive-OR tree for pattern generation. The cell implements cellular automata Rule 150, but we can convert it to a Rule 90 cell simply by removing the Exclusive-OR gate that feeds back the previous value of the cell.

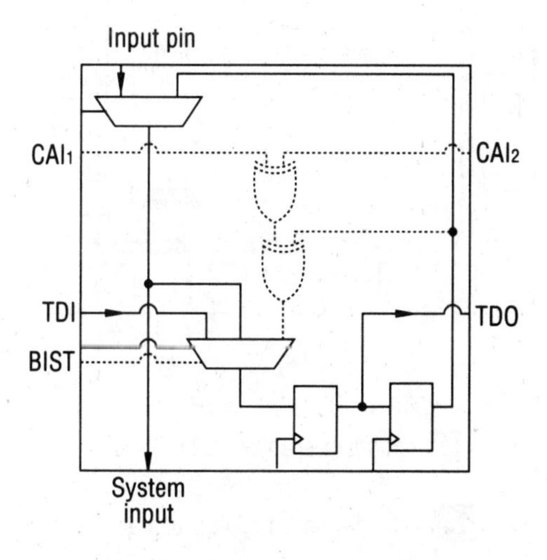

Figure 4. *Realization of a Rule 90/150 cell; TDI = test data in, TDO = test data out.*

While the implementation overhead of a cellular automata register is generally higher than that of an LFSR, a CAR has the advantage of modularity.

COSTS

In determining the overhead involved in adding boundary scan with BIST to an existing design, we used the largest unidirectional cell. To verify the functions of this cell, we used three Exclusive-OR gates, two multiplexers, one demultiplexer, and a scannable flip-flop. The implementation required 13 logic gates or standard cells. The design used 91 transistors and, with loose wiring, occupied 255 × 255 sq. μm in a 3-μm CMOS technology. After we optimized the largest cell, we decreased the number of transistors to 68. Custom design of these cells will decrease the area also, but we believe that the resulting need for feedthroughs will offset any decrease. For this reason, we used the conservative area estimate of 255 × 255 sq. μm in analyzing the chip area required.

Since there will be a boundary-scan cell for each primary I/O, we suggest placing the cells adjacent to the pins of the design. The cells are then on the periphery of the design.

Figure 5 shows the projected location of all required hardware to enable boundary scan and BIST. The template consists of the boundary-scan cells as well as some additional control logic. In some instances, part of the area in the shaded region, nominally reserved for boundary-scan cells, will accommodate additional control logic. If we include the boundary scan cells in the gray region of Figure 5, we must ensure that the width of the cells is less than the distance between adjacent pins. This requirement is not difficult to satisfy. Even our largest cell fits beside a pin with area left for routing.

We analyzed several pad frames to estimate the cost of testability in terms of chip area. Table 2 gives the results. The maximum usable area before boundary scan is the frame area minus the area of the pads. We calculated our maximum usable area after boundary scan by placing our largest cell beside each I/O pin. For large frames, the decrease in usable area is relatively small.

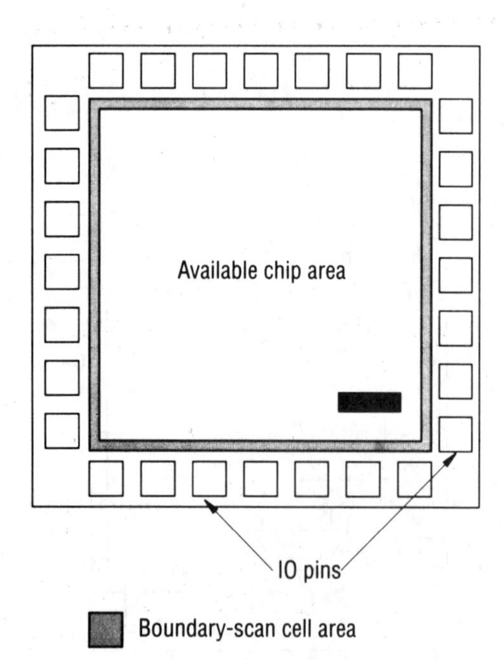

Figure 5. *An effective location for boundary scan with BIST.*

Table 2. *Projected overhead including built-in self-test for pad frames of different sizes.*

Pins	Frame Size	Usable Area Before Boundary Scan (mm^2)	Usable Area After Boundary Scan (mm^2)	Calculated Overhead (%)
28	S	11.8	10.1	16.96
40	S	11.8	10.1	16.96
40	M	25.7	23.2	11.15
40	L	40.2	37.0	8.55
64	M	25.7	23.2	11.15
64	L	40.2	37.1	8.55
64	XL	64.3	60.2	6.70
84	L	40.2	37.1	8.55
84	XL	64.3	60.2	6.70

By merging an input and output cell, we can get a bidirectional cell. Because of the constraint in pin spacing, however, these cells would have to be rectangular and would thus increase overhead.

As we mentioned earlier, all our projections are based on a 3-μm CMOS technology. With a 1-μm CMOS technology, we can place even the most complex cell on the chip boundary.

EVALUATING OPTIONS WITH BIST

We experimented with using a register based on cellular automata principles, called CAR, and a traditional linear-feedback shift register as sources to generate random patterns. We used the two CAR configurations in Table 1 and several LFSR polynomials tabulated in work by Bardell et al.[14]

THE BIST MODEL

In the BIST mode, the boundary-scan input register is reconfigured into a CAR or a LFSR of length *s*. Either register can serve as a source of *s*-bit wide $2^s - 1$ random patterns. These patterns are distributed in parallel to *n* primary inputs and serially to *m* interior scannable latches, as Figure 6 shows. In fact, *n* of the *s* latches from the source register are primary inputs. We add *s* – *n* register latches to the input register only when the random pattern testability of the circuit under test requires such an addition. Given that the number of uniformly distributed random patterns required to test the circuit in the scan mode is N_{TEST}, then we need to maintain $s > \log(N_{TEST}+1)/\log(2)$ with some margin.

TEST-PATTERN GENERATION

We formed a pattern *n*+*m* wide by clocking the source register for *m* cycles to serially load the interior register. We then applied the pattern in a single clock cycle to the circuit under test (Figure 7). Table 3 shows an exhaustive set of trial patterns that the CAR source register (configuration 1 in Table 1) can generate for parameters (*s*=4, *n*=3, *m*=2).

In generating trial patterns, we traversed $2^s - 1$ source patterns *m* times. In this example, the period of the trial patterns is the

Table 3. *Trial pattern generation for parameters n=3, m=2, s=4.*

	Source Patterns	Trial Patterns Trav. 1	Trav. 2
0	1111		
1	1100		11010
2	1010	10101	
3	0001		00000
4	0011	00110	
5	0110		01111
6	1011	10101	
7	0010		00110
8	0101	01001	
9	1101		11010
10	1001	10011	
11	0111		01111
12	1000	10011	
13	0100		01001
14	1110	11100	
15	1111		11100

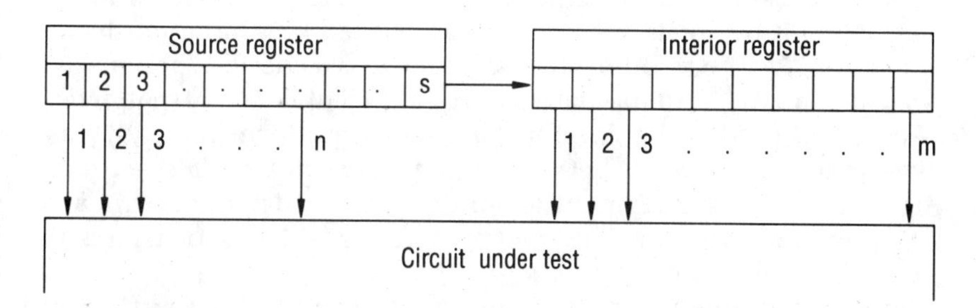

Figure 6. *Characteristic parameter set (n,m,s); N_{TEST} = the number of random patterns to cover 100% of stuck-at faults.*

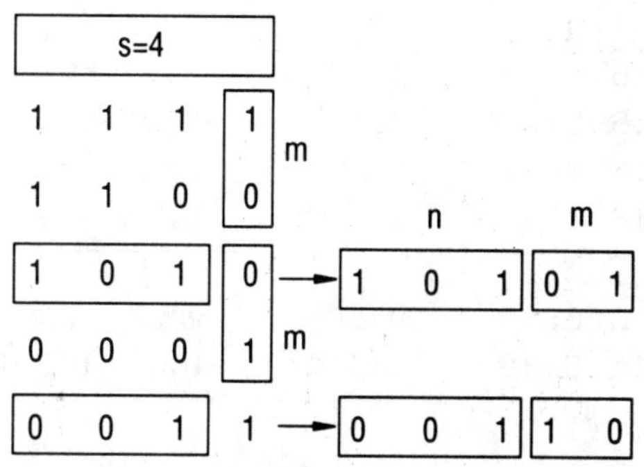

Figure 7. *Test-pattern generation.*

same as that of the source, but this is not always the case. If the number of interior latches, m, and the source period $2^s - 1$ have a common prime factor, say r, then the period of trial patterns becomes $(2^s - 1)/r$. Trial patterns begin repeating after the first traversal of the source patterns. The source period in Table 3 is 15, so by choosing m=3, for example, we reduce the period of trial patterns to 5.

Note also that only eight of the 15 patterns in the table are unique. If we change the source to a cellular automata register with configuration 2 (see Table 1) or to an LFSR, we would generate 15 unique trial patterns. For this reason, we compare trial pattern generation on the basis of trial pattern coverage, which is

$$\text{trial pattern coverage} = \frac{\text{no. of unique trial patterns}}{2^s - 1}$$

We exhaustively analyzed trial patterns for coverage with several values of s, n, and m, using both CARs and LFSRs as the source register. Table 4 summarizes the results. We divided the data into five groups to represent the aspects of pattern generation. Group A represents a case in which the period of the source is 63. The period has several prime factors in common with several choices of m, so the period of trial patterns varies. In Group B, pattern coverage starts at less than 100% when $m<s$. When the source is based on a CAR instead of an LFSR, pattern coverage rises toward 100% much faster. Group C shows the requirements for an exhaustive test, given these values of m and n. To achieve 128 unique patterns, we need a CAR source with value of s=9 and an LFSR source with a value of s=11. Group D conveys the same message as Group B except that register lengths are in a somewhat more practical range. Group E uses the source register that relates to an actual design.

***Table 4.** Trial pattern coverage for a set of cellular automata registers and linear-feedback shift registers.*

n	m	s	Period	Maximum Coverage CAR	LFSR
			A		
4	2	6*	63	100	50.8
4	3	6	21	33.3	28.6
4	4	6	63	100	50.8
4	5	6	63	100	100
4	6	6	21	33.3	33.3
4	7	6	9	14.3	14.3
4	8	6	63	100	100
			B		
4	3	7	127	50.4	25.2
4	4	7	127	50.4	25.2
4	5	7	127	50.4	25.2
4	6	7	127	100	100
4	7	7	127	100	100
4	8	7	127	100	100
			C		
4	3	8*	255	27.3	27.3
4	3	9*	511	25.04	12.5
4	3	10*	1,023	12.5	12.3
4	3	11*	2,047	6.25	6.25
			D		
16	1	17*	131,071	100	50
16	2	17*	131,071	100	50
16	3	17*	131,071	100	100
16	4	17*	131,071	100	100
			E		
16	6	18	87,381	33.3	20.7
16	6	19	524,287	100	100

*CAR configuration 1 from Table 1 (all others are CAR configuration 2).

WHY DISTINCTIVE PATTERN COVERAGE?

As Table 4 shows, trial patterns repeat at different rates for a CAR or an LFSR source. A plot of pattern coverage as a function of trial patterns is shown in Figure 8 for a set of parameters. Trial patterns repeat for a number of reasons. First, a pattern from the n-bit segment, illustrated in Figure 7 repeats itself 2^{s-n} times. The exception is the (000...00) pattern, which repeats $2^{s-n} - 1$ times. This repetition does not depend on the order of patterns, and we can verify it by sorting the patterns in ascending binary order.

Second, for the m-bit segment, we have two cases. When $m < s$, we can have 2^m unique patterns from a total of $2^s - 1$ patterns. We cannot predict the distribution of these patterns as readily as we can for the patterns of the n-segment. For $m \geq s$, all $2^s - 1$ patterns in the m-bit segment are unique, so all trial patterns are unique. Thus, we can generate fewer than $2^s - 1$ unique trial patterns only when m < s.

The pattern distribution of the m-bit segment for the CAR is similar to that for the LFSR. Therefore, we conclude that the single most important influence on pattern coverage is the order in which the n-bit and m-bit segments combine into $2^s - 1$ trial patterns. Trial patterns generated with a CAR as a source register more readily produce trial patterns with higher coverage. We believe this higher coverage occurs because the adjacent bit correlation with a CAR is lower than that with an LFSR.

ON PATTERN AND FAULT COVERAGE

For some circuits, we must generate many random patterns before 100% of the stuck-at faults are covered. The effectiveness of BIST depends on how well we can match the source of random patterns to the testability requirements of the circuit under test. We must ensure that BIST hardware will deliver patterns that have sufficient coverage.

We applied boundary scan and BIST techniques to an existing scan-based chip design[15] that has a small number of interior scannable latches (m=6) relative to the number of inputs (n=16). Let N_{TEST} be the number of random tests we must apply to fully test the circuit. In fault simulation with computer-generated random patterns,[16] we found that N_{TEST} = 131,040 covers 100% of the single stuck-at faults in this design.

We chose a source register of s=18 to match the random testability requirements of this design. However, we did not realize at that time that m=6 and $2^s - 1$ = 262,143 have a common prime factor, 3. This factor reduced the period of the trial patterns to 87,381. Despite this shorter period, the pattern coverage of the CAR was higher than that of LFSR, 33.3% vs. 20.7%, as shown in Table 4.

This pattern coverage correlates well with the fault coverages we attained when we simulated CAR-based and LFSR-based patterns. Figure 9a shows the results of this simulation. With 41,888 CAR-based trial patterns, we reached 100% coverage, but the LFSR-based test flattened at 98.28% after about 40,000 trials.

The last entry in Table 4 shows that for (s=19, n=16, m=6), both CAR and LFSR achieve 100% pattern coverage. We also easily covered 100% of the single stuck-at faults in both cases. However, if we consider a two-pattern test and measure transition fault coverage,[17] the test patterns are not equivalent between the CAR and the LFSR. As Figure 9b shows, when we use a CAR as a pattern source, we cover 99.7% of the transition faults in 500,000 patterns. With an LFSR as a pattern source, we cover only 93.4% of the faults in the same number of patterns.

We can realize boundary scan with a variety of test modes for high-performance boards, including a mode for built-in self-test, while keeping overhead to an acceptable level. A pattern generator based on the principles of cellular automata is a viable alternative to the more conventional LFSR-based generator in terms of pattern coverage as well as for transition fault testing. While the implementation overhead of a CAR is generally higher than that of an LFSR, a CAR has the advantage of modularity. Since only adjacent neighbor communication is required, we can readily change the length

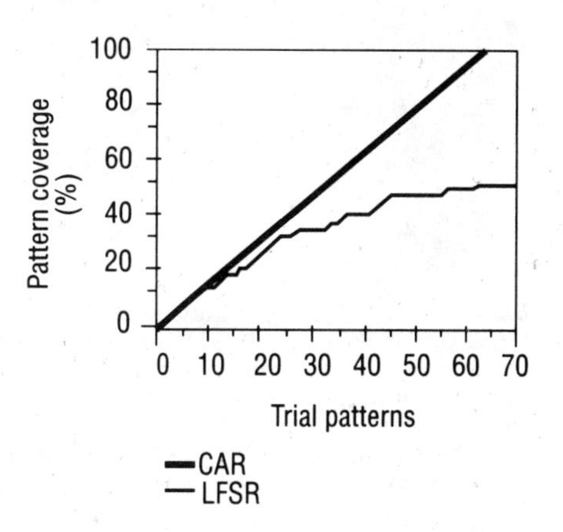

Figure 8. *Trial pattern coverage with a cellular automata register and a linear-feedback shift register for (n=4, m=2, s=6).*

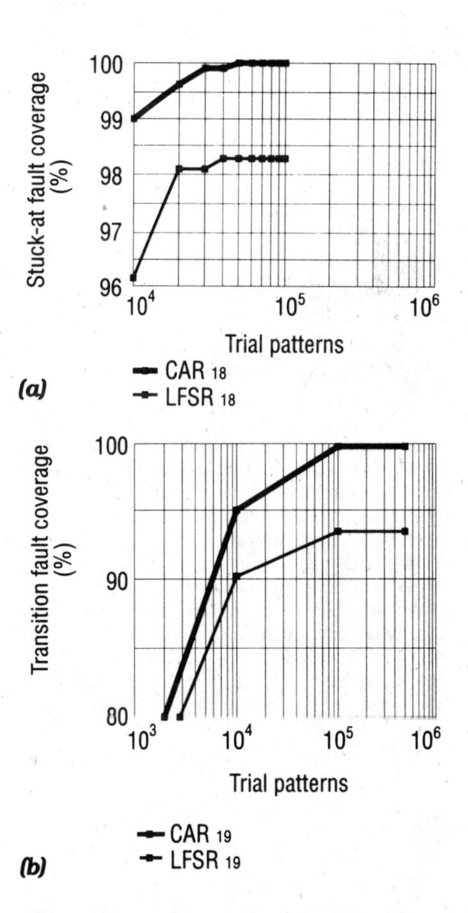

Figure 9. *Fault coverage curves for the CAR and LFSR: stuck-at fault coverage **(a)** and transition fault coverage **(b)**.*

of the generator by simply adding or removing adjacent cells. Thus, we do not have the major redesign effort involved with LFSR-based generators.

We have begun work on a tool to automate boundary-scan layout. The tool characterizes a universal mask-programmable register that we can reconfigure into either a CAR or an LFSR. We are also investigating properties of CARs and LFSRs in weighted random test-pattern generation and test-pattern compaction.

ACKNOWLEDGMENTS

We gratefully acknowledge Rod Tulloss from AT&T Engineering Research Center for keeping us updated with versions of JTAG's recommendations, Peter Hortensius from the University of Manitoba for sharing with us early results from his PhD dissertation on random number generation with cellular automata, BNR for supporting a summer student position at the Microelectronics Center of North Carolina in 1987, and NTI for supporting an Industrial Affiliate position at MCNC in 1988. In addition, Harold Martin from North Carolina A&T State University has been a source of constant encouragement throughout this effort.

REFERENCES

1. F. Beenker and C. Maunder, Boundary-Scan, "A Framework for Structured Design-For-Test," *Proc. Int'l Test Conf.*, Sept. 1987, pp. 724-729.
2. C. Maunder, F. Beenker, and C. Vivier,. *A Standard Boundary Scan Architecture Version 1.0*, June 1987 (available by writing to R. Tulloss, AT&T Eng. Res. Ctr., PO 900, Princeton, N.J. 08540).
3. *JTAG Boundary-Scan Architecture Standard Proposal Version 2.0*, Mar. 1988 (available by writing to R. Tulloss at the address above).
4. J. LeBlanc, "LOCST: A Built-In Self-Test Technique," *IEEE Design & Test of Computers*, Vol. 1, No. 4, December 1984, pp. .
5. P. Bardell and W. McAnney, "Self-Testing of Multichip Logic Modules," *Proc. Int'l Test Conf.*, Nov. 1982, pp. 200-204.
6. D. Komonytsky, "LSI Self-Test Using Level-Sensitive Scan Design and Signature Analysis," *Proc. Int'l Test Conf.*, Nov. 1982, pp. 414-424.
7. C. Gloster, Jr., and F. Brglez, "Boundary Scan with Cellular-Based Built-In Self-Test," *Proc. Int'l Test Conf.*, Sept. 1988, pp. 138-145.
8. F. Brglez et al., "Automated Synthesis for Testability,"*IEEE Trans. Industrial Electronics* (to be published).
9. C. Gloster and F. Brglez, *Integration of Boundary Scan with Cellular-Based Built-In Self-Test for Scan-Based Architectures, Version 1.0*, tech. rpt. TR87-18, Microelectronics Center of North Carolina, Research Triangle Park, N.C., Aug. 1987.
10. P. Hortensius, *Parallel Computation of Non-Deterministic Algorithms in VLSI*, PhD dissertation, University of Manitoba, Winnipeg, 1987.
11. B. Koenenmann, J. Mucha, and G. Zwiehoff, "Built-In Logic Block Observer," *Proc.Int'l Test Conf.*, Oct. 1979, pp. 37-41.
12. D. Lancaster, *TTL Cookbook*, Howard W. Sams & Co., New York, 1974.
13. L.-T. Wang and E. McCluskey, "Hybrid Designs Generating Maximum-Length Sequences," IEEE Trans. Computer-Aided Design of ICs and Systems, Vol. 7, No. 1, Jan. 1988, pp. 91-99.
14. W. McAnney, P. Bardell, and J. Savir, *Built-In Test for VLSI: Pseudorandom Techniques*, John Wiley & Sons, New York, 1987.
15. G. Kedem and J. Ellis, "The Ray Casting Machine," *Proc. Int'l Conf. on Computer Design*, Oct. 1984, pp. 533-538.
16. J. Calhoun, D. Bryan, and F. Brglez, *Automatic Test Pattern Generation (ATPG) for Scan-Based Digital Logic: Version 1.0*, tech. rpt. TR87-17, Microelectronics Center of North Carolina, Research Triangle Park, N.C., Aug. 1987.
17. M. Schulz and F. Brglez, "Accelerated Transition Fault Simulation," *Proc. Design Automation Conf.*, 1987, pp. 237-250.

Clay Gloster, Jr., is graduate student in the Department of Electrical and Computer Engineering at North Carolina State University, where he is working towards a PhD in electrical engineering. His research interests are boundary-scan and built-in self-test architectures in conjunction with random pattern generation and pattern compaction. Previously, he worked on superconducting DC motors for David Taylor Naval Ship Research. Gloster holds an MSEE from North Carolina A&T State University.

Franc Brglez is with Bell-Northern Research and is a resident professional at Microelectronics Center of North Carolina, where he manages a research team in design synthesis and testability. He is also an adjunct professor in the Department of Electrical and Computer Engineering at North Carolina State University. His research interests have been in digital testability analysis, fault simulation, and automatic test-pattern generation. Currently, he is focusing on work in bridging logic synthesis, verification, and testability.

Direct comments or questions on this article to F. Brglez, MCNC, 3021 Cornwallis Rd., Research Triangle Park, NC 27709.

Boundary Scan And Its Application To Analog-Digital ASIC Testing In A Board/System Environment

Patrick P. Fasang

National Semiconductor Corporation
ASIC Division
2900 Semiconductor Drive
M/S 10-165
Santa Clara, CA 95052-8090

Abstract

This paper first introduces the concept and motivations for developing Boundary Scan (BS), then explains the input BS cell, the output BS cell, and the bidirectional BS cell. Then the paper explains the application of Boundary Scan to the testing of analog-digital ASICs in a board/system environment. An example is given to illustrate the concept and the application.

1. Introduction

Design For Testability (DFT) at the chip level for the purpose of making sure that the chip can be tested in a stand-alone manner is not new. Companies have been using this approach for some time. However, DFT used at the chip level by itself does not mean that such chips when used on a board will make it easier for the board or for that matter the system to be tested. The reason is that new packaging technologies for ICs, while allowing more ICs to be mounted on a given size of board, have made it difficult or impossible to use current in-circuit board testing techniques. Examples of new packaging technology of this type are surface mounted package and pin-grid array. The former requires no plated through holes on the printed circuit board (PCB), and the latter has its pins inaccessible from the device side of a PCB. One additional problem is that the spacing between adjacent pins has decreased to the point that physical probing of package pins is impossible or requiring too expensive probes. To solve this problem, the Joint Test Action Group (JTAG) developed a specification that became a proposed IEEE standard (P1149.1) test interface and Boundary Scan architecture for increasing the testability of PCBs, and hence systems, by adding a scan path around the periphery of ICs.

2. The Boundary Scan Concept

Boundary scan is the application of a scan path to the internal periphery of the signal pins of an IC to provide controllability and observability to the pins when the IC is mounted on a PCB and the pins are not physically accessible for probing. Figure 1 shows an IC with BS cells placed next to the signal pins. During normal application or mission function, a signal travels from the Mission Input through the BS cell into the Application Circuit. The response from the Application Circuit travels out to the Mission Output pin through a second BS cell. When the IC is mounted on a PCB and if the pins are not physically accessible, test data can be applied to the Application Circuit in a serial manner via the Test Data Input (TDI) pin. Likewise the response from the Application Circuit can be captured into the BS cell on the output port and serially shifted out via the Test Data Output (TDO) pin. Figure 1 shows only one Mission Input pin and one Mission Output pin. However, the concept can be extended whereby the IC may have n Mission Input pins and m Mission Output pins, where n and m are some arbitrary numbers. In that case, the IC would have n BS cells on the input port and m BS cells on the output port. The IC needs to have, however, only one TDI and only one TDO. The TDO of the first BS cell on the input port would be connected to the TDI of the second BS cell on the input port, etc. Likewise the TDO of the first BS cell on the output port would be connected to the TDI of the second BS cell on the output port, etc. The general concept of Boundary Scan as seen from the board level is depicted in Figure 2. Note that the TDI of the first IC in the BS path is connected to the signal called Scan In on the edge of the PCB, and the TDO of the last IC in the BS path is connected to the signal called Scan Out on the edge of the PCB. From the above information, one sees that functionally there is a need to distinguish those BS cells on the input port from those on the output port. Hence the BS cells on the input port are called input BS cells, and those on the output port are called output BS cells. For bidirectional signals, bidirectional BS cells are needed. Likewise, for Tri-state output signals, Tri-state output BS cells are needed.

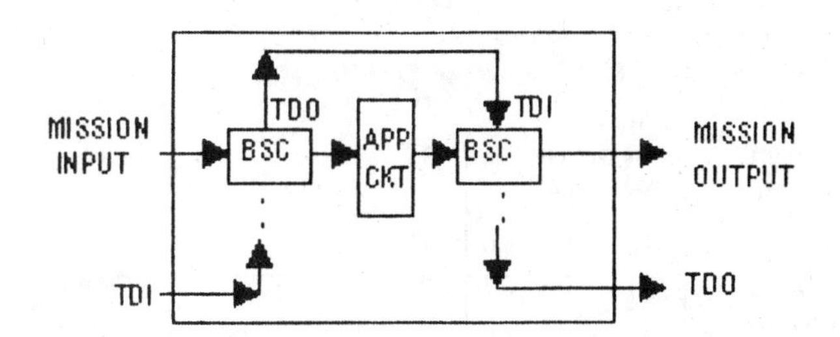

FIGURE 1. BOUNDARY SCAN IN AN INTEGRATED CIRCUIT

3. Input BS Cell, Output BS Cell, & Bidirectional BS Cell

Figures 3, 4, and 5 show the block diagrams of one implementation of the input BS cell, the output BS cell, and the bidirectional BS cell, respectively. JTAG also specifies a Test Access Port (TAP) controller which generates the various control signals used in the boundary scan architecture (1). The BS cells for each of the signal pins of an IC are interconnected to form a scan path around the

Reprinted from *IEEE 1989 Custom Integrated Circuits Conference*, pages 22.4.1-22.4.4.

EH0321-0/90/0000/0275$01.00 © 1989 IEEE

border of the design, and this path is provided with serial input and output connections and appropriate clock and control signals. For further information on the BS concept, specification, and applications, see references (1 - 6).

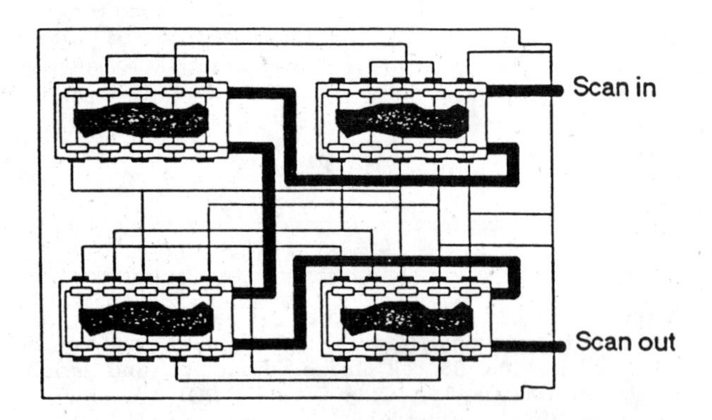

FIGURE 2. A BOUNDARY-SCAN BOARD

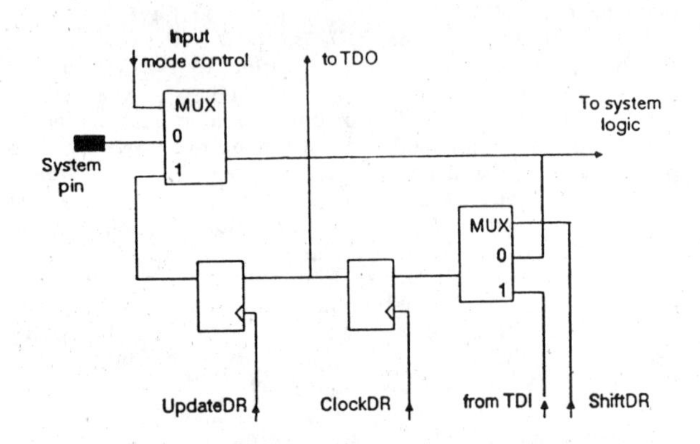

FIGURE 3. INPUT BOUNDARY-SCAN CELL

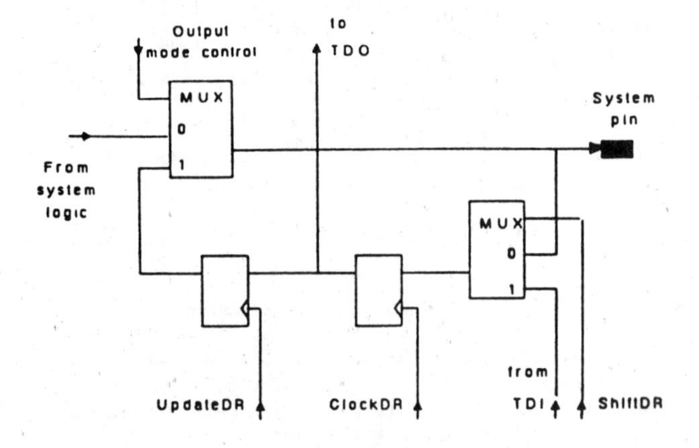

FIGURE 4. OUTPUT BOUNDARY-SCAN CELL

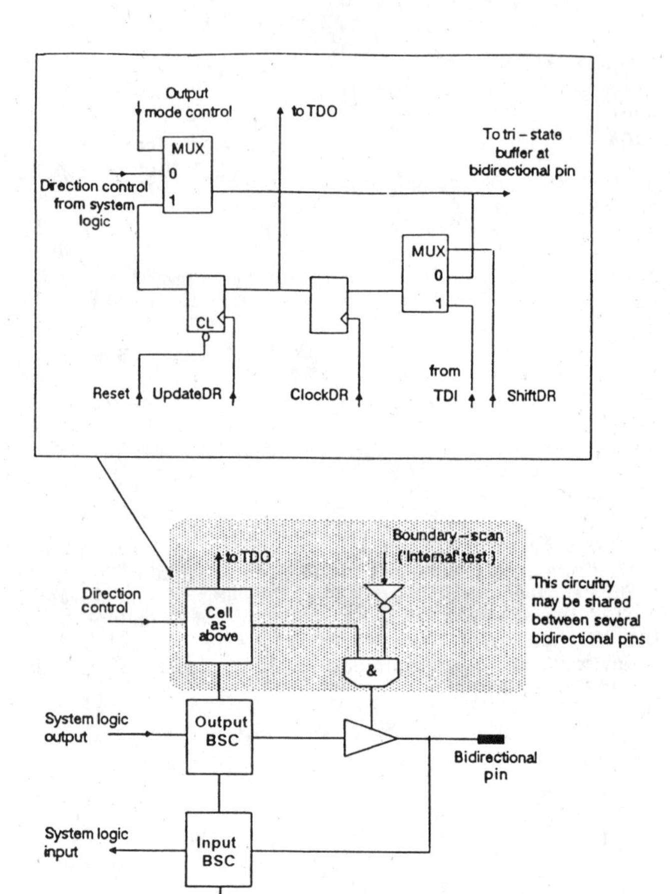

FIGURE 5. BIDIRECTIONAL BOUNDARY-SCAN CELL

4. Analog-Digital ASICs

Some semiconductor companies offer analog as well as digital circuits in their ASIC libraries (7). This type of mixed signal ASICs are difficult to test even when they are stand-alone devices (8). When placed on a PCB, the testing problem becomes more difficult. The reason is due to the fact that analog signals do not lend themselves to shifting via scan paths as digital signals do, and mixed-signal simulators are not yet capable of providing a complete set of analog and digital input and output data for testing as in the case of purely digital circuit simulator.

5. Application Of BS To Analog-Digital ASIC Testing

To manage the testing problem of analog-digital ASICs, the following procedure is suggested:

A. Partition the analog circuit from the digital circuit

B. Add demultiplexers (DMUX) to observe the digitized analog input signal(s) via output BS cells during testing

C. Add multimplexers (MUX) to control the digital circuit with digital test patterns applied during testing via the input BS cells

D. Perform logic simulation of the digital circuit without using the analog input signal(s) but use the digital test patterns applied during simulation (and testing) via the input BS cells

E. Test the analog and digital circuits separately by applying analog test signals at the analog input pins and observing the digitized analog outputs at the output BS cells; and by applying digital test patterns to the digital circuit via the input BS cells and observing the digital circuit outputs via the output BS cells associated with the digital circuit output pins.

F. With one or two analog input values, check the overall behavior of the ASIC chip to make sure that the link between the analog circuit and the digital circuit is not faulty. This particular step is not meant to be an exhaustive test but only to ensure that the analog-digital link is not broken.

Note that the above procedure is useful even when the IC is not mounted on a PCB because each BS cell has the property that it aids in performing an external interconnection test or as a scan element to allow internal testing of the application circuit or simply a transparent (buffer) element. So when an IC is in a stand-alone mode, meaning not mounted on a PCB, its BS cells can be controlled so that they are in the transparent mode and allow the testing procedure described above to be performed.

6. An Example

Figure 6 shows an example of an analog-digital ASIC with DMUXs, MUXs, input BS cells, and output BS cells added for the purpose of making the IC testable both at the chip level as well as at the board or system level. In this example, the analog circuit consists of an amplifier and an analog-to-digital (A/D) converter. The analog input is a primary input pin to the ASIC chip. When mounted on a PCB, this analog primary input pin needs to be connected to a dedicated analog signal pin (finger) on the edge of the PCB containing this ASIC chip. The output BS cells associated with the DMUXs allow the digitized analog signal to be observed when the ASIC chip is in a stand-alone mode as well as when the ASIC chip is mounted on a PCB. In the stand-alone mode, these output BS cells can be placed in the transparent mode, and the digitized outputs from the DMUXs can be observed in parallel at their corresponding output pins. When the ASIC chip is mounted on a PCB, these output BS cells can be controlled such that they are linked together to form a scan path around the border of the ASIC chip, and the digitized analog-signal bits can be captured into these BS cells and then shifted along the boundary scan path on this ASIC chip, and via other boundary scan paths in other ICs if the PCB is so designed, and finally to the Scan-Out pin (finger) on the edge of the PCB on which this ASIC chip is mounted. In the stand-alone mode, the digital circuit can be tested by applying, in parallel, the digital patterns used in the logic simulation of the digital circuit at the pins associated with the input BS cells which feed into the MUXs. The responses from the digital circuits can be observed in parallel at the pins associated with the output BS cells on the output port of the digital circuit. When mounted on a PCB, the digital circuit can be tested by having the input test pattern bits shifted in serially from the Scan-In pin (finger) on the edge of the PCB, and via other boundary scan paths in other ICs if the PCB is so designed, and then loaded from the input BS cells into the MUXs and then into the digital circuit. The responses from the digital circuit are captured into the output BS cells on the output port of the digital circuit and then serially shifted along the boundary scan path in this ASIC chip, and via other boundary scan paths in other ICs if the PCB is so designed, and then to the Scan-Out pin (finger) on the edge of the PCB. Note that under normal mission mode, the DMUXs and the MUXs would be controlled in such a way that the BS cells associated with these DMUXs and MUXs would not be selected. To make Figure 6 simple to follow, various control signals such as system clock, test clock, and boundary scan shift and update signals are not shown.

7. Conclusions

New IC packaging technologies which allow more ICs to be mounted on a given size of PCB also make it increasingly difficult to gain physical access to the pins of the ICs for testing purposes. JTAG whose membership includes some 30 to 40 systems companies as well as semiconductor companies developed a concept and specification to address this particular issue of testing boards/systems containing state-of-the-art ICs. The JTAG solution is known as boundary scan. The boundary scan concept entails embedding a BS cell next to each signal pin on an IC to, in effect, emulate electrical in-circuit testing without requiring physical access to each signal pin of an IC when mounted on a PCB. For ASIC chips with both analog and digital circuits, partitioning the whole circuit so that the analog circuit is separate from the digital circuit and adding DMUXs, MUXs, and BS cells make the chip testable both in the stand-alone chip environment as well as in a board/system environment.

8. References

(1) Boundary-Scan Architecture Standard Proposal, Version 2.0, JTAG Technical Committee, March 30, 1988

(2) Expanding Beyond Boundary Scan Techniques And JTAG, Pete Fleming, WESCON'88 Professional Program Session Record 13, November 15-17, 1988, Anaheim, California

(3) A Proposed Standard Test BUS And Boundary Scan Architecture, Lee Whetsel, WESCON'88 Session Record 13

(4) Boundary Scan -- A User's Point Of View, Ulrich Ludemann and Heinz Vogt, WESCON'88 Session Record 13

(5) Merging BIST And Boundary Scan At The IC Level, Scott Davidson, WESCON'88 Session Record 13

(6) A Method For Using JTAG Boundary Scan For Diagnosing Module Level Functional Failures, John Sweeney, WESCON'88 Session Record 13

(7) National Semiconductor ASIC Analog Cells, 1987, National Semiconductor Corporation, Santa Clara, CA 95052-8090

(8) Design For Testability For Mixed Analog/Digital ASICs, Patrick P. Fasang, et. al., Proceedings of the 1988 IEEE Custom Integrated Circuit Conference, May 1988, Rochester, N.Y.

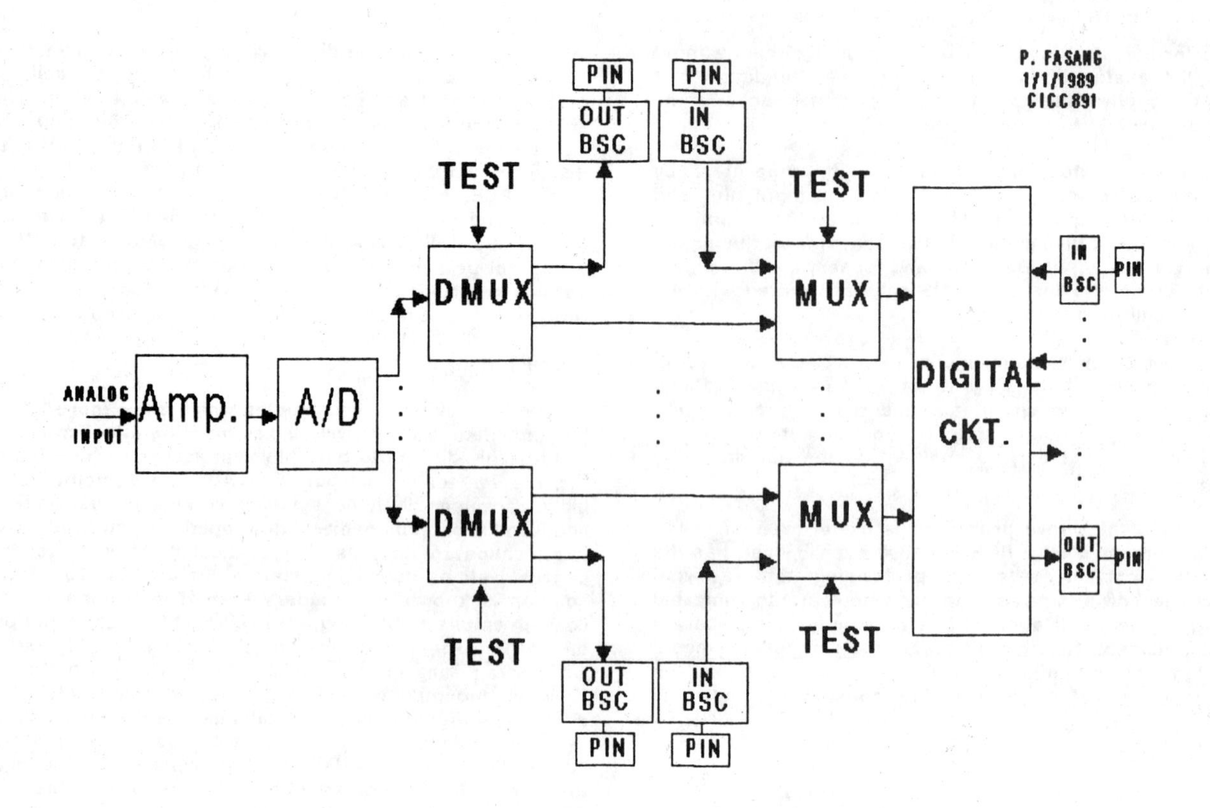

FIGURE 6. ANALOG – DIGITAL ASIC WITH BOUNDARY SCAN CELLS

Reprinted from *IEEE Transactions on Industrial Electronics*, Volume 36, Number 2, May 1989, pages 231-240.

A Universal Test and Maintenance Controller for Modules and Boards

JUNG-CHEUN LIEN, STUDENT MEMBER, IEEE AND MELVIN A. BREUER, FELLOW, IEEE

Abstract—**The design of a Module test and Maintenance Controller (MMC) is presented. Driven by structured test programs, an MMC is able to test every chip in a module or PCB via a test bus, such as the JTAG boundary scan bus. More than one test bus can be controlled by an MMC. The proposed MMC is quite versatile. It can support several bus architectures and many modes of testing. The differences between MMC's on different modules are the test programs which they execute, the number of test buses they control, and the expansion units they employ. A simple yet novel circuit, called a *test channel*, is used in an MMC. The MMC processor can control a test channel by reading/writing its internal registers. Once initialized by the MMC processor, a test channel can carry out most of the testing of a chip. Thus the processor need not deal with detailed test-bus control sequences since they are generated by the test channel. This strategy greatly simplifies the development of test programs. The proposed MMC can be implemented as a single-chip ASIC or by off-the-shelf components. Some self-test features of the MMC are also presented.**

I. INTRODUCTION

DESIGNING testable chips which can be connected to standard test buses has recently drawn much attention [1]–[3], [12], [16]–[18]. This is due primarily to two major initiatives dealing with testable designs, which have emerged over the last few years. One is the ETM-BUS protocol proposed by the VHSIC committee [13]; another is the boundary scan protocol proposed by the JTAG committee [21], which has attracted considerable industrial support. Recently, the IEEE Testability Bus Standard Committee (TBSC) developed several test-bus protocols for board-level testing, known as P1149.x (x = 1, 2, 3, 4) [22]. The serial test bus, namely P1149.1, was adopted from the JTAG proposal. As a result of these IEEE proposals, the ETM-BUS protocol will probably be abandoned.

The main objective of these efforts is to support the design for testability (DFT) of a module (or a board). An acceptable degree of testability is not always achievable by simply using a set of testable chips unless they are properly integrated at the module level. Similar problems exist at both the subsystem level and system level. For a system to have a high degree of testability and maintainability, the system must be testable at every level of integration. Examples of such systems are described in [7] and [11].

Manuscript received November 16, 1988; revised December 10, 1988. This work was supported by the Defense Advanced Research Projects Agency and monitored by the Office of Naval Research under Contract N00014-87-K-0861. The views and conclusions considered in this document are those of the authors and should not be interpretted as necessarily as representing the official policies, either expressed or implied, of the Defense Advanced Research Projects Agency or the U.S. Government.

The authors are with the Department of Electrical Engineering-Systems, University of Southern California, Los Angeles, CA 90089.

IEEE Log Number 8926814.

A hierarchical system design methodology to support test and maintenance, known as an HTM, has recently been reported [7]. In this methodology, a hierarchy of test controllers is embedded into a target system's physical hierarchy. In an HTM system, each testable chip contains an on-Chip test and Maintenance Controller (CMC); each testable module contains a Module test and Maintenance Controller (MMC); each testable subsystem contains a Subsystem test and Maintenance Processor (SuMP); and each system has a System test and Maintenance Processor (SMP). These controllers participate in all system test and maintenance activities and communicate via test buses. Fig. 1 shows part of the test hierarchy with these four levels of controllers. Different buses may be used for communcation at different levels. The SMP communicates with SuMP's through a Level-2 bus (L2-bus); a SuMP communicates with MMC's through a Level-1 bus (L1-bus); and an MMC communicates with CMC's through a Level-0 bus (L0-bus).

Bus interfaces are required for both the controlling party (master) and controlled parties (slaves) on a bus. For example, each CMC contains an L0-slave to interface to an L0-bus. Each MMC contains an L0-master to control an L0-bus and an L1-slave to communicate with an L1-bus. If MMC's, SuMP's, and SMP's are all designed to be testable chips, they each should contain a CMC. It is possible for an MMC to have more than one L0-master and thus control more than one L0-bus.

Suitable L0-bus designs are the JTAG boundary scan bus, the VHSIC ETM-BUS, and the IEEE P1149.x (x = 1, 2, 3, 4) bus. The TM-BUS [14] is suitable for an L1-bus design. The TM-BUS or a system functional bus, such as the pi-bus, can be used for the L2-bus. In this paper, we employ the JTAG bus and the L0-bus; hence, every L0-slave contains a test access port (TAP) controller. This bus consists of a data line input (TDI) to the chip, a data line output (TDO), a control input (TMS), a clock (TCK), and an optional interrupt output line (INT).

This paper deals primarily with the design of an MMC. An MMC is able to control the self-test process of a module (or board) by accessing each chip's BIT structures through an L0-bus. The proposed MMC is universal in that the same basic design is used for all modules. MMC's differ by the test programs they execute, the number of these buses they control, and the expansion units they employ. Test programs direct the processor in an MMC in the execution of the built-in self test (BIST) process for the entire module. The test results are then reported to a SuMP via an L1-bus. A SuMP can initiate the self-test process of a module by sending a "begin

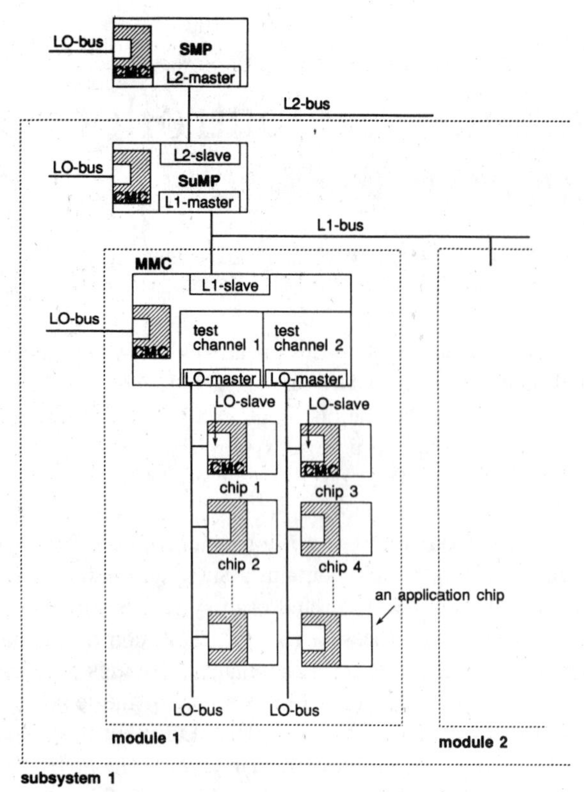

Fig. 1. Test hierarchy for a module.

test'' commmand to the MMC on that module. The MMC then reports the ''health status'' of that module to a SuMP.

An MMC contains bus interface units, such as an L1-slave and an L0-master, a processing unit such as a processor, a memory unit consisting of RAM's and ROM's, one or more test channels, a bus driver/receiver, one or more expansion units such as testability registers and analog test interface, and a CMC. Only bus interface units are shown in Fig. 1.

A simple yet novel design, called a *test channel*, is used in an MMC. Since every testable chip has an L0-slave in its CMC, a test channel, which contains an L0-master, can communicate over an L0-bus with a CMC. The MMC's processor can control a test channel by reading or writing its internal registers. Once initiated by the processor, a test channel can completely control an L0-bus and the testing of a chip. The separation of processor and test buses provided by test channels prevents the processor from dealing with detailed bus timing activities. A test channel translates processor instructions into proper timing sequences for an L0-bus. A test process can now be represented as high-level processor instructions.

Budde reported on the design of the Testprocessor [9], which is similar to our MMC. The Testprocessor is intended to carry out some of the functions of the CMC and the MMC. Since it may be part of an application chip, it must be simple. The Testprocessor is programmed at the microinstruction level. All peripherial devices are controlled directly by the control signals provided by the microinstructions. The number of expansion units is limited by the total number of control signals the control unit can provide. Data can be moved directly between the test-pattern RAM and the test interfaces without going through the processor register. Obviously, this is an efficient approach for data movement. However, due to the limitation of the bus, only one serial interface can run at a time. Comparisons are done by a fault-secure comparator. There is no other data processing unit in the Testprocessor. Due to the limited processing capability, diagnostic programs cannot run on the Testprocessor.

In Section II a control model for a testable chip is presented. The design requirements of an MMC are presented in Section III, followed by its architecture in Section IV. Major building blocks such as a test channel, processor, and memory are described in turn. Some self-test aspects of the MMC are presented in Section V.

II. A Control Model for Testable Chips

A test controller for the DFT and/or BIST hardware on a chip must be able to: 1) provide data to the circuit under test (CUT), such as test vectors or seed values; 2) switch between test and functional clocks; 3) provide required control signals; 4) count the number of tests executed; and 5) execute and process test results. More details on DFT and BIST test controllers can be found in [5] and [6]. Both control signals and data required to test a chip are supplied, to some extent, by the MMC. Thus the hardware for carrying out this test process can be distributed between the MMC and the CMC.

An MMC can transmit two types of information to a chip, namely instructions and data. Instructions are sent to the instruction register in the L0-slave to control and/or configure the test function of a chip, while data are set to a selected scan chain in the chip. Two types of information are sent from a chip to an MMC, namely status and results. The status consist of the values of important signals monitored by the chip, while results come from a selected scan chain in the chip.

Fig. 2(a) shows a typical testable chip employing the JTAG boundary scan architecture. Components within dashed boxes are optional. The original portion of the chip, denoted as the application circuit, has been modified to have *n* scannable data registers. Everything outside of application circuit, which is added for the purpose of testing the chip, is called the CMC. The CMC consists of an L0-slave and a BIT controller (see Fig. 2(b)). Assuming the L0-bus is a JTAG boundary scan bus, then the L0-slave consists of a TAP controller, an output buffer, an instruction register, two multiplexers, a bypass register and, optionally, an interrupt circuit.

In the control of the CMC by the MMC, two control schemata exist, namely *centralized* and *distributed*. In the centralized control schema, the MMC and CMC are tightly coupled during the entire test process of a chip, and the test bus is thus tied up during this time. The CMC cannot execute a test process without the help of the MMC. In the distributed control schema, the test bus is used only to initialize the test process. The CMC then executes the test process without any help from the MMC. During this time, the test bus can be used to communicate with other CMC's. At the termination of the test, the bus is used for the transmission of test results from the CMC to the MMC.

III. MMC Design

An MMC must be able to respond to request from a SuMP, to carry out tests for every chip on the module, and to report

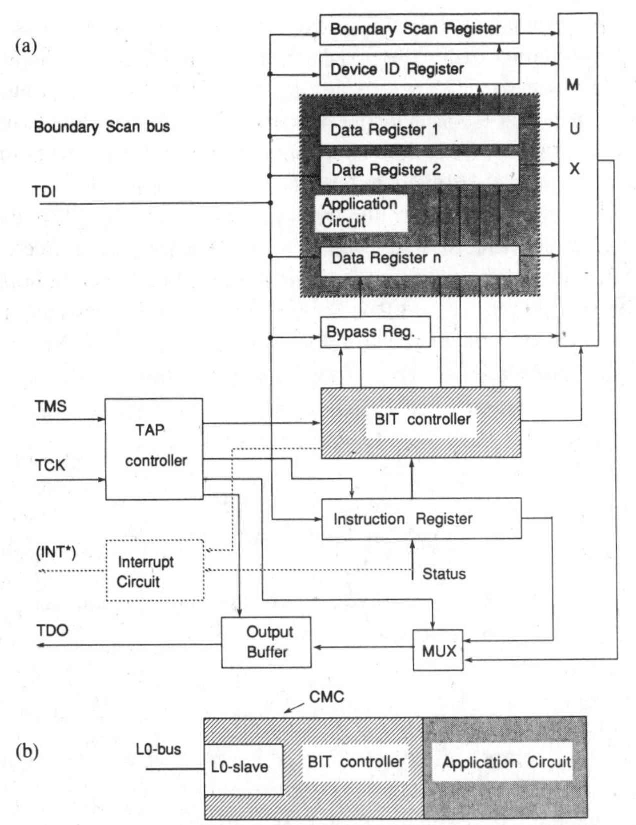

Fig. 2. Control model for a testable chip: (a) JTAG architecture, and (b) abstract model.

test results to a SuMP. The requirements for an MMC are outlined next, followed by a description of its architecture in Section IV.

A. Requirements for an MMC

Based on the test control model presented, one can design an MMC to satisfy all requirements for testing a module containing testable chips. An MMC should be able to support the following functions:

1) access the on-chip BIT structures via an L0-bus;
2) provide proper control sequence for the execution of a chip's BIT structures;
3) provide test data and collect test results if necessary;
4) analyze test results to decide on the health status of chips;
5) test the interconnection among different chips on the module via the boundary scan registers;
6) provide controllability and observability for nontestable chips and analog circuits; and
7) interface with a SuMP or the control console.

An MMC must have memory to store test data and/or test results if deterministic test data are used. For random or exhaustive test methodologies, much less memory is required since only seed data and signatures need to be stored.

IV. MMC Architecture

Fig. 3 shows the architecture of an MMC. It consists of a 16-bit general- or special-purpose processor, a ROM, a RAM, a test channel, a CMC with an L0-slave, an L1-slave, and a

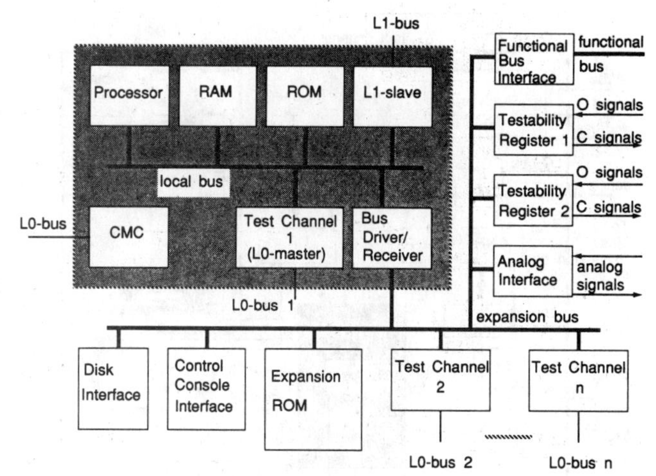

Fig. 3. Architecture of an MMC.

bus-driver/receiver (BDR), which support an expansion bus. Extra units can be added to the MMC via the BDR. For example, a functional bus interface, two testability registers, an analog test interface, several test channels, an expansion ROM, a control console interface, and a disk interface are shown in the figure. The components shown within the dashed line box are required for every MMC. This unit can be implemented as a single ASIC chip. All other units on the expansion bus can be designed for one or more ASIC chips. CMC's for these chips are not shown.

All units on the local and expansion bus are accessed by the processor in a *memory-map* schema. That is, every accessible register of each unit occupies one location in the global address space. The processor can read from or write into these registers by first addressing the appropriate registers. Each unit must be able to decode the address lines. Once a register is selected, an enable signal is generated to initiate a READ or WRITE operation.

A. Test-Channel Design

A CMC may have a pseudorandom test-pattern generator (TPG) and a signature analyzer (SA), which can be implemented using linear feedback shift registers (LFSR's) [19]. In this case, only control signals need be supplied by a test bus during self-test. An example of such a design is presented in [1]. However, if the chip does not have these registers and is to be tested using pseudorandom test data, then a TPG and an SA must be made a part of the MMC. For chips tested by deterministic test vectors, an MMC must be able to provide test vectors and obtain test results via a test channel.

Once initialized by the processor, the primary function of a test channel is to control an L0-bus autonomously. The processor can then be used for other tasks. Thus, high test parallelism can be achieved through running several test channels at the same time.

The major functions of a test channel are listed below:

1) serve as an L0-master;
2) transmit instructions to and receive status from chips;
3) generate and transmit pseudorandom test data and receive and compact test results;
4) transmit deterministic test vectors to and receive test results from chips;

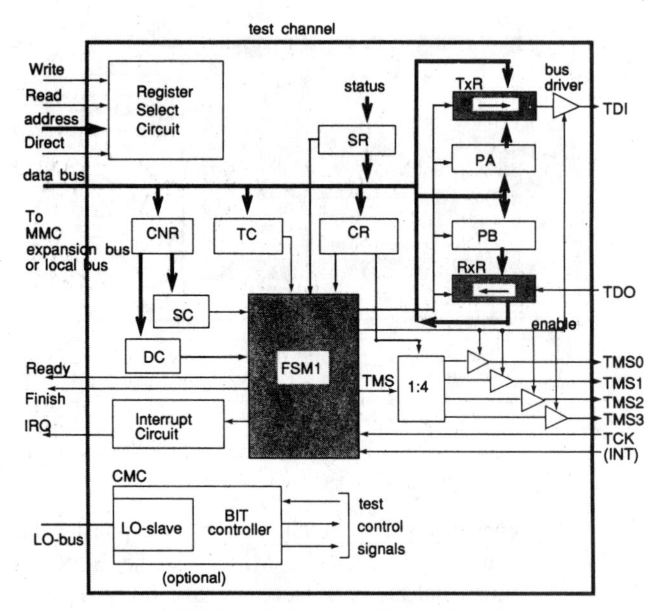

Fig. 4. Architecture of a test channel.

5) generate interrupts and also direct interrupts from chips to the processor; and
6) keep count of the number of tests applied and the number of bits of each test or instruction transmitted.

Organization of a Test Channel: Fig. 4 shows a block diagram of a test channel. Solid lines represent data flow paths, and dashed lines represent control flow paths. A test channel consists of a transmitter register (TxR) for transmitting data over the TDI line; a receiver register (RxR) for receiving data on the TDO line; two polynominal control and buffer registers PA and PB; a control register (CR), which specifies operation mode, selection, and function enabling information; a status register (SR), which contains the current chip status; three counters, namely a test counter (TC), which stores the total number of test vectors to be sent, a scan counter (SC), which keeps track of the number of bits in a test vector which have been transmitted, and a delay counter (DC), which keeps track of the elapse idle time between two vectors; a register count number register (CNR), which contains the initial values for SC and DC; a register select circuit for processor READ/WRITE control; an interrupt circuit to request service from the processor; and a control unit FSM1, which implements the L0-master protocol and is used to send and receive information via an L0-bus under the control of the CR and the three counters. If a test channel is implemented as a stand-alone unit, then it should also have a CMC.

Output signals, such as *TDI* and *TMS*, are all driven through a tri-state buffer thus allowing two or more test channels to be connected to an L0-bus. This enhances the reliability of the test process and makes external testing of a module by another MMC feasible [7]. A more detailed description of the major blocks follows.

1. TxR (Transmitter Register): The TxR is a 16-bit register with parallel LOAD, SHIFT, and TPG capabilities. It is used to transmit data over the *TDI* line. During pseudorandom data transmission, the TxR acts as a TPG. The feedback polynominal of the TPG is controlled by PA. Any feedback polynominal can be realized since PA is directly writable by the processor. The seed value for the TPG also can be loaded by the processor. During instruction or deterministic data transmission, TxR acts as a shift register. It must be loaded with a new word of data before transmission is initiated. The PA serves as a buffer for transmission. Once TxR is empty, the next word of data, which is already PA, is copied into TxR. Processor service is then requested in order to load a new word of data into PA. Transmission over the L0-bus is not interrupted during the 16-clock-cycle window in which PA may receive a new data word. If the data transfer rate is not fast enough, or when TxR is empty and PA does not contain a new word of data, the L0-bus enters a pause state until PA is loaded.

2. RxR (Receiver Register): The RxR is a 16-bit register with parallel READ, SHIFT and SA capabilities. It is used to receive data from the TDO line. Received data are either read by the processor or compressed into a signature. During pseudorandom data transmission, RxR acts as an SA. The feedback polynominal is controlled by PB. The final signature in RxR can be read out via a processor READ operation. During transmission of status or deterministic results, data on the *TDO* line are shifted into RxR. PB serves as a buffer. Once the RxR is full, its content is copied into PB. A service request is generated to signal the processor to read PB and store the data in the RAM. If the previous results in PB have not yet been read, the L0-bus enters a pause state. Transmission cannot start again until PB is read and RxR transfers its data to PB.

3. PA, PB (Polynominal Control Registers): Both registers are 16 bit and have parallel LOAD capability. They can be accessed by the processor via the data bus. Their functions have already been described.

4. CR (Control Register): CR is a 7-bit register. Symbolic names used for the CR bits are *FSMen*, *INTen*, *MS*0, *MS*1, *BS*0, *BS*1, and *Scan*. *FSMen* and *INTen* are used to enable FSM1 and the interrupt circuit, respectively; *MS*0 and *MS*1 are used to specify operation modes; *BS*0 and *BS*1 address one of the *TMSi* ($i = 0, 1, 2, 3$) signals, and *Scan* is for the selection operation type.

5. SR (Status Register): SR is a 4-bit register consisting of bits *Finish*, *IRQ*, *Ready*, and *Wait*. The *Ready* bit is cleared whenever the content of PA is copied into TxR and is set whenever the processor loads new data into PA. The *Finish* bit is set only when the required information has been transferred or TC reaches 0. The *IRQ* bit is set when the *INT* line from the test bus is active. The *Wait* bit is set when both the TxR and PA are empty and is cleared when the TxR is loaded. A processor SR READ operation also reads the contents of CR, i.e., 11 bits are read. This operation can be performed independent of the state of the FSM1. Bits *Finish* and *IRQ* are cleared whenever the SR is read.

6. TC (Test Counter): TC is used to keep count of the number of test vectors transmitted during the execution of one test session. The TC is a 22-bit down counter and requires two processor WRITE operations to load. One of the WRITE operations loads part of this counter and part of the CR. This counter is able to count down to 0 from any number between 1 and 4 194 303.

7. SC (Scan Counter): SC is used to keep count of the number of bits of a test vector or instruction which have been transmitted. SC is a 10-bit down counter and can count down to 0 from any number from 1 to 1023. Its initial value is loaded from the CNR. A terminal count signal will be activated whenever the value in SC reaches 0, and the value *s* in CNR will be copied into SC. In transmitting *t* test vectors to a chip during one test session, SC must be re-initialized (to the value *s*) *t* times.

8. DC (Delay Counter): DC is a 5-bit down counter and is used to count the number of clock cycles between the transmission of two consecutive test vectors. Its initial value can be loaded from the CNR. The DC can count down to 0 from any number from 1 to 31. A terminal count signal will be activated whenever DC reaches 0, and the value *d* in CNR will be copied into DC.

9. CNR (Count Number Register): This buffer is used to store the initial value of the constants for both SC and DC, i.e., *s* and *d* referred to above. These counters destroy their original contents after a test vector is transmitted. Thus, this register is used to restore the value of both SC and DC so that the next vector can be transmitted. The CNR is 15 bits long. It can be loaded by a single processor WRITE operation.

10. Register Select Circuit: This circuit is driven by the processor and is used so that the processor can write into and/or read from various registers in the test channel. Registers CNR, TC, CR, SR, TxR, RxR, PA, and PB are accessible to the processor. When the *Direct* signal here is inactive, the registers are selected by address. When the *Direct* signal is active, this circuit interprets a processor READ operation as a WRITE to PA operation, thus ignoring the address lines. In addition, the address and READ signals are used to read a word from the memory unit. Thus, a word of data is transferred from the memory unit to the PA of the selected test channel. Similarly, when *Direct* is active, a processor WRITE operation is interpreted as a READ from PB operation. The address and WRITE signals are used to write the contents of PB into the memory unit.

11. FSM1: This circuit controls the operation of a test channel and acts as an L0-master. It receives control signals from CR and conditional signals from counters TC, SC, and DC. When the *FSMen* bit is set, a processor-generated WRITE operation is used to generate a *Start* signal, which in turn initiates the FSM1.

Operation of the Test Channel: The operation of a test channel is controlled by its FSM1. The FSM1 controls the state of a test bus via signal line *TMS* (see Fig. 4). For the possible JTAG bus states, the reader is referred to [21].

A test channel provides for two types of operation, namely *RunTest* and *Scan*. During *RunTest*, the test bus enters the *Idle/RunTest* state for a predetermined number of clock cycles. The TC counter keeps tracking of this number. No data is transmitted on either the *TDI* or *TDO* lines. This type of operation is used when a CUT has BIST capability, and the BIST hardware has been properly initialized through the test bus. The chip's BIST controller runs the self-test as long as the bus stays in the *Idle/RunTest* state.

TABLE I.
COUNTER USAGE

	PTD	DTD	DRC	INS	RunTest
TC	no. of tests	no. of tests	no. of tests	set to 1	no. of clock cycles
SC	no. of bits	no. of bits	no. of bits	no. of bits	—
DC	elapsed clock cycles	set to 15	set to 15	set to 15	—
TxR	TPG	SHIFT	SHIFT	SHIFT	—
RxR	SA	SHIFT	SA	SHIFT	—

During *Scan* operation, the test channel tranfers either pseudorandom test data (PTD), deterministic test data without results compression (DTD), deterministic test data with results compression (DRC), or instruction (INS). The operation of the test channel is controlled by the CR and three counters. These counters are used for all types of information transfer. During different operations, a counter may be used for different purposes. For example, in PTD transmission, TC keeps track of the number of test vectors applied, SC keeps track of the number of bits transmitted, and DC keeps track of the number of elapsed clock cycles between two consecutive test vectors. Table I indicates how these counters are used. The operation modes of the TxR and RxR are also shown in the table.

Fig. 5 shows the state transitions carried out by a test channel. Dashed rectangles represent a wait for processor service. The operations indicated in the solid rectangles execute in one clock cycle. The protocol corresponding to this state-transition diagram is consistent with the JTAG boundary scan protocol.

The *FSMen* bit is cleared during the power-up process, and the test channel enters the *idle* state at this time. The processor can read from and write into internal registers of a test channel while in this state. After initializing the appropriate set of registers, setting the *Start* signal and *FSMen* bit will initiate the operation of the FSM1. Depending on the setting of bits *Scan*, *MS*0, and *MS*1, the FSM1 follows one of the five major branches as shown in Fig. 5(a).

The branch labeled PTD is followed when pseudorandom testing is used. Registers PA, PB, TxR, RxR, CNR, and TC are assumed to have been initialized to appropriate values, such as *pa, pb, seed1, seed2,* (*s*, *d*) and *t*. The TxR acts as a TPG with *pa* selecting the feedback polynominal and *seed1* as its initial value; RxR acts as an SA with *pb* selecting the feedback polynominal and *seed2* as its initial value. The test channel then autonomously transmits *t* random test vectors generated by TxR to TDI and compresses *t* test results in the RxR. Each test result is *s* bits long, and *d* clock cycles of delay exist between two consecutive test vectors. No service from the processor is required during pseudorandom testing. The *Finish* bit is set to signal the processor that the process has completed. The processor then reads the signature stored in RxR to determine the test result.

The branch labeled DTD (see Fig. 5(b)) is used when deterministic test data are employed. Registers CNR and TC contain the values (*s*, *d*) and *t*. Note that *d* is always equal to 15 for the DTD process. Its purpose is to clear the *Ready* bit after every 16 bits transmitted. For a test vector longer than 16 bits, TxR is loaded with the first 16 bits of deterministic test data before the *Start* signal is activated. After 15 shift operations, TxR contains the last bit of the test data. One clock cycle later, RxR is full. Two possible situations exist. After these shift operations have occurred, it is possible that PA is

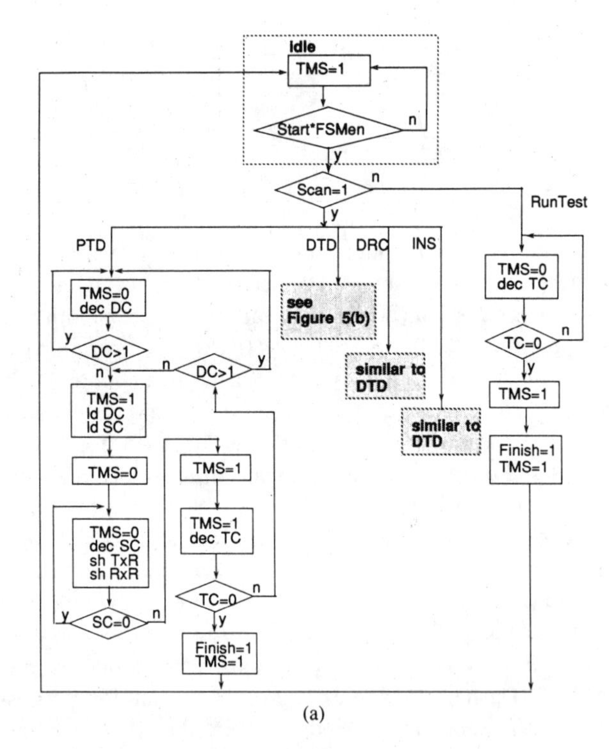

(a)

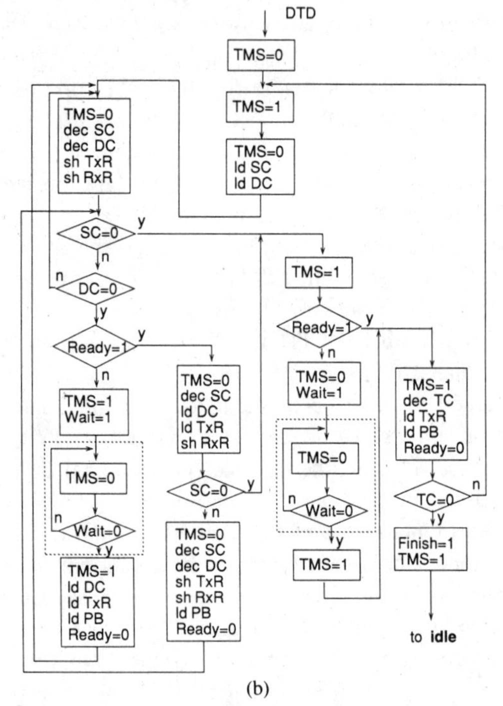

(b)

Fig. 5. State transition diagram for a test channel: (a) overall diagram, and (b) the DTD case.

full (*Ready* = 1). Then the content of PA is copied into TxR and one clock cycle later the content of RxR is copied into PB. The *Ready* bit is cleared, and transmission over TDI and TDO is not interrupted. The processor then has another 16 clock cycles to load PA, read PB, and set the *Ready* bit.

Another possibility is that PA is empty (*Ready* = 0). Transmission is then interrupted, and the *Wait* bit is set to request service from the processor. Waiting for the processor to read the RxR and load TxR is indicated by a dashed rectangle in Fig. 5(b). The test bus is in the pause state during the wait period. Once the processor finishes the READ/WRITE process, it clears the *Wait* bit to allow the FSM1 to transfer another 16 bits of information. The *Finish* bit is set upon the completion of the DTD test, i.e., when TC reaches zero.

The branch labeled DRC is followed when deterministic test data are used and test results are compressed in RxR. The volume of information flow between the memory unit and test channel is reduced by half over the DTD operation.

The branch labeled INS is followed when transmitting instructions. The content of TC is set to 1. The operation of the test channel is similar to that for DTD operations. The only difference is that the sequence of values on the *TMS* line is different.

The branch labeled RunTest is followed when the *RunTest* operation is used. The test channel transmits a specific sequence as specified by the JTAG protocol over the *TMS* line such that all L0-slaves connected to the selected signal *TMSi* will enter the *Idle/RunTest* state for t clock cycles. The *Finish* bit is set before returning to the idle state again.

The loop conditions depend on condition signals ($TC = 0$, $SC = 0$, $DC = 0$, and $DC > 1$) generated by counters TC, SC, and DC, respectively. The processor can stop or disable the operation of the FSM1 by loading a new word into CR through a processor write operation. Resetting the *FSMen* bit will halt the operation of the FSM1. In order to maintain consistent operation, modification of all other registers, except PA, PB, TxR and RxR, is prohibited until the *Finish* bit is set or an interrupt has occurred.

B. Bus Driver/Receiver

The BDR is a bidirectional interface to the local bus of the MMC. It provides driving capability for the signals to/from the expansion bus. Fig. 6 shows the basic architecture of the BDR. Signals IN and OUT control the flow of information between the local bus and the expansion bus. These two signals are decoded from the address and control buses, which are subbuses of the local bus. When the addressed unit is not directly tied to the local bus, the BDR is used to allow searching for the appropriate unit on the expansion bus. To allow interrupts from units tied to the expansion bus to reach the local bus, the expansion bus interrupt signals can also assert the IN signal.

C. Functional Bus Interface

The funtional bus interface (FBI) allows communications between the module's functional bus and the MMC's expansion bus. Through the FBI, the MMC can execute functional tests for the module. Details of this interface will not be presented here. Further information on related interfacing techniques can be found in [4].

D. Testability Register

This is a 16-bit register used to increase the testability of modules containing chips which are either not designed to be testable or do not have a test bus interface. The boundary scan registers on testable chips can be used to increase the testability of nontestable chips. However, in many cases, no boundary scan registers can be found to access signals between

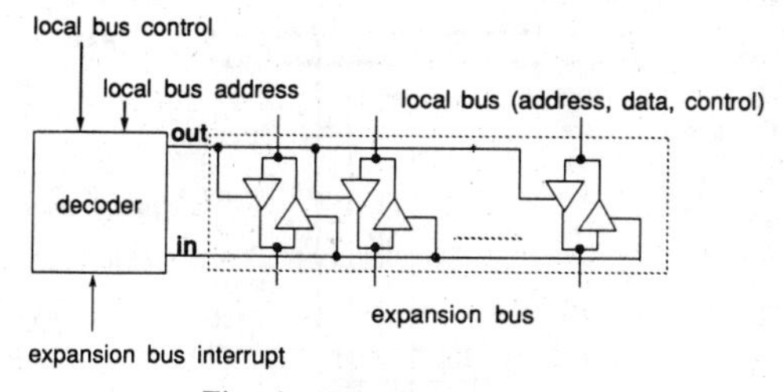

Fig. 6. Bus driver/receiver.

nontestable chips. The testability register can be used to increase the testability of these chips and their signals in the following way. Signal points which need to be controlled (C) and/or observed (O) are cut and fed into the testability register. The O signals are connected to the C signals during normal operation (see Fig. 7). In test mode, the processor writes a word to the testability register which in turn applies this data to the C signals. O signals are loaded into the testability register and then read by the processor. Thus, both the controllability and observability of these cut points are enhanced. A technique for selecting these signal points is presented in [10].

E. Analog Test Interface

This circuit is used when there are analog circuits on the module under test (see Fig. 8). To generate an analog signal, the processor writes a word to the analog test interface, and the D/A converter then converts this data into an analog signal. For observability, an analog signal is converted into a digital word, which can then be read by the processor.

F. L1-slave

The MMC communicates with its higher level SuMP controller via an L1-bus; thus, it must have an L1-slave. The design of a TM-slave is given in [8].

G. Processor

Processor functions can be classified into five categories: 1) transfer data between memory and test channels; 2) transfer data between memory and an L1-slave; 3) compare test results with good results; 4) transfer data between memory and expansion units; and 5) execute test and/or diagnostic programs.

A general- or special-purpose 16-bit processor can be used in the MMC. It controls all other units in the MMC. Through READ/WRITE operations, the processor can access internal registers of a peripheral device, such as the L1-slave and test channels. Operations of a peripheral device can thus be controlled by a processor WRITE to the CR of the peripheral device. Data exchange between memory and a peripheral device are controlled by processor READ/WRITE operations. Any processor having the following instructions is powerful enough for the application of an MMC.

instruction	meaning
LDA R_i	Load Acc with R_i.
LDA M	Load Acc from memory location M.
STA R_i	Store Acc to R_i.
STA M	Store Acc to memory location M.
ADD R_i	Add R_i to Acc.
AND R_i	Bitwise AND R_i with Acc.
CMP R_i	Compare Acc with R_i.
NEG	Complement Acc.
CLA	Clear Acc.
BRZ R_i	Branch to location R_i if Acc not zero.
JMP R_i	Jump to location R_i.
PUSH	Push Acc onto stack.
POP	Pop Acc from stack.
NOOP	No operation.
HALT	Halt the processor.

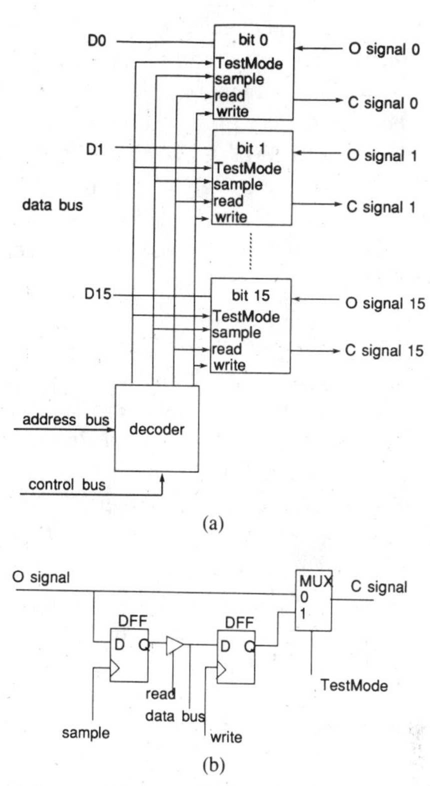

(a)

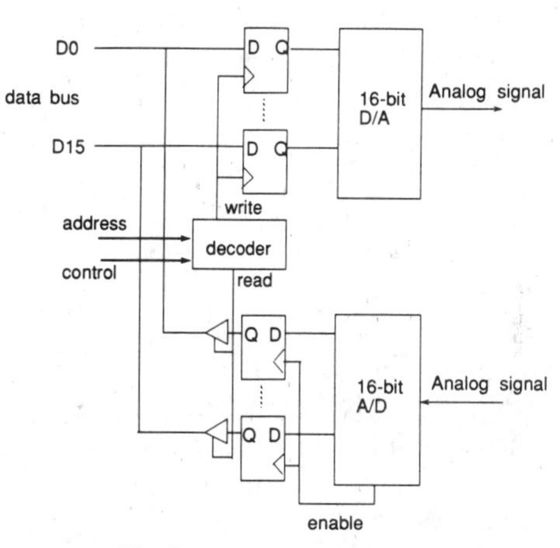

(b)

Fig. 7. Testability register: (a) block diagram, and (b) circuitry for bit i.

Fig. 8. Analog test interface.

The minimal architecture for a processor which is able to execute the above instruction set consists of an accumulator, four general-purpose registers, an ALU, a program counter, a program status word, a stack with at least four words, an interrupt circuit, and a microprogrammed control unit.

If an MMC is implemented as a single-chip ASIC, two additional instructions are useful to increase the data transfer efficiency between the memory unit and a test channel. The added instructions are MULTIPLE READ (MR) and MULTIPLE READ and MULTIPLE WRITE (MRMW). The signal lines *Direct, Finish,* and *Ready* are used exclusively to support these two instructions (see Fig. 9). Signal *Direct* is active when the microcontroller is executing any one of these two instructions. Signals *Finish* and *Ready* are used as conditional signals for the microcontroller of the processor. All *Ready* (*Finish*) signals from test channels are wired-*OR*ed together.

When executing an MR instruction, the processor waits until the *Ready* signal is cleared and then issues a READ operation to the memory location addressed by general-purpose register *R*0. Meanwhile, the test channel with *FSMen* bit set and operation mode being either DTD, DRC, or INS can generate a load PA signal using signals *Direct* and *Read*. Thus, a data word is moved directly from memory to a test channel. The value of *R*0 is increased by one after each READ. The processor waits for the *Ready* signal to be activated and then issues another READ operation. This process is repeated until the *Finish* signal is set. Thus, a block of information can be moved from the memory unit to the selected test channel and transmitted to a chip without any interruption.

When executing an MRMW instruction, the processor waits until the *Ready* signal is deactivated and then issues a READ operation to the memory location addressed by *R*0; meanwhile, the enabled test channel generates a load PA signal, and the data word from memory is loaded into the PA. The value of *R*0 is incremented by one. The processor then issues a WRITE operation to the memory location addressed by *R*1; meanwhile, the enabled test channel generates a READ PB signal, and a data word is read out of the PB and sent to the memory. The value of *R*1 is incremented. The processor waits for the *Ready* signal to be deactivated again and then issues another READ/WRITE operation. This process is repeated until the *Finish* signal is set. Thus, a block of deterministic test data is moved from the memory unit to the selected test channel, and a block of test results is moved from the selected test channel to the memory unit.

H. Memory

The memory unit in an MMC is composed of a RAM unit and a ROM unit. The ROM unit contains test programs to test the entire module. These programs are compiled separately before testing. Some crucial information about the chips on the module is stored here such as the number of chips to be tested, ordering of chips along the test bus ring, number and length of scan chains for each chip, number of random test vectors to apply to each chain, test instructions for each chip's CMC, TPG seeds, and good signature for each test session. MMC functional self-test programs can also be stored in this unit. If the MMC is implemented using commercial IC's, then these programs are essential for MMC self-test. The expansion ROM can be added where a module requires a large test program.

The RAM unit provides scratch pad memory for test program execution. Response signatures are stored here for latter evaluation. The RAM also provides storage for the Go/NoGo status for all chips, as well as for the entire module.

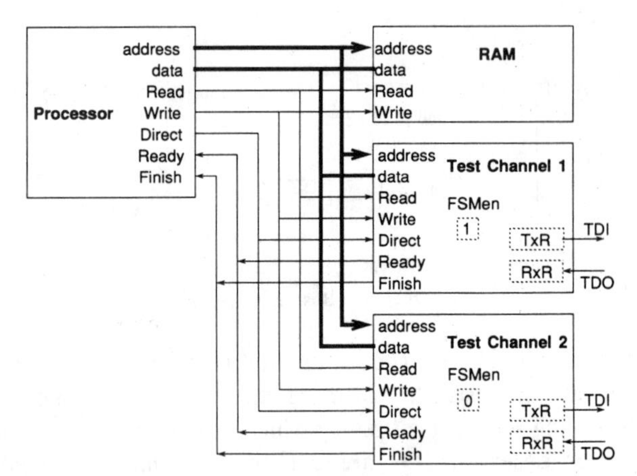

Fig. 9. Control signals for MR and MRMW instructions.

I. Stand-alone MMC

The MMC can be used as a stand-alone mini ATE, provided extra storage and console capabilities are added. For this application, a console control interface and disk interface can be added to the MMC.

V. MMC SELF-TEST

If the MMC is implemented with an off-the-shelf "nontestable" processor, ROM, and RAM, then some form of functional self-test is required. After finishing self-test, the MMC then reports its status to the control console or to a SuMP.

An MMC can also be tested by either an ATE or by another MMC. In the first case, an ATE can access the expansion bus of the MMC under test. The ATE invokes the self-test program of the MMC under test and waits until its completion. The test results, which are stored in the RAM, are then read by the ATE. In the second case, an MMC uses its FBI to access the expansion bus of the MMC under test. Again, self-test programs can be invoked. Test results can be read and interpreted by the monitoring MMC.

If the MMC under test is implemented as a custom testable ASIC, then we assume it has a CMC. The MMC can thus be tested by another MMC. All units in the MMC, such as the processor, RAM, ROM, test channel, and L1-slave must be designed to be testable, and their BIT structure must be accessible via the L0-slave.

The testable design features of a test channel are shown in Fig. 10. Major combinational logic blocks are indicated by rectangles having dotted lines. Registers are indicated by rectangles having solid lines. Some logic is associated with these registers, since some are counters and LFSR's. Normal functional connections are not shown. Instead two scan chains formed during self-test are shown. Scan chain 1 is the boundary scan chain. All I/O signals can be controlled and observed by shifting test data or results along this chain. All other registers make up scan chain 2. The state of the test channel is controlled by shifting data along this scan chain. If a

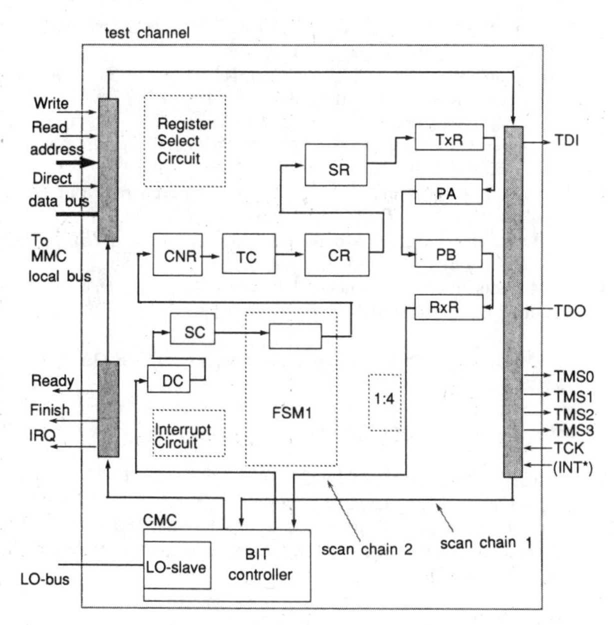

Fig. 10. Testable design features for a test channel.

functional clock is then activated, the next state of the test channel also can be observed by shifting out the content of this chain.

During testing, scan chain 1 is first loaded with test data which is held in place while the logic associated with scan chain 2 is tested. The module I/O is tested using the boundary scan chains of this chip and those to which it is connected.

VI. Conclusions and Discussions

We have described an MMC design suitable for controlling the self-test process of a module. The design uses the concept of test channels, which can run a test autonomously (in PTD case) once it is initialized by the processor. Because of the test channel, the processor need not deal with detailed control sequences over the JTAG boundary scan bus. Test execution sequences for chips can be generated in terms of processor READ/WRITE operations, which greatly simplifies the development of test programs.

The MMC architecture is expandable. More test channels can be added so that more chips can be tested in parallel. In addition, the MMC supports the functional testing of a module, the testing of clusters of chips which are not designed to be testable, and the testing of analog devices.

Clock Synchronization: Four or more clocks may be applied to an MMC, viz. *TCK* for the CMC, *FCK*1 for the L1-slave, *FCK*2 for each test channel connected to an L0-bus, and *FCK*3 for the operation among processor and other peripheral devices. Synchronization problems will occur in a test channel where both *FCK*2 and *FCK*3 may access the same component, such as TxR and RxR. Techniques to solve this problem can be found in [4] and [15]. In the design presented here, a common clock is used to drive all the clocks mentioned above, thus avoiding the clock synchronization problem.

Portable tester: The proposed MMC is designed to be part of an HTM system. It is assumed that each module contains an MMC, which under request from a SuMP can test all chips on the module and report back test results. However, it is possible to build an MMC as a portable stand-alone unit. In this case, the L1-slave can be replaced by a control panel. A stand-alone MMC can test any module having an L0-bus. The module's built-in MMC is tested first through its L0-slave. Application chips on the module can be tested either by the built-in MMC or by the stand-alone MMC. For the latter case, the built-in MMC must be disabled to allow the stand-alone MMC to take control the module's L0-bus. An operator can start the test process via the control panel. Test programs stored in the ROM then take over control. After all chips have been tested, test results are shown on the control panel to indicate the Go/NoGo status of the module under test.

Overhead: There are several ways of implementing an MMC. One or more test channels can be built on an ASIC chip. The processor, RAM, and ROM can be implemented using standard chips. The other functions, which are optional, can be implemented using standard parts or an ASIC chip, excluding the expansion ROM. The application chip requires overhead to support testability, such as scan registers, as well as a L0-slave. For double-latch designs, scan area overhead usually varies from 2.8 to 6.3 percent, depending on the ratio of gates to latches [20]. The overhead for the L0-slave depends on the length of the instruction register and the number of I/O pins. Assuming each shift register latch (SRL) is equivalent to ten gates, an L0-slave with a 16-bit instruction register and a 60-bit boundary scan register requires about 1600 gates. For a 50 000-gate ASIC chip, the total overhead for testability will typically be between 5 and 10 percent.

The JTAG bus consists of four wires. Assuming 60 pins/chip prior to adding the bus, the routing overhead to support testability will be at least $4/60*100 = 6.7$ percent. This is a lower bound since most pins on a chip are tied to only two to three point nets, while the JTAG bus goes to all IC's. We estimate the wiring overhead to be closer to 10 percent.

Fault Isolation: One of the important attributes of boundary scan is the ability to test the interconnect between chips. Assuming chips are also designed to be testable via DFT or BIST techniques, the MMC should be able to accurately isolate hardware faults to a chip or interconnect.

Analog Performance: Since the A/D and D/A conversion time is much smaller than the data transfer rate over the bus, the speed of observing or controlling an analog signal is determined by the data-bus bandwidth. For example, an Intel 80186 processor running at 8-MHz clock rate can transfer 4-MByte of data from memory to the analog interface in 1 s.

CAD Tool Support: Clearly, a great deal of binary data flows between an MMC and test application chips. It is not feasible to develop these data manually. Thus, CAD tools are required so that test programs for chips can be written in a high-level language and these programs compiled into code which is executed by an MMC. We are currently working on developing such tools and associated test languages.

References

[1] L. Avra, "A VHSIC ETM-BUS compatible test and maintenance interface," in *Proc. Int. Test Conf.*, 1987, pp. 964–971.
[2] F. Beenker, "Systematic and structured methods for digital board testing," *VLSI Syst. Des.*, vol. 8, pp. 50–58, Jan. 1987.
[3] F. Beenker, K. Eerdewijk, R. Gerritsen, F. Peacock, and M. van der

Star, "Macro testing: Unifying IC and board test," *IEEE Design & Test Computers*, pp. 26–32, Dec. 1986.

[4] G. Borriello and R. H. Katz, "Synthesis and optimization of interface transducer logic," in *Proc. Int. Conf. CAD*, 1987, pp. 274–277.

[5] M. A. Breuer, "On-chip controller design for built-in-test," Dept. EE-Systems, Univ. of Southern Calif., Techn. Rep. CRI-88-04, Dec. 1985.

[6] M. A. Breuer, R. Gupta, and J. C. Lien, "Concurrent control of multiple bit structures," in *Proc. Int. Test Conf.*, 1988, pp. 431–442.

[7] M. A. Breuer and J. C. Lien, "A methodology for the design of hierarchically testable and maintainable digital systems," in *Proc. 8th Digital Avionics Systems Conf.*, 1988, pp. 40–47.

[8] M. A. Breuer and J. C. Lien, "A test and maintenance controller for a module containing testable chips," in *Proc. Int. Test Conf.*, 1988, pp. 502–513.

[9] W. O. Budde, "Modular testprocessor for VLSI chips and high-density PC boards," *IEEE Trans. Computer-Aided Des.*, vol. 7, no. 10, pp. 1118–1124, Oct. 1988.

[10] K. K. Chua and C. R. Kime, "Selective I/O scan: A diagnosable design tehnique for VLSI systems," *Comput. Math. Applic.*, vol. 13, no. 5/6, pp. 485–502, 1987.

[11] J. E. Haedtke and W. R. Olson, "Multilevel self-test for the factory and field," in *Proc. Annual Reliability Maintainability Symp.*, 1987, pp. 274–279.

[12] C. L. Hudson, Jr. and G. D. Peterson, "Parallel self-test with pseudo-random test patterns," in *Proc. Int. Test Conf.*, 1987, pp. 954–963.

[13] IBM, Honeywell, and TRW, "VHSIC Phase 2 interoperability standards," ETM-BUS specification, Dec. 1986.

[14] IBM, Honeywell, and TRW, "VHSIC Phase 2 interoperability standards," TM-BUS specifications, Dec. 1986.

[15] S. Y. Kung, S. C. Lo, S. N. Jean, and J. N. Hwang, "Wavefront array processors—Concept to implementation," *IEEE Computer,* vol. 20, pp. 18–33, July 1987.

[16] D. van de Lagemaat and H. Bleeker, "Testing a board with boundary scan," in *Proc. Int. Test Conf.*, 1987, pp. 724–729.

[17] J. J. LeBlanc, "LOCST: A built-in self-test technique," *IEEE Design & Test Computers*, pp. 45–52, Nov. 1984.

[18] C. Maunder and F. Beenker, "Boundary-scan: A framework for structured design-for-test," in *Proc. Int. Test Conf.*, 1987, pp. 714–723.

[19] E. J. McCluskey, "Built-in self-test techniques," *IEEE Design & Test Computers*, Apr. 1985, pp. 21–28.

[20] M. J. Ohletz, T. W. Williams, and J. P. Mucha, "Overhead in scan and self-testing designs," in *Proc. Int. Test Conf.*, 1987, pp. 460–470.

[21] Technical Subcommittee of Joint Test Action Group (JTAG), "Boundary-scan architecture standard proposal," Version 2.0, Mar. 1988.

[22] J. Turino, "IEEE P1149 proposed standard testability bus—An update with case histories," in *Proc. Int. Conf. Comput. Design* (ICCD), 1988, pp. 334–337.

Reprinted from *IEEE Design & Test of Computers*, August 1989, pages 18-30.

THE IMPACT OF BOUNDARY SCAN ON BOARD TEST

KENNETH P. PARKER
Hewlett-Packard

Boundary scan, which began as a proposal from Joint Test Action Group, is now IEEE proposed standard P1149.1. This technology for incorporating design for testability into ICs can actually benefit several levels of manufacturing from IC fabrication through boards and into system test. Boundary scan's impact seems particularly noticeable in production-board testing. Pure boundary-scan implementations, in which all ICs are scannable, are not likely to appear in the near future, but the benefits of partial implementations are still significant. While definitely not a replacement for ATE, boundary scan can still reduce test complexity and cost, and increase accuracy. Those not willing to incorporate boundary scan at the IC level must be prepared to balance costs at that level with the costs of board test, which are escalating in the face of growing complexity.

Testing technology has been examined formally and in depth for many years. This year marks the 20th International Test Conference, the 12th Design for Testability Workshop, and the Eighth Built-In Self-Test Workshop. However, despite the fact that design-for-testability technology has been widely disseminated, the industry overall—except for large, vertically integrated companies—has taken remarkably little advantage of it.[1,2] There are only a few examples of industrial DFT and built-in self-test applications, such as the level-sensitive scan design approach developed by IBM.[3]

One source of this resistance is designers, who tend to be offended by the overhead of extra circuitry and possible performance degradation from adding testability. When taken in the narrow view, that of designing a small part of a product, the overhead does appear onerous. If, however, we look at the larger view, that of the entire design/manufacturing process, the costs are made up in reduced testing costs at all levels, and decreased development time. Thus, as Dave Ballew of AT&T put it, we must avoid being "silicon wise and system foolish."

Manufacturers of ICs have also been reluctant to provide testability features for users even though for VLSI it is absolutely mandatory to incorporate testability to economically produce such parts. Some reasons for their hesitancy are

- Testability features must be documented and maintained just like any other IC feature.
- Security of designs may be compromised.
- Fault coverage may be embarrassingly low (if known at all).
- An implied warranty could be attached to these features making the IC vendor liable if the device contains an uncovered fault.
- Test requirements conflict from customer to customer.

An article in a 1986 issue of *D&T*,[4] examined these and other sources of resistance to testability. At that time, another reason was the lack of standards. Since then we've seen a push for boundary scan, once referred to as the JTAG proposal now called IEEE proposed standard P1149.1.[5] This effort began as an attempt to develop a standard that can be embraced at many levels of digital-circuit test, from IC fabrication through system

This article is based on the keynote presentation given at the BIST Workshop, Kiowah Island, South Carolina, March 1989.

test. It has been so well-received that the proposed standard, as of publication date, has been issued for balloting.

A standard such as P1149.1 relieves many of the problems that have caused resistance to design for testability. The IC vendor is free to use this boundary-scan standard as a gateway to public BIST features while leaving proprietary tests undisclosed. Boundary scan allows board and system designers to more easily test their respective products independently of the content of the ICs. It lets IC vendors off the hook legally, since all they have to do is adhere to an official standard and then blame it if the users complain.

There is a great deal of interest now in integrating the design and test phases of a product-development cycle.[7] The implementation of boundary scan within ICs will have a two-fold effect: It will make digital designs more testable and producible, and it will take pressure off designers who might otherwise have to pursue ad hoc testability modifications to their designs. This will make cooperation between design and test departments easier to achieve.

> ***We must look at savings in the larger view and avoid being silicon wise and system foolish.***

BOARD-TEST PHILOSOPHIES

It's fair to ask why we go to the trouble of testing boards at all. The main reason is economics. It is relatively easy to test boards and much more difficult to test systems. If boards used to build systems are nearly perfect, then system turn-on success rates will be acceptable. If boards are somewhat less than perfect, system turn-on success decreases very rapidly.

What are some of the approaches to board test in manufacturing today? Broadly categorized, they are

- board test as a sorting process
- board test as a repair driver
- board test as a process monitor

In the following discussion, bear in mind that any approach to board test is heavily influenced by such factors as product mix, volume, complexity, reliability requirements, quality requirements, capital budgets, and available skills as well as inertial factors such as "it's always been done this way" and "we have to use existing processes and equipment."

A SORTING PROCESS

Sorting in this context means to separate good boards from faulty boards. It is a go/no-go approach. Essentially, only one bit of information is generated about a board during test. Finding and repairing faults in such an environment can be quite difficult, but we can justify this approach if we simply discard faulty boards. For example, if a board-manufacturing process enjoys very high yields, then why invest in diagnostic test and repair processes when it costs less to discard the few bad boards? In another case, seen in military applications, extremely high technology boards are built in very low quantities and are technically challenging to repair without introducing new failures (some

An all-too-familiar practice is to sort good boards for immediate shipment, leaving piles of dog boards to be dealt with later.

latent). Consequently, manufacturers simply discard bad boards, while taxpayers cringe, and keep making new ones until they accumulate enough good boards.

In yet another scenario, schedule pressures dictate simply sorting good boards for immediate shipment, leaving piles of dog boards behind to be dealt with later. This result is unfortunately all too familiar across the electronics industry. It is yet another indication of how forces other than testing issues can influence efficiency.

A REPAIR DRIVER

Using board test as a repair driver is perhaps the most common board-test approach today. Here, testing is more thorough and delivers more bits of information about failures, which forms the basis for a failure diagnosis. More sophisticated techniques are used during test, such as in-circuit isolation, guided probes, current tracing, and fault dictionaries. Information about faults, in the form of symptoms such as failed outputs and test numbers, is collected into a fault syndrome that can be interpreted to help repair. This information is important because we need enough correct data on the fault to enable a complete repair. It is unproductive to send a bad board back without complete information, since partially repaired boards inevitably return for rework.

A PROCESS MONITOR

The newest view of board test is as a process monitor. A well-conceived and thorough board test that emphasizes diagnosis can provide a wealth of information about the various processes that go into a board's manufacture. For example, soldering is a well-known point for faults. Solder problems show up at board test as shorts in the form of blobs, opens in the form of SMT tombstones (small low-mass SMT components like resistors and capacitors that have moved during soldering because of the surface tension of the liquid solder) or incomplete flow, or thermally damaged devices. When such problems are exposed, we know, in the long run it will be more fruitful to fix the soldering process than to fix the solder defects after the fact.

We can use this type of board test to change the structure of the manufacturing process. For example, we may discover fabrication errors, like a wayward solder process, or errors in the design of the board. We can also improve the tests themselves to get a better diagnosis. With time, we gain expertise in the board's manufacture and can begin to improve diagnosis by correlating data from independent tests and previous repairs.

Another use of this type of board test is to observe and control all the system-level process changes. Machines wear out and require calibration. Different people are involved, depending on the shift and who's on vacation. Vendors of components may change what they send, perhaps because their own process changes. Manufacturers may add new people and machines, update production technology, or change the locale of production.

The design group may implement changes to the board's design. All these changes have an impact on the manufacturing process, and using board test as a process monitor provides a way of controlling them efficiently.

This philosophy of board test is not easy to implement quickly, however. It requires skill, experience, and teamwork to a degree that some organizational structures will not readily allow. Management involvement is a key ingredient. Managers need to recognize that process control is becoming more and more important with each new round of technical evolution in board manufacture, and is becoming the dominant reason for using board test. Without process control, a manufacturer cannot hope to attain world-class quality.

Managers need to recognize that process control is becoming the dominant reason for using board test.

WHAT ARE THE COSTS?

Everyone knows that testing is costly and that it is grabbing a larger percentage of manufacturing costs every year. It is phenomenally expensive to let a failure escape to later stages of product manufacture or support. The rule is fix it now or suffer far costlier problems downstream.

Each board-test philosophy has an associated cost. In general, more testing for better coverage and diagnosis implies more test-development costs in time and talent and more time actually spent testing on test heads. Testers themselves become more expensive. The driving forces here are increased tester versatility, complex test-preparation software, increased operating frequencies and higher active tester channel counts. It is not unusual today to see board testers with five times more channels operating 10 times faster than they did just a few years ago. Ironically, while boards are often the same size, many board testers have increased in volume, mass, and power consumption. Especially in the IC test world, mass ratios between the tester and the IC are nearly unbelievable. Channel counts and operating frequencies are two main contributers to this trend.

As mentioned earlier, controlling a process requires teamwork. The people involved have many disciplines. Purchasing agents enforce quality controls on vendors. Inspectors inspect incoming parts that must meet precisely identified parameters. Stock/inventory personnel need to take care that parts are not damaged before they are used—even if it's in today's Just-In-Time manufacturing process. Those controlling phases of fabrication must be alert to their own controls. The test department has to correlate and communicate failures to points upstream in the process. The design team has to have design for manufacturability as a measured goal. And finally, management has to see how these disciplines interrelate. Teamwork is a challenge.

INTERNAL THREATS TO SUCCESS

A number of elements in the overall testing and manufacturing process can sabotage the success of a board-test operation, which depends on sophisticated monitoring and control techniques.

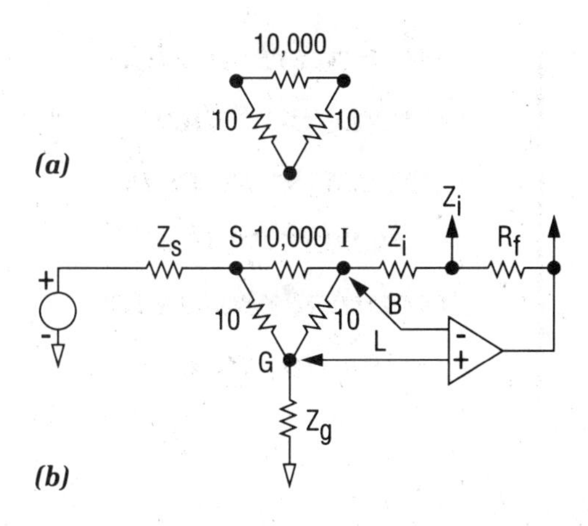

Figure 1. *The resistor network in **(a)** with an extreme ratio of component values is tested using the six-wire guarding scheme in **(b)**, where error impedances, labeled Z_x, are sensed at points A, B, and L for mathematical correction. Accuracy to 2% is easily achieved.*

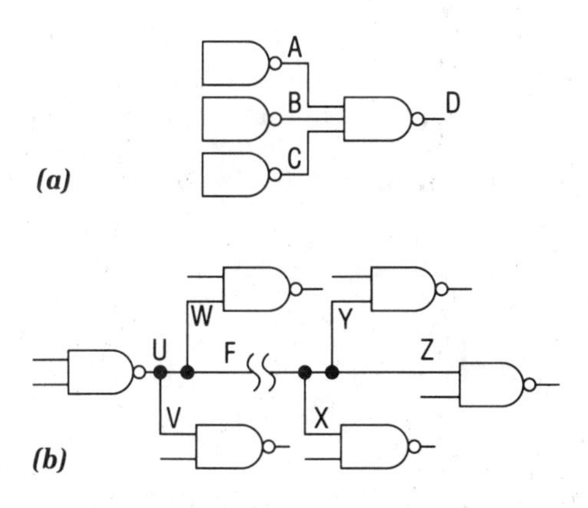

Figure 2. *In **(a)**, three stuck-at-0 faults at A, B, or C will have identical behavior to the output D stuck-at-1 during test. We need to consider only one fault (usually D) when developing the test, but we must consider all four faults when attempting to diagnose a failing circuit. In **(b)**, a simulator analyzes single stuck-at faults at nodes U through Z, but none will produce the syndrome caused by an open at point F. Though it is likely (but not guaranteed) that we will detect F during testing, its syndrome will be missing from the fault dictionary produced from the simulation, so diagnosis will be poor.*

When these factors intrude on the control procedures, the object of board testing—the exchange of diagnostic information to prevent future failures—is greatly impaired. Elements that can sabotage a successful board-test operation include test inaccuracy, misdiagnosis, loops in the test-repair process, and failures from extra handling.

Another threat, which is very real but not always taken into account, is the psychological factor of confidence in the control procedures. If we lose confidence in our control of a process, regardless of reality, matters will rapidly degenerate from tight control to a bare-minimum position of simply sorting good boards from bad. Like biological homeostasis, a well-controlled process requires the coordinated balance of myriad variables and feedback relations. Without this balance, a system enters shock.

Accuracy. If a test is not accurate, it will not give results that we can trust in monitoring and controlling the process. For example, look at the analog test in Figure 1. A three-resistor network has a delta configuration of two 10-ohm resistors and a 10,000-ohm resistor. In-circuit testers find this circuit challenging because of the ratio of resistances and the need to remove error terms.[8] A simple three-wire guarding scheme can measure the 10,000-ohm resistor in this configuration, but without proper consideration of error sources, the measurement can yield 900 ohms. This error is large enough to invalidate testing the 10,000-ohm resistor altogether. A six-wire guarding scheme[8] can account for errors due to voltage drops in the measurement setup (Z_g, Z_i, and Z_s) and mathematically reduce their effects to achieve a 2% accuracy on the measurement.

This analog example illustrates that real-world board-test is not simply for digital testing problems. Harder problems may actually be more prevalent, and not just in the analog domain. The digital bus-fault problem has a number of complexities, for example. We don't know which driver sources a bus at any time, so we have to isolate the source(s) of current on the bus. Test measurement must then move from voltage measurement to current measurement, or we are forced to make complex deductions about the states of the drivers on the bus. Such interactions can profoundly affect digital testing accuracy.

Misdiagnosis. This problem occurs when we detect a failure but do not properly resolve it. For example, we may make an invalid assumption in fault modeling (Figure 2), or several failure modes with identical syndromes may be represented by only one mode. Misdiagnosis leads to erroneous and unnecessary repairs that pollute the information stream we are using to observe the process. The box on the righthand page elaborates on fault isolation. Misdiagnosis also makes the other structural threats even more damaging.

Test-repair looping. Looping occurs when testing is not able to completely or accurately resolve all the faults on a board in one pass across the test head. This inability could be due to tester inadequacies but more often is due to poorly designed or immature test programs. Looping can also be due to handling-induced failures. A loop occurs when the board is sent to repair with an incomplete or inaccurate list of defects. Subsequent testing will fail, and another repair cycle will start. This cycling consumes time and talent. Even though paperless test-repair networks have reduced some of the bookkeeping overhead associated with multiple repairs, repair looping still wastes tester time and uses skilled technicians to poor advantage.

Handling-induced failures. Failures can result from the extra handling and rework in a repair cycle. Clearly, each additional repair cycle compounds the risk of new failures. But another kind of risk is of even greater concern. Handling may weaken the circuitry enough to accelerate the time to failure. Thus, handling increases the risk of infant mortality, the failure of the board in early stages beyond manufacturing.

Repair looping wastes tester time and uses skilled technicians to poor advantage.

EXTERNAL THREATS TO SUCCESS

Industry trends can threaten a manufacturer's ability to control the board-manufacturing process. Some of these are ASICs,

ISOLATING INTERCONNECTION FAULTS

One of the prime advantages of boundary scan at board test is the ability to test board interconnections for integrity. (Algorithms and analysis for this will be presented in detail at the 1989 International Test Conference, held in Washington D.C. August 27-31.) In a pure boundary-scan implementation, in which every IC is scannable, each source and destination of a node is connected to a scannable point. The figure below shows a simplified circuit of six boundary-scan devices with a node driven by point *U* and received at points *V* through *Z*. This figure is simply a scannable version of the circuit in Figure 2b on the facing page. We no longer show the NAND gates since these are not relevant. We also do not need a simulator to help us develop an interconnection test for this circuit.

We can test any single stuck-at fault on the six labeled points, *U* through *Z* by driving *U* to 1 (through the scan path) and latching bits *V* through *Z* for scan out, followed similarly with *U* set to 0. If a single bit in *V* through *Z* is incorrect, it indicates a single stuck value at the corresponding receiver. If all five bits, *V* through *Z* are incorrect, there may be a stuck problem with driver *U*. Suppose a circuit is open at point *F*, and the test of *V* and *W* pass and *X*, *Y*, and *Z* fail (say to 1). We then know that the problem with the interconnection affects only part of the net, and we can deduce the significant topological clues necessary to isolate the physical defect.

The simulator-based test for the circuit in Figure 2b may misdiagnose the fault at point *F* because its effects were not modeled. Because of this, the test fails but the observed syndrome does not match any syndromes predicted by the simulation. When this occurs, we must supply meaningful diagnostic information with additional measurements. For example, with a handheld probe guided by the actual syndrome data and deductions on the cause of the fault derived from circuit topology, we can backtrace to the region of the physical defect. However, we've invested a great deal in the simulation for nothing. Moreover, physical access constraints may apply to the guided probe as well, which may limit its effectiveness.

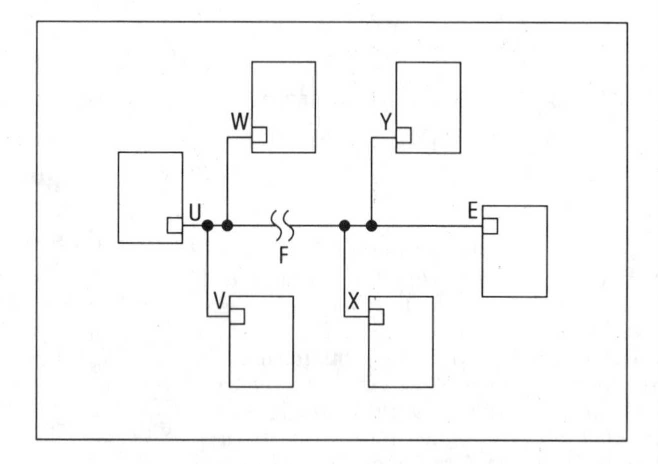

This circuit is the same as the circuit shown in Figure 2b, except NAND gates are replaced with scannable NAND gates. The output U and inputs V through Z are scannable cells. We do not have to know that the devices are NANDs. The complexity is arbitrary. The open-circuit defect is at point F.

In-circuit testing, probably the most popular board-test methodology, is the most threatened by access problems.

surface-mount technology/tape automated bonding, miniaturized components, and node counts.

ASICs. Each application-specific IC—unique as a fingerprint—is an adventure in test. There is nothing inherently untestable about ASICs, although a conjunction of forces makes them seem that way. They are often developed in parallel with the board they will be placed on, which constricts test-development time. Along with this, the ASIC design database is commonly either nonexistent or incompatible with the board test. ASICs are often used to garbage collect random gating or glue logic into one package, which makes their function appear random and undecipherable to a test engineer. Last-minute design changes often destroy any test written for the ASICs—and you can bet that the ASICs *will* be affected.

SMT/TAB. Surface-mount technology and tape automated bonding are two newer packaging technologies that present new board-test problems. They introduce severe test-access problems and are difficult to rework in repair situations. Repairing such boards is a time-consuming task that requires a lot of skill and may elevate the risk of collateral damage that could result in scrapping the board entirely. Also, these technologies are more sensitive to physical damage. Their fragile nature makes them particularly vulnerable to misdiagnosis and test-repair looping. Each pass through the loop increases the probability of irreparable damage.

Access problems. Boards are becoming much more densely laden with higher complexity devices. But even generic resistors and capacitors are presenting test problems,[9] all because of miniaturized packaging and surface-mount configurations. In-circuit testing, probably the most popular board-test methodology, is the most threatened, since it relies heavily on direct electrical access to all nodes through bed-of-nails probing. In talking about access, an important distinction is that access is required to all nodes (also known as nets or circuits), not to each device pin on each node.

We are now seeing boards with components on both sides without the through-holes that enabled test nails to see every node from one side of a board. This packaging technology is also producing components with large numbers of pins on much closer centers. Some boards that are SMT redesigns of existing through-hole boards are 30% to 50% of the original size. New designs are often done on familiarly sized boards to reduce board count in products or to add functionality without increasing physical size.

Two problems arise with this increased number of components and absence of through-holes. First, we may need to place nails much closer together, which means they will be much smaller, more fragile, and more expensive. Second, the targets the nails must hit are much smaller. Since boards are not vastly smaller (on average) today, we need tight control of mechanical tolerances for these nails across thousands of square centimeters. Other access problems are presented by hybrids on ceramic substrates, conformal coatings, and similar applications.

Node counts. Because of these packaging technologies, the node counts of boards are rising, although node counts per board can actually fall if the level of integration within packages is high enough to reduce interpackage data flow on parallel buses. An example is the recently introduced microprocessors that contain on-chip memory management and floating-point units.

Generally, though, node counts are rising. Entire systems now fit on single boards. For in-circuit testing, this trend means adding more nails to ensure access to all nodes. For functional testing (based on edge connectors), the ratio of functions per I/O pin increases. The already difficult problem of test preparation for functional test will become legendary in the future.

Boundary scan promises to enhance fault diagnosis, reduce test-repair looping, and enable standardized testing, and the reuse of tests.

THE IMPACT OF BOUNDARY SCAN

Boundary scan promises to relieve at least some of the difficulties in board test, regardless of the philosophy adopted—in-circuit, functional, or combined (called combinational) testing. These benefits include enhanced diagnosis, reduced test-repair looping, standardized testing, and reuse of tests.

Enhanced diagnosis. The scan port provides access to hundreds of additional control and observation points. We can access the publicly available BIST features of the ICs in the scan chain, for example. The BIST functions run independently from the board and do not require tedious programming. There is also less sensitivity to the initial state of the board. Of course, this access is not without cost. Because it is serialized, we have to consider the impact on test length and time, which could be considerably greater.

The scan chain allows us to perform many tests without great concern about synchronizing or homing sequences, since we can effectively ignore the logic within scanned ICs. In doing so, however, we are assuming a certain amount of luck as to the integrity of the scan chain itself.

Another benefit to diagnosis is the increased stability of fault syndromes, which is due to the insensitivity of tests to initial states. A syndrome becomes unstable when it depends on an initial state that cannot be reliably achieved. It is difficult to isolate faults with unstable syndromes when we must run a test several times to collect isolation information, as in backtracing.

Reduced test-repair looping. Because the scan path offers additional control and observation points, we can isolate more faults per pass across a test head. Thus, we need to make fewer passes across the test head, and we decrease the time and handling involved. The improved immunity to initial-state problems and more stable fault syndromes makes each pass yield higher quality diagnoses, again resulting in fewer passes.

Standardized tests. Tests in a boundary-scan environment are prewritten, or they are easily derived from the topological structure of the circuit. The content of ICs within the boundary-scan

Analog components are a reality, and the large number of analog or hybrid boards being tested is not going to get any smaller.

perimeter may not be exhaustively tested as is true today at board-level test. Of course, if the scanned ICs possess BIST functions, they can be accessed with a standard RUNBIST protocol without much programming effort. The problem of last-minute changes to ASICs is no longer of great concern because the internal workings of the ASIC have been removed from consideration in developing the board test.

Reuse of tests. Tests that work in concert with the scan protocol, such as BIST functions, will be accessible at several stages of manufacture. This is not true in a non-scan environment. For example, we may not be able to use an IC test developed along with the IC during in-circuit board test if there are simple constraints such as tying some pins to ground.[7]

Reduced access problems. Fixturing, that is, connecting the tester to the board for testing, is traditionally troublesome. This problem is particularly true for in-circuit testing, in which we use hundreds or thousands of nails to access a board's internal nodes. This access gives exceptional test control, observability, and fault isolation. Further, it allows comprehensive analog testing along with digital testing. For this reason alone, in-circuit testing will remain popular. Analog components are a reality even on so-called digital boards, and the large number of analog or hybrid boards being tested is not going to go away. The access problems described earlier are indeed a threat to board test, and particularly aggravating to in-circuit access. Boundary scan offers some relief, even though not all the ICs will be scannable initially.

IN-CIRCUIT TEST

Boundary scan implementations—even partial ones—will have a number of benefits to in-circuit testing. They will reduce the need for a 1:1 nail-to-node ratio, for example. We can test any digital node composed completely of scannable sources and destinations from the scan port without a nail. Another benefit is less need for close-centered probing, which is done with thin, fragile, expensive nails necessitating small target pads and fine mechanical tolerances. By reducing the number of nails on a device, even if we don't eliminate them entirely, we reduce the crowding, which is the reason we do close-centered probing in the first place.

Boundary scan will also reduce the need for two-sided probing. If devices mounted on the nonprobed side of a board are testable through scanning, we can avoid nails on this side, but we must consider this strategy in the design-for-testability specifications. Reducing the nail count will, in turn, reduce the flexing of boards caused by the uneven concentrations of closely spaced nails. Each nail presses against the board with roughly two newtons of spring force. When the nails are spaced too closely—that is, concentrated in groups—the force is no longer uniform across the board, which causes it to flex. Flexing can cause the board to misalign with the nails, which may make vacuum-activated fixtures unreliable without some mechanical augmentation. Worse, flexing may cause open connections to close during test, masking their existence and making the test inaccurate.

Pure boundary-scan may never be widespread because the increase in fault diagnosis isn't enough to justify the expense of putting testability in every chip.

FUNCTIONAL TEST

Boundary scan will also benefit functional (edge-connector) testing. The ability to observe and control the circuit is greater, which helps mitigate the unfavorable impact of high gate-to-pin ratios. Boundary scan also allows us to conduct topologically derived tests for interconnection faults. These faults are currently modeled as single stuck-at faults at the IC pin level, when preparing functional tests.

The ability to conduct topologically derived tests is perhaps boundary scan's greatest potential contribution because it reduces the pressure to do board-level simulation to develop tests for common manufacturing faults. Of course, the success of this strategy will depend on how pure (what percentage of the ICs are scannable) the boundary-scan implementation is in the board design.

There is some hope that scannable devices can incorporate aids for performance testing, also known as at-speed testing. This testing attempts to run a board at its native clock speed during test to excite timing problems or other marginal conditions. We can use the INTEST and SAMPLE modes of P1149.1 to set up test experiments and examine the results. As is true today, the technical challenges of such testing are likely to be high.

FUTURE TRENDS

Partial boundary-scan implementations are beginning to crop up in board designs. Partial implementations are likely to be the most we can expect for some time because the number of ICs available using the discipline is limited. Actually, pure implementations may never be the rule because the increase in fault diagnosis isn't significant enough to justify the expense of incorporating testability in every chip.

The big question is will boundary scan die out before reaching its potential because of a lack of critical mass? I am confident that the discipline will achieve acceptance in short order. But today's board-level ATE will not be disappearing. Indeed, some people equate boundary scan with the total absence of ATE, but this is a fallacy. We should avoid overselling the impact of this new technology, which will only create credibility problems.[10]

In fact, ATE and boundary scan are mutually beneficial. In the real world of manufacturing, we have to test economically, so we still need the enhanced diagnosis that our current ATE systems can give. Boundary scan, in turn, allows ATE to reduce the testing cost, since the nail count goes down. The nice thing about boundary scan is that we don't need it in every chip to make it effective. We can use this technology with ATE to get a more in-depth fault diagnosis without an impossible rise in cost and complexity. We still get the additional test capability, but bound-

Board-level ATE in its present form will not vanish. It will span the gap between current test problems and the new boundary-scan test environment.

ary scan allows us some flexibility in determining how far we want to take the implementation.

In digital testing, the upward spiral of board-level ATE costs will slow and even decline with the advent of boundary scan, and the cost of test preparation will be less. The capital (purchase) cost of an ATE system is heavily driven by the cost of its electronics and the cost of the software. With unconstrained board designs, both these costs will escalate as board circuitry becomes denser.

Boundary scan will also improve hybrid analog/digital testing because the digital portion will have greater testability. Analog testing problems will remain largely unaffected, however, so in-circuit approaches must solve new analog test problems without it. Thus, board-level ATE in its present form will not vanish because we still need it to span the gap between current test problems and the new boundary-scan test environment.

IN-CIRCUIT TEST

With boundary scan, in-circuit test will see the nail-to-node ratio drop below one, (see box at right on in-circuit testing) because some tests can be accomplished without nails on the scannable nodes. As a result, many of the aforementioned fixturing difficulties will be relieved. Also, the number of simultaneously active nails (connected to independent drive/receive channels) will stabilize or decline. An advantage to test serialization (there is one) is that we can examine connectivity between high pin-count devices with just four active channels. This decline allows higher nail-to-channel multiplex ratios which means more nails can be serviced with fewer active (expensive) electronics channels. Test preparation is eased since scanned nodes can be tested with software-derived tests based on board topology, and there is no need to write IC interconnect tests that sensitize pathways through the internal regions of the ICs.

FUNCTIONAL TEST

For functional test, we see control and observability rise for internal areas of a board. This makes test preparation easier, and may reduce or even eliminate the requirement for fault simulation. (Simulation for design verification may still be required, but that is a very different problem.[7]) Each scannable IC is a candidate for elimination from simulation which is attractive if simulation models are unavailable or inadequate, or if the IC is complex and expensive to simulate. As a result, the cost of test preparation is reduced. Hardware costs for functional testers has been increasing due to the depth of pattern storage required behind all pins. With higher levels of boundary scan adherence, only those tester channels dedicated to scanning may actually generate long bit streams with the general I/O channels only active a small fraction of time.

In whichever case, in-circuit or functional or the combined approach, the increased usage of boundary scan will have a cost savings impact that can be readily demonstrated. This savings will make it easier to measure the value of a level of adherence to the boundary scan discipline.

SETTING UP AN IN-CIRCUIT TEST

How does one set up an in-circuit test for a digital circuit that has some boundary scan components? Consider the simple circuit in the figure below. This circuit has two scanned components *U1* and *U2* and three glue gates, which are not scannable. For simplicity, the figure shows NAND gates, but the devices could be much more complex. The nodes are labeled in alphabetical order from *A* to *T*. The triangular pointers show the location of in-circuit test nails.

We need nails *A* through *I* at the circuit's main inputs to test for connectivity to *U1* and to drive tests for gates *U3* and *U5*. We need nails *P* through *T* at the circuit's main outputs to test for connectivity to *U2* and to monitor gate *U5*. We need nail *N* to test the second input to gate *U4*. The channel driving this nail requires overdrive capability. We need nail *O* to test the first input to gate *U5* and to monitor gate *U4*. The channel driving this nail also requires overdrive capability. Nails *A, E, F* are the boundary-scan inputs TDI, TCK and TMS. Nail *P* is the scan-chain output TDO. These nails allow us to control the entire chain.

Nodes *B, C, D, K, L,* and *M* on *U1* are scannable through the boundary scan chain, as are nodes *K, L, M, Q, R,* and *S* on *U2*. Therefore, all interconnections to *U1* and *U2* are completely testable without any knowledge (or concern) for the internal logic of these devices. Interconnection here includes printed trace integrity, solder integrity, the existence of IC bond wires, and bare-bones silicon integrity—that is, the devices can at least perform boundary-scan protocols. *U2*'s input node *N*, however, is not scannable because gate *U3* is not a boundary-scan device. Instead, we use nail *N* to overdrive gate *U3* as if node *N* were a primary input.

Gate *U4* has nails on nodes *N* and *O*, but we can control node *M* only by scanning through *U1*. Think of node *M* as a virtual nail that we can control from *U1* to provide inputs while directly driving node *N* and receiving on node *O*. It is this idea of a virtual nail that allows the nail-to-node ratio to drop below one.

Nail *J* is an interesting case. It sits on the scan data (TDI-TDO) path between *U1* and *U2* and must have overdrive capability. This path is supposedly tested by the procedure to test boundary-scan integrity run first to ensure that the path works. But what if a fault has damaged the path? Say, for example, that device *U2* is completely dead. We can still test *U3*, *U4*, and *U5*, plus all board-input interconnections because nail *J* lets us see the results scanned by *U1*.

Stopping the test because the scan path is faulty can cause test-repair looping, especially if a large board with many scannable devices fails the test for scan-path integrity. In this case, there may be several faults on the scan path, which ruins our ability to isolate the culprits. The result is misdiagnosis, which may cause several iterations through the test-repair loop before real testing can begin. To reduce distress due to a faulty scan path, we can add a TDI-TDO nail on every package.

When the board is 100% scannable, we need the nails on all of the following: all board inputs, *I*; all board outputs, *O*; all TDI-TDO signals (one per device pair), *D*; and all analog nodes, *A*. The nail count is then $I + O + D + A$, which has essentially saved us placing nails on all the internal digital-only nodes. When not all the ICs on a board are scannable, we need to add nails for all nonscanned glue-gate inputs and outputs such as nodes *O* and *N* in the figure.

Thus, we begin to see the benefits of putting boundary scan into a device. If we balk at the cost of incorporating boundary scan in ICs, we must be prepared to balance the savings at the IC level against the costs of fixturing at board test. There are also the benefits to test preparation. Either *U1* or *U2* can be horribly complex internally, but we don't care. Boundary scan has isolated us from this complexity. Likewise, if either device is difficult to initialize internally, we are not affected. If either device has a public BIST function for self-test, we use the RUNBIST instruction and read out the result using canned routines. This simplicity is understandably exciting for test programmers.

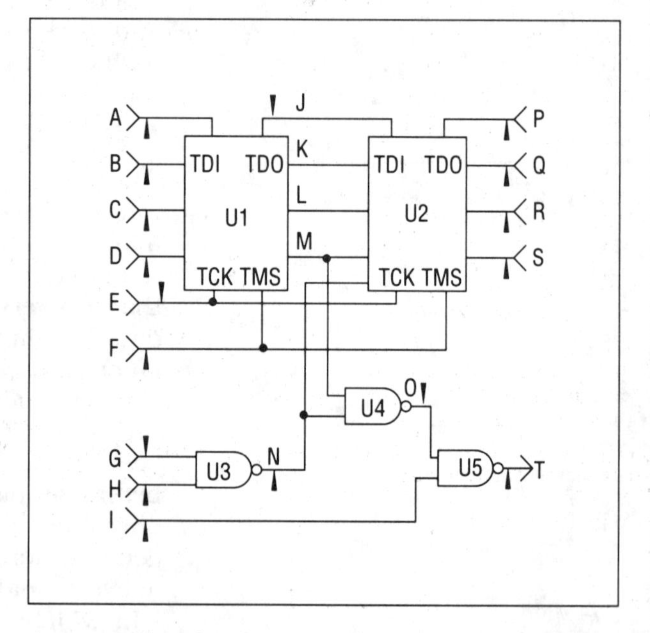

An example of a circuit with mixed standard and boundary-scan components showing the location of nails for in-circuit test. Note that not all nodes need a nail.

Some prognostications for boundary scan have been nothing less than euphoric. Some say boundary scan will eliminate the need for today's testers in manufacturing. A new generation would appear consisting of a four-wire interface I/O card for a personal computer and a few floppy discs of software. The reality lies somewhere between the past and this happy outcome. Just as a carpenter, upon receiving a new power saw, does not discard his collection of older tools, ATE systems that control a board-production process will not discard the capabilities they have today. The future can bring many pleasant developments if boundary scan (as well as other BIST/DFT technologies) is accepted, and we could all enjoy the reduced cost and complexity. This hesitancy should go away as we quantify the benefits of this new technology, and its use should become more widespread. Perhaps the largest obstacle will then be deciding how to use what we have developed, and that will be up to management. D&T

REFERENCES

1. T.W. Williams and K.P. Parker, "Design for Testability: A Survey," *Proc. IEEE*, Vol. 71, Jan. 1983, pp. 98-112.
2. C.C. Timoc, *Selected Reprints on Logic Design for Testability*, IEEE Computer Society Press, Los Alamitos, Calif., 1984.
3. E.B. Eichelberger and T.W. Williams, "A Logic Design Structure for LSI Testability," *Proc. Design Automation Conf.*, 1977, 77CH1216-1C, pp. 462-468.
4. K.P. Parker, "Testability: Barriers to Acceptance," *IEEE Design & Test of Computers*, Vol. 3, Oct. 1986, pp. 11-15.
5. C. Maunder and F. Beenker, "Boundary Scan: A Framework for Structured Design-for-Test," Proc. Int'l Test Conf., 1987, pp. 714-723
6. IEEE Standard 1149.1-1989/D4, "IEEE Standard Test Access Port and Boundary-Scan Architecture," IEEE Standards Board, 345 East 47th Street, New York, NY 10017, Draft D4, May 5, 1989.
7. K.P. Parker, *Integrating Design and Test: Using CAE Tools for ATE Programming*, IEEE Computer Society Press, Los Alamitos, Calif., 1987.
8. D.T. Crook, "Analog In-Circuit Component Measurements: Problems and Solutions," *Hewlett Packard J.*, Mar. 1979, pp. 19-22.
9. M. Bullock, "Designing SMT Boards for In-Circuit Testability," *Proc. Int'l Test Conf.*, 1987, pp. 606-613.
10. "Built-In Self-Test: Are Expectations Too High?" *IEEE Design & Test of Computers*, Vol. 6, No. 3, June 1989, 66-74.

Ken Parker is a member of the technical staff in Hewlett-Packard's Manufacturing Test Division, where he is involved in the design and development of systems to test circuits. Previously, he worked for HP in California and for NASA/Ames Research Center, where he was involved in the Illiac IV project.

Parker is cofounder of the IEEE Subcommittee on Design for Testability. He holds a BS in computer engineering from the University of Illinois, and an MS and a PhD in electrical engineering from Stanford University. His address is Hewlett-Packard, PO 301, Loveland, CO 80537.

An Optimal Test Sequence for the JTAG/IEEE P1149.1 Test Access Port Controller

Anton T. Dahbura
AT&T Bell Laboratories
Murray Hill, NJ 07974

M. Ümit Uyar
AT&T Bell Laboratories
Holmdel, NJ 07733

Chi W. Yau
AT&T Bell Laboratories
Engineering Research Center
Princeton, NJ 08540

ABSTRACT

A test sequence is given for the Test Access Port (TAP) controller portion of the boundary-scan architecture proposed by the Joint Test Action Group (JTAG) and IEEE Working Group P1149.1 as an industry-standard design-for-testability technique. The resulting test sequence, generated by using a technique based on Rural Chinese Postman tours and Unique Input/Output sequences [1], is of minimum cost (time) and rigorously tests the specified functional behavior of the controller. The test sequence can be used for detecting design faults for conformance testing or for detecting manufacture-time/run-time defects/faults.

I. Introduction

The Joint Test Action Group (JTAG), an *ad hoc* committee comprised of major semiconductor users in Europe and North America, together with IEEE Working Group P1149.1, has proposed a framework for standardized design-for-testability of integrated circuits for module-level (e.g., board-level) testing. The so-called *boundary-scan architecture* consists of circuitry which allows the inputs and outputs of the digital logic of the integrated circuit to be accessed from outside the module [10]. The advantage of the boundary-scan approach is that the controllability and observability of a module containing many components is vastly improved while the input/output overhead of the module consists of only three extra inputs and one extra output.

In most cases, boundary-scan components which have been designed and produced by different manufacturers reside within the same module. Thus, it is paramount that the implementation of the boundary-scan portion of each component *conforms* to the set JTAG/IEEE P1149.1 standard to ensure that the component can be successfully integrated into a module-level design-for-testability scheme. While design verification is necessary at virtually every step of the design process, it is ultimately desired to check that, in the *physical implementation* of the component, the functionality of the boundary-scan portion is as expected, based on its specification.

Of course, any test sequence which checks the conformance of many different designs must, by its nature, be implementation-independent. This means that test sequence generation techniques which are based on logic-level information such as stuck-at-faults are of little value. Furthermore, since it is applied to a physical implementation, the test sequence must be able to overcome the severe observability and controllability constraints which arise. As a result, a high-level approach is required for generating a test sequence which is 1) implementation and fault-model independent, 2) of compact length, and 3) able to detect an extremely high percentage of faults which occur as a result of design faults.

In this paper, a minimum-cost (time) conformance test sequence, based upon such a high-level approach, is presented for checking the joint functional behavior of the TAP controller and associated registers in a boundary-scan implementation. The test sequence has been derived using optimization techniques for managing the observability and controllability limitations which arise in testing in a "black box" environment such as this. While the test sequence has been designed for conformance testing (design verification), it is also an effective manufacture-time and/or run-time test for the boundary-scan portion of an implementation.

In Section II, the JTAG/IEEE P1149.1 architecture is described. The model used to derive the test sequence is given in Section III. In Section IV, the test sequence genera-

Reprinted from *IEEE Proceedings 1989 International Test Conference*, pages 55-62.

EH0321-0/90/0000/0302$01.00 © 1989 IEEE

tion technique is discussed. Finally, the resulting test sequence is described and the fault coverage results for an implementation of the boundary-scan architecture are presented in Section V.

II. JTAG/IEEE P1149.1 Architecture

The boundary-scan technique consists of placing a *boundary-scan* cell adjacent to each component input/output pin in order to observe and control the component's signals at its boundaries [10]. Each boundary-scan cell is able to either capture data from an input pin or from the component logic, and can load data either into the component logic or onto a component output pin. The boundary-scan cells are interconnected as a shift-register chain and, if desired, several components can be connected as a single chain. The boundary-scan cells can be used to test the interconnections among various components (*external test*) or to isolate a component while an *internal test* is performed. Also, the boundary-scan cells can be used to *sample* values at a component's input and output pins.

The overall JTAG/IEEE P1149.1 boundary-scan architecture is shown in Figure 1. The primary elements are as follows:

Test Access Port (TAP) controller: a sixteen-state circuit (Figure 2) which receives the test clock signal (TCK) and test mode select (TMS) control input and generates clock and control signals for the remainder of the architecture. The actions initiated by the TAP controller occur on the rising edge of TCK, when the con-

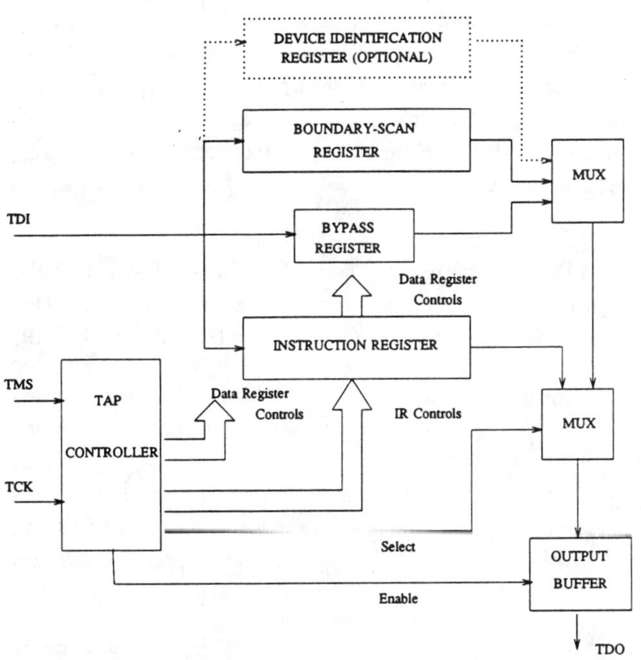

Figure 1. The JTAG/IEEE P1149.1 boundary-scan architecture [10].

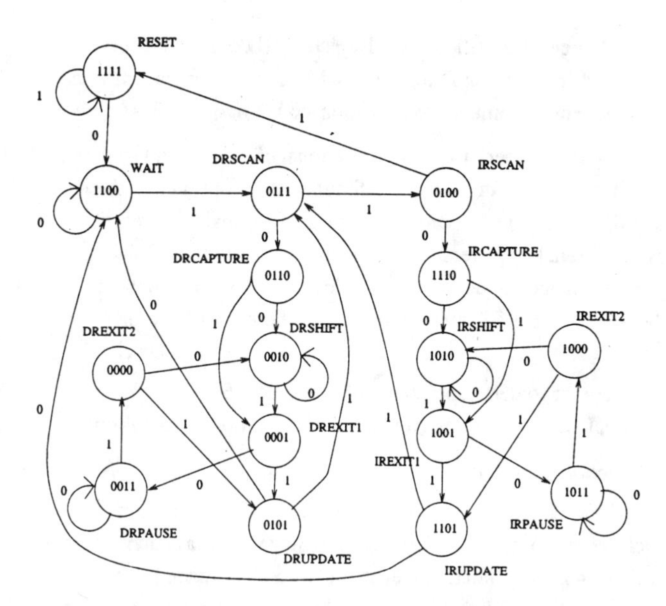

Figure 2. State diagram of the TAP controller [10].

troller *leaves* the corresponding state, except for the reset operation (state 0000), which occurs asynchronously. A block diagram of the TAP controller is shown in Figure 3.

Instruction Register (IR): stores an instruction, shifted into it through the TDI input, which selects the test to be performed (external, internal, sample) and/or the data register (boundary-scan, bypass, or device identification) to be accessed.

Boundary-Scan Register (BSR): a multiple-bit shift-register consisting of the boundary-scan cells interconnected in serial fashion with access to the component's input/output pins and internal logic.

Bypass Register (BPR): a single-bit connection from TDI to TDO to allow test data to flow through to other components with a single TCK period delay.

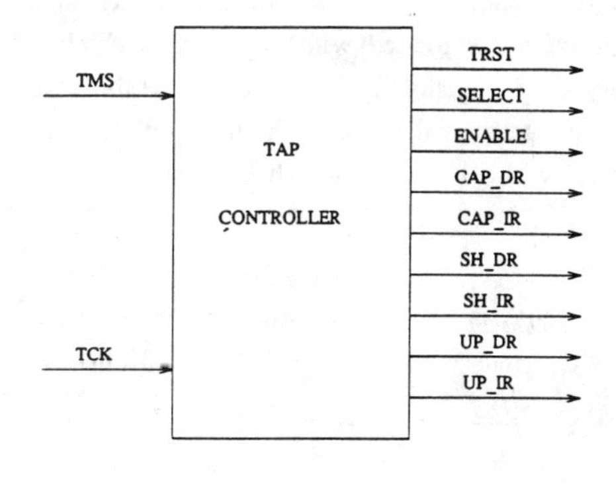

Figure 3. Block diagram of the TAP controller.

Device Identification Register (IDR): an optional multiple-bit shift-register which contains a device-dependent binary identification code.

In this paper, two configurations of the boundary-scan architecture, with and without the optional device identification register, are considered. Optional features, such as additional registers, are not included, although the test generation technique is also applicable. For more details on the JTAG/IEEE P1149.1 architecture, the reader is referred to [10].

The detailed operation of the IR, BSR, BPR, and optional IDR in the design under consideration is as follows.

Instruction Register

The IR consists of three IR cells, IR0, IR1, and IR2, each consisting of a shift-register stage and a latch stage (Figure 4). The latch stage of an IR cell is loaded with the corresponding shift-register value upon receiving an Update_IR (UP_IR) signal from the TAP controller (Figure 3). Upon receiving a Capture_IR (CAP_IR) signal from the TAP controller, IR0 retains its previous value, the shift-register stage of IR1 is loaded with a "0", and the shift-register stage of IR2 is loaded with a "1". Upon receiving a Shift_IR (SH_IR) signal from the TAP controller, the value on the TDI input is stored in the shift-register portion of IR0, the old value of the shift-register portion of IR0 is shifted to IR1, and IR1's old value is shifted to IR2, which is then observable on the TDO output. Upon receiving a Test-Logic Reset (TRST) signal from the TAP controller, 1) in the configuration without the IDR, the latch stages of IR0, IR1, and IR2 are each loaded with a "1", which corresponds to the "bypass-select" instruction, and the shift-register portions of the cells retain their previous values; 2) in the configuration with the IDR, the latch stages of IR0, IR1, and IR2 are each loaded with a design-specific bit corresponding to the "IDR-select" instruction and the shift-register portions of the cells retain their previous values. In this study, the two instructions of relevance are that in which the latch stages of the three IR cells are loaded with "0" (000), which selects the boundary-scan register and places it in the external test mode, and that in which the latch stages of the three IR cells are loaded with "1" (111), which selects the bypass register;

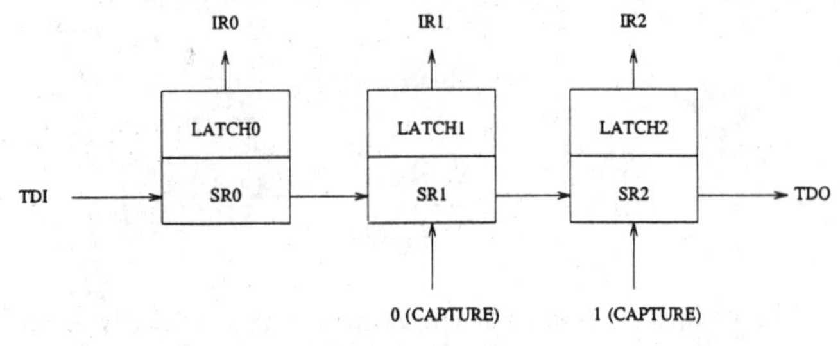

Figure 4. Functional diagram of the instruction register.

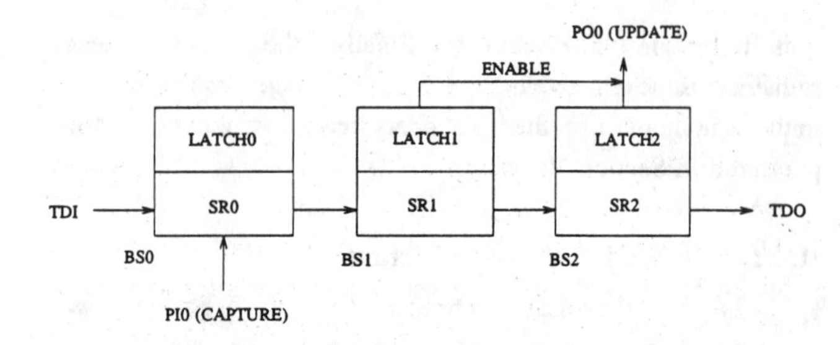

Figure 5. Functional diagram of the boundary-scan register.

in addition, the implementation-specific "IDR-select" instruction is relevant in a design which includes the IDR.

Boundary-Scan Register

The considered design consists of three cells in the BSR: an input cell (BS0), an output cell (BS2), and an output enable control cell (BS1) which enables the PO0 pin of the output cell (Figure 5). Each cell consists of a shift-register stage and a latch stage. Upon receiving a Capture_DR (CAP_DR) signal from the TAP controller and when the BSR is selected in the IR (instruction 000), the shift-register stage of BS0 is loaded with the value of input pin PI0; BS1 and BS2 retain their previous values. Upon receiving a Shift_DR (SH_DR) signal from the TAP controller when the IR is in external test mode (instruction 000), the values of the shift-register portions of the BSR cells shift one cell to the right (from BS0 to BS1 to BS2), the value on the TDI input is stored in the shift-register portion of BS0, and the new value of the shift-register portion of BS2 is then observable on the TDO output. Upon receiving a Test-Logic-Reset (TRST) signal from the TAP controller, the latch stage of BS1 is loaded with a "0" to disable the PO0 signal and BS0 and BS2 retain their previous values. In the external test mode and upon receiving an Update_DR (UP_DR) signal from the TAP controller, the latch stages of BS0, BS1, and BS2 are loaded with the values of their corresponding shift-register stages.

Bypass Register

The bypass register consists of a single-stage shift-register cell (Figure 6). Upon receiving a Capture_DR (CAP_DR) signal from the TAP controller and when the IR

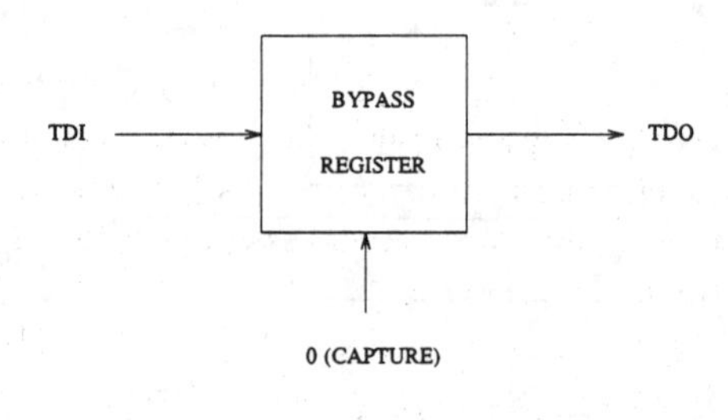

Figure 6. Functional diagram of the bypass register.

is in bypass mode (instruction 111), a "0" is stored in the BPR cell. Upon receiving a Shift_DR (SH_DR) signal from the TAP controller when the IR is in bypass mode, the value on the TDI input is stored in the BPR cell and is consequently observable on the TDO output. An Update_DR (UP_DR) signal from the TAP controller when the IR is in bypass mode produces no effect on the bypass register or any other register. Finally, a Test-Logic-Reset (TRST) signal from the TAP controller produces no effect on the BPR.

Device Identification Register (optional)

The IDR consists of N (normally N=32) single-stage shift-register cells (Figure 7). Upon receiving a Capture_DR (CAP_DR) signal from the TAP controller and when the IR is in "IDR-select" mode (an implementation-specific instruction), each bit of the device identification code is stored in the corresponding IDR cell. Upon receiving a Shift_DR (SH_DR) signal from the TAP controller when the IR is in "IDR-select" mode, the the values of the IDR cells shift one cell to the right (from IDR1 to IDR2, and so on, to IDRN), the value on the TDI input is stored in IDR1, and the new value of IDRN is then observable on the TDO output. An Update_DR (UP_DR) signal from the TAP controller when the IR is in "IDR-select" mode produces no effect on the IDR or any other register. Finally, a Test-Logic-Reset (TRST) signal from the TAP controller produces no effect on the IDR.

The instruction stored in the IR ("000", "111", or the optional implementation-dependent "IDR-select" instruction) controls the multiplexer determining which of the shift-register outputs of the data register cells (BS2, BYPASS, or IDRN, respectively) is to be observable at the TDO output. The appropriate data register value is observable at the TDO output only when the TAP controller enables the output buffer (Figure 1) in states DRCAPTURE, DRSHIFT, DREXIT1, DRPAUSE, and DREXIT2 (Figure 2). The value of the shift-register portion of the IR2 cell is observable only when the TAP controller enables the output buffer in states IRCAPTURE, IRSHIFT, IREXIT1, IRPAUSE, and IREXIT2 (Figure 2).

III. Finite-State Machine Model of the JTAG/IEEE P1149.1 Architecture

As described in the previous section, the TAP controller is a sixteen-state finite-state machine (FSM); since its output depends solely on its present state, it is a *Moore circuit* [6]. If the values on the TAP controller outputs could be directly measurable then several of its states would be essentially directly observable. For instance, if the SH_DR line is active (Figure 3) and all other lines are inactive, then, by definition, the TAP controller is in the DRSHIFT state (state 0010 in Figure 2).

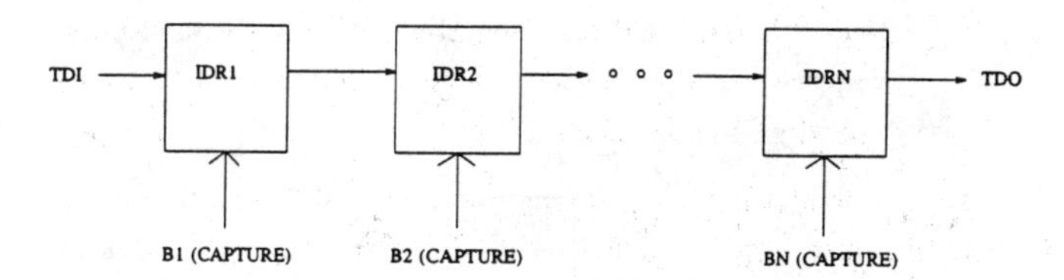

Figure 7. Functional diagram of the device identification register (optional).

In fact, however, the implementation of the complete JTAG/IEEE P1149.1 boundary-scan architecture is such that the TAP controller outputs are not directly measurable, but control the instruction register, decoding logic, boundary-scan, bypass, and (optional) device identification registers, multiplexers, and buffers (Figure 1). The contents of the instruction register control the behavior of the data (BS, BP, and optional ID) registers. Thus, determining the levels of the TAP controller outputs consists of observing the effects of these outputs on the TDO and primary outputs as a function of the contents of the instruction, boundary-scan, bypass, and optional device identification registers and the values applied to the TMS, TDI, and primary inputs. The FSM representation of the control portion of the JTAG/IEEE P1149.1 boundary-scan architecture, therefore, describes the joint behavior, or *composition* of the TAP controller with the contents of the instruction register. This composition is shown in Figure 8 for a design which does not include a device identification register, and in Figure 9 for an implementa-

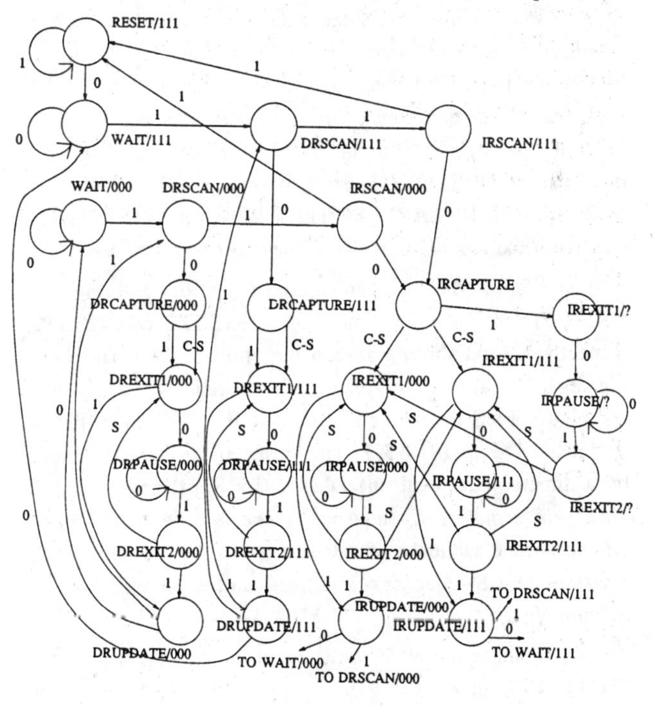

Figure 8. Finite-state machine showing the joint behavior of the TAP controller and instruction register.

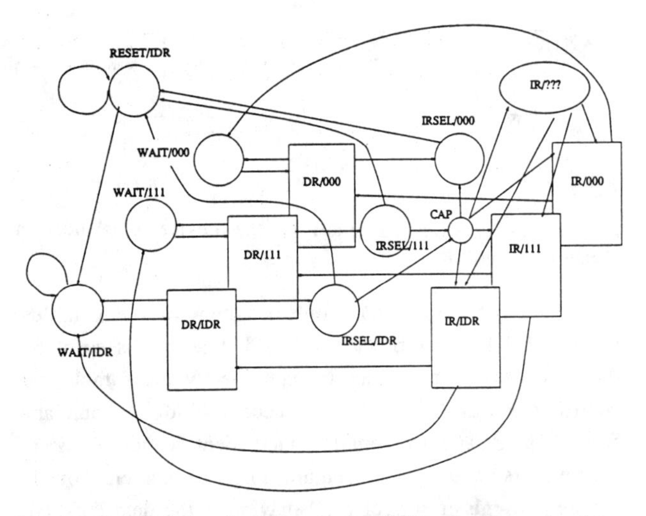

Figure 9. High-level view of the FSM showing the joint behavior of the TAP controller and IR (IDR included).

tion which does. For instance, in Figure 8, the DRSHIFT/000 state corresponds to the TAP controller SH_DR line active and "000" in the instruction register (external test mode); in this state, the value on the TDI input is shifted through the boundary-scan register. On the other hand, in the DRSHIFT/111 state, the value on the TDI input is shifted through the bypass register.

In this model, the operation consisting of capturing, followed by shifting several bits through any of the registers, is represented by a single directed edge, and is said to be a *capture-shift sequence* (CS-sequence). For example, in Figure 8, the operation consisting of capturing a "0" in the bypass register, followed by shifting through a certain sequence of bits and then completing the operation by applying a "1" to the TMS input, thereby putting the TAP controller in the DREXIT1/111 state, is represented by a directed edge from the DRCAPTURE/111 state to the DREXIT1/111 state with the label C-S.

Similarly, the operation consisting of entering the DRSHIFT or IRSHIFT state from DREXIT2 or IREXIT2, respectively, and shifting several bits through the corresponding register and then completing the operation by applying a "1" to the TMS input, thereby putting the TAP controller in the DREXIT1 or IREXIT1 state, respectively, is represented by a directed edge with the label S (for Shift) and is said to be a *shift sequence* (S-sequence). The issue of determining an appropriate sequence of values to shift through the various registers in CS-sequences and S-sequences is addressed in Section V.

Note that, as intended to be used in this paper, the IRCAPTURE state (CAP in Figure 9) is entered with the intention of placing either "000" or "111" (or the "IDR-select" instruction) in the instruction register latch, which occurs when either the IRUPDATE/000 or IRUPDATE/111 states (or IRUPDATE/IDR), respectively, are exited. Thus, there are three states, IREXIT1/?, IRPAUSE/?, and IREXIT2/?, which may be entered from the IRCAPTURE state but which may not lead directly to the IRUPDATE/000 or IRUPDATE/111 (or IRUPDATE/IDR) states without placing a "000" or "111" (or the "IDR-select" instruction) in the instruction register cells.

IV. Test Generation Technique

Verifying that the boundary-scan portion of a given implementation of a component conforms to the JTAG/IEEE P1149.1 boundary-scan standard is equivalent to verifying the functional behavior of the underlying sequential logic circuit. Several techniques for automatically generating test sequences for sequential logic circuits have been published in the literature and, for the most part, can be categorized into *structural testing* for logic-level fault coverage and *functional testing* for design verification.

Structural testing approaches, including the D-algorithm [7] and its variants [3],[4], aim at detecting gate-level faults such as single stuck-at-1 and stuck-at-0 faults in an electronic device. While these techniques can be adapted to generating test sequences for sequential circuits, they are implementation-specific; furthermore, they are designed to detect only a very limited subset of the possible faults that can occur in the design, manufacture, and operational stages of a device.

The classical *functional testing* approach consists of the design of so-called *checking experiments* [5], which produces a set of input and output test sequences which are used to verify that the behavior of the "black box" under test is exactly as specified by the given state transition table. A checking experiment tests the implementation of an *m*-state finite-state machine (FSM) for the correctness of every specified transition; that is, it verifies that each specified input for state s_i, $i=1,...,m$, in the implementation produces the expected output and takes the implementation to the expected state.

In the past, checking experiments were based on the existence of an input sequence called a *distinguishing sequence*, which produces a distinct output sequence for each initial state of the FSM [5]. Unfortunately, only a very limited number of FSMs have a distinguishing sequence [2,5]. A new approach was proposed in [1], based on the concept of *Unique Input/Output (UIO)* sequences [8]: a UIO sequence of a given state s_i in an FSM is an input/output sequence of minimum length starting from state s_i which could not be produced by starting at any other state. Thus, a UIO sequence can be used to verify that the initial state of an input/output sequence is that which is expected. The difference between a UIO sequence and a distinguishing sequence is that if the FSM is not in the expected initial state, the actual initial state can be deduced from the output sequence of a distinguishing sequence but not from that of a UIO sequence; however, this information is unnecessary for the purposes of testing.

The advantage of UIO sequences over distinguishing sequences is twofold. First, while few FSMs have a distinguishing sequence, almost all FSMs have a UIO sequence, or a variant described below, for each state. Second, the length of a UIO sequence is at most that of a distinguishing sequence and usually much less, so that UIO sequences are the method of choice for checking that an implementation is not in an expected state.

In the technique described in [1], after computing the UIO sequences for each state of the FSM specification, a *test* is formed for each transition of the FSM. The test of a transition consists of placing the FSM in the initial state of the transition, applying the appropriate input for the transition and observing that the output is that which is expected, and then applying the UIO sequence for the final state of the transition to ensure that the final state of the given transition under test is that which is expected.

The set of tests is then assembled in an optimal manner, using a network flow algorithm based on the Chinese Postman problem of graph theory [9], such that the resulting test sequence is a continuous tour of the FSM which 1) contains a test for each transition of the FSM, 2) begins and terminates at a designated *start state* of the FSM, and 3) is of minimum total cost. In the case of the TAP controller, each transition requires the same time to realize, so that the test sequence generated, described in the following section, consists of the minimum number of transitions necessary to rigorously test the FSM in the manner described above. (For more information on the UIO sequence/Chinese Postman test sequence generation approach, the reader is referred to [1],[2],[8],[9]).

V. Test Sequence Description

UIO Sequences

For the implementation with no device identification register, a UIO sequence was computed for 28 of the 33 states shown in Figure 8. For example, the UIO sequence for the DRSCAN/000 state consists of the following: 1) starting in the DRSCAN/000 state, apply a "0" to the TMS input and observe the previous value of the shift-register portion of BS2 on the TDO output; 2) apply a "0" to the TMS input, hence capturing the value of PI0 in BS0; 3) shift a *CS-signature* (for *capture-shift signature*) to be described below, during the capture-shift sequence, from the TDI input, through the boundary-scan register, to the TDO output so that it may be deduced that the values from the TDI input are actually shifted through the boundary-scan register. It can be observed that this output behavior could have been produced only if DRSCAN/000 had been the initial state. Similar UIO sequences exist for all of the other states except for IREXIT1/?, DREXIT1/000, DREXIT1/111, IREXIT1/000, and IREXIT1/111. These five states are considered in the following.

It is easily observable that state IRPAUSE/? is *weakly equivalent* [2] to state IREXIT1/? in that there is no effect on the functionality of the boundary-scan architecture if the TAP controller and instruction register composition is in the IRPAUSE/? state when it is expected to be in the IREXIT1/? state. This is because applying a "0" to the TMS line when the TAP controller is in the IREXIT1/? state has the same effect as applying a "0" to the TMS line when the TAP controller is in the IRPAUSE/? state. Note that a "1" input to the TMS when the TAP controller is in the IREXIT1/? state is undefined since that would ultimately place an undefined instruction in the instruction register.

For the last four of the states, a set of input/output sequences, henceforth called a set of *partial* UIO sequences, is used instead of UIO sequences, since the four states do not initiate any unique input/output behavior themselves and yet are not weakly equivalent to any other states. For example, note that the result of applying a "0" to the TMS input when the TAP controller is in state DREXIT1/000 is identical to that produced when the TAP controller is initially in state DRPAUSE/000; from that point on, all actions are identical since the two transitions lead to the same state. Also, the result of applying a "1" to the TMS input when the TAP controller is in state DREXIT1/000 is identical to that produced when the TAP controller is in state DREXIT2/000 and the two transitions also lead to the same state, so that further actions are identical. The first partial UIO sequence for DREXIT1/000 distinguishes DREXIT1/000 from all other states except DRPAUSE/000. The second partial UIO distinguishes DREXIT1/000 from all other states except DREXIT2/000. Each transition leading to DREXIT1/000 is tested twice as in the manner described in the previous section, once using each partial UIO sequence. Together, the partial UIO sequences yield the same diagnostic power as a UIO sequence. Analogously, sets of partial UIO sequences can be easily derived for the other three states.

For the implementation with an IDR, 34 of the 41 states have UIO sequences and six have partial UIO sequences as described above. Also, state IRPAUSE/? is weakly equivalent to state IREXIT/?.

CS-Signatures and S-Signatures

The process of capturing and shifting a sequence of values through the boundary-scan, bypass, instruction, or device identification register during a capture-shift sequence or shift sequence must serve two functions: 1) to ensure that the shift operation is occurring by means of the intended register, and 2) to exercise as fully as possible the targeted register. Such a sequence is said to be a *CS-signature* (if the shift operations are preceded by a capture operation) or *S-signature* (if there is no capture operation before shifting) of the appropriate register.

In addition, specific logic values must be placed on all of the input lines to ensure that the proper operation is taking place. For example, when the BS0 cell captures a value from the PI0 primary input, the complement of that value is placed at the TDI input to distinguish the capture operation from an ordinary shift. Also, the complement of each CS-signature and S-signature is used at least once in the test sequence to exercise the register adequately.

The CS-signatures and S-signature chosen are as follows, where each bit corresponds to a TMS input on consecutive TCK pulses. Note that capture operations apply only for CS-signatures, and are omitted for S-signatures:

boundary-scan register: (capture 0 in BS0 register), 1,0,0,1,1, followed by 0,1,1;

bypass register: (capture 0 in BP register), 1,0,0,1,1;

device identification register (optional): (capture ID), 1,0,0,1,1, followed by complement of ID;

instruction register (external test mode): (capture 0 in IR1 register, capture 1 in IR2 register), 1,0,0,1,1, followed by 0,0,0;

instruction register (bypass mode): (capture 0 in IR1 register, capture 1 in IR2 register), 0,1,1,0,0, followed by 1,1,1;

instruction register ("IDR-select" mode): (capture 0 in IR1 register, capture 1 in IR2 register), 1,0,0,1,1, followed by "IDR-select" instruction.

The left-most five bits of the signatures, which are the first to be applied to the respective register, exercise the register logic. Note that each value (0 and 1) is stored at least once in each register cell and that each transition (0→0, 0→1, 1→0, and 1→1) occurs at least once in each register cell. Finally, the right-most bits, which remain in the corresponding register after the signature is applied, are such that when they are ultimately shifted out, it is evident which of the registers is being exercised. The CS-signatures and S-signatures can easily be generalized for use with boundary-scan architecture implementations consisting of arbitrary-length boundary-scan, device identification, and instruction registers; in general, only the right-most bits which remain in the register must be changed.

Length of Test Sequence

The generated test sequence for the implementation under consideration without the device identification register consists of 694 input/output operations, corresponding to 694 pulses of TCK, including CS-sequences. The boundary-scan CS- and S-sequences are used 12 times, the bypass register CS- and S-sequence is used 20 times, the instruction register CS- and S-sequence (external test mode) is used 12 times, and the instruction register CS- and S-sequence (bypass mode) is used 11 times. In general, given that the number of cells in the boundary-scan register is N_{bs} and that the number of cells in the instruction register is N_{ir}, the overall length of the test sequence for the implementation without the device identification register is

$$589 + 12N_{bs} + 23N_{ir};$$

for an implementation which includes the device identification register, where the number of cells in the device identification register is N_{id}, the overall length of the test sequence is

$$986 + 12N_{bs} + 44N_{ir} + 21N_{id}.$$

As an example, for a device with 200 I/O pins, a four-cell instruction register, and a 32-cell device identification register, the test sequence is of length 4234.

Coverage

The test sequence for the implementation without the device identification register was evaluated by means of a fault simulator, using a gate-level description of the boundary-scan architecture. Recall that the goal of the test sequence is to be a conformance test for the TAP controller portion of the boundary-scan architecture. Any faults detected beyond the boundaries of the TAP controller is a desirable, yet optional, feature. The test sequence detected 100% of the non-redundant single stuck-at-faults associated with the TAP controller (151 out of 151); with the judicious selection of the CS- and S-signatures, it also detected a very high percentage (85%-95%) of the 546 single stuck-at-faults associated with the registers, multiplexers, and buffers. It is not possible to give an exact figure for the fault detection in the circuitry external to the TAP controller because: 1) many of the faults can only be detected by entering the sample or internal test modes, which do not have standard, reserved instructions, and 2) the number of undetectable faults outside the TAP controller are design-dependent and therefore, unpredictable.

Discussion

The figures for the single stuck-at-fault coverage do not adequately quantify the capabilities of the test sequence because the UIO sequence/Chinese Postman test generation technique, unlike other approaches for testing sequential circuits, is designed to detect *functional* faults, of which stuck-at-faults are but a small subset. Therefore, it seems reasonable to believe that the robustness of the test sequence using UIO sequences should exceed that of other known techniques based on the stuck-at-fault model. At present, however, there is no way to verify this using simulation techniques.

The test sequence generated by the UIO sequence/Chinese Postman technique avoids needlessly detecting so-called *operationally redundant* faults, that is, faults which do not affect the specified operation of the implementation. This further reduces the overall length of the test sequence. An example of an operationally redundant fault is a fault in the TAP controller which can only be detected by updating the instruction register with an undefined instruction.

Finally, another advantage of the UIO Sequence/Chinese Postman-generated test sequence is that it is not circuit-dependent. Unlike other techniques, the test sequence generated here is based solely upon a functional description of the circuit. Therefore, the same test sequence can be used as a manufacture-time and/or run-time test as well as a conformance (design-time) test for the many different boundary-scan gate-level implementations expected to be designed by the various manufacturers that plan to incorporate the JTAG/IEEE P1149.1 boundary-scan architecture into their chip designs.

VI. Conclusions

In this paper, a novel technique for generating test sequences has been applied to the TAP controller portion of the JTAG/IEEE P1149.1 boundary-scan architecture. The resulting test sequence is based on a functional-level, finite-state machine description of the circuit and has in its initial analysis indicated impressive capabilities in detecting design inconsistencies and run-time faults in boundary-scan implementations.

Acknowledgements

The authors would like to thank J.D. Sutton and A.B. Sharma for their valuable contributions which made this paper possible. The work reported in this paper was first suggested by R.E. Tulloss.

References

[1] A.V. Aho, A.T. Dahbura, D. Lee, and M.U. Uyar, "An optimization technique for protocol conformance test generation based on UIO sequences and Chinese postman tours," in *Proc. 8th. Int. Symp. on Protocol Specification, Testing, and Verification*, North Holland, ed. S. Aggarwal and K. Sabnani, 1988.

[2] A.T. Dahbura and K.K. Sabnani, "An experience in estimating the fault coverage of a protocol test," in *Proc. IEEE INFOCOM '88*, pp. 71-79.

[3] H. Fujiwara & T. Shimono, "On the acceleration of test generation algorithms", *IEEE Trans. on Computers*, vol. C-32, no. 12, pp. 1137-1144, Dec. 1983.

[4] P. Goel, "An implicit enumeration algorithm to generate tests for combinational logic circuit", *IEEE Trans. on Computers*, vol. C-30, no. 3, pp. 215-222, March 1981.

[5] Z. Kohavi, *Switching and Finite Automata Theory*. New York: McGraw-Hill, 1978.

[6] E.F. Moore, "Gedanken-experiments on sequential machines," Automata Studies, *Annals of Mathematical Studies*, no. 34, Princeton Univ. Press, Princeton, NJ, pp. 129-153, 1956.

[7] J. P. Roth, *Computer Logic, Testing, and Verification*. Rockville, MD: Computer Science Press, 1980.

[8] K.K. Sabnani and A.T. Dahbura, "A protocol test generation procedure," *Computer Networks*, vol. 15, no. 4, pp. 285-297, 1988.

[9] M.U. Uyar and A.T. Dahbura, "Optimal test sequence generation for protocols: the Chinese postman algorithm applied to Q.931," in *Proc. IEEE Global Telecommunications Conference*, pp. 68-72, 1986.

[10] Boundary-Scan Architecture Standard Proposal, Version 2.0, JTAG, 31 March, 1988.

A New Framework for Analyzing Test Generation and Diagnosis Algorithms for Wiring Interconnects

Najmi Jarwala and Chi W. Yau

AT&T Bell Laboratories
Princeton, NJ

Abstract

Increasing complexity of circuit boards and surface mount technology has made it difficult to test them using traditional in-circuit test techniques. A design-for-testability framework has been proposed as the IEEE Standard 1149.1, Test Access Port and Boundary-Scan Architecture. This architecture simplifies board test by providing an electronic bed of nails. It also provides access to other test features that may be present on a chip.

Because of the serial nature of the tests that use Boundary-Scan, it is important to minimize the test size while maintaining diagnosability. This has renewed interest in exploring efficient test algorithms and implementation techniques. This paper presents a new framework for analyzing the algorithms proposed for testing and diagnosing wiring interconnects. Using this framework, the algorithms proposed in the literature are analyzed, clearly identifying their capabilities and limitations. A new optimal adaptive algorithm that can reduce test and diagnosis complexity is also presented.

Keywords: Boundary-Scan, Board Test, Test Generation, Diagnosis, Interconnect Test.

1. Introduction

Higher levels of system integration have resulted in circuit boards shrinking onto devices, while systems are packed onto circuit boards. The increase in both the number of integrated circuits and their complexity has made testing circuit boards difficult and expensive. Traditionally, manufacturers use two techniques to test boards: in-circuit test and functional test. In the in-circuit test technique, the devices on a board are accessed by a "bed-of-nails" — probes on the ATE that directly make contact with the device I/O pins from pads on the surface of the circuit board. This makes it possible to test each device and the interconnects between devices. Note that this technique requires extensive access to the circuit under test.

Functional tests are applied through a board's normal terminations—for example, edge connectors. The objective is to test the board as a single entity. However test generation, fault simulation, and test application costs are excessive to achieve acceptable fault coverage. For some products it may be impossible to either verify or achieve the desired fault coverage.

Design-for-testability (DFT) and Built-in Self-Test (BIST) make the test problem more tractable. ATE's have also evolved to cope with the growing complexity of circuit boards. However surface mount technology, silicon-on-silicon, etc. further reduce the access that the in-circuit test methodology needs.

1.1 Review of the Boundary-Scan Architecture

To solve the problems discussed above, IEEE Std. 1149.1 has proposed the Boundary-Scan Architecture. This architecture basically consists of a Test Access Port (TAP) which consists of a four or five signal interface, a controller, an instruction register, and two or more test data registers (Figure 1). One of these test data registers is the Boundary-Scan (B-S) Register. This register is formed by serially linking latches (each of which is part of a B-S cell) that are placed at each device I/O so that the signals at the I/O can be controlled and observed. The B-S data registers in the parts of a board are linked into one or more serial paths through the assembled product. Such a path allows one to test the interconnects, apply tests to each device on this path, apply tests to clusters of logic that are not on the B-S path, access BIST and other testability features within devices, and take "snapshots" of the system state in real time. More details about this architecture can be obtained from [3].

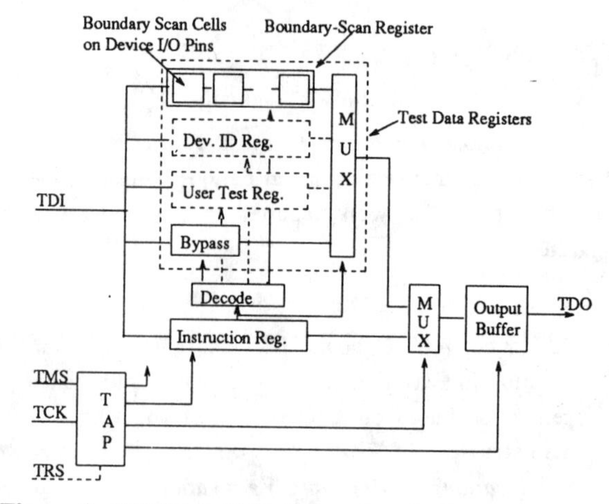

Figure 1. IEEE Std. 1149.1 Boundary-Scan Architecture.

1.2 Testing Circuits with Boundary-Scan

If all the devices on a board implement B-S, then the board test

Reprinted from *IEEE Proceedings 1989 International Test Conference*, pages 63-70.

procedure can be divided into the following test sequence:

- Test B-S chain integrity
- Test board interconnects
- Activate BIST and scan out the resultant signatures

Testing the integrity of the B-S chain is supported by the B-S architecture standard and is described in detail in [3]. Activating BIST is done by scanning in the RUNBIST instruction, waiting for it to complete, and scanning the signature out of the B-S Register. Testing the board interconnect is the subject of the remainder of the paper. While the emphasis of this work is testing in the IEEE Std. 1149.1 scan environment, the algorithms and results discussed here are also applicable to the general problem of testing wiring interconnects.

Section 2 presents the notations and definitions that will be used in the remainder of this paper. Section 3 defines the fault model which is followed by a review of fault detection algorithms in Section 4. Section 5 discusses fault diagnosis. In this section the faulty response is analyzed and classified, and the new framework is presented. The diagnostic capabilities and limitations of existing algorithms are analyzed under this new framework. Section 6 proposes a new, optimal adaptive test and diagnosis algorithm. Sections 7 and 8 discuss stuck-at faults and opens on wire/3-state nets respectively. Section 9 reviews some implementation issues, and Section 10 offers conclusions.

2. Notations and Definitions

In this section we present the notations and definitions that will be used in the remainder of the paper.

Net: A net on a circuit board is defined as an equipotential surface, formed by a physical wire connecting a set of input buffers and a set of output buffers.

The most general form of a net is shown in Figure 2.

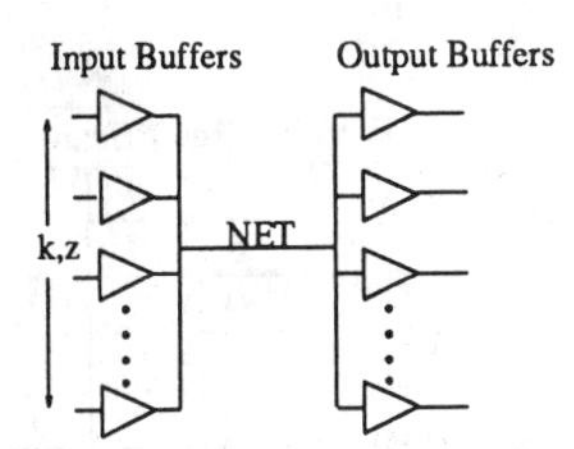

Figure 2. A Wire or 3-state Net.

A net can be driven by one or more buffers. Each net can be fanned out to one or more buffers. If a net is driven by a single buffer, we refer to it as a *simple net.* A net which is driven by more than one buffer is either a wire-AND/OR or 3-state and is referred to as a *wire net* or a *3-state* net respectively.

Net Degree: The degree of a wire net, denoted by k, is defined to be the number of buffers that drive that net. The degree of a 3-state net, denoted by z, is also defined to be the number of buffers that drive that net.

Let:

- L: Length of the B-S chain
- N: Number of nets
- n_i: A net identified by a unique number (ID), i
- n_i^*: A wire net identified by a unique number, i
- k_i: Degree of a wire net n_i^*
- K: $\max(k_i)$
- n_i^t: A 3-state net identified by a unique number i
- z_i: Degree of a 3-state net n_i^t
- Z: $\max(z_i)$

We also review the following definitions:

Union of Vectors, denoted by $\bigcup(v_1, v_2, \cdots)$, is the vector that results from a bit-wise OR of the component vectors.
Example: $v_1 = 0\,0\,0\,1$; $v_2 = 1\,1\,0\,0$. $\bigcup(v_1, v_2) = 1\,1\,0\,1$.

Intersection of Vectors, denoted by $\bigcap(v_1, v_2, \cdots)$, is the vector that results from a bit-wise AND of the component vectors.
Example: $v_1 = 1\,0\,0\,1$; $v_2 = 1\,1\,0\,0$. $\bigcap(v_1, v_2) = 1\,0\,0\,0$.

Parallel Test Vector (PTV), is the vector applied to all nets of an interconnect network in parallel.

Sequential Test Vector (STV), is the vector applied to a net, over a period of time, by a number of PTVs.
Note that the STV also represents a net ID. In Table 1, the PTVs are represented by the columns while the STVs are represented by the rows.

Sequential Response Vector (SRV), is the response of a net to an STV. If the net is fault free, its STV and SRV will be identical. A faulty net will differ in its STV and SRV.

Fault Syndrome, denoted by s^f, is the serial response of faulty or potentially faulty net(s). It is a SRV that is either different from its corresponding STV or a SRV that is common to two or more nets.

Vector Set, denoted by S is the set of all STVs. Note that $|S| = N$.

3. Fault Model

We consider the following classes of faults:

Multi-Net Faults. These are bridging faults that create a short between two or more nets. The behavior of the nets is a function of the driver characteristics of the individual nets involved in the short. This behavior can be either deterministic or non-deterministic. Deterministic behavior can be characterized as follows:

- *OR-type Short.* If the drivers are such that a '1' dominates, then the resultant logic value is an OR of the logic values on the individual nets.
- *AND-type Short.* If the drivers are such that a '0' dominates, then the resultant logic value is an AND of the logic values on the individual nets.
- *Strong-Driver Short.* If a specific driver in the short dominates, then the value of the net follows that of the driver regardless of the output of the other drivers.

The logic value on the net can also be non-deterministic or undefined. *This behavior is not included in this fault model and is not considered in the remainder of this paper.*

Single-Net Faults. These are stuck-at-one, stuck-at-zero, and open faults on single nets. Note that in the case of wire or 3-state nets, stuck-at faults affect the net as a whole while open faults may affect only part of the net.

The fault model allows for single or multiple occurrences of either multi-net faults and/or single-net faults with deterministic behaviour.

4. Fault Detection

In this section we discuss detection of stuck-at faults, shorts between nets and opens on simple nets. Note that opens on simple nets are equivalent to a stuck-at '0' or stuck-at '1' at the receiving buffer(s), depending on the technology. Opens on wire and 3-state nets are more complex and are dealt with in section 8.

Counting Sequence Algorithm

Some of the earliest work in this area was reported by Kautz [5]. In this paper he showed that $\lceil \log(N) \rceil$ PTVs are optimal for detecting all shorts in a network of N unconnected terminals. The test requirement for detecting shorts is very simple; a unique STV must be applied to each net. If the board is fault-free, each response must be unique. In case of a short, the nets involved will have the same response and hence the short will be detected. The requirement of applying a unique STV to each net can be easily met by following a simple counting sequence.

Modified Counting Sequence Algorithm

The vector set proposed by Kautz contained the all '0' and all '1' STVs. This was extended to $\lceil \log(N+2) \rceil$ PTVs by Goel and McMahon. This vector set eliminated the all '0' and all '1' STVs so that every STV has at least one '0' and one '1'. This modification permits stuck-at fault testing. This vector set is also a counting sequence with the all '0' and all '1' vectors excluded.

Nets	Parallel Test Vectors				Sequential Test Vectors
	v_1^T	v_2^T	v_3^T	v_4^T	
n_1	0	0	0	1	v_1
n_2	0	0	1	0	v_2
n_3	0	0	1	1	v_3
n_4	0	1	0	0	v_4
n_5	0	1	0	1	v_5
n_6	0	1	1	0	v_6
n_7	0	1	1	1	v_7
n_8	1	0	0	0	v_8
n_9	1	0	0	1	v_9
n_{10}	1	0	1	0	v_{10}

TABLE 1. Test Set defined by Modified Counting Sequence Algorithm.

Consider a circuit with ten nets. The four test vectors that would be applied are shown in Table 1. As defined previously, the *Parallel Test Vectors* are represented by the columns of Table 1 and the *Sequential Test Vector* are represented by the rows of Table 1. Each STV applies at least a '0' and '1' to each net so that stuck-at faults can be detected. Also every STV is unique so that a short between any pair of nets can be detected.

Structure Independent Algorithm

The test proposed by Hassan et al. [2] further generalizes the test proposed by Goel and McMahon. Their goal is to generate a test that is independent of the structure of the scan path so that test generation is simplified. To do this they propose using $\lceil \log(L+2) \rceil$ test vectors (L is the length of the B-S path) instead of $\lceil \log(N+2) \rceil$. The logic is that while this is non-optimal, it does not need formatting and is more suitable for BIST implementation. This test, however has a serious limitation: It only works if there are no 3-state nets. This constraint is not realistic for complex, bus-oriented circuit boards. In general, for scan paths based on the IEEE Standard 1149.1, order independent test sets are not possible. This is because the 3-state control signals and the outputs they control lie on the same scan path; and, to prevent conflicts, the relationship between the control signal and the outputs must be deterministic.

Walking One's Algorithm

Nets	Walking One's Sequence									
n_1	1	0	0	0	0	0	0	0	0	0
n_2	0	1	0	0	0	0	0	0	0	0
n_3	0	0	1	0	0	0	0	0	0	0
n_4	0	0	0	1	0	0	0	0	0	0
n_5	0	0	0	0	1	0	0	0	0	0
n_6	0	0	0	0	0	1	0	0	0	0
n_7	0	0	0	0	0	0	1	0	0	0
n_8	0	0	0	0	0	0	0	1	0	0
n_9	0	0	0	0	0	0	0	0	1	0
n_{10}	0	0	0	0	0	0	0	0	0	1

TABLE 2. Walking One's Test Sequence.

This is a very common test sequence used for testing memories, etc (Table 2). This has also been discussed by

Hassan et al. [2]. We consider the properties of the original sequence *without* response compression. The sequence has N PTVs. Each PTV is applied and the response stored for analysis. If applied in parallel this sequence is $O(N)$. In a scan environment this test is $O(N^2)$. Note that this sequence satisfies the minimum requirement for stuck-at and short detection as discussed above. This sequence also has a unique property that guarantees diagnosis, as discussed in the next section.

5. Diagnosis of Short Faults

In this section we discuss diagnosis of short faults. Diagnosis of stuck-ats and detection/diagnosis of other single-net faults will be discussed in sections 7 and 8.

Diagnostic resolution is of two types: The first identifies, without ambiguity, a list of nets that have a fault. The second type further identifies the sets of nets affected by the same short, the nets that are stuck-at zero or one, or the net that is open. When we use the term *diagnose,* we refer to the second type of diagnosis. This is important for rapid repair during manufacture.

Further, there are two test and diagnostic techniques. The first we call the **One Step Test and Diagnosis** where a set of test patterns are applied and the response is analyzed for fault detection and diagnosis. The other technique we call **Adaptive Test and Diagnosis** where the test is applied, response analyzed and then one or more additional tests may be applied to aid diagnostics. The implication of these techniques and their suitability for different test and repair environments is discussed in the Section 4.

Traditional testing applies PTVs to a circuit-under-test, receives a response and then analyzes one or more failing PTVs for diagnosis. We use a different approach. We assign a unique ID (STV) to each net and then consider the test procedure as 'requesting' each net to respond with its ID. Fault free nets respond with their correct IDs; faulty nets with IDs that differ from their assigned IDs. These incorrect IDs (or SRVs) have been defined earlier as *fault syndromes* and their analysis leads to diagnosis.

5.1 Syndrome Behavior

We analyze the relationship of the syndrome to the STVs of the nets involved in a short.

Let v_i represent the STV applied to net n_i. The SRV for a fault free net is same as the STV applied — v_i. Let the faulty response be represented by v_i^f. This notation implies that net n_i has responded with an ID f instead of i. All nets involved in a short will have the same faulty SRV. This SRV is called the fault *syndrome* s^f. If nets $n_i, n_j, n_k, \ldots$ are shorted together, then each will have the response $s^f = v_i^f, v_j^f, v_k^f, \ldots$

Let $(v_i, v_j, v_k, \ldots)$ be the STVs that were assigned to the nets involved in a short. Based on the type of short the syndrome can be characterized as follows:

- If $s^f = \bigcup(v_i, v_j, v_k, \ldots)$, then the short is of OR-type and the syndrome is called a *Disjunctive Syndrome.*
- If $s^f = \bigcap(v_i, v_j, v_k, \ldots)$, then the short is of AND-type and the syndrome is called a *Conjunctive Syndrome.*
- If the syndrome is neither conjunctive nor disjunctive and $s^f \in (v_i, v_j, v_k, \ldots)$ then the short is strong-driver short and the syndrome is called an *Identity Syndrome.*

5.2 Syndrome Classification

Based upon the above characterization, we classify a syndrome into two types:

- **Aliasing Syndrome.**
  Let S^F be the set of nets which respond with the syndrome s^f. If $s^f \in S$ (S is the set of STVs) and $s^f = \bigcup(S^F - s^f)$ or $s^f = \bigcap(S^F - s^f)$, then the syndrome s^f is called an *aliasing syndrome.* If this happens then the faulty response of a set of failed nets is the same as the fault-free response of another net. It cannot be determined whether or not this net is also involved in the short.

  Consider the test in Table 1. Assume all shorts are of type OR. If nets n_3 and n_4 are shorted then both will have a syndrome 0 1 1 1. However the fault-free response of net n_7 is also the same. Therefore it is not possible to distinguish whether n_3, n_4, n_7 are shorted or only nets n_3, n_4 are shorted. The syndrome 0 1 1 1 is an aliasing syndrome. Clearly, if nets n_1, n_2 and n_7 were shorted then the syndrome would be 0 1 1 1. However this syndrome would not be aliasing because $s^f \neq \bigcup(n_1, n_2)$. (Note that it is possible for a syndrome to be conjunctive or disjunctive and the corresponding short be a strong-driver short. For example, if net n_5, n_6, n_7 are shorted and the driver associated with n_7 is a strong driver, then the syndrome will still be disjunctive and aliasing. However for diagnosis, this situation is not relevant and will not be considered further.)

- **Confounding Syndrome.**
  A syndrome is called a *Confounding Syndrome* if the syndromes that results from multiple independent faults are identical, that is $s^i = s^j = \ldots = s^k$. Therefore it cannot be determined if these faults are independent.

  Consider (Table 1) two independent faults: nets n_4, n_{10} are shorted and nets n_6, n_8 are also shorted. Both shorts have the same syndrome 1 1 1 0. Consequently it cannot be determined if the faults are independent or one fault, a short between n_4, n_{10}, n_6, n_8, has occurred. The syndrome 1 1 1 0 is a confounding syndrome.

Degree of Confounding, denoted by c, of a syndrome is defined as the maximum number of potentially independent faults which all have the same syndrome. In the above example the syndrome 1 1 1 0 has $c = 2$.

Note that a syndrome can be both confounding and aliasing. Full diagnosis (as defined earlier) is possible if and only if no aliasing or confounding syndromes can exist.

5.3 Diagnostic Capabilities of One-Step Algorithms

Using the syndrome analysis framework introduced in the previous section, we now analyze the diagnostic capabilities and limitations of the algorithms proposed in the literature.

Modified Counting Sequence Algorithm

This algorithm can diagnose all short faults, provided the syndromes are neither aliasing nor confounding. It is impossible to predict the nature of the syndrome and consequently Algorithm 2 has very limited diagnostic capability.

True/Complement Test and Diagnosis Algorithm

To resolve the ambiguity caused by aliasing syndromes, Wagner [6] proposed a technique that we refer to as the True/Complement Test sequence.

Nets	True Vectors				Complement Vectors			
n_1	0	0	0	1	1	1	1	0
n_2	0	0	1	0	1	1	0	1
n_3	0	0	1	1	1	1	0	0
n_4	0	1	0	0	1	0	1	1
n_5	0	1	0	1	1	0	1	0
n_6	0	1	1	0	1	0	0	1
n_7	0	1	1	1	1	0	0	0
n_8	1	0	0	0	0	1	1	1
n_9	1	0	0	1	0	1	1	0
n_{10}	1	0	1	0	0	1	0	1

TABLE 3. True/Complement Test Sequence.

The technique applies $2\lceil \log(N+2)\rceil$ patterns. The additional $\lceil \log(N+2)\rceil$ patterns are obtained by complementing the first set of patterns. This test can diagnose all shorts with unique syndromes which are not confounding. Consider Table 3, which shows the $2\lceil \log(N+2)\rceil$ patterns applied to the same ten nets. If now nets n_3, n_4 are shorted, the complement test set gives a syndrome of 1 1 1 1 while the fault free response of n_7 is 1 0 0 0. Therefore the combined syndrome is no longer aliasing and the short can be diagnosed. However confounding syndromes cannot be diagnosed. This can be seen by analyzing the pairs of shorts, n_8, n_6 and n_{10}, n_4. The diagnostic capabilities of this technique is summarized by the following lemma:

Lemma 1. *The True/Complement Test and Diagnosis algorithm will not generate aliasing syndromes. It cannot diagnose syndromes that confound.*

Proof: Obvious. □

Walking One's Algorithm.

This algorithm is unique in that it is the only known algorithm that guarantees complete one step diagnosis of shorts for unrestricted faults. It has a property that we call Diagonal Independence. (Note that the definition that follows applies for OR-type shorts, that is the definition is actually for *Disjunctive Diagonal Independence*. Its dual, *Conjunctive Diagonal Independence* would apply to AND-type shorts. For clarity, OR-type shorts are assumed in the discussion that follows.)

Diagonal Independence of a Vector Test Set.

Let $S_{N\times M}$, $M \geq N$ denote the matrix of the vector test S. Let b_{ij}, $0 \leq i \leq N-1$, $0 \leq j \leq M-1$ be an element of S. If S, or the matrix obtained from S by successive row and/or column interchanges, has the form:

$$b_{ij} = \begin{cases} 1 & \text{for all } i=j \\ 0 & \text{for all } i>j \\ \times & \text{for all } i<j \end{cases}$$

where $\times \in \{0,1\}$, then S is said to be *Diagonally Independent.*

The general form of a matrix of test vectors that is Diagonally Independant is:

$$\begin{bmatrix} 1 & x & x & x & x & x \\ 0 & 1 & x & x & x & x \\ 0 & 0 & 1 & x & x & x \\ 0 & 0 & 0 & 1 & x & x \end{bmatrix}$$

The following are two other examples of Diagonally Independant test vector matrices:

$$\begin{bmatrix} 1 & 0 & 1 & x & x & x \\ 0 & 1 & 0 & 1 & x & x \\ 0 & 0 & 1 & 0 & 1 & x \\ 0 & 0 & 0 & 1 & 0 & 1 \end{bmatrix} \quad \begin{bmatrix} 1 & x & x & 0 & 0 & 0 \\ 0 & 1 & x & 0 & 0 & 0 \\ 0 & 0 & 1 & 0 & 0 & 0 \\ 0 & 0 & 0 & 1 & 0 & 0 \\ 0 & 0 & 0 & x & 1 & 0 \\ 0 & 0 & 0 & x & x & 1 \end{bmatrix}$$

Theorem 1: *Unrestricted shorts of the nets whose STVs* $\in$ ***S*** *are diagnosable if S is Diagonally Independent.*

Proof: Let Q be the matrix that is obtained from S after successive row and/or column interchanges, so that Q is in the form defined by the Diagonal Independence property. Let $v_1, v_2, \ldots, v_N$ represent the N rows of the matrix Q or the N STVs of the corresponding test set. Let each STV be represented by $v_i = b_{i0}, b_{i1}, \ldots, b_{i(M-1)}$. Consider any two STVs v_i, $v_j \in Q$. Let b_{il} and b_{jn} be the lowest bit positions that are '1', of the two vectors v_i, v_j respectively. From the definition of Diagonal Independence, it is clear that $i \neq j \Rightarrow l \neq n$. In other words, no two row vectors of a matrix that is Diagonally Independent can have identical lowest bit positions that are '1'.

Further, diagnosis is possible if and only if the syndromes do not alias or confound.

Let us assume that the syndromes can alias. Therefore there exists a set of vectors $R = \{v_i, v_j, \ldots, v_k\}$, $R \subset Q$ such that $\bigcup(v_i, v_j, \ldots, v_k) = v_l$, $v_l \notin R$, $v_l \in Q$. This implies that there is at least one vector in R which has a '1' in the same lowest bit position as v_l. This implies that the matrix Q has two unique vectors that have the same lowest bit position as '1', and consequently cannot be Diagonally Independent. This contradicts the given fact that Q is Diagonally Independent and hence it is impossible for the syndromes to alias.

Similar it can be shown that the syndromes cannot confound. For two syndromes to confound, there must exist two independent vector sets, such that their respective unions result in the same vector. This implies that each vector set must have at least one vector which has '1' in the same lowest bit position as the syndrome. This further implies that the matrix has at least two unique vectors with a '1' in the same lowest bit position which leads to the same contradiction as above. Hence syndromes cannot confound.

Since syndromes can neither alias nor confound, full diagnosis is possible. □

Note that this condition is sufficient, but not necessary, for a test vector set to avoid aliasing and confounding. However this leads to a systematic method of generating vector sets that guarantee diagnosability.

Corollary 1: *The Walking One's algorithm can diagnose unrestricted shorts with N test vectors.*

Proof: The Walking One's test set is Diagonally Independent and consequently the proof follows directly from Theorem 1. □

5.4 Diagnostic Capabilities of Adaptive Algorithms

As defined previously, adaptive test refers to the process of applying test vectors to a CUT, analyzing the response and then applying one or more tests to perform diagnosis. Note that each of the previous one-step algorithms has its adaptive dual. We present a new algorithm, called the One-Test Adaptive Algorithm, that is the dual of the True/Complement Algorithm. We then analyze the *W*-Test Adaptive algorithm, proposed by Goel and McMahon [1], which is the dual of the Walking One's Algorithm.

One-Test Adaptive Algorithm

This algorithm is the equivalent of the True/Complement Algorithm. That is, they have the same diagnostic capability. The algorithm is as follows:

1. Apply the $\lceil \log(N+2) \rceil$ tests for fault detection.
2. Analyze the syndromes. If the syndromes are neither aliasing nor confounding, then diagnosis is immediate.
3. If the syndromes are aliasing (but not confounding) then we need to resolve whether or not the vector, which is aliased to, has also failed. This can be accomplished by one additional test. A PTV is applied to the interconnects in which the bits applied to the nets whose STVs are aliased to, are set to '1'. The remaining bits are set to '0'. Clearly if a net is part of a short, then the response bits of the nets in the same short will be driven to '1'. Otherwise they will remain '0'. This test can be performed for all unique aliasing syndromes in parallel and hence only one additional test is required.

Note that like the True/Complement Algorithm, One-Test Adaptive Algorithm also cannot diagnose confounding syndromes.

Lemma 2. *The One-Test Adaptive Algorithm can diagnose shorts with syndromes that are aliasing but not confounding with no more than* $1 + \lceil \log(N+2) \rceil$ *tests.*

Proof: Follows from the discussion above. □

***W*-Test Adaptive Algorithm**

Goel and McMahon [1] have proposed a two-step test and diagnosis procedure. This algorithm is equivalent to the Walking One's Algorithm and, in the limit, reduces to the Walking One's Algorithm. In the first step they apply the $\lceil \log(N+2) \rceil$ vectors discussed in the Modified Counting Sequence Algorithm. From analyzing the response, it is possible to identify a set of vectors $R \subseteq S$ which have produced faulty response vectors. Let $W = |R|$. In the second step the procedure applies a Walking One Test to the set R to diagnose the failures. This algorithm has requires $W + \lceil \log(N+2) \rceil$ PTVs.

In case there are a large number of faults, $W \rightarrow N$. However in practice $W \ll N$ and so this will require fewer vectors than the Walking One's Sequence.

Lemma 3. *The W-Test Adaptive Algorithm can diagnose all unrestricted shorts with* $W + \lceil \log(n+2) \rceil$ *vectors, where W is the number of faulty nets.*

Proof: Follows directly from discussion above. □

6. Optimal *C*-Test Adaptive Algorithm

This section describes a new, optimal diagnostic algorithm. It has the same capability as the Walking One's Algorithm, however the analysis stage uses the analytical framework developed earlier to avoid the potential inefficiency the *W*-Test Adaptive Algorithm. Instead of looking at the set of all faulty nets, this algorithm analyzes the syndromes to determine their nature and then decides if additional tests are required.

The Algorithm is as follows:

1. Apply the $\lceil \log(N+2) \rceil$ tests for fault detection.

2. Analyze the syndromes. If they are neither aliasing nor confounding, then, diagnosis is immediate.

3. If syndromes are only aliasing, then full diagnosis requires one additional test, as described in the One-Test Adaptive Algorithm.

4. The remaining syndromes are either confounding, or both aliasing and confounding. The confounding is resolved as follows: Let $C = \max(c_i)$ be the largest degree of confounding of these syndromes. No more than $C-1$ tests are required to resolve the confounding of the fault whose syndrome has a degree of C. Since the diagnosis of faults with unique syndromes can be done in parallel, $C-1$ tests suffice to completely resolve all confounding syndromes.

5. Finally aliasing (if exists) needs to be resolved. This requires one more vector. Therefore if the syndromes are confounding and aliasing, then $C-1$ tests suffice to resolve confounding and one more test resolves aliasing. Therefore, in general, at most C tests suffice to completely diagnose all shorts.

Lemma 4. *The C-Test Adaptive Algorithm can diagnose unrestricted shorts with no more than* $C+\lceil \log(N+2) \rceil$ *tests, where C is the highest degree of confounding.*

Proof: Follows directly from the discussion above. □

7. Diagnosis of Stuck-at Faults

Diagnosis of a stuck-at fault is relatively simple, since the affected receiving buffer reads a constant value. However to achieve complete diagnosis, care must be taken to ensure that the resultant all-one or all-zero syndrome is not disjunctive or conjunctive respectively. If that is the case then the set of nets reporting a constant '0' or '1' syndrome could all be stuck-at or shorted together. This can be a potential problem for one-step diagnosis. For example, consider a short of nets n_7,n_8 in Table 1. The SRV of both these nets will be 1 1 1 1, which is same if both the nets had been stuck at '1'. This ambiguity can be resolved by adding a all-zero (for OR-type shorts) and all-one (for AND-type shorts) PTV to the test vector set. If Table 1 had another all PTV, v^{T_5}, which is all '0', the the last bit of the SRVs of n_7 and n_8 will distinguish between a stuck-at-1 fault and a short. A stuck-at-1 fault will drive this bit to a '1' while a short will result in this bit having a value '0'. This will clearly distinguish between a stuck-at and short.

8. Testing and Diagnosing Opens on Wire and 3-State Nets

Wagner [6] presents detailed algorithms for testing wire nets. There are three types of wire-nets: wire-AND, wire-OR and 3-state nets. Testing wire-AND and wire-OR nets is simple and is equivalent to testing an AND or OR gate. Note that stuck-at and bridging faults affect the net as a whole and are detected by the previous test procedures. The principle interest in these tests is to test for and diagnose opens which affect only a subset of the pins in a net.

Wire-AND/OR Nets Test and Diagnosis Algorithm
Consider a wire-AND net of degree k. Testing it is analogous to testing a k input AND gate and the test set consists of k tests formed by 'walking' a '0' across the k output buffers (the remaining $k-1$ output buffers are held at '1') and one additional test which consists of all ones. The dual applies to wire-OR nets. Therefore a net of degree k can be tested with $k+1$ tests. Since multiple wire-nets can be tested in parallel, $K+1$ tests suffice to test all AND/OR wire nets.

3-State Nets Test and Diagnosis Algorithm
To test a 3-state net, we have to ensure that each buffer can independently drive the net to both a '0' and a '1' and that this value is correctly received by the receiving buffers. This implies that both a '0' and a '1' has to walked across the input of the output buffers, (with the other output buffers being held at the complementary state and disabled). Therefore a 3-state net of degree z can be tested with $2z$ test vectors. Further since multiple 3-state nets can be tested in parallel, 2Z tests suffice to test all 3-state nets.

Lemma 6. *All stuck-ats and open faults in 3-State, Wire-AND and Wire-OR nets can be detected and diagnosed by the Wire-AND/OR and 3-State Test and Diagnosis Algorithms using at most max(K + 1, 2Z) test vectors.*

Proof: Follows directly from the observation that all wire-AND, wire-OR and 3-state nets can be tested in parallel. Note that by observing the output response, it is trivial to diagnose which driver or receiver is isolated from the net by an open. □

9. Implementation Issues

There are several implementation issues that are important when generating tests for a circuit board. The objective is to minimize the test/repair time (and consequently the cost) while achieving high fault coverage and diagnostic resolution. Some of the parameters to be considered are:

- ATE Capabilities
- Test/Repair Environment
- Board Yield

9.1 ATE Capabilities

One of the motivations for using B-S is that low cost ATE's (e.g. one that is PC-based) can be used for board test. These ATE's may have some limitations. Of principal concern is the maximum length of the test sequence that the ATE can apply between successive disk accesses. This may constrain the algorithm that can be used. An ATE with limited test length capability may make it impractical to use the Walking One's

Algorithm. Another limitation concerns the 'openness' of the architecture. In some cases it is not possible for a test engineer to gain direct access to the response for the purpose of diagnosis; in other cases, the ATE provides a compressed failure report. Other factors are the ATE's handling of the failed response and the number of failures permitted before test termination. If few such failures are permitted, then insufficient failure information might be obtained and this may impact diagnostic resolution.

Another important consideration is the computational capability of the ATE and the ease of generating and applying test vectors in real time. These factors determine the feasibility of adaptive testing. Some ATE's may have very limited computational capabilities making diagnosis difficult. Others may require extensive processing before test vectors can be generated making it impossible to apply tests in real time.

9.2 Test/Repair Environment

A good understanding of this factor is crucial in designing efficient tests. Frequently this may be pre-determined by existing equipment and practices. If a shop set-up initially performs a go/no-go test and failing boards are sent to a repair station, then to maximize throughput, the Modified Counting Sequence algorithm can be used for the initial test, and the ***C***-Test Adaptive Algorithm for diagnosis and repair. If however test and repair are done at the same station, and the ATE does not have the capability to do adaptive testing, then the Walking One's Algorithm may be required. This problem is addressed in another paper by the authors [4], where design and process information is used to constrain the size of the test produced by one-step algorithm without sacrificing diagnostic resolution.

9.3 Board Yield

If the process is mature and high yields are being obtained, even the Modified Counting Sequence or the True/Complement Algorithms may provide enough diagnostic resolution. On the other hand for a new product/process the number of failures are likely to be large and and consequently the Walking One or C-Test Adaptive may be required. Note that in such an environment the W-Test Adaptive Algorithm may suffer reduced effectiveness

10. Conclusions

This paper makes several contributions. A new framework for analyzing test generation and diagnosis algorithms for wiring interconnect have been presented. A property of test vector sets, called Diagonal Independence, has been identified which guarantees the diagnostic resolution of the vector test set. The failing responses or syndromes have been classified into aliasing and confounding syndromes, and this classification permits precise analysis of the diagnostic capabilities of different test algorithms. Using this framework, all the algorithms that have been proposed for board interconnect test are analyzed. Their capabilities and limitations are clearly defined. A new, optimal adaptive test and diagnosis algorithm is proposed.

An important aspect of test design is to take into account the test/repair environment and its relationship with the product being tested. This relationship is not static and it changes as the process matures. Ideally we need a design procedure that takes into account the ATE capabilities, test/repair strategies, product yield and the fault data from Failure Mode Analysis so that an efficient, cost-effective test can be developed. To do this we not only need a good understanding of the entire test/repair process but we also need good test algorithms that permit a tradeoff between diagnostic resolution and test complexity. Presently, if adaptive tests are not possible, then we basically have two choices: The Modified Counting Sequence Algorithm with $O(\log N)$ test size or the Walking One's Algorithm with $O(N)$ test size. If Boundary-Scan is used to apply the test vectors then the test time, because of the serialization of the PTVs, is $O(N \log N)$ and $O(N^2)$ respectively. In this environment, the difference between $O(N \log N)$ and $O(N^2)$ test application times may be too high to make the algorithm with $O(N^2)$ complexity practical.

Using the framework established in this paper, the authors propose [4] a family of One-Step diagnosis algorithms that use design and process information to generate tests of $O(\log N)$ without sacrificing diagnostic resolution. The algorithms (Modified Counting Sequence, Walking Sequence) that are discussed in this paper are shown to be special cases of the general theory that is used to generate these new algorithms.

References

[1] P. Goel and M. T. McMahon, "Electronic Chip-in-Place Test," *Proceedings International Test Conference* 1982, pp. 83-90

[2] A. Hassan, J Rajski, and V. K. Agarwal, "Testing and Diagnosis of Interconnects using Boundary Scan Architecture," *Proceedings International Test Conference* 1988, pp. 126-137.

[3] JTAG Boundary Scan Architecture Standard Proposal, Version 2.0, published March 1988.

[4] N. Jarwala and C. W. Yau, "A Unified Theory for Designing Optimal Test Generation and Diagnosis Algorithms for Board Interconnects," *Proceedings, International Test Conference* 1989.

[5] W. K. Kautz, "Testing of Faults in Wiring Interconnects," *IEEE Transactions on Computers,* Vol C-23, No. 4, April 1974, pp. 358-363.

[6] P. T. Wagner, "Interconnect Testing with Boundary Scan," *Proceedings, International Test Conference* 1987, pp 52-57.

A Unified Theory for Designing Optimal Test Generation and Diagnosis Algorithms for Board Interconnects

Chi W. Yau and Najmi Jarwala

AT&T Bell Laboratories
Princeton, NJ

Abstract

To test wiring interconnects in a printed circuit board, especially one equipped with boundary-scan devices, it is important to minimize the test size while maintaining diagnostic capability. This has provided the motivation for research work that explores efficient test generation and diagnosis algorithms. In this paper, we propose a unified theory for designing various types of interconnect test algorithms. We demonstrate that the algorithms proposed in the literature are special cases of the general algorithms presented in this paper. The new algorithms are shown to be optimal or near-optimal for a given set of design and process parameters. They increase the designer's flexibility by offering a full range of solutions (i.e. test vector sets) based on various trade-off criteria such as test compactness and diagnostic accuracy. Parameters for quantifying the quality of the tests are described. The significance and limitations of the proposed algorithms are also discussed.

Key Words: Board testing, boundary-scan, interconnect test, design-for-testability.

1. Introduction

The problem of test generation for wiring interconnects has been extensively studied. Several algorithms have been proposed which assure detection of all opens, shorts, and stuck-at faults [1-4]. Some of them [1, 2] are optimal in the sense that they produce tests that are most compact. That is, the test size is $O(\log N)$, and, in the boundary-scan environment, the test time is $O(N \log N)$. However, as described in the accompanying paper [5], these tests are inadequate in terms of their diagnostic capability. An algorithm based on the walking patterns [4] has been proposed for fault diagnosis. Although it guarantees complete diagnosis, the test size and the test time are $O(N)$, and $O(N^2)$ respectively. This may be intolerable for high-density boards.

Clearly, it is advantageous to develop a general approach for designing test generation and diagnosis algorithms which enable the designer to *gradually* give up compactness while still maintaining maximal diagnostic resolution. In this paper, we present a unified theory which will allow us to accomplish this goal. In particular, the designer will have the freedom to choose from a wide variety of tests, ranging from those primarily designed for fault detection to those primarily designed for fault diagnosis.

We assume that the reader is familiar with the boundary-scan test architecture [6], as well as some basic interconnect test algorithms. (Detailed review and analysis of these algorithms can be found in the accompanying paper.) In addition, this work is based on the framework and some basic concepts described in the accompanying paper from which we also adopt all the necessary notations and definitions. For clarity, we assume throughout this paper that all shorts exhibit wire-OR behavior. By duality, all results presented can be easily extended to handle shorts with wire-AND behavior. Additionally, strong-driver shorts [5] exhibiting deterministic behavior can also be handled easily.

In the following section, we present the unified theory for designing test generation and diagnosis algorithms. Two new algorithms are proposed. It is shown that both algorithms produce the fault-detection test (the Modified Counting Sequence) and the fault-diagnosis test (the Walking-1 Sequence) as special cases. Section 3 discusses some important characteristics of the proposed algorithms. Some directions for future work are given in Section 4. The last section provides some concluding remarks.

2. A Unified Theory

In this section, we generalize the results from the previous works [1-4] by proposing a unified theory for designing optimal test generation and diagnosis algorithms. This theory is based on a general concept: Suppose, for an N-net board, that the number of parallel test vectors (PTVs) which we can "afford" to apply is p, where $p \geq \lceil \log(N+2) \rceil$. Then, the problem of generating a test vector set with optimal diagnostic capability is equivalent to that of "intelligently" assigning a unique (p-bit) sequential test vector (STV) to each of the N nets such that the overall diagnostic ambiguity of the test is minimized. Since, excluding the all-0 and all-1 STVs, there are 2^p-2 possible STVs, the solution space is defined by

$$\binom{2^p-2}{N}$$

To circumvent combinatorial explosion, we will describe the *optimality* of a solution in *heuristic* terms only. Among possible solutions, we will propose two heuristic algorithms. These algorithms can generate test vectors with complete fault detection capability, *and* also "good" fault isolation capability. Since both algorithms produce test vectors with full fault detection capability, we will describe their "goodness"

Reprinted from *IEEE Proceedings 1989 International Test Conference*, pages 71-77.

EH0321-0/90/0000/0318$01.00 © 1989 IEEE

only in terms of their fault isolation capability. In general, we attempt to increase the diagnostic capability of a test vector set by reducing its potential for producing aliasing and/or confounding syndromes [5].

All algorithms proposed in the literature provide the designer with individual, *ad hoc* solutions which fall within two extremes—the test generated by the Modified Counting Sequence Algorithm [2], which is most compact, but least helpful to diagnosis ($p = \lceil \log(N+2) \rceil$); and the test generated by the Walking-One Algorithm [4], which is most helpful to diagnosis, but least compact ($p = N$). In contrast, both of our algorithms share an important property: They enable the designer to select from a *full range* of solutions ($\lceil \log(N+2) \rceil \le p \le N$) based on such trade-off criteria as test compactness and diagnostic capability.

Now, we will describe the two heuristic algorithms in detail. The first one assumes that no physical design information is available; while the second one assumes that certain design and process information can be used.

2.1 The Min-Weight Algorithm

This algorithm can be used when no design and process information is available. Typically, the designer specifies the total number p of parallel test vectors (PTVs) to be produced in advance. During test generation, the Min-Weight Algorithm sequentially assigns a unique (p-bit) STV of minimum weight to each of the N nets (hence the name Min-Weight Algorithm). Since, the number of unique STVs which can have a weight of w is

$$\binom{p}{w}$$

the maximum weight w_{max} of the N assigned STVs is given by the minimum value of k for which the following is true

$$\sum_{i=1}^{k} \binom{p}{i} \ge N \quad (1)$$

A necessary consequence of the Min-Weight Algorithm is that w_{max} is always greater than or equal to the weight of any of $(2^p-2)-N$ unassigned STVs (excluding the all-1 STV). Also observe that if $N=2^p-2$, then $w_{max}=p-1$ and all (2^p-2) possible STVs will be assigned. A sample test produced by this algorithm is shown in Table 1, where $p = 4$ and $N = 12$

Intuitively, one can see that if all shorts exhibit wire-OR behavior, the test generated by the Min-Weight Algorithm is less likely to produce aliasing syndromes than that by the counting methods [1, 2]. This follows from the observation that shorting of two or more nets often produces a syndrome whose weight is greater than those of all the STVs assigned to the shorted nets. Since all assigned STVs have minimum possible weights, the (heavy) syndrome is less likely to alias with one of them.

Nets	p-Bit STVs				Weights
n_1	1	0	0	0	1
n_2	0	1	0	0	1
n_3	0	0	1	0	1
n_4	0	0	0	1	1
n_5	1	1	0	0	2
n_6	1	0	1	0	2
n_7	1	0	0	1	2
n_8	0	1	1	0	2
n_9	0	1	0	1	2
n_{10}	0	0	1	1	2
n_{11}	1	1	1	0	3
n_{12}	1	1	0	1	3

TABLE 1. Test Vectors Produced by the Min-Weight Algorithm

It is interesting to observe the lower and upper boundary conditions of this algorithm: If $p = \lceil \log(N+2) \rceil$, the test is spatially most compact, but diagnostically least helpful; its diagnostic capability is marginally better than, or even identical to, that generated by the counting methods. On the other hand, if $p = N$, the test is diagnostically most helpful, but spatially least compact; it is the same as that generated by the Walking-One Algorithm.

2.2 The Max-Independence Algorithm

The Max-Independence Algorithm minimizes the size of a test without sacrificing its diagnostic accuracy. This is achieved by using net adjacency (i.e. wire routing) information as well as certain process-related information—in particular, the maximum size of expected shorts or solder defects. (The *size* σ of a short or solder defect is the number of nets affected by the defect.) The following definitions are needed before describing the Max-Independence Algorithm.

Definition 2.1. Given a binary vector $v = (b_0, b_1, \ldots, b_n)$, let i and j be the lowest and highest bit positions respectively such that $b_i = b_j = 1$. The **potential weight** $\tilde{w}$ of v is equal to $j-i+1$ for all non-zero v; otherwise, $\tilde{w} = 0$. For example, vectors (0,1,1,1,0) and (0,1,0,1,0) both have a potential weight of 3. Further, the number $N_{\tilde{w}}$ of unique (p-bit) STVs that can have a potential weight of $\tilde{w}$ is given by

$$N_{\tilde{w}} = \begin{cases} 1 & \text{for } \tilde{w} = 0 \\ p & \text{for } \tilde{w} = 1 \\ (p-\tilde{w}+1)\cdot 2^{\tilde{w}-2} & \text{for } \tilde{w} > 1 \end{cases} \quad (2)$$

Definition 2.2. A set of N nets $\{n_1, n_2, \ldots, n_N\}$ is an **adjacency-ordered** set if n_i is more adjacent to, or more likely to be shorted with, n_{i+1} than n_{i+2} for $1 \le i \le N-2$, and if n_i is more adjacent to n_{i-1} than n_{i-2} for $3 \le i \le N$. Note that a net on a circuit board can be physically adjacent to its neighboring nets in more than one dimension. Therefore, only a *partial*, or approximate, adjacency-ordering is achievable in practice. Fortunately, bare board testing eliminates many faults (e.g. inter-layer shorts) which weaken the "net-adjacency assumption." We will show that even a partial ordering is more useful than a random one. A "good" partial adjacency ordering can be easily obtained by approximating net adjacency with device pin adjacency. (Here, we assume that most common shorts are caused by solder bridges affecting physically adjacent device pins.)

Definition 2.3. Given an adjacency-ordered set of nets, let n_i be the lowest ordered net, and n_j the highest ordered net affected by a given short. The **extent** e of the short is then defined as $j-i+1$. In a physical sense, e is related to σ, the number of nets affected by a short (e.g. the number of nets shorted together by a single "solder blob"), and to the ordering of the nets: for a complete adjacency-ordering, $e = \sigma$; for a partial adjacency-ordering, $e \geq \sigma$. Clearly, e is bounded by $2 \leq e \leq N$.

Finally, we assume that the reader is familiar with the **diagonal independence** property of a test vector set which is detailed in the accompanying paper. Having provided sufficient background material, we are ready to describe the Max-Independence Algorithm in detail.

The Max-Independence Algorithm consists of the following steps:

1. Find the minimum number p of PTVs that are required for unambiguous diagnosis of all expected shorts, given that the extents of these shorts will never exceed some predetermined limit E. (As seen in Theorem 2.1 below, p is given by

$$p = \begin{cases} \lceil E + \log(N+1) - \log E - 1 \rceil & \text{if } E > 2 \text{ or } \log(N+1) < \lceil \log(N+1) \rceil \\ \lceil \log(N+2) \rceil & \text{if } E = 2 \text{ and } \log(N+1) = \lceil \log(N+1) \rceil \end{cases}$$

   where $2 \leq E \leq N$).

2. Generate an adjacency-ordered list of all N nets. (Partial adjacency-ordering is acceptable if complete adjacency-ordering is impractical.)

3. Form unique subsets of p-bit STVs (excluding the all-0 and all-1 STVs) such that each subset is made of all possible STVs which have the same potential weights *and* the same Hamming weights. (Observe that each STV subset contains a maximum number of unique STVs that are diagonally independent—thus the name Max-Independence Algorithm.)

4. Use the STV subset with the smallest potential weights (i.e. 1) to form an initial ordered set of STVs.

5. Concatenate, repeatedly, a new STV subset with the next smallest potential weights to the ordered STV set until it contains at least N STVs. (If two or more STV subsets have equal potential weights, pick the subset with the smallest Hamming weight.)

6. Assign, sequentially, an STV from the ordered STV set to the next unassigned net in the adjacency-ordered net list until all N nets have been assigned a unique STV.

A sample test generated by this algorithm is shown in Table 2, where $p = 4$ and $N = 12$.

Adjacency-Ordered Nets	STVs (p bits)					Potential Weights
n_1	1	0	0	0	0	1
n_2	0	1	0	0	0	1
n_3	0	0	1	0	0	1
n_4	0	0	0	1	0	1
n_5	0	0	0	0	1	1
n_6	1	1	0	0	0	2
n_7	0	1	1	0	0	2
n_8	0	0	1	1	0	2
n_9	0	0	0	1	1	2
n_{10}	1	0	1	0	0	3
n_{11}	0	1	0	1	0	3
n_{12}	0	0	1	0	1	3

TABLE 2. Test Vectors Produced by the Max-Independence Algorithm

It is important to point out a unique property of the STV set produced by the Max-Independence Algorithm. First, notice in Table 2 that the vector set exhibits a very regular pattern. Specifically, it is made of (successively smaller) STV subsets ($\{n_1,n_2,n_3,n_4,n_5\}$, $\{n_6,n_7,n_8,n_9\}$, etc.) which are diagonally independent. Also, the unique ordering of the STVs guarantees that *as long as the extent of any given short (see Definition 2.3) never exceeds a certain upper bound (4 in this case), the fault can always be unambiguously diagnosed.* This is because any 4 consecutive STVs in the vector set possess the diagonal independence property. One can be easily convinced of this assertion by observing a number of examples. The most obvious example involves the STVs assigned to nets n_1, n_2, n_3 and n_4. A less obvious example consists of the STVs associated with nets n_4, n_5, n_6 and n_7, whose diagonal independence property becomes evident upon realizing that the STVs of n_4 and n_5 and the STVs of n_6 and n_7 can be interchanged [5].

Table 3 summarizes some important characteristics associated with the Max-Independence Algorithm. As seen in this table, STVs of successively larger potential weights are assigned to each of the N nets, and the maximum potential weight of all assigned STVs is denoted by k (column 1). In addition, as the potential weight $\tilde{w}_i$ of an STV subset increases, the maximum defect extent e_i for which full diagnosability is still maintained,

$\tilde{w}_i$	e_i	No. of STVs (Eq. 2)†
1	p	p
2	p	$(p-1)\cdot 2^0$
3	$p-1$	$(p-2)\cdot 2^1$
4	$p-2$	$(p-3)\cdot 2^2$
.	.	.
.	.	.
.	.	.
$k-1$	$p-(k-3)$	$[p-(k-2)]\cdot 2^{k-3}$
k	$p-(k-2)$	$[p-(k-1)]\cdot 2^{k-2}$

† The all-0 and all-1 STVs are excluded to detect all stuck-at faults.

TABLE 3. The STV Assignment Sequence of the Max-Independence Algorithm

decreases (column 2). Specifically, upon completion of test generation, the maximum allowable defect extent E of the test vector set becomes $\min\{e_i\}$ or $p-(k-2)$. Finally, since the total number of STVs to be assigned is N, we should be able to equate N to the sum of the terms in column 3.

At this point, it has become apparent that the larger the maximum defect extent E, the larger the number p of PTVs that are required to guarantee unambiguous diagnosis of all expected shorts. Obviously, given E, it is desirable to compute the minimum p which still assures complete diagnosis. The following theorem enables us to do precisely that.

Theorem 2.1. *Let E be the maximum extent of all expected shorts on a board with N adjacency-ordered nets, and let the test generation algorithm be the Max-Independence Algorithm. Then, the minimum number p of parallel test vectors required to unambiguously diagnose all expected shorts is given by*

$$p = \begin{cases} \lceil E + \log(N+1) - \log E - 1 \rceil & \text{if } E > 2 \text{ or } \log(N+1) < \lceil \log(N+1) \rceil \\ \lceil \log(N+2) \rceil & \text{if } E = 2 \text{ and } \log(N+1) = \lceil \log(N+1) \rceil \end{cases} \quad (3)$$

where $2 \le E \le N$.

Proof: Obviously, the total number of STVs assigned by the Max-Independence Algorithm (i.e. the sum of the terms in column 3 of Table 3) must equal N. Therefore,

$$N = p\,(1 + \sum_{i=0}^{i=k-2} 2^i) - \sum_{i=0}^{i=k-2} (i+1)2^i \quad (4)$$

Simplifying the two series on the right hand side of (4), we get

$$N = p\cdot 2^{k-1} - [(k-2)\cdot 2^{k-1} + 1]$$

or

$$N+1 = [p-(k-2)]\cdot 2^{k-1} \quad (5)$$

We know that unambiguous diagnosis of all expected shorts is guaranteed if the minimum of e_i, the maximum allowable defect extents associated with the STV subsets (column 2 of Table 3), is equal to E. Since $\min\{e_i\} = p-(k-2)$,

$$E = p-(k-2) \quad (6)$$

must hold. Solving (5) and (6) for p, we obtain

$$p = \lceil E + \log(N+1) - \log E - 1 \rceil \quad (7)$$

Note that the term $\log(N+1)$ in (7) clearly signifies the exclusion of the all-0 STV from the test vector set. The Max-Independence Algorithm automatically avoids assigning the all-1 STV to the last net n_N as long as $E > 2$ or $\log(N+1) < \lceil \log(N+1) \rceil$. Otherwise, when the boundary condition that $E = 2$ *and* $\log(N+1) = \lceil \log(N+1) \rceil$ is true, we must subtract 1 from the right hand side of (4) to account for the exclusion of the all-1 STV. This, after solving for p again, will yield

$$p = \lceil E + \log(N+2) - \log E - 1 \rceil \quad (8)$$

Note that the term $\log(N+2)$ in (8) reflects the omission of both the all-0 and all-1 STVs from the test vector set. Finally, putting $E = 2$ in (8), we get

$$p = \lceil \log(N+2) \rceil \quad (9)$$

□

Note in (3) that p is a function of N and E. While N is always known, E can only be estimated or empirically obtained for a given board and manufacturing process. Also, it can be verified that p satisfies the two well-known boundary conditions. That is, for $2 \le E \le N$,

$$\lceil \log(N+2) \rceil \le p \le N$$

Further, as seen in Figure 1, p is essentially a *linear* function of E for a given N.

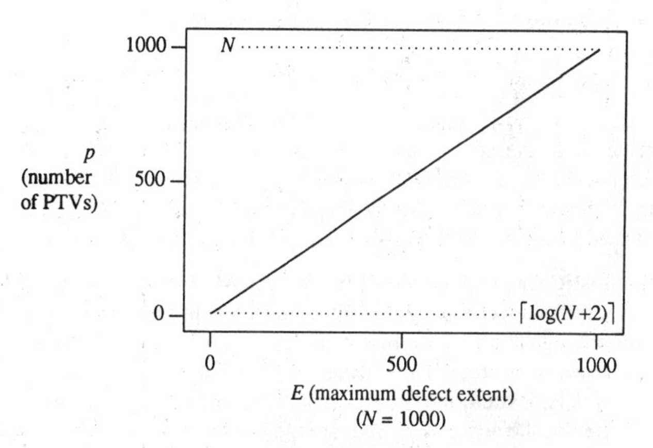

Figure 1. The Effect of Maximum Defect Extent on the Number of PTVs

Note that although in the worst case (i.e. when *all possible* shorts are considered, including that which affects all N nets) $E = N$, in practice, $E \ll N$ holds for most of the faults that will *actually* occur. For example, $E = 20$ maybe an upper bound on the extents of, say, 99% of the actual shorts encountered by a particular board.

We now illustrate the significance of Theorem 2.1 with a simple example. Given that a board has 1000 nets ($N = 1000$), which are (completely) adjacency-ordered, and that the number of nets affected by any given short never exceeds 20 ($E = 20$), the minimum number of PTVs required to fully diagnose all expected shorts, according to Eq. (3), is $p = 25$. Note that p is significantly smaller than the upper bound $N = 1000$, and that full diagnostic capability of the test is still maintained. Of course, this is possible only because we have prior information regarding net adjacency and the maximum size of expected short/solder defects.

3. Discussion

In this section, we will discuss some important aspects of the algorithms proposed in the last section.

3.1 The Min-Weight Algorithm

The characteristics of this algorithm will be described in terms of its trade-off criteria, measure of goodness, and diagnostic capability.

Trade-off Considerations: Virtually no design and process specific information is needed by the Min-Weight Algorithm. Moreover, the fault model includes all theoretically possible shorts. The primary trade-off criterion offered by the algorithm is p, the number of PTVs that the test engineer is willing to apply to the board under test, given certain spatial and temporal constraints (i.e. vector storage space and test throughput). Once p is determined, the algorithm generates an STV set whose maximum Hamming weight is minimal (assuming only OR-type shorts are possible).

Measure of Goodness: The qualitative justification for the Min-Weight Algorithm is that the probability of the test to produce aliasing syndromes (whose weights are generally greater than the constituent STVs) is likely to diminish if the weights of the assigned STVs are minimized. In general, the larger the p (where $\lceil \log(N+2) \rceil \leq p \leq N$), the smaller the chance of aliasing, and thus the "better" the test vector set.

Diagnostic Capability: The Min-Weight Algorithm does not totally prevent the test vector set from producing aliasing and confounding syndromes although the probability of their occurrence is reduced. Therefore, when unambiguous diagnosis is desired, the optimal adaptive diagnostic algorithm presented in the accompanying paper [5] can be used to resolve any ambiguity.

3.2 The Max-Independence Algorithm

Our discussion on the Max-Independence Algorithm willl cover four aspects: trade-off criteria, measure of goodness, diagnostic capability, and test complexity.

Trade-off Considerations: This algorithm typically requires the knowledge of net adjacency and maximum defect extent. Using this knowledge, the size of a test can be minimized without compromising its diagnostic capability. Similar to the Min-Weight Algorithm, this algorithm allows the test engineer to use the maximum allowable p as the trade-off criterion. Given a predetermined value of p, a set of p-bit STVs is generated. This test set is less likely to cause diagnostic ambiguities because it consists of maximum STV subsets which are diagonally independent.

The Max-Independence Algorithm gives the designer another trade-off option. Assuming that unambiguous one-step diagnosis [5] is required for all shorts whose extents do not exceed a predetermined upper bound E, this algorithm allows one to compute the corresponding minimum value of p which guarantees the full diagnosis of those shorts. Typically, the trade-off parameter E can be estimated using accummulated statistical data pertaining to the sizes of solder defects.

Measure of Goodness: In most applications, the designer uses the Max-Independence Algorithm to derive p from E, the maximum extent of all (or most of) the expected shorts. Therefore, the larger the E (and hence the p), the "better" the test vector set. This is because the probability that the extent of a short exceeds E (or equivalently the probability of diagnostic ambiguity) decreases as E increases. (Recall that aliasing and/or confounding syndromes can be encountered only when the extent of a short is greater than E.)

Diagnostic Capability: As mentioned previously, as long as the extent of a short does not exceed E, the Max-Independence Algorithm guarantees complete diagnosis. Strictly speaking, however, the only value of E that assures full diagnosis of *all possible* shorts is N, the total number of nets on the board. Fortunately, in reality, the "equi-probable assumption" about all shorts never holds. That is, *realistic* shorts do not occur with the same frequency. For example, a 5-net short is far more likely to occur than a 50-net short. This implies that it is possible to select a proper E (e.g. 50) which is greater than the extents of a great majority of realistic shorts. Of course, in the unlikely event where the extent of a short exceeded E, and aliasing and/or confounding syndromes were encountered, we can always resort to the optimal adaptive algorithm described in [5] to achieve full diagnosis.

Test Complexity: Previously, we have shown that p, the number of PTVs generated by the Max-Independence algorithm varies almost linearly with E, the maximum defect extent (Figure 1). Fortunately, E in reality is much smaller than N the total number of nets, and can be treated as a constant parameter indicative of a particular character of the manufacturing process (e.g. maximum solder blob size). Therefore, we can assume that p is essentially $O(\log N)$, especially when E is small relative to $\log N$ (see the upper diagram in Figure 2). However, as E gets much larger than

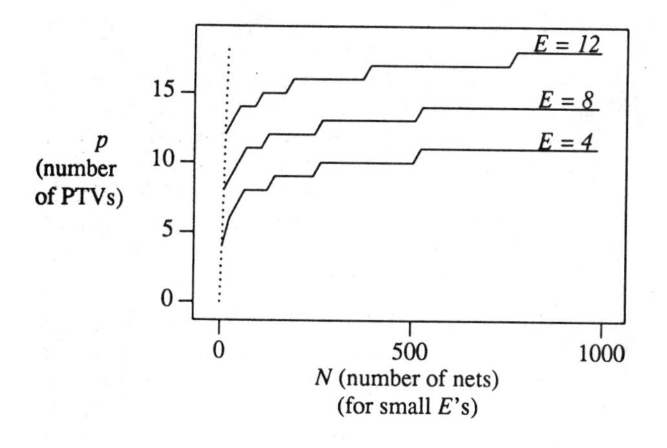

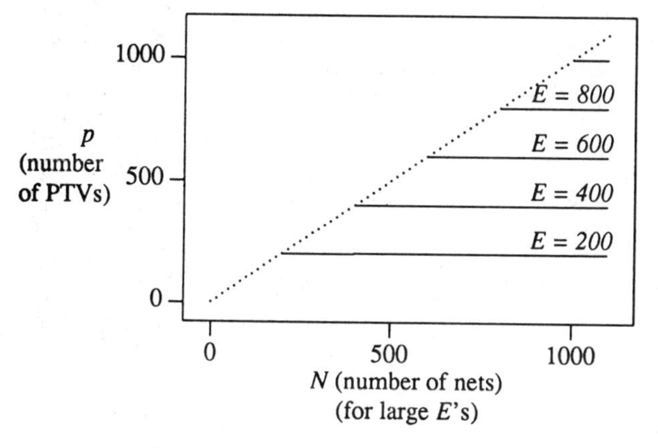

Figure 2. The Effect of Total Net Count on the Number of PTVs

$\log N$, p becomes almost insensitive to variations in N. That is, p remains nearly constant in spite of changes in N (see the lower diagram in Figure 2). Note that in Figure 2 each solid line plots p as a function of N for a typical value of E (e.g. 12). The dotted lines represent $p{=}N$. The values of p are plotted only to the right of the dotted line, where $N \geq E$. (Clearly, the maximum defect extent should never exceed the total number of nets.)

To summarize, in terms of test compactness, the Max-Independence Algorithm is comparable to the Modified Counting Sequence Algorithm; however, in terms of diagnostic capability, it is equivalent to the Walking-One Algorithm.

4. Future Work

The unified theory proposed in this work has transformed test generation for wiring interconnects into a more general problem. This problem involves assigning unique p-bit vectors (STVs) to a set of N nets such that the diagnostic capability of the resultant test is maximized. As mentioned previously, the solution space of this problem is extremely large, and various *heuristic* techniques must be employed to make the problem computationally tractable. Potentially, families of heuristic algorithms can be developed. We have proposed two such algorithms which are based on different assumptions about the availability of certain information (e.g. design and process information). In the future, we intend to explore other heuristic algorithms including those based on the techniques of binary-tree search.

We are convinced that the Max-Independence Algorithm is a very powerful test generation technique for wiring interconnects. However, its effectiveness directly depends on our ability to obtain a "good" adjacency-ordered net set. For such a net set, the extent of a short tracks very closely to the size of the short (e.g. the size of the solder blob). On the other hand, for a poorly adjacency-ordered net set, the maximum defect extent E could become proportional to the total number of nets N. As a result, p, the number of required PTVs may become $O(N)$ instead of $O(\log N)$. Therefore, we intend to develop effective schemes for obtaining good, partially adjacency-ordered net sets. Realizing that most solder bridges affect the adjacent pins of the same device, an obvious and good approximation of net-adjacency-ordering is pin-adjacency-ordering. Note that the latter can be easily derived from existing device libraries.

5. Conclusion

In this paper, we addressed the need of minimizing the test size while maintaining its diagnostic capability. This need is particularly pressing if the board under test is equipped with boundary-scan devices, and the interconnect test vectors have to be applied through the serial scan chain. We proposed a unified theory which reduces interconnect test generation to a more general problem. This problem involves assigning unique sequential test vectors (STVs) to all the nets on the board so that the overall diagnostic accuracy of the test is maximized. With this new approach, it is possible to develop *families* of interconnect test algorithms. We demonstrated that the algorithms reported in the literature are special cases of the two new algorithms proposed in this paper: the Min-Weight Algorithm and the Max-Independence Algorithm. These algorithms were shown to be optimal or near-optimal for a given set of design parameters. In particular, the Max-Independence Algorithm can achieve virtually full diagnosis of shorts with $O(\log N)$ parallel test vectors (PTVs). The new algorithms increase design flexibility by providing a full range of solutions (i.e. test vector sets) based on various trade-off criteria such as test compactness and diagnostic accuracy. We also described some of the trade-off parameters as means for quantifying the quality of the tests. Finally, we discussed the significance and the limitations of the proposed algorithms, and provided some directions for future work.

References

[1] W. K. Kautz, ''Testing of Faults in Wiring Interconnects,'' *IEEE Transactions on Computers*, Vol C-23, No. 4, April 1974, pp. 358-363.

[2] P. Goel and M. T. McMahon, ''Electronic Chip in Place Test,'' *Proceedings International Test Conference 1982*, pp. 83-90.

[3] P. T. Wagner, ''Interconnect Testing with Boundary Scan,'' *Proceedings International Test Conference 1987*, pp. 52-57.

[4] A. Hassan, et al., ''Testing and Diagnosis of Interconnects using Boundary Scan Architecture,'' *Proceedings International Test Conference 1988*, pp. 126-137.

[5] C. W. Yau and N. Jarwala, ''A New Framework for Analyzing Test Generation and Diagnosis Algorithms for Wiring Interconnects,'' *Proceedings International Test Conference 1989*.

[6] *JTAG Boundary Scan Architecture Standard Proposal, Version 2.0*, published March 1988.

A Self-Test System Architecture for Reconfigurable WSI

David L. Landis

University of South Florida
Center for Microelectronics Research
Department of Electrical Engineering
Tampa, FL 33620

ABSTRACT

Progress in Wafer Scale Integration (WSI) has brought the problem of electronic system testing into the semiconductor manufacturing arena. The problem is complicated by the reduced controllability and observability implicit at the full wafer integration level. Structured methods must be employed to generate and apply tests in a hierarchical fashion at the function, chip, and system levels. This paper describes a methodology under development within the WSI program at the University of South Florida which addresses these problems for both the manufacturing and field test environments. A uniform testing interface is defined for each functional chip (cell), with built-in self-test incorporated whenever possible on all new designs. Use of a standard interface will reduce test complexity and costs by allowing entire wafer probing by a common standardized probe card, irrespective of the number of different species of functional cells. Details are provided for the function (cell), chip, and wafer level testing standards as well as for the procedures to be followed at wafer level restructuring and test. The proposed methods will allow current generation wafer restructuring methods to be applied to the next generation of WSI designs requiring numerous cell types and increasing on-wafer complexity.

I. Background

As Very Large Scale Integration (VLSI) technology grows at a rapid rate, the problem of testing state-of-the-art devices is growing even faster. Difficulties associated with chip testing (cost, time, test data volume, tester complexity) have grown because advances in IC technology must slightly precede advances in test technology. Furthermore the trend toward larger levels of on-chip integration have not been matched by increasing package pin count, worsening the accessibility of on-chip circuitry. The size of monolithic integrated circuits has generally been limited by the acceptable yield loss associated with defects within the manufacturing process. Thus while advancing technology has reduced feature size and defect densities to allow higher levels of on-chip integration, the defect, yield, and cost relationships always place a physical limit on the maximum economical chip size. Traditional large scale system designs continue to be implemented using multiple packaged chip assemblies.

* This research is being supported by the Defense Advanced Research Projects Agency under DARPA Grant No. MDA 972-88-J-1006.

An alternative to traditional assembly methods is to develop an IC design and fabrication technology which is capable of tolerating defects. This is accomplished through the careful use of redundant components, along with a means to restructure each fabricated device to circumvent its unique defect pattern. One successful method for providing defect avoidance at the wafer level has been demonstrated at the MIT Lincoln Laboratories. This Restructurable VLSI (RVLSI) technique uses a laser to configure wafer level interconnections, following wafer probe tests which identify defective components [1]. Both additive and deletive interconnections can be made using this technology through the use of fuse and anti-fuse connections on the wafer. Under DARPA support, this technology is being transferred to the University of South Florida in support of wafer scale research which includes technology, architecture, and test activities.

Recent progress in WSI at Lincoln Labs and elsewhere has brought the problems of system level testing into the semiconductor manufacturing arena. The problem is complicated by the reduced controllability and observability implicit at the full wafer integration level. Structured methods must be employed to generate and apply tests in a hierarchical fashion at the function, chip, and system levels. Extensive CAD tool support is mandatory because each manufactured wafer is potentially unique due to its personal defect map. Cells must be tested prior to restructuring to determine the cell level defect map. For WSI designs containing multiple species of cells, this requires multiple probe cards and extensive wafer handling during test. During the restructuring process, the interconnections must also be tested to assure proper connectivity. Only after the restructuring task has been completed can traditional system level functional testing be performed. Failures found at the final system test level are traditionally the most expensive to find and repair, and this becomes even more expensive for RVLSI and WSI because of the inherent difficulties associated with repair at the wafer level.

II. Reconfigurable WSI System Test Requirements

Many of the inherent test problems associated with Wafer Scale Integration could be reduced through limitations upon allowable system designs. It is well accepted that there are substantial benefits associated with restricting the system building blocks to one or a few simple cells [2]. For example, a totally homogenous design (only a single building block) could be wafer probe tested using only a single probe card. Furthermore, the

Reprinted from *IEEE Proceedings 1989 International Test Conference*, pages 275-282.

assignment of logical functions to physical cells is simplified because all cell instances are functionally interchangeable. Unfortunately such homogeneity in system design is virtually unattainable. For even the simplest homogenous processor system design, several different cell types will be required to account for input and output requirements. Restructurable WSI designs currently under development have as many as five cell types [3], and it is easy to envision system requirements for even larger numbers of cell species. Obviously the system level design must balance the requirements of efficient overall design using many function specific cell designs, against efficient wafer level layout and redundant resource utilization which dictates only a few unique cells. This problem will become more acute as the development and acceptance of WSI ushers in an era of application specific wafer scale designs, which will likely require numerous cell species to allow for a wide range of customization.

Given the previously identified problems and requirements for restructurable WSI, the following subsections outline the procedures which must be followed during the three testing phases: *i.) silicon processing/manufacturing test; ii.) reconfiguration and restructuring test; iii.) system test (field test and verification / fault tolerance).*

i.) Processing/manufacturing test: Silicon processing for WSI is generally performed in much the same manner as that for conventional VLSI chip fabrication. The ability to create WSI by placing multiple chip types on a single wafer follows the Mead and Conway pioneered multi-chip project wafer methodology [4]. This technology is readily available to universities and defense contractors through the DARPA funded MOSIS silicon foundry. Upon completion of processing, each unique chip or cell must be independently verified on a VLSI tester. This requires a separate probe card and test vector set for each cell type. A typical procedure would involve setting up a probe card and test program for a particular chip, and then probing all instances of that chip on all untested wafers. Once this test is complete, a new probe card and test vector set is loaded for the next cell type, and the defective devices of this type are identified. This process must be repeated until a complete wafer defect map is obtained. The extreme amount of wafer handling (and corresponding yield loss) associated with designs containing multiple cell species should be obvious. Following the cell tests, a wafer scale interconnect verification test must be performed to identify the defective wafer level tracks which cannot be used for global interconnections.

ii.) Restructuring test: Following test step i.) described above, those wafers with sufficient cell and track yield are packaged to enable electrical circuit connections to be monitored during the laser restructuring process. Given the defect map of a specific wafer and the logical description of the target system design, a logical to physical mapping must be performed to route the wafer scale interconnections. This process is typically assisted by a CAD tool [5] which makes the assignment of logical cells to good cells on the wafer, and then performs the routing of required cell-to-cell interconnections. Once this assignment has been made, the laser restructuring process is used to physically attach all the required system interconnections. Because there is less than 100% yield associated with this restructuring process, it is desirable to perform incremental tests during the restructuring process. Such testing allows defective interconnection links to be quickly identified and re-routed while the capability and wiring resources still exist to perform such an operation. This is currently done in the RVLSI system by using the laser to illuminate a junction within a cell, and then measuring the generated photo-current at an external pad which must pass through the link or cut under test [2]. For interconnection signals which are completely internal to the wafer, extra links and wires must be connected to temporarily route these signals to a wafer pad for test purposes. Following a successful test, these extra connections are deleted using the laser to blow link fuses.

iii.) System test: In a conventional WSI system, complete wafer functional tests are performed totally under external I/O control. Where provisions are made to use extra internal redundancy for field level fault tolerance, the testing and identification of bad cells (and the swapping in of good cells), must be done under the control of off-wafer resources.

III. WSI Self-Test System Architecture

In an attempt to reduce the number and magnitude of wafer scale test problems (as described in sections I and II of this paper), a WSI self-test system architecture is under development at the University of South Florida. This architecture utilizes standardized system test interfaces for Wafer Level test coordination within the framework of a system level Built-In Self-Test (BIST) strategy. Several maintenance network standards have been developed which potentially address the wafer scale test coordination problem. These include the Test & Maintenance (TM) and Element Test & Maintenance (ETM) networks [6], developed primarily for military applications, and the Test Access Port (TAP) defined in the Joint Test Action Group (JTAG) boundary-scan architecture standard proposal [7], endorsed by commercial manufacturers. Under the auspices of the Test Technology Committee of the IEEE Computer Society, a testability bus standardization committee has been formed to develop an IEEE standard for a testability bus. The minimum serial signal subset of this proposed standard (P1149.1) corresponds to the current JTAG test access port (version 2), and provides a standardized serial interface for ATE (automatic test equipment) as well as BITE (built-in test equipment) access. Work is already underway at Honeywell [9], Texas Instruments [10] and elsewhere, to develop chips which can provide a standardized test-bus interface based upon the VHSIC, JTAG, and/or IEEE P1149.1 test bus standards.

Utilizing the P1149.1 (JTAG TAP) standard, an evolutionary path has been defined within our WSI self-test architecture to allow existing components to be added to a WSI design by adding only a simple set of boundary scan I/O pads in addition to the standardized self-test interface. However, supplemental BIST circuitry is recommended for all new cell designs. The basic elements and requirements of this architecture are summarized in section III.a. below. Several WSI cell designs are currently in development which incorporate this architectural standard, and their functional architecture and test features are briefly summarized in the results (section IV).

A. Elements of the WSI Self-Test Architecture

- All WSI cell designs include a standardized test interface {based upon the proposed IEEE P1149.1 (JTAG v.2) standard [8]}
- A common probe pattern is defined for all cells, utilizing a standardized placement of probe pads and test interface signals. This allows a single probe card to be used to test an entire wafer irrespective of the number of different cell species.
- All cells incorporate boundary scan [7] to facilitate internal and inter-chip testing. Where possible, Built-In Self-Test (BIST) will be incorporated on new designs to reduce testing time, cost, and data volume; and to simplify the test generation problem.
- A "standardized" maintenance processor is proposed for future wafer level system designs to facilitate laser restructuring testing and system verification testing. This maintenance processor could also provide the basic support necessary for field level fault tolerance given an appropriate underlying system architecture.

B. Application of the WSI Self-Test Architecture

This section describes how the features defined in section III A.) will be used to facilitate wafer level testing and system verification. Figure 1 illustrates the overall WSI self-test system concept. As indicated, a common seven pin interface is included on each wafer level cell (or chip). By placing the test interface in the corner of the chip layout, a common probe card can be used to probe this interface, irrespective of the actual cell dimensions. Thus a complete wafer test can be performed without physically removing the wafer from the tester or having to interchange multiple probe cards. Naturally, the test process must be organized as a sequence of tests of individual cell types, with delays between these tests to load the unique test program associated with each cell type. Note that the incorporation of BIST within cells can drastically reduce the test time, test program length, test cost, and test data storage requirements.

As indicated, the standard interface includes both power and ground pins for the unit under test. This is because the physical power and ground connections are made during the restructuring process, so that devices with

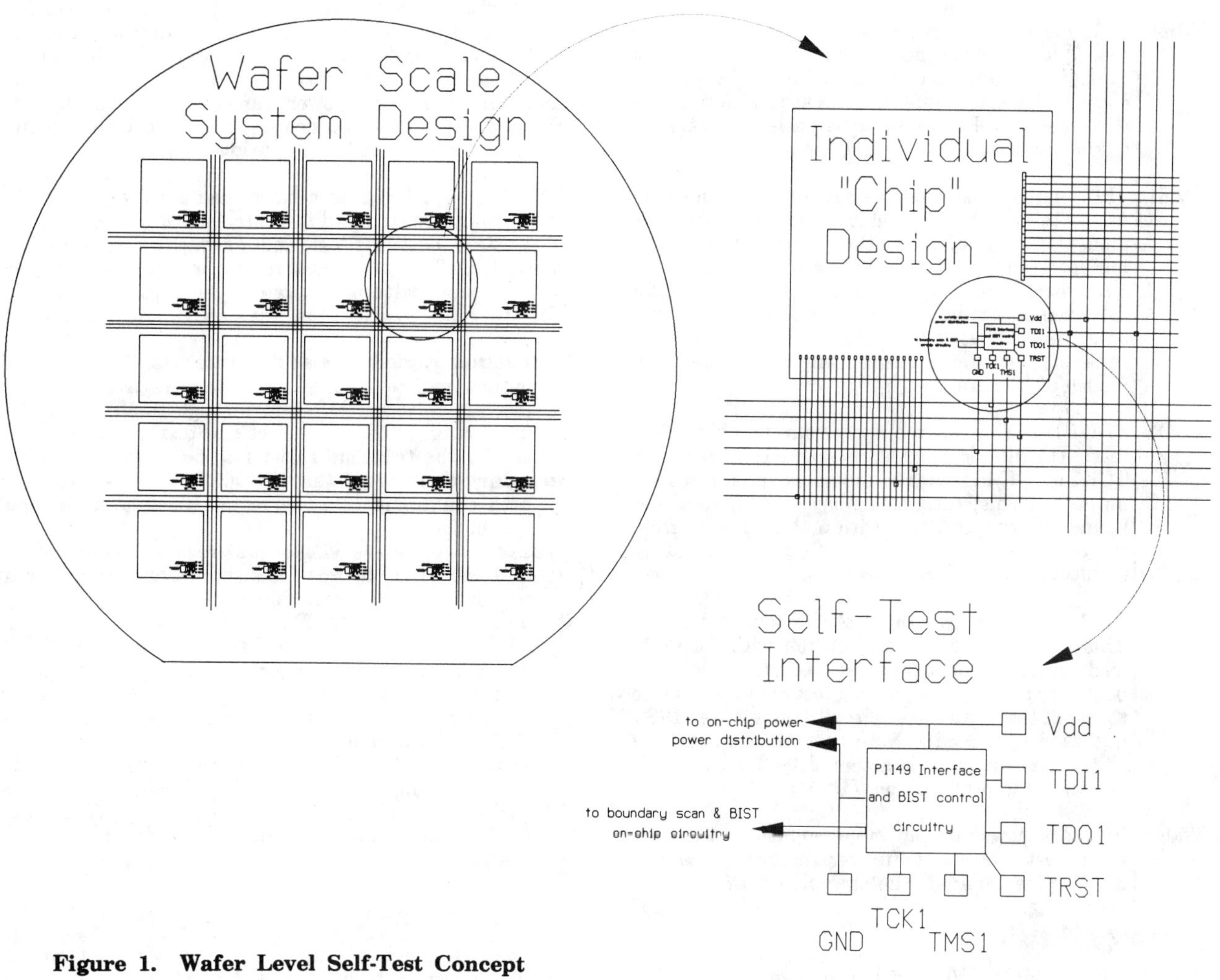

Figure 1. Wafer Level Self-Test Concept

catastrophic Vdd - Gnd shorts do not disable an entire wafer. For the simple WSI cells currently being developed (<10K devices), a single pair of Vdd and Gnd pins will be adequate to provide power during wafer probe testing. However, for more complex devices which require high power / high speed testing, multiple Vdd and Gnd pins will be required at wafer probe. This can be accommodated within the standard by placing additional power pin pairs adjacent to the existing power pins at the edge of the standard test probe card. The signal designations on the self-test interface follow the proposed IEEE P1149.1 standard, with signal definitions as given in Table 1 below.

TABLE 1. Self-Test Interface Pin/Signal Definitions

Signal	Functional Description:(including P1149.1 - JTAGv.2 standard signal definitions)
GND*	GND provides a ground connection for the entire chip under test. Note that after restructuring, this (and other pins around the periphery of logically assigned good chips) will be permanently connected to ground.
TCK1	TCK1 is the test clock input to the chip under test. This signal not only provides the clock control for the test circuitry itself, but under control of the test interface circuitry also provides those chip level system clock signals necessary for test / self-test.
TMS1	TMS1 is the test mode select input for the chip under test. This signal is used to control the modes of operation of the testability circuitry incorporated in the self-test interface. In particular, it is used to enable boundary scan for interconnect testing, on-chip test pattern input and result output, as well as for initiation of built-in self-tests for those devices which include such modes of operation.
TDO1	TDO1 is the test data output from the chip under test. This line is activated in conjunction with the TCK1 and TMS1 lines to provide a serial test data output. This test data may come from the boundary scan registers within the chip, or from the test interface circuitry itself (eg. the signature resulting from a BIST operation).
TDI1	TDI1 is the test data input to the chip under test. This input is used in conjunction with TCK1 to load either functional test patterns into the boundary scan path; or to load test instructions into the self-test interface circuitry (eg. a BIST initiation command). Note that for each cycle of TCK1, one input bit is accepted by TDI1 and one output bit is produced on TDO1.
Vdd*	Vdd is a power supply input voltage to the chip under test. Note that after restructuring, this (and other pins around the periphery of logically assigned good chips) will be permanently connected to Vdd.

* *note: multiple GND and Vdd pins are permitted to accommodate high speed / high power cells*

In addition to a standardization of the self-test interface itself, there are several other advantages associated with the use of this interface. It is not uncommon for a VLSI design to have its overall chip dimensions dictated by the pad size and pitch, especially where large numbers of I/O are required on designs of moderate complexity. Reductions in pad size and pitch are limited by the accuracy and repeatability limitations of wafer probing. However, in the WSI self-test strategy just defined, only those seven self-test interface pads need to be full size probe pads, with the remaining I/O connections only directly connected to global wafer interconnections and the boundary scan path. Thus in some cases there would be a substantial wafer area savings associated with the use of the standardized self-test interface.

An additional advantage is found in the potential for performing closer to at-speed testing using this architecture. Output drivers for WSI designs are typically sized to drive the worst case on-wafer capacitance. However the capacitive loading associated with driving a probe card and test head can be orders of magnitude larger than that encountered on-wafer. In the WSI self-test architecture, only the five common self-test interface signals must logically communicate with the probe card. Furthermore, only one of these signal is an output, and its driver can be appropriately sized to drive the test head load. Consequently, the on-chip test / built-in self-test can proceed at operational speeds, or as limited by the test clock generation and distribution circuitry. Furthermore, additional area and power can be saved through the reduced size of off chip drivers, which no longer need to drive off-wafer capacitances during test.

Figure 2 provides a block diagram level illustration of the complete standard test interface, which is comprised of registers, decoding logic and a sequential state machine controller. The five registers are: instruction, boundary-scan, bypass, pattern generator, and signature analyzer. The individual register functions are defined as follows:

Instruction register - a shift-register stage and a parallel output register. The instruction register allows an instruction to be shifted in through the TDI pin. The instruction is used to select the test to be performed and/or the test data register to be accessed.

Boundary-Scan register - a single shift-register-based path containing cells connected to all module inputs and outputs.

Bypass register - a single shift-register stage between TDI and TDO. It provides a short circuit route for the test-data during a scanning cycle.

Pattern Generator - the Pattern Generator is constructed using a 17-bit Linear Feedback Shift Register (LFSR) which is configured to generate test patterns for the built-in self-test mode. It may also be configured into a shift-register so that an initial seed value may be shifted in through TDI.

Signature Analyzer - forms a signature for the test results during the built-in self-test mode using an LFSR circuit. A seed can be shifted in through TDI and the final signature may be shifted out through TDO after the self-test is completed.

The scan path connections provide access to the internal registers of the circuit, allowing all internal parallel registers to be operated as shift-registers. The number of scan paths may vary according to the structure

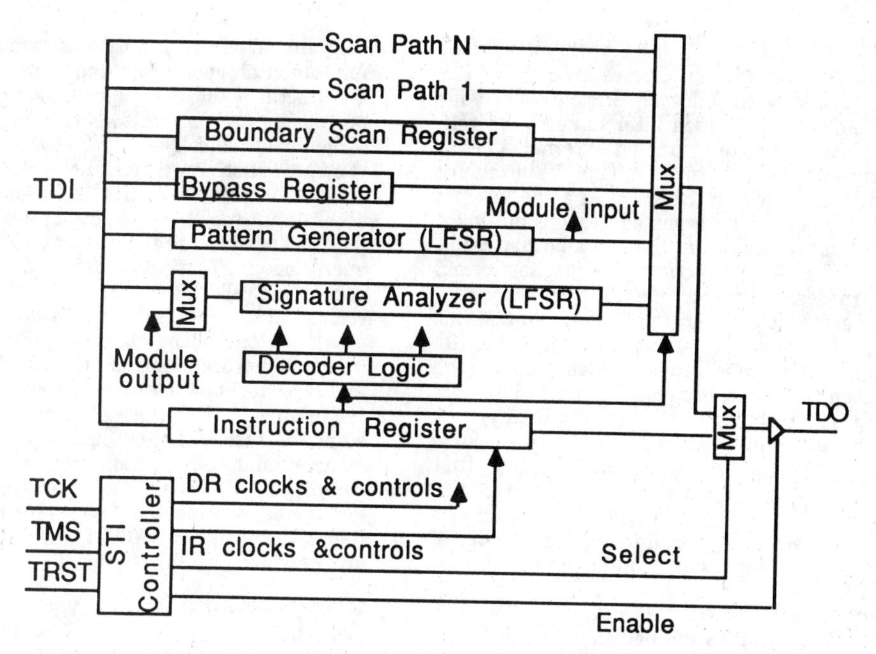

Figure 2. Standard Self-Test Interface

of the design and the total number of on-chip registers. Decoding logic identifies a selected test data register according to the instruction in the instruction register. The unselected registers maintain their previous values. The controller is a synchronous finite state machine which sequences through its various operations under the control of the TMS and TCK signals. This design follows the JTAG Test Access Port Controller specification [7].

The contents of the instruction register and the state of the test interface controller determine the mode of operation of the cell and test circuitry. The various operating modes are generically defined as follows:

Functional - On power-up reset, the instruction register is initialized to the functional command mode. This mode continues until a different instruction is clocked in. Pulling the TRST low for one clock cycle, or holding TMS high for more than 5 clock cycles will also force the controller into the functional mode of operation.

External test - In this mode the interconnects of the wafer scale system design are tested by means of the boundary-scan registers. This is accomplished by scanning data into the output boundary cells of a module, and then observing the inputs of all attached modules using their boundary scan input features.

Sample test - This may occur during the functional mode of operation. The boundary-scan registers sample the input and/or output of the module without interfering with functional operation.

Internal test - In this mode the module is isolated from the other modules on the wafer by means of the boundary-scan registers, and the internal circuitry is tested. This test may be conducted by means of BIST or by shifting in external data scan path data.

Serial Scan - Instructions or test-data are shifted through the 'daisy chain' connection of TDOs to TDIs. However, the destination of the data may be different in each module.

Using the self-test system architecture and interface just described, the WSI manufacturing, restructuring, and system level test procedures, as previously defined in section II, are modified and enhanced as follows:

i.) Processing/Manufacturing test: The 7-pin standardized wafer probe card will be loaded onto the probe station, and the test vector set for the first chip type will be loaded into the ATE memory. For the case of a chip containing BIST, the test program could be as simple as providing a test initiation command, and then reading back the good - bad test result. Optionally, the tester could be required to check the resultant BIST signature against a known good value. In the worst case of a device containing only boundary scan circuitry on each I/O pad, conventional ATPG test data would be loaded in serial fashion into the boundary registers to exercise the device. In this case, the worst case test time would be N times longer that of a conventional broadside (parallel) I/O test; where N is the maximum of the number of bits in the input / output boundary registers. At the completion of a test, the chip position is marked as either good or bad, and the wafer table is stepped to the next instance of the cell whose test program is currently loaded in the tester. Following completion of a particular cell type, the test program for the next cell type is loaded into the tester, and the procedure is repeated for the wafer sites corresponding to that particular cell type. This entire process is repeated until all cell types have been tested and a complete wafer map has been obtained. The defect map for wafer scale interconnections would be created at this time using conventional capacitance measurements, as is currently done [2].

ii.) Restructuring test: Testing during the laser restructuring phase could be performed in a similar fashion to that described in section II. However, an evolutionary goal of the WSI self-test architecture is to allow logic driven interconnect testing concurrent with the restructuring process. This would be performed through coordination of the restructuring laser with an off-chip ATE controller or an on-chip maintenance processor. The process sequence would be as follows: First, a wafer wide test bus would be configured which interconnects the self-test interfaces of each chip to either an external ATE or an on-chip maintenance processor. The Maintenance Processor would be included on the wafer as a special chip instance. This chip would be assigned responsibility for control of the testability bus connections to all other chips, and have direct off-wafer connections for diagnostic and maintenance purposes. An example of a maintenance processor test interface is given in the single chip test-bus interface unit under design reported by Honeywell [9]. Using the boundary scan capabilities of each chip on the wafer, the maintenance processor can be utilized to test each chip-to-chip interconnection as it is restructured. A more realistic and efficient method might be to perform this interconnect testing in a staged fashion, first restructuring logical groups of signals, and then testing the individual groups. These signal groups would be partitioned in such a way as to maximize the probability of being able to reroute a faulty signal interconnection.

iii.) System test: The incorporation of the Maintenance Processor chip as defined in ii.) above can also be used to facilitate system level testing. For an advanced system containing all self-testing chips, the maintenance processor would be used to initiate a self-test of all chips, collect the self-test responses, and signal complete wafer self-health assessment to the external world. Furthermore, if given sufficient intelligence, it could be used to generate test patterns and compress results for those chips which do not have on-chip BIST. In addition, it could be used as a repository for supplemental patterns which would be applied to BIST devices as a means to improve the fault coverage provided by the self-tests.

A complete wafer scale system containing the above described features would be an ideal candidate for the incorporation of field level fault tolerance. Given that redundancy of components is implicit in the wafer scale concept, intelligent partitioning of hard restructured and soft restructured (electrically switchable) redundant resources could be made. If all redundant resources are not consumed during the initial restructuring process, then appropriate architectures could permit the extra resources to be available for field level fault tolerance. If an error is observed (for example, via parity), the entire wafer system could be configured into the self-test mode by the maintenance processor. Following a self-health assessment, the maintenance processor could determine if all assigned resources are functioning correctly (the error was transient), or whether an assigned resource was actually faulty (permanent error). In the later case, if a redundant resource is available via soft switching to replace the failed resource, then this reconfiguration would be performed under the auspices of the maintenance processor. Such an operation could happen at power-on self-test time in a manner which is totally transparent to the system user.

IV. Summary and Status

Under the design, architecture, and applications task of our WSI research project, we are developing cells to support Fast Fourier Transform (FFT) and related signal processing operations. Figure 3 provides a system block diagram of our reciprocal cell which provides one of the fundamental operations required in a high speed array architecture for LU decomposition (a common signal processing task). This is the first of our chip designs targeted specifically for WSI system implementation, and it includes both on-chip Built-In Self-Test and standardized self-test interface circuits. As indicated in the figure, the 15-bit input register includes a pseudo-random 15-bit linear feedback shift register pattern generator mode of operation. This provides exhaustive self-testing by generating all (32K - 1) possible input sequences (the reciprocal of zero is considered separately as a special case). At the output boundary register, a 20-bit multiple input signature register (MISR) mode of operation is included for test result compression (again using LFSR techniques). The figure also indicates the inclusion of the standard test interface circuitry which provides for test control and observation via the test bus. Design complexity for this cell is approximately 7000 transistors, and speed is predicted to be 120ns for a 3 micron CMOS fabrication process. Additional details of this chip design are provided in [11].

Application of the WSI self-test interface to a radically different cell type is illustrated in the block diagram of Figure 4. This MSA (Multiply-Subtract-Add) component provides the primitive computation necessary for the implementation of a pipelined FFT algorithm. Due to the large number of inputs and pipelined nature of this design, exhaustive self-test cannot be practically implemented. Thus a serial self-test is provided which is initialized and controlled from the standard test interface. Additional details of the MSA cell design and on-chip self-test circuitry can be found in [12] and [13] respectively.

Summarizing the costs and benefits of the proposed self-test system architecture for WSI:

- As with all structured test and self-test methodologies, additional circuitry is required on each chip. For the levels of integration characteristic of WSI cells (on the order of 10K devices), the self-test interface represents a minimum percentage overhead (generally < 4%), and the boundary scan register requirements should only add approximately 3% additional penalty [14].

- The addition of boundary registers in all inter-chip signal paths could reduce system throughput, especially in asynchronous data driven system designs. However, all candidate WSI systems currently under evaluation involve synchronous sequential design techniques. In such designs, the use of boundary register is commonplace to provide synchronization and re-timing for global wafer level signals, and no time penalty is incurred.

- Use of a common 7-pin probe card for all WSI designs will significantly reduce the probe card cost (\$0 for new designs after the initial investment). While the use of a serial test interface can produce a test slow-down by a factor of N over conventional wafer probe parallel testing, this can easily be offset in future designs by the incorporation of a significant fraction of self-testing devices on the wafer. Equally important is the reduction in wafer handling (complete wafer probe testing without replacement of probe card) which could have a

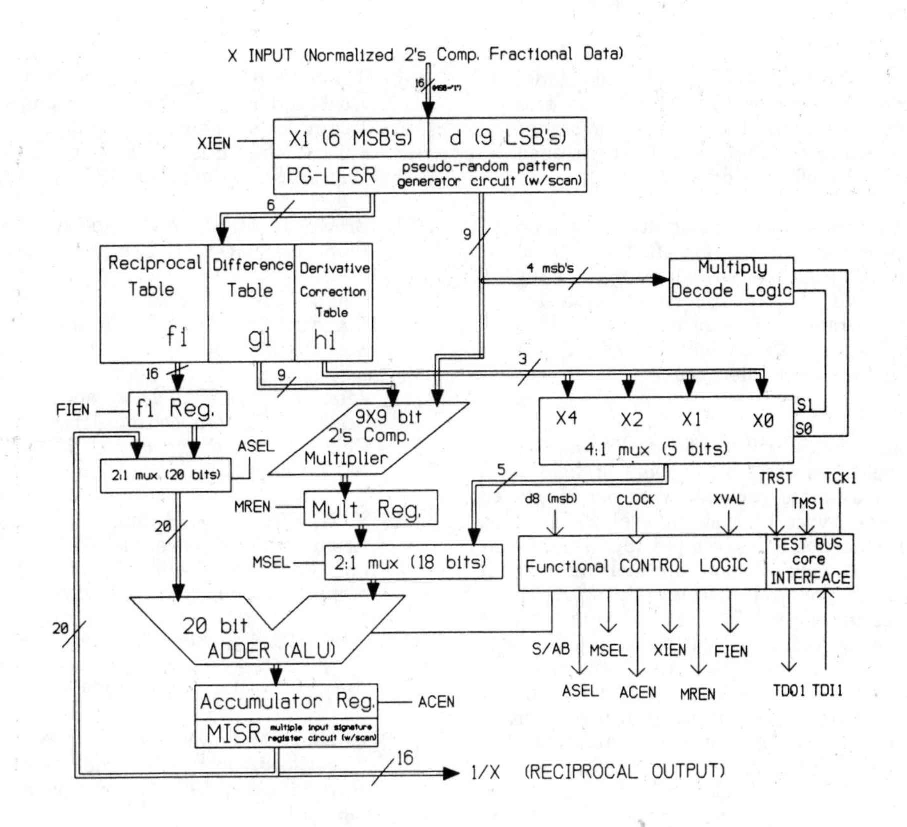

Figure 3. High Speed Reciprocal Cell

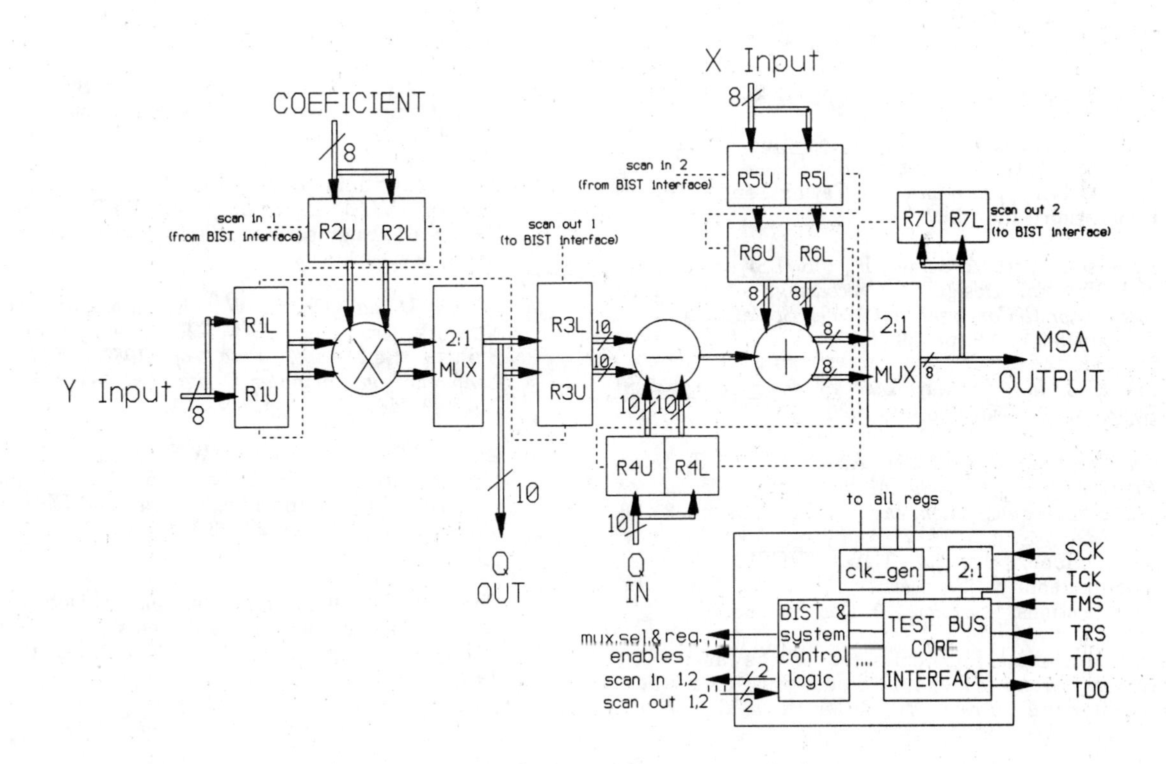

Figure 4. MSA (multiply-subtract-add) Cell

substantial impact on final wafer yield. Additional benefits include area savings (due to I/O pad and driver area reductions), and the potential for higher speed testing because test head capacitances need not be driven by the off-chip drivers.

- The on-wafer standardized testability bus and maintenance processor provide for efficient logical interconnect testing during restructuring. They additionally provide an easy migratory path for including fault tolerance in fielded systems. This is very appealing because of the built-in redundancy implicit in WSI for defect tolerance.

In addition, future work will be directed toward more efficient use of wafer resources in support of field level fault tolerance. For example, previous work has identified optimal amounts and types of system level BIST using computational performance measures [15,16]. Extensions of this work have considered the optimal number of maintenance processors for a self-testing architecture using similar performance measures [17]. Related work has also considered the use of redundant maintenance network connections (redundant standardized test interfaces) to improve overall system reliability and performance [15,16,17]. Results from each of these previous research activities could be included in a second generation WSI self-test system architecture targeted for highly reliable fault-tolerant digital system applications.

References

[1] Wyatt, P. A., & Raffel, J. I., "Restructurable VLSI - A Demonstrated Wafer Scale Technology", *Proceedings of the 1989 International Conference on Wafer Scale Integration*, January 3-5, 1989, pp. 13-20.

[2] Jesshope, C. ed, Wafer Scale Integration, section 5.3, Anderson, A. H., "Computer Aided Design and Testing for RVLSI",pp.216-222, Taylor and Francis Publications, Inc, 1987.

[3] Anderson, A. H. & Berger, R., "RVLSI Applications and Physical Design", *Proceedings of the 1989 International Conference on Wafer Scale Integration*, January 3-5, 1989, pp. 39-45.

[4] Mead, C. and Conway, L., Introduction to VLSI Systems, Addison Wesley, 1980

[5] Frankel, R. et. al., "SLASH - An RVLSI CAD System", *Proceedings of the 1989 International Conference on Wafer Scale Integration*, January 3-5, 1989, pp. 31-38.

[6] IBM, Honeywell and TRW, "VHSIC Phase 2 Interoperability Standards TM / ETM Bus Specifications," *version 2.0*, Dec. 31, 1986.

[7] The Technical Sub-Committee of the Joint Test Action Group (JTAG), "A Standard Boundary Scan Architecture", *version 2.0*, 30 March, 1988.

[8] P1149 T-Bus Standardization Committee (TBSC) Working Group of the IEEE Computer Society's Test Technology Technical Committee, "IEEE Standard for a Testability Bus", *IEEE Std P1149-1989/D8 (Draft Version 8)*, January 31, 1989.

[9] Brown, D. et. al, "A Single-Chip Test-Bus Interface", *Proc. 1988 GOMAC Conference*, pp 565-568.

[10] Mokhoff, N. & Weitzner, S., "TI Boosts Testability in Standard Cell ASICs", *Electronic Engineering Times*, May 8, 1989, pp. 1,8.

[11] Jain, V.K., Landis, D.L., and Alvarez, C. "A Wafer Scale L-U Decomposition Array with a new Reciprocal Cell", *Proceedings of the 1989 IFIP Workshop on Wafer Scale Integration*, Milano Italy, June 6-8, 1989

[12] V. K. Jain, H. A. Nienhaus, D. L. Landis, S. A. Al-Arian, and C. E. Alvarez, "Wafer scale architecture for an FFT processor," *Proc. IEEE International Symposium on Circuits and Systems*, May 7, 1989.

[13] S. A. Al-Arian, "The BIST Structure of the WSI MSA Cell", *Univ. of South Florida - CMR Technical Report*, WSI-T8, May 1989.

[14] Ohletz, M.J., Williams, T.W., & Mucha, J.P., "Overhead in Scan and Self-Testing Designs", *Proceedings of 1987 International Test Conference*, pp. 460-470.

[15] Landis, D., Check, W., & Muha, D., "Influence of Built-In Self-Test on the Performance of Fault Tolerant VLSI Multi-Processors", *Proc. 1987 Intl. Conference on Parallel Processing*, Aug. 17-21, 1987, St. Charles, Illinois, pp. 114-116.

[16] Landis, D. and Muha, D., "Evaluation of System BIST using Computational Performance Measures", *Proc. 1988 Intl. Test Conference*, Sept. 1988, pp 531-536

[17] Muha, D. "Built-In Self-Test Resources for Fault Tolerant VLSI Environments", Ph.D. Thesis in Electrical Engineering, *Penn State University*, December 1988.

[15] Landis, D. and Check, W., "Essential Maintenance Network Issues for Highly Reliable System Level Built-In Self-Test", *Proc. of 1987 IEEE Intl. Conference on Computer Design*, Oct 5-8, 1987, Port Chester, New York, pp. 458-461.

[16] Landis, D. and Check, W., "Built-In Self-Test Maintenance System Impact on VLSI Computer Performability", *Proceedings of the 1988 IEEE VLSI Test Workshop*, March 22, 23 1988, Atlantic City, N.J., pp. 31-40.

[17] Check, W. "Fault-Tolerant Maintenance Networks for Highly Reliable Self-Testing Systems", Ph.D. Thesis in Electrical Engineering, *Penn State University*, Aug. 1988.

DESIGNING AND IMPLEMENTING AN ARCHITECTURE WITH BOUNDARY SCAN

Reprinted from *IEEE Design & Test of Computers*, February 1990, pages 9-19.

R.P. VAN RIESSEN
H.G. KERKHOFF
A. KLOPPENBURG
University of Twente, The Netherlands

The authors describe a standardized, structured test methodology based on the boundary-scan proposal from the Joint Test Action Group, which is now IEEE proposed standard P1149.1. The architecture ensures testability of the hardware from the printed-circuit- board level down to integrated-circuit level. In addition, the architecture has built-in self-test at the IC level. The authors have implemented this design using a self-test compiler.

The problems of testing increasingly complex digital integrated systems continue to challenge the design and test community. At the printed-circuit-board, or PCB, level, these problems led to the formation of JTAG, short for Joint Test Action Group, a collaborative organization of major semiconductor users in Europe and North America. JTAG subsequently developed the boundary-scan standard[1] with the goal of improving the controllability and observability of an IC's primary inputs and outputs. Because of this standard, which is now IEEE proposed standard P1149.1, we can now easily implement testability hardware using computer tools, which reduces overall design time.

However, boundary scan does not address testability at the IC level—primarily because there is no standard for designing BIST circuits. At this level are many approaches to adding testability, but the one that seems most promising for future VLSI and ULSI circuits is built-in self-test, or BIST.[2] In BIST, test data is generated and evaluated on the chip.

In this article, we present an architecture called hierarchical testable, or H-testable, architecture for integrating boundary scan at the PCB level and BIST at the IC level. We believe that this integration is important because the boundary-scan standard defines access to the IC during the IC test. The extra test pins let us control on-chip testability hardware.

A digital system has several levels of hierarchy. First, we have the PCB level, which contains such items as a Winchester control board. The second level is the IC level, where we have units such as a microprocessor chip. The third level, called the macro level, allows us to make finer distinctions between functional modules like PLAs and ALUs, for example—the so-called macros.[3] We use these three levels to define the H-testable architecture. With this hierarchical structure, we can use BIST for a macro at the higher levels and so more completely integrate the testability features at the IC and PCB levels.

As we show in more detail later, the H-testable architecture can be implemented using a self-test compiler.[4] This compiler automatically generates the layout of a macro, including hardware to generate data for and evaluate the results of self-test.

Test interface elements are located at the primary inputs and outputs of both macros and ICs. Each TIE contains boundary-scan cells and serial control registers.

ARCHITECTURAL REQUIREMENTS

To be H-testable, an architecture must have certain characteristics. First, it must be hierarchical because, as we just mentioned, the hierarchical approach allows us to use test results from a lower level in higher levels. The results of a macro test can be used for the IC test, for example. Second, the architecture has to be standard because PCB manufacturers use ICs from different vendors on a single board. If we have well-defined test-interface rules and control definitions for every level of hierarchy, we can use standard test approaches. Third, the architecture has to be structured also to reduce extra design time. With a structured approach, test hardware is developed only once and is reusable. A structured approach also facilitates the design by allowing us to use computer tools. Finally, to be H-testable, the architecture must incorporate the BIST facilities of different macros.

INTEGRATION WITH BOUNDARY SCAN

Figure 1 shows the JTAG boundary-scan architecture for PCB testing. The behavior and architecture of all blocks in this figure are defined in the standard.[1] We use this architecture to define our H-testable architecture.

Figure 2 is a schematic block diagram of the H-testable architecture at the IC level. In this diagram, we can distinguish two levels of hierarchy: the macro level and the IC level. At the IC level are (self-)testable macros, connections between these macros, and additional testability hardware. The macro level consists of a (self-)testable macro with additional testability hardware. Both levels of the testability hardware incorporate test interface elements, or TIEs; a test processor; and a scan path.

The TIEs separate a macro (IC) from the connections with other macros (ICs). Therefore, TIEs are located at the primary inputs and primary outputs of both macros and ICs. Each TIE contains boundary-scan cells and serial control registers. Test processors provide parallel control of the TIEs.

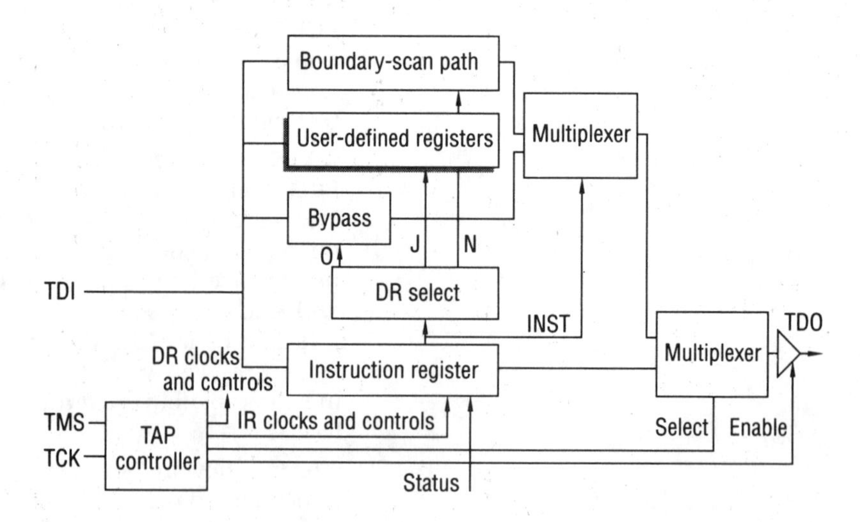

Figure 1. *Architecture of the Joint Test Action Group boundary-scan standard.*

The IC test processor provides the TIEs at the IC level and the macro test processors with parallel control. The macro test processor provides the TIEs at the macro level with parallel control. The macro test processor can also control a macro self-test.

We can merge the JTAG boundary-scan and H-testable architectures without any changes to either. Thus, the H-testable architecture is already defined at the IC level.

ARCHITECTURE AT THE IC LEVEL

At the IC level, the H-testable architecture is compatible with the boundary-scan architecture and its behavior. Therefore, we have in effect merged the H-testable architecture with the JTAG boundary-scan architecture as evidenced by the following structural characteristics:

- The JTAG boundary-scan path in Figure 1 is part of the boundary-scan cells of the TIEs at the IC's input and output in Figure 2.
- The JTAG instruction-register path is implemented in the IC-level test processor. The registers in this path provide the serial control data for the IC-level TIE.
- The JTAG test-access port (TAP) controller is implemented in the IC-level test processor of the H-testable architecture. The TAP controller generates the parallel control signals for the IC-level TIEs and the macro-level test processors.
- The JTAG user-defined register path is used to implement the local scan path in Figure 2.

We can merge the architecture of the JTAG boundary-scan standard and the H-testable architecture without any changes to either. Consequently, at the IC level, the H-testable architecture has already been defined.

ARCHITECTURE AT THE MACRO LEVEL

The test hardware for the H-testable architecture at the macro level consists of TIEs and a macro test processor, as Figure 2 shows.

A TIE in the local scan path forms the link between a macro and the macro interconnection. We add this element only to enhance testability. The TIEs are located at both inputs and outputs of a macro and do not affect the functional behavior of the IC during normal operation.

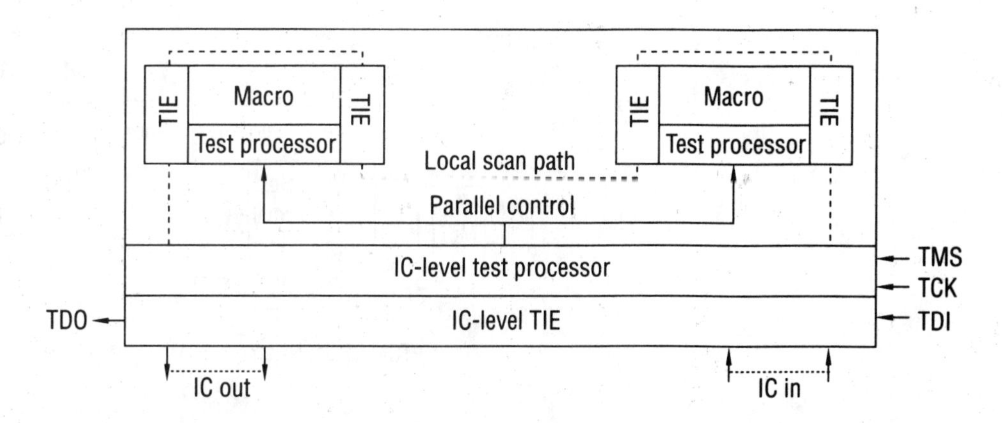

Figure 2. *Block diagram of the hierarchical testable (H-testable) architecture at the IC level.*

Data-register cells form the interface of the macro and macro interconnections. DRC modes vary according to the control signals applied to the multiplexers.

During an IC test however, the TIEs are able to separate macros from their interconnections, which allows an independent test of both. Test patterns are shifted serially into the TIE via the local scan path, and the TIE applies the patterns in parallel to the macro or to the interconnection of macros. Results from a macro self-test are loaded in parallel into the TIEs at the output of the macro. Results from the macro interconnection test are loaded in parallel into the TIEs at the input of a macro. Next, data m the TIE's will be shifted out serially via the local scan path. Control signals for the TIE are applied serially via the local scan path and in parallel via the control signals from the macro-level test processor.

Figure 3 shows the implementation of the TIE. A TIE consists of data-register cells *D*, two control-register cells (*M* and *S*), a bypass path, and a multiplexer.

Data-register cells form the interface of the macro and the interconnections to other macros. Figure 4a is a block diagram of one of these cells. This cell consists of two multiplexers and a master-slave register. The macro-level test processor provides the signals Mode, DRC1, and DRC2. The data-register cell is used in different modes, which vary according to the control signals applied to the multiplexers. Figure 4b shows the truth table of multiplexer 2. The first mode

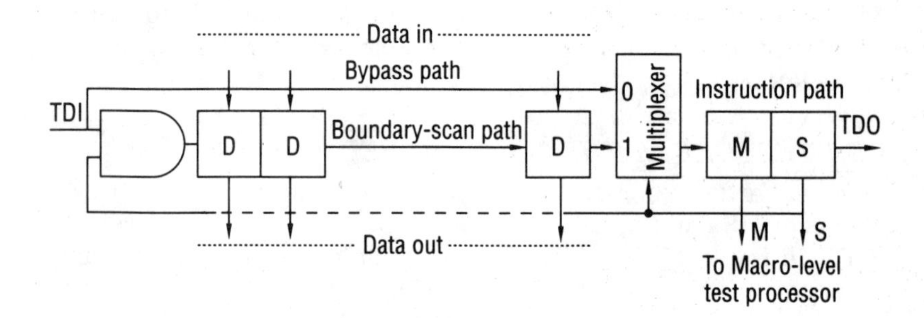

Figure 3. *Implementation of a test interface element, or TIE; D is the data-register cell and M and S are control-register cells.*

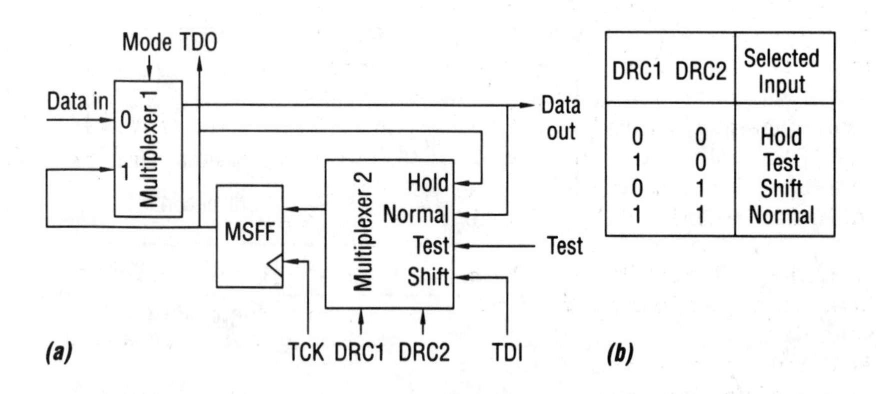

DRC1	DRC2	Selected Input
0	0	Hold
1	0	Test
0	1	Shift
1	1	Normal

Figure 4. *Block diagram of a data-register cell **(a)** and the truth table for data-register multiplexer 2 **(b)**.*

of the data-register cell is the hold mode (DRC1=0, DRC2=0), in which data in the register remains unchanged. The second mode (DRC1=1, DRC2=0) is the test mode, in which the input Test is used for BIST. The third mode (DRC1=0, DRC2=1) is the scan mode, in which the cell is placed in the local scan path at the IC level. Figure 2 shows this path. We can now shift data into input TDI and towards output TDO. The fourth mode (DRC1=1, DRC2=1) is the normal functional operation. Data enters the cell via the input Data-In and propagates through the cell with minimal delay to Data-Out.

At the PCB level, the JTAG boundary-scan architecture requires two scan operations to initiate the test interface elements: shift in instruction bits and shift in data bits.

The control-register cells in the instruction path of a TIE (*M* and *S* in Figure 3) provide its serial control. These registers consist of a shift register (L2) and an output latch (L1). Figure 5a is a block diagram of an instruction-register cell. The TAP controller of the IC-level test processor supplies the control signals Update-IR, IRC1, and IRC2. At the rising edge of Update-IR, the contents of shift register L2 are loaded into the output latch L1. The signals IRC1 and IRC2 control which input is selected by the multiplexer. Figure 5b shows the truth table of this multiplexer. The input Hold (IRC1=0, IRC2=0) is selected to retain the data in the output latch L1. The input Status (IRC1=1, IRC2=0) is required to load a signal into the shift register. The input Shift (IRC1=0/1, IRC2=1) is the serial scan input. This input is connected to the output TDO of the previous shift-register cell.

Because TIEs are at both the input and output of a macro, there are two mode registers—Ml at the input, M2 at the output—and two select registers—S1 at the input and S2 at the output. These four instruction registers can define 16 modes for the data-register cells.

The select register *S* in Figure 3 controls the bypass of the data-register cells. The data-register cells in a TIE are placed in the local scan path if *S*=1. If *S*=0 the scan path of a TIE contains only the instruction-register cells.

The value in the mode register *M* is decoded in the macro-level test processor and, together with parallel control signals from the IC-level test processor, controls the two functions of the data-register cells. In Figure 4a the data-register cells transmit data if mode=1 and receive data via input Data-In if mode=0.

Figure 3 shows, in contrast with the JTAG architecture, that the boundary-scan path and the instruction path are connected serially. With this architecture at the macro level, we can use a simple multiplexer to select either the bypass mode or the boundary-scan mode. Because both modes include the instruction path, a data scan will always contain data bits and instruction bits. We need only one scan operation to initialize the TIEs for a macro test. At the PCB level, the JTAG boundary-scan architecture requires two scan operations to initiate the TIEs. In the first stage, the instruction bits are shifted in. In the second stage, the data bits are shifted in.

Another difference between the macro-level TIE and the IC-level TIE is the number of modes that a data-register cell has. The boundary-scan data-register cell at the IC level has three modes of operation. At the macro level, it has four modes. As we mentioned earlier, this additional mode is the test mode, which allows the data-register cell to be used for BIST. This mode does not require an extra control signal as compared with the boundary-scan register cell.

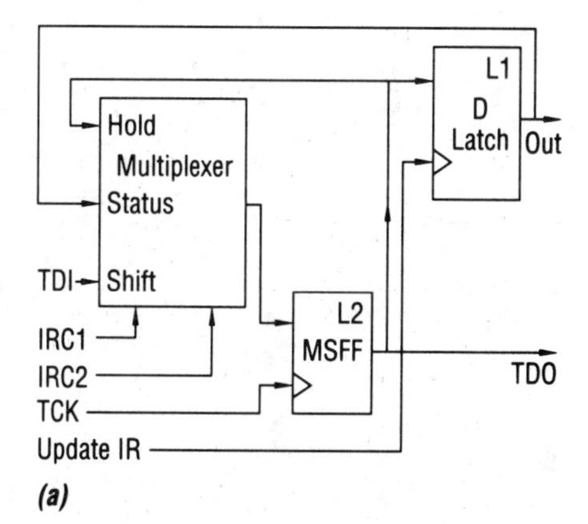

IRC1	IRC2	Selected Input
0	0	Hold
1	0	Status
0	1	Shift
1	1	Shift

(b)

Figure 5. *Block diagram of an instruction-register cell **(a)** and the truth table for the instruction-register multiplexer **(b)**.*

Central to the H-testable architecture is the self- testable macro, which has only combinational logic.

THE MACRO-LEVEL TEST PROCESSOR

The test processor forms the control part of the H-testable architecture. At the macro level, the processor has to perform a macro self-test and apply the parallel control signals to the data-register cells of the TIEs at both the input and output of the macro.

To carry out BIST, we must generate test patterns and compact them using some hardware implementation of a test-pattern generation/compaction algorithm. Test patterns are applied in parallel to the macro inputs by loading the test patterns in the data-register cells via the extra Test input, as Figure 4a illustrates. The test result is loaded in parallel into the TIE at the output of the macro.

During the self-test, the macro test processor generates the parallel control signals for the data-register cells of the TIE. Figure 6 is

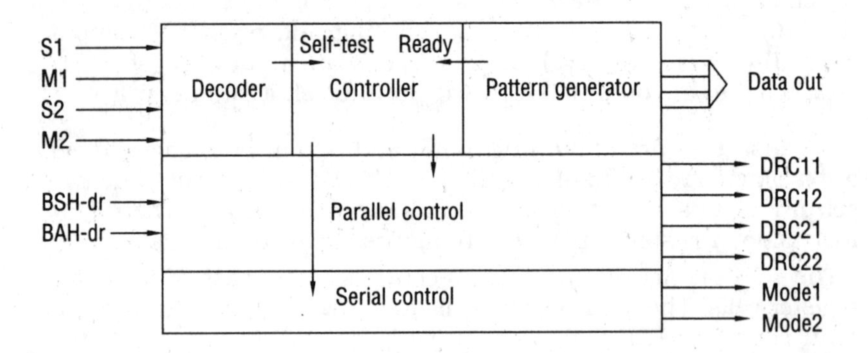

Figure 6. *Block diagram of a macro-level test processor.*

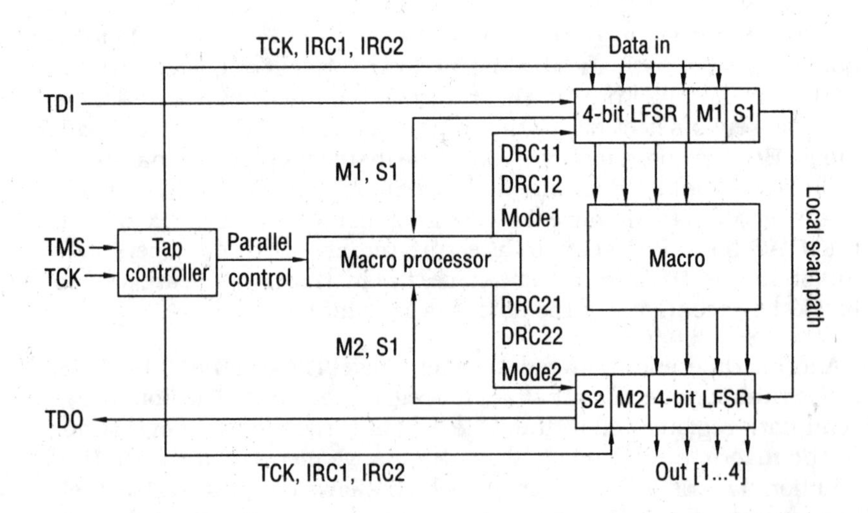

Figure 7. *Example of the hierarchical testable, or H-testable architecture.*

a block diagram of the macro-level test processor. We briefly describe the main parts of the test-processor architecture. A more detailed description is available elsewhere.[5]

Parallel and serial control logic supply the data-register cells with control signals. The signals DRC11, DRC12, and Mode1 form the signals DRC1, DRC2, and Mode for the TIE at the input of a macro. The signals DRC21, DRC22, and Mode2 form the signals DRC1, DRC2, and Mode for the TIE at the output of a macro. These signals depend on the state of the controller and on the state of the IC-level TAP controller (BSH-dr, BAH-dr).[1]

A decoder signals the controller to start a macro self-test. The self-test is activated by the contents of the registers in the instruction path at the input (S1 and M1) and the output (S2 and M2) of the macro. A controller, which is, in fact, a synchronous state machine controls the macro self-test. We can, however, implement a macro self-test in many ways, depending on the type of macro. Therefore, we have a dedicated controller for each macro. Every controller must be able to start the self-test, indicate the end of a self-test, and control the registers involved.

A pattern generator, which is governed by the controller, generates the test patterns for the macro. The pattern generator uses the TIE's data-register cells at the input of the macro. The generated patterns (Data-Out) are applied to the Test inputs of the data registers. The pattern generator generates a signal (Ready) for the controller to indicate the completion of a self-test.

A pattern generator, which is governed by the controller, generates the test patterns for the macro.

AN H-TESTABLE ARCHITECTURE

The best way to illustrate the features of the H-testable architecture is to describe an actual implementation. Figure 7 shows the example we have used. Our intent is primarily to show the integration of boundary-scan hardware with BIST at the IC level. Our example incorporates two TIEs, a macro test processor, one TAP controller, and a simple macro.

The central part in the architecture is the self-testable macro, which has four inputs and four outputs. This macro contains only combinational logic and is tested with pseudorandom patterns. A signature analyzer compacts the test results. We have added some hardware to the data-register cells of the TIEs so that we can use the data register as a building block for pseudorandom pattern generation and signature analysis. Figure 8 illustrates the additional hardware.

We form the pattern generator/compactor by connecting a number of modified data-register cells as a linear-feedback shift register. To do this, we feed Data-Out of the last register cell back to the Feedback terminal of specific data-register cells. The connections are determined by the feedback polynomial.[6] We can use the structure as a pseudorandom pattern generator when (DRC1, DRC2)=(1,0) and mode=1. The circuit operates as a signature analyzer when (DRC1, DRC2)=(1,0) and mode=0.

In our example, the TIEs form an LFSR during the test mode that has a feedback polynomial of $1+X+X^4$. Figure 9 shows the data register of a TIE realizing this LFSR.

The macro-level test processor also incorporates the logic to start and complete the self-test. The Ready signal, which indicates the completion of the self-test, is true when a specific test pattern is generated.

We have added some hardware so that we can use the data register as a building block for pseudorandom pattern generation and signature analysis.

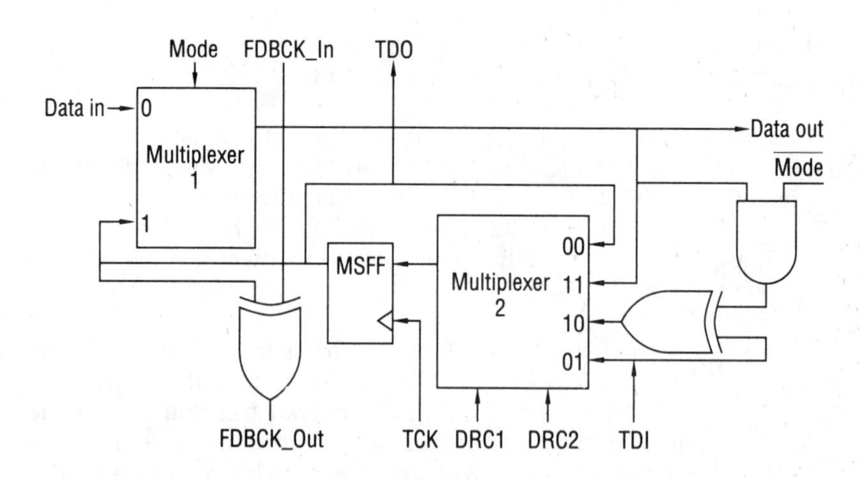

Figure 8. *Block diagram of a modified data register.*

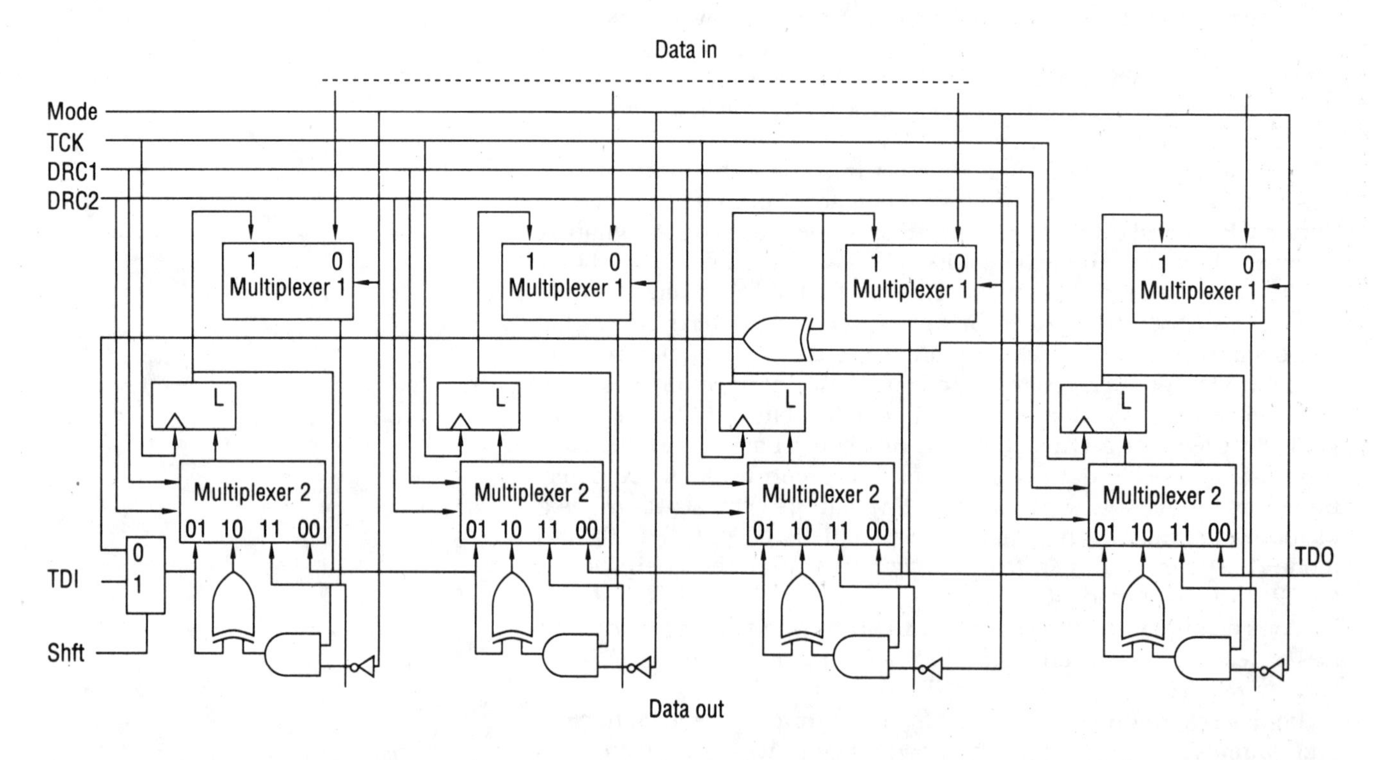

Figure 9. *Block diagram of the data register part of the test interface element, or TIE, that forms a four-bit linear feedback shift register (LFSR) with feedback polynomial $1+X+X^4$.*

The TAP controller is identical to the TAP controller as described in version 2.0 of the boundary-scan standard.[1]

SIMULATION OF THE SAMPLE CIRCUIT

Table 1 shows the scan actions applied to the example circuit during simulation. Scan action 1 initializes the instruction-register cells. Four clock cycles shift the values for the initialization path into the instruction-register cells: (M1,S1,M2,S2) = (1,1,0,1). Because S1=1 and S2=1, we can initialize the data-register cells at both the input and output.

Scan action 2 initializes the data-register cells for a macro self-test. Both data-register cells are initialized with the value (1,1,0,0). Scan action 3 indicates that during a macro self-test, we can still shift data through the TIEs. While the two TIEs perform a macro self-test, a pattern is shifted via the input TDI to the output TDO.

We need scan action 4 to place the data-register cells in the scan path after the macro self-test has been completed. Finally, with scan action 5, the signature in the output TIE appears at the serial output TDO.

We simulated the test process for this sample circuit using a switch-level description.[7] The results of the simulation[5] show the correct operation of the H-testable architecture. A layout for the individual blocks of the H-testable architecture has since been designed and will be used in our self-test compiler.

With the self-test compiler, designers define only the type and size of the macro to be realized and the fault coverage they desire for the self-test.

A SELF-TEST COMPILER

As we mentioned earlier, the purpose of the H-testable architecture is to develop a standard, hierarchical test approach to ease the burden of test development. Towards that end, we implemented our architecture in a self-test compiler.[4] The compiler automatically generates the layout of the most appropriate on-chip test hardware for self-testing along with the functional macro. Designers define only the type and size of the macro to be realized, along with the fault coverage they desire for the self-test. Using the described architecture, the compiler generates self-testable macros that we can control in a standardized format. The H-testable architecture defines the signals to initialize, control, and verify a macro self-test from the macro level to the PCB level.

Table 1. *Tests applied to the sample circuit in Figure 9.*

No.	Scan Action	Instruction M1	S1	M2	S2
1	Select initialization path	1	1	0	1
2	Initialize data and instruction register	1	0	0	0
3	Scan operation during self-test	1	0	0	0
4	Select result path	0	1	0	1
5	Verify test result and scan in pattern for external test	0	1	1	1

For small macros of say, 10 to 20 I/O ports, the controllers will probably determine the overhead of the extra test hardware.

Figure 10 shows part of the layout of a self-testing carry-save array multiplier, which was generated by the self-test compiler. The self-test, performed in this particular structure, is an exhaustive test. We used a signature analyzer to evaluate the test responses. The bottom row of cells in the figure shows the layout of some data-register cells used for data compaction.

The overhead needed for the extra test hardware varies with the size of the array multiplier. For a 16×16-input carry-save array multiplier, for example, the overhead is about 20%. For a 32×32-bit array multiplier, the overhead is about 12%.

The H-testable architecture we have described will ease the problems of testing ICs on printed-circuit boards. It is hierarchical, structured, and compatible with the JTAG boundary-scan standard for PCB testing. Using this architecture, we can initialize, control, and verify a macro self-test from the IC level up to the PCB level. During a macro self-test, the IC-level scan path can still be used, which implies that we can test different macros in

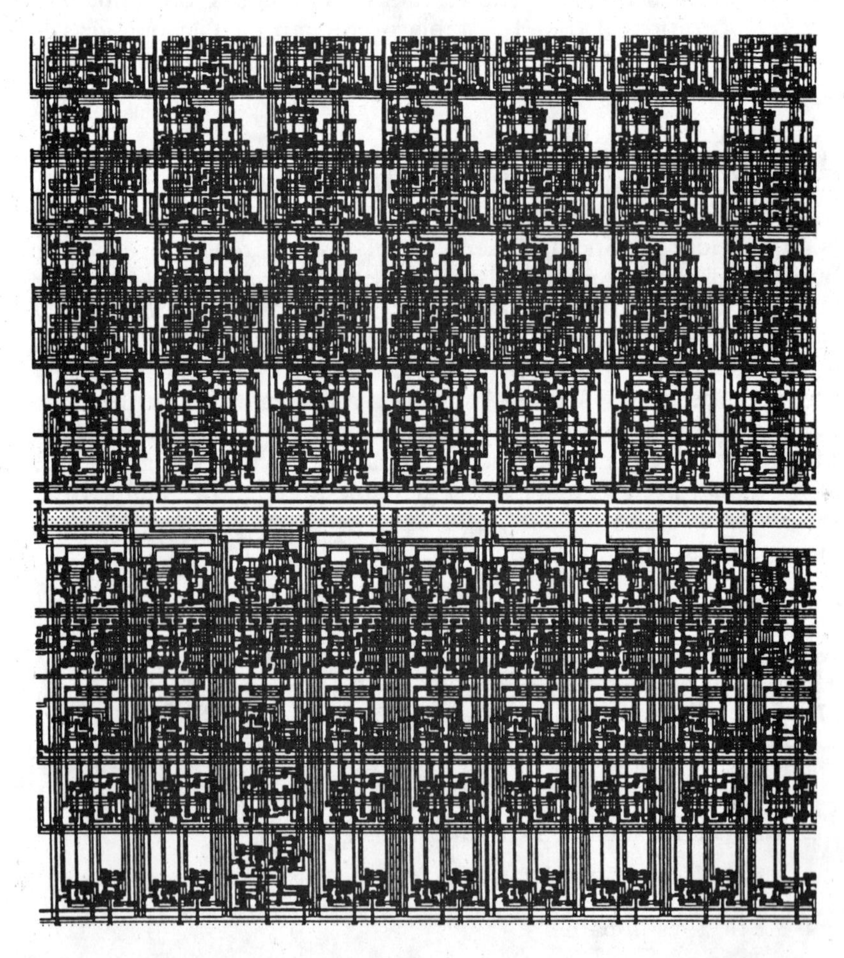

Figure 10. *Part of the layout of a self-testing carry-save array multiplier, which was generated by the self-test compiler. The bottom row of cells shows the layout of some data-register cells used for data compaction.*

parallel with the H-testable architecture. We have implemented this architecture in a self-test compiler. An example circuit, generated by this compiler, shows the possibilities of this architecture.

The overhead of the extra test hardware remains a problem that needs more research. For small macros of say, 10 to 20 I/O ports, we expect the controller parts of the H-testable architecture to determine the overhead. Therefore, we advise the use of only one macro test processor for a set of small self-testable macros. D&T

ACKNOWLEDGMENTS

This research is supported by the Innovative Research Program (IOP) on IC technology under HTO-049/1, part testing. We thank F. Beenker and his group at Philips Research Laboratories, Eindhoven, for their contribution to this research.

REFERENCES

1. *A Test Access Port and Boundary Scan Architecture*, JTAG standard, Version 2.0, Draft 2, Jan. 1988.
2. B. Koenemann, "Built-In Logic Block Observation Techniques," *Proc. Int'l Test Conf.*, 1979, pp. 37-41.
3. F. Beenker et al., "Macro Testing: Unifying IC and Board Test," *IEEE Design & Test of Computers*, Vol. 4, No. 6, Dec. 1986, pp. 26-32.
4. R. van Riessen, H. Kerkhoff and A. Kloppenburg, "Design of a Compiler for the Generation of Self-Testable Macros," Proc. European Solid-State Circuits Conf., 1988, pp. 194-197.
5. A. Kloppenburg, *Investigations on the Possibilities of Using Boundary-Scan Techniques in Silicon*, MSc thesis, Univ. of Twente, 060-7001, Enschede, The Netherlands, July 1988.
6. A. Miczo, *Digital Logic Testing and Simulation*, John Wiley & Sons, New York, 1987.
7. A. van Genderen, A. de Graaf, "SLS: A Switch-Level Timing Simulator," in *The Integrated Design Book*, Delft Univ. Press, 1986, The Netherlands, pp. 2.93-2.145.

Ronald P. van Riessen is working towards a PhD in electrical engineering at the University of Twente, Enschede, The Netherlands, where his research interests include automatic design for testability and BIST of CMOS VLSI systems, particularly boundary scan with BIST in a module compiler. He holds an MSc in electrical engineering from the University of Twente.

Hans G. Kerkhoff is a staff member of the IC Technology and Electronics Group of the University of Twente, where his interests are in testable CAD for VLSI digital signal processors. He holds an MSc in electronical engineering from the Technical University of Delft and a PhD in microelectronics design from the University of Twente.

Ad Kloppenburg is a staff engineer at Witteveen en Bos, an engineering consultancy in Deventer, The Netherlands. He holds an MSc in electrical engineering from the University of Twente.

Direct questions or comments on this article to R. van Riessen, University of Twente, EF Bldg., PO Box 217, 7500 AE Enschede, The Netherlands.

Reprinted from *IEEE Proceedings 1990 International Test Conference*, pages 222-234.

A Language for Describing Boundary-Scan Devices

Kenneth P. Parker and Stig Oresjo
Hewlett-Packard Company, Manufacturing Test Division
P. O. Box 301, Loveland Colorado, 80537

ABSTRACT

Boundary-Scan (IEEE Standard 1149.1-1990) technology is beginning to be embraced in chip and board designs. One key need is a way to simply and effectively describe the feature set of a Boundary-Scan compliant device in a manner both user friendly and suitable for software to utilize. A language subset of VHDL is proposed here for this purpose. As with any new standard, the industry is learning how to apply its rules and mistakes will occur. A derivative effect of the language proposed here is that if a device is not describable by the language, then that device does not comply with the 1149.1 standard. While the converse is not true, the language still allows a syntactic check for compliance as well as a number of semantic checks.

INTRODUCTION

IEEE Standard 1149.1-1990[1] was approved in February 1990, and is now available from the IEEE. The Boundary-Scan concept was formally investigated by the Joint Test Action Group, a consortium of European and North American companies starting in 1985, and is often refered to as the JTAG Standard. The standard is rich in options and is open-ended in that user defined features are provided for. This richness can be a source of complication that must be accounted for while utilizing the standard. The testability enhancing attributes of the standard are quite powerful. Many of the barriers that have slowed the adoption of testability technology[2] are directly overcome by Boundary-Scan[3]. For these reasons, expect to see widespread application of the standard. In this paper it is assumed the reader has a passing knowledge[1][3][4] [5] of Boundary-Scan.

As new products become available to support Boundary-Scan designs, each will have the problem of how to describe a designer's unique application of the standard. Some sort of description will be necessary for each device containing Boundary-Scan. This paper describes a language that captures the essential features[6] of an implementation. This language is called the Boundary-Scan Description Language (BSDL) and is written within a subset of the VHSIC Hardware Description Language (IEEE Std 1076-1987 VHDL[7]). It has two criteria to meet: first that it be 'user friendly', since people will have to create the files; and second, it should be simply and unambiguously parsable by computer. This proposal is intended to be a 'straw man' or 'Version 0.0', illustrating a structure and illuminating needs.

It is important to note that the language described here is necessarily evolving. However, it represents a consensus developed from discussions[6] with many individuals within various sectors of the electronics industry as noted in the acknowledgements. Several groups had already begun their own development efforts on proprietary languages suited to their individual needs; of note, AT&T, Hewlett-Packard, Philips, and Texas Instruments. In particular, the Philips work is part of an effort supported by the multinational European Commission ESPRIT Project 2478. It is now the intention that this European activity will merge with this proposal. This process is now underway and in this respect, this proposal reflects both North American and European thinking. While this language definition is expected to change as applications develop, it is our hope that the resulting evolution will differ in minor ways, with a goal of upward compatibility. Thus, software tool developers can make use of this proposal now rather than continue to wait. In so doing they will benefit from compatibility with other segments of the industry. Ultimately, this language should be taken over and maintained by a body devoted to standards, such as the IEEE.

THE SCOPE OF THE LANGUAGE

The BSDL language allows description of the *testability features* in IEEE Std 1149.1-1990 compliant devices. This language can be used by tools that make use of those testability features. Such tools include testability analysis, test generation and failure diagnosis. Note that BSDL itself is not a general purpose hardware description language. With a BSDL description of a device *and knowledge of the standard*, it is possible for tools to completely understand the data transport characteristics of the device. With additional capabilities provided by VHDL, it is possible to perform simulation, verification, compliance analysis, and synthesis functions. Support for these functions is beyond the scope of BSDL alone.

A key characteristic of a BSDL description of the parameters of an implementation is orthogonality to the rules of the standard. As a result, elements of a design absolutely mandated by the standard are not included in BSDL descriptions. For example, the BYPASS Register is not described in BSDL because it is completely described by the standard itself, without option. This eliminates both redundancy and the opportunity for error.

BOUNDARY-SCAN CHARACTERISTICS

What are the characteristics of any Boundary-Scan device that need description? All such devices must have two major features; a Test Access Port (TAP) and a Boundary Register. The parameters of these features are described by BSDL.

The parallel/serial Boundary Register is made up of Boundary Cells which are associated with device inputs, device outputs, device bidirectional signals, and specific embedded device control signals. A great deal of the flexibility of the standard is reflected in the Boundary Register rules.

The TAP possesses either four or five dedicated signals, familiarly labeled TCK, TMS, TDI, TDO and, optionally, TRST*. It must contain an instruction register and a BYPASS register. The TAP implements a minimum set of mandatory instructions which control operation of the Boundary-Scan facility. These instructions operate in conjunction with the dedicated TAP signals in a precisely prescribed way. The TAP may also contain optional data registers and optional instructions as specified by the 1149.1 standard. Additionally, the TAP may also be endowed by a device designer with additional user-specified data registers and instructions beyond those specified by the standard, but governed by rules of implementation within the standard.

Notice by conspicuous absence that the TAP state diagram is not described here. This information is inherent in the 1149.1 standard itself and does not need to be specified as part of a device adhering to the standard. In essence, stating "1149.1-1990" implies a great deal of information common to any such device. The proposed language is intended to specify those parameters necessarily unique to a given Boundary-Scan device implementation.

As further context, a device should be thought of as a black-box with terminal connections. Inside is the TAP and the *system logic*[1] surrounded on its perimeter by the Boundary Register logic. We want to describe the properties of the Boundary Register and terminal connections without need for describing the system logic. This independence recognizes a major contribution of Boundary-Scan; we can decouple problems such as board test from the system logic of the ICs.

LANGUAGE ELEMENTS

The language consists of a case-insensitive free-form multiline terminated syntax which is a *subset* of VHDL.[7] Comments are any text between a "--" symbol and the end of a line, syntactically terminating that line. Some of the information is conveyed in VHDL strings; sequences of characters between quote marks. This information is associated with a VHDL attribute and has a BSDL syntax requirement. Obviously, this is not checked by VHDL itself, but by applications that consume this information. (This is one reason the BSDL name is retained.) In practice, this information will be used in two environments. The first is a full VHDL-based system. It passes a BSDL description through its VHDL analyser into a compiled design library. From there, VHDL design library based tools can extract Boundary-Scan data by referencing the appropriate attributes. The second environment is a non-VHDL system capable of parsing a limited set of VHDL syntax (simply skipping items it doesn't recognize) to find and parse the BSDL information. In support of these systems, we constrain the full power of VHDL into a *standard practice*. Standard practices will be indicated as they are used in this text. Thus, BSDL is a "subset and standard practice" of VHDL.

BSDL is composed of three sections. These are: **Entity, Package,** and **Package Body.** An entity is the basis for describing a device within VHDL and an example for a real device is shown in Appendix A. Within the entity, the Boundary-Scan parameters of a device are described. The 1149.1 related definitions come from a pre-written, standard VHDL package (and related package body). The definitions for a 1149.1-1990 package and package body are given in Appendix B. The package information is directly related to the 1149.1 standard and is only expected to change when the standard itself is changed. Typically, this information would be write-protected. The development of new standards in the future would require new packages to be created.

A user may add an additional package (and package body), to contain user-specific design information. An example of this would be to contain a library of cell definitions unique to the users application, perhaps dependent upon the silicon technology in use. The reason for breaking out package bodies as seperate units is to allow the updating of the data within these without causing the need for recompilation of all entities that reference the corresponding package.

A simple Backus-Naur Form (BNF)[8] is used to describe the syntax of BSDL data within VHDL strings (see also Appendix C). Where the meaning is obvious without the use of BNF, the description is given by example. Since many of these strings are potentially long, the concatenation operator '&' is used to break them into managable pieces. The syntax descriptions will not show this, and, the concatenation operation may be thought of as a lexicographical pre-processing step before parsing.

THE ENTITY DESCRIPTION

An *entity* describes a device's I/O port and important *attributes* of the device. For BSDL, an entity has the following structure:

```
entity My_IC is          -- an entity for my IC

  [generic parameter]
  [logical port description]
  [use statement(s)]
  [package pin mapping]
  [scan port identification]
  [TAP description]
  [Boundary Register description]

end My_IC;               -- End description
```

The order of the elements within the entity as shown above is a required standard practice to simplify non-VHDL applications. The next few sections will examine each element of the entity.

Generic Parameter

The generic parameter is a VHDL construct used to pass data into a VHDL model. In BSDL it is intended as a method for selecting among several packaging options that a device may have. Each option may have a different mapping between the pins of the package and the bonding pads of the device. Even devices manufactured in a single package will be tested before packaging, with a different mapping possible. We call this the *logical-to-physical* relationship of the signals of the device. The description of the Boundary-Scan architecture of the device is done with the logical signals. Applications such as board testing will need to know how the logical structure of the device maps onto a set of physical pins. A VHDL **generic** parameter is used for this. It must have the name shown in order for software to seperate it from other parameters that might be passed to the entity. It has this form:

1. The 'system logic' is the same refered to by the 1149.1 document. However, important 'null' logic cases must also be treated as will be discussed.

generic(PHYSICAL_PIN_MAP:string := "undefined");

Note the string is initialized to an arbitrary value ("undefined") that will not allow a package selection if the parameter is not bound to a value, i.e., not passed. The use of this parameter will become clear shortly.

Logical Port Description

The port description uses the VHDL **port** list in a standard practice. Here, we are assigning meaningful symbolic names to the device's system terminals. These symbolic names are used in subsequent descriptions. This allows the majority of the statements to be 'terminal independent'; that is for example, independent of a renumbering or other reorganization of the terminals of the device. It also allows description of devices which may be packaged in several different forms. It is optional to include non-digital pins such as power, ground, no-connects, or analog signals, but these should be included for completeness. Non-digital pins will not be referenced later in the description, but all pins referenced in the description must have been defined here. The form is:

port(*<PinID>*; *<PinID>*; ... *<PinID>*);

<PinID> ::= *<IdentifierList>*: *<Mode>* *<PinType>*

<IdentifierList> ::= *<Identifier>* |
<IdentifierList> , *<Identifier>*

<PinType> ::= *<PinScaler>* | *<PinVector>*

<PinScaler> ::= *<Identifier>*

<PinVector> ::= *<Identifier>*(*<Range>*)

<Mode> ::= in | out | inout | buffer | linkage

<Range> ::= *<number>* to *<number>* |
<number> downto *<number>*

The *<Mode>* identifies the system usage of a device pin, with *in* for a simple input pin, *out* for an output pin that may participate in buses, *buffer* for an output pin that may not participate in buses, *inout* for a bidirectional signal pin, *linkage* for other pins such as power, ground, analog, or no-connect. A *<PinVector>* is a shorthand for grouping related signals. For example, *Data(1 to 8)* indicates there are 8 signals named *Data* indexed from 1 to 8, like *Data(3)*. A *<PinScaler>* is a single signal. Note, every pin must have a unique name, so if there are several ground pins for example, they must have different names such as *GND1*, *GND2*, etc, or be expressed as a vector. An example of a port statement for a 22 pin device is:

port(CLK:in bit; CLEAR:in bit; Q:out bit_vector(1 to 8);
DATA:in bit_vector(1 to 8); VCC, GND:linkage bit);

Bit and *bit_vector* are type names known to VHDL.

Use Statement(s)

The **use** statement identifies a VHDL package needed for defining attributes, types, constants, and other items that will be referenced. The following statement is mandatory in BSDL. Others may also be added to support user defined Boundary Register cells. The content of this package and its associated package body is shown in Appendix B.

use STD_1149_1_1990.all; -- Get 1149.1 information

Package Pin Mapping

VHDL **attribute** and **constant** statements are used to show the package pin mapping. These are shown by example:

attribute PIN_MAP of My_IC:entity
is PHYSICAL_PIN_MAP;

constant dw_package:PIN_MAP_STRING :=
<MapString>;

Attribute PIN_MAP is a string that is set to the value of the parameter PHYSICAL_PIN_MAP, already described. VHDL constants are then written, one for each packaging variation, that describe the mapping between the logical and physical pins of the device. (The BSDL syntax for *<MapString>* is given in Appendix C.) In a VHDL design library, the value of PIN_MAP can be used to identify the constant (by name) that contains the mapping of interest. In a non-VHDL implementation, the parse phase would look for the constant with a name matching the value of PIN_MAP. Note, the type of the constant must be PIN_MAP_STRING. This allows such parsers to ignore constants of other types. An example of a mapping is:

"CLK:1, DATA:(6,7,8,9,15,14,13,12), CLEAR:10, " &
"Q:(2,3,4,5,21,20,19,18), VCC:22, GND:11"

Notice it is the concatenation of two smaller strings. This is arbitrary; a string is the result after all concatenations are performed. A BSDL parser will read the content of the string. It matches signal names like *CLK* with the names in the port definition. The symbol on the right of the colon is the physical pin associated with that port signal. It may be a number, or an alphanumeric identifier because some packages such as Pin-Grid Arrays (PGAs) use coordinate identifiers like A07, or H13. If signals like *DATA* are *<PinVector>*'s in the port definition, then a matching list of pins enclosed in parenthesis are required. The physical pin mapped onto *DATA(5)* is pin 15 in the above example.

Scan Port Identification

Next we give the 5 attributes that define the scan port of the device. These are shown by example:

attribute TAP_SCAN_IN of TDI:signal is true;
attribute TAP_SCAN_OUT of TDO:signal is true;
attribute TAP_SCAN_MODE of TMS:signal is true;
attribute TAP_SCAN_RESET of TRST:signal is true;
attribute TAP_SCAN_CLOCK of TCK:signal
is (17.5e6, BOTH);

Here, signal names TDI, TDO, TMS, TRST and TCK must have appeared in the port description. The names chosen here match the 1149.1 standard, but may be arbitrary. The TAP_SCAN_RESET attribute is optional but the others must be specified for a correct implementation. The boolean assigned is arbitrary; the statement is used to bind the attribute to the signal. The TAP_SCAN_CLOCK attribute is a record with a real number field (the first) that gives the maximum operating frequency for TCK. The second field is an enumerated type with values LOW and BOTH which specify which state(s) the TCK signal may be stopped in without data loss in Boundary-Scan mode.

TAP Description

The next major piece of Boundary-Scan functionality that must be described is the device dependent characteristics of the TAP. It may have four or five control signals, already identified. It may have a user specified instruction set (within the rules) and a number of data registers and options. The following sections show how this is described.

The TAP Instruction Register may have any length 2 bits or longer and is required to support certain opcodes and some (but not all) of these have mandatory bit patterns. A designer may add 1149.1-identified optional instructions and/or new instructions with completely dedicated functions. An instruction may have several bit patterns. Unused bit patterns will default to the BYPASS instruction. Upon resetting the TAP or passing through the *Test-Logic-Reset* state, the instruction register is jam-loaded with a specific instruction. The standard provides for 'private' instructions which need not be documented *except* if their access could create an unsafe condition such as a board level bus conflict. Our language must easily denote these characteristics and take advantage of opportunities for semantic checks.

The characteristics of the instruction register that we capture with the language are *length, opcodes, capture, disable, private* and *usage*. Since these are basically simple, they are introduced by example.

attribute INSTRUCTION_LENGTH of My_IC:entity
is <integer>;
attribute INSTRUCTION_OPCODE of My_IC:entity
is <OpcodeTable>;
attribute INSTRUCTION_CAPTURE of My_IC:entity
is <Pattern>;
attribute INSTRUCTION_DISABLE of My_IC:entity
is <OpcodeName>;
attribute INSTRUCTION_PRIVATE of My_IC:entity
is <OpcodeList>;
attribute INSTRUCTION_USAGE of My_IC:entity
is <UsageString>;

Example:

```
attribute INSTRUCTION_LENGTH of My_IC:entity is 4;

attribute INSTRUCTION_OPCODE of My_IC:entity is
        "Extest (0000), " &
        "Bypass (1111), " &
        "Sample (1100, 1010), " &
        "Preload (1010)," &
        "Hi_Z  (0101), " &
        "Secret (0001)  ";

attribute INSTRUCTION_CAPTURE of My_IC:entity is
        "0101";

attribute INSTRUCTION_DISABLE of My_IC:entity is
        "Hi_Z";

attribute INSTRUCTION_PRIVATE of My_IC:entity is
        "Secret";
```

The **instruction_length** attribute defines the length that all opcode bit patterns must have. The **instruction_opcode** attribute is a BSDL string (syntax defined in Appendix C) containing the opcode identifiers and their associated bit patterns. The rightmost bit in the pattern is closest to TDO. The standard mandates the existence of **EXTEST**, BYPASS, and SAMPLE instructions, with mandatory bit patterns for the first two. Note that other bit patterns may also decode to these same instructions.

The **instruction_capture** attribute string states what bit pattern is jammed into the shift register portion of the instruction register when the TAP passes through the *Capture-IR* state. This bit pattern is shifted out whenever a new instruction is shifted in, and the standard mandates the least 2 significant bits must be a "01". Note, this bit pattern may be design-specific data. Since it is possible, by traversing from *Capture-IR* to *Exit1-IR* to *Update-IR,* to cause this pattern to become the effective instruction, it will act as some instruction (if not simply BYPASS) when it becomes effective. This bit pattern is **not** the instruction loaded into the instruction register when passing through the *Test-Logic-Reset* state. The standard states that on passing through the reset state, the *effective* instruction is jammed either to BYPASS, or IDCODE if it exists.

The optional **instruction_disable** attribute identifies an opcode that makes a Boundary-Scan device "disappear". In this mode, the 3-state outputs are disabled and the BYPASS register is placed between TDI and TDO. This is not (yet) a specified behavior in the 1149.1 standard, but many devices have this capability today because it is very useful for testing purposes. This attribute allows the opcode to be identified for software use.

The optional **instruction_private** attribute identifies opcodes that are private and potentially unsafe for access. By definition, the results of these instructions are undefined to the general public and should be avoided. Software can monitor the instruction register to issue warnings or errors if a private instruction is loaded during run time.

The optional **instruction_usage** attribute is a BSDL string with the syntax given in Appendix C. The usage concept will be covered in its own section later.

ID Register Values

Next, we need to identify standard prescribed optional registers. These are the IDCODE and USERCODE registers. Note, if an IDCODE instruction exists, an IDCODE register must also exist. Further, the existence of USERCODE implies the existence of IDCODE. To describe these instructions we need two attributes.

```
attribute IDCODE_REGISTER of My_IC:entity is
        "0011" &                  -- 4  bit version
        "1111000011110000" &      -- 16 bit part number
        "00000000111" &           -- 11 bit manufacturer
        "1";                      -- mandatory LSB

attribute USERCODE_REGISTER of My_IC:entity is
        "10xx" & "0011110011110000" &
        "00000000111" & "1";
```

The bit patterns must be 32 bits long. The rightmost bit is closest to TDO. In the examples above, concatenation is used to delimit fields within the codes. The "X" values specify a don't-care for that bit position. This is used to nullify subfields within a code that are not important for testing purposes.

Register Access

All TAP instructions must place a shiftable register between TDI and TDO. User-defined instructions may access data registers mandated by the standard; the Boundary Register, the IDCODE register, and the BYPASS register. The standard allows a designer to place additional data registers in the design. These are referenced by user-defined TAP instructions. It is important for software to know the existence and length of these registers and their associated instruction(s). Therefore we need to express these associations in the language. The attribute for this is:

attribute REGISTER_ACCESS of My_IC:entity
is <RegisterString>;

The syntax for <RegisterString> is in Appendix C. Example:

attribute REGISTER_ACCESS of My_IC:entity is
"Boundary (Secret, User1), " &
"Bypass (Hi_Z, User2), " &
"MyReg[7] (LoadSeed, ReadTest)";

In this example, *Secret, Hi_Z, User1, User2, LoadSeed* and *ReadTest* must be previously defined user instructions. Note that a seven bit user-register *MyReg* has been added to the TAP, with two instructions that access it. The 1149.1 standard itself defines the following relationships implicitly, so these need not be given.

attribute REGISTER_ACCESS of My_IC:entity is
"Boundary (Extest, Sample, Intest)," &
"Bypass (Bypass), " &
"Idcode (Idcode, Usercode)";

This ability to identify register access allows software to know the length of a scan sequence, which is dependent on the currently effective instruction. The mandatory Boundary Register, Bypass Register and Instruction Register are known from other statements, as well as their relationship to TAP instructions. Note that a semantic check can be made here ensuring that each instruction has an associated data register as required by the standard. Exceptions to this are the instructions marked 'private' since they are not to be accessed, nor do their target registers need to be identified.

The standard also allows user-instructions to reference several registers at once in a concatenated mode, but also requires them to have a new name in this mode. Here, we would treat this concatenation as if it were a new register with a distinct name and length. The reason is that in any case, the data flowing out of any register after passing through the *Capture-DR* state is not known to BSDL because *it is not a simulation language.* We are not attempting here to completely characterize the entire design so that its behavior is simulatable. This is more properly the domain of VHDL itself. We are simply trying to capture the relevant characteristics of Boundary-Scan devices so that we can intelligently manipulate chains of such devices. Other software can predict what must be coming out of various registers. This allows us to divide testing problems into two parts: calculating tests at an abstract level and manipulating the chain to deliver them. The language described here deals mainly with this second task. This division is important since there are several configurations (even proposed within the standard itself) for setting up Boundary-Scan chains. The abstracted test can be independent of these configurations.

Boundary Register Description

The Boundary Register is an ordered list of Boundary Cells, numbered 0 to N where $N+1$ is the number of cells in the register. Cell 0 is closest to TDO. There are cells of varying design and purpose. The standard, in chapter 10[1], shows fifteen such designs as examples. Others are possible as well. In discussing cell structures we will make heavy use of reference to the standard and its figures depicting cell designs to save space in this paper. To avoid confusion with figure references, a symbol such as *f10-16*[1] will refer to figure 10-16 in the standard. Symbols such as *f10-19c*[1] and *f10-19d*[1] refer to the *control* and *data* cells that make up the structure shown in figure 10-19 of the standard.

Cells must be identified before they are referenced in the Boundary Register description that follows. However, since the standard does give a number of examples that will likely be adopted in a design, we have constructed the language to have *intrinsic* or predefined cells that may be referenced via a simple nomenclature. Cell names are listed in Table 1 and their definitions are shown in the VHDL package body given in Appendix B. However, there will still be a need to define other cells not covered by the intrinsics. The details of these definitions are defered until later. If the intrinsic definitions contain the cells one needs for a description, then no cell definitions are required at all.

BC_1	*f10-12*[1], *f10-16*[1], *f10-18c*[1], *f10-18d*[1], *f10-21c*[1]
BC_2	*f10-8*[1], *f10-17*[1], *f10-19c*[1], *f10-19d*[1], *f10-22c*[1]
BC_3	*f10-9*[1]
BC_4	*f10-10*[1], *f10-11*[1]
BC_5	*f10-20c*[1]
BC_6	*f10-22d*[1]

Table 1. List of intrinsically defined cells and the figures covered in the standard.

Numerous rules must be observed when using the cells to create a Boundary Register as covered in chapter 10[1] of the standard. Some of these may be checked during compilation of a device's description. For example, some cell designs may only be used on a device input. Some will not support the *INTEST* instruction, which is allowable if the device TAP description does not list that instruction. Some cells require the aid of another cell to control 3-state enables. Checks can be performed and problems discovered as soon as a device's Boundary-Scan behavior has been specified and described, which may be well in advance of device fabrication.

A very general cell design from the standard (*f10-16*[1]) is shown in Figure 1. In Figure 2a we show a symbol that captures the essence of this cell needed for discussion. The design in Figure 1 is comprised of a *parallel input*, a *parallel output*, a multiplexer controlled by a *Mode* signal, and two *flip-flops*. The *Mode* signal is a function of the currently effective instruction. Yes, there are other elements such as the signals shifted in from the last cell and to the next cell. Yes, there is a second multiplexer controlled by signal ShiftDR. Yes, there are two clock signals ClockDR and UpdateDR. *But,* all these additional elements are always

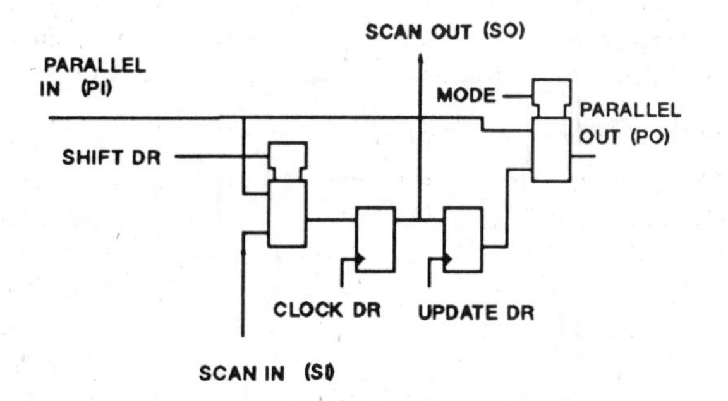

Figure 1. Cell design from f10-16[1].

precisely prescribed by 1149.1 and as such, *may be omitted from our consideration* in this language. This leads us to the symbol in Figure 2a which is simple.

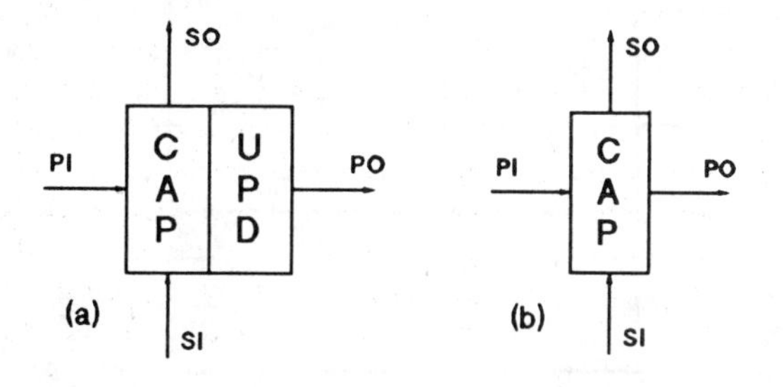

Figure 2. Two symbols for a typical Boundary Cell, one (a) with an UPD flip-flop and one without (b).

The parallel input and output are shown, and these are connected to various places depending on the application. The two flip-flops are labeled CAP and UPD to symbolize their use; the CAP flip-flop captures data in the *Capture-DR* state. The UPD flip-flop captures data in the *Update-DR* state. The shift path is shown because many such cells will be linked together in a shift chain that makes up the Boundary Register. The shift path links *only* the CAP flip-flops. Now, one cell design shown in *f10-11[1]* has a symbol (Figure 2b) with no UPD flip-flop.

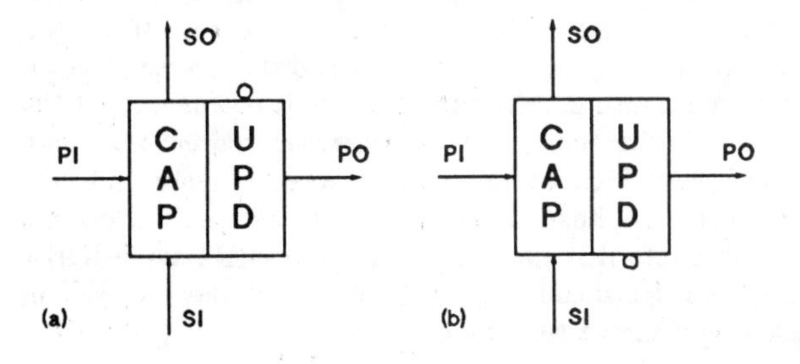

Figure 3. Symbols for a Boundary Cells with preset (a) and clear (b) on the UPD flip-flop.

The symbols in Figure 3 show bubbles on the top or bottom of the UPD flip-flop to indicate that flip-flop may be preset (1) or cleared (0) when passing through the *Test-Logic-Reset* state, as the standard allows in *f10-21c[1]*). No signal connection is made to these bubbles.

Now we show the three attributes needed to define the Boundary Register:

attribute BOUNDARY_CELLS of My_IC:entity is
<CellList>;

attribute BOUNDARY_LENGTH of My_IC:entity is
<integer>;

attribute BOUNDARY_REGISTER of My_IC:entity is
<CellTable>;

The <CellList> and <CellTable> are strings with syntax given in Appendix C. An example of a 3 cell Boundary Register is:

```
attribute BOUNDARY_CELLS of My_IC:entity is
        "BC_1, MyCell";
attribute BOUNDARY_LENGTH of My_IC:entity is 3;
attribute BOUNDARY_REGISTER of My_IC:entity is
  -- num cell   port function safe [ccell disval rslt]
   " 0 (BC_1,   IN,     input,   X), "   &
   " 1 (BC_1,   *,      control, 0), "   &
   " 2 (MyCell, OUT,    output3, X,   1,   0,   Z)";
```

The first attribute shows the cells used in the register; *BC_1* from the standard package, and *MyCell*, which must have been described in a user defined package. A semantic check can occur here; do these cells support the standard instructions that are listed for the TAP opcodes? For example, the TAP may support INTEST, but does *MyCell*?

The second attribute defines the number of cells in the Boundary Register. This number must match the number actually found in the third attribute, the register itself. This attribute is a string containing a list of elements, each with two fields. The first field is merely the cell number, which must be between 0 and LENGTH-1. (They may be listed in any order.) The second is a set of subfields within parentheses. There are either four or seven subfields. They are labeled, as in the comment above, **cell, port, function, safe, ccell, disval,** and **rslt.** All cells have the first four subfields. Only cells providing data for device outputs that can be disabled have the remaining three subfields. These three specify how to disable the output.

The **cell** subfield identifies the cell design used. It must match a cell given in the **boundary_cells** attribute.

The **port** subfield identifies the port signal actively driven or received by this cell. A cell serving as an output control or internal cell will have an asterisk in this position.

The **function** subfield shows the primary function of the cell. Table 2 shows the values this subfield may have:

input	a simple input pin receiver (*f10-8[1]*)
clock	a cell at a clock input (*f10-11[1]*)
output2	supplies data for a 2-state output (*f10-16[1]*)
output3	supplies data for a 3-state output (*f10-18d[1]*)
control	controls 3-state drive or cell direction (*f10-18c[1]*)
controlr	a *control*, disables at *Test-Logic-Reset* (*f10-21c[1]*)
internal	captures internal constants (*see page 10-7[1]*)
bidir	reversible cell for a bidirectional pin (*f10-22d[1]*)

Table 2. Function subfield values, meanings, and a figure reference of a representative cell in the standard[1].

Shortly, we discuss cells with more than one function. Note that the function is with respect to the boundary cell and **not** the device pin. This reflects the fact that two cells may service a single pin, for example, one serving as an input receiver and the other serving as an output driver, on a bidirectional pin (*f10-21*[1]). *Internal* cells are used to capture 'constants' (0's and 1's) within a design. They are specifically *not* allowed to be surrounded by system logic (*f10-7*[1]). One proposed use of this is to capture an encoded value (perhaps in the first few bits of the Boundary Register) as an informal identification code. Another was proposed in[9] where sense amplifiers monitor redundant power connections and place the measured results in internal Boundary Register cells. If the power connections are good, the data loaded will be constant. Finally, there may be "extra" cells unused in a programmable device (see page 10-7 of the standard[1]).

The **safe** subfield gives the value that a designer prefers to be loaded into the UPD flip-flop of the cell when software would otherwise choose a value randomly. Two examples are; the value that an output should have that is safe to overdrive during In-Circuit testing; or, the value to present to on-chip logic at a device input during EXTEST. An 'X' signifies that it doesn't matter.

The **ccell** subfield identifies the cell number of the cell that serves as an output enable. The **disval** subfield gives the value the **ccell** must have to disable the output driver. The **rslt** subfield gives the state the disabled driver goes to; a high impedance state (Z), a weak '0' (Weak0), or a weak '1' (Weak1). The last two values correspond to asymetrical drivers such as TTL open-collector drivers or ECL open-emitters. The functions in effect when these three subfields exist must be *output2, output3* or *bidir.* If it is *bidir,* then disabling the driver implies the cell is a receiver.

An Example Boundary Register Description

We now use the device shown in Figure 4 to illustrate a Boundary Register description and how special cases are handled. These special cases arise because the standard allows cells to be merged when *the system logic between them is null.* (See for example, *f10-4*[1], *f10-5*[1].) Cells may be merged if the logic between them is simply a non-inverting data path, like a wire or buffer. When merging is done, the resulting cell must obey a combination of the rules of the merged cells. Here is the definition of the Boundary Register for Figure 4.

```
attribute BOUNDARY_CELLS of Figure4:entity is
        "BC_1, BC_2, BC_6";
attribute BOUNDARY_LENGTH of Figure4:entity is 10;
attribute BOUNDARY_REGISTER of Figure4:entity is
-- num cell  port      function  safe  [ccell disval rslt]
 "0 (BC_1, *,          control,  0)," &
 "1 (BC_1, OUT2,       output2,  1,    1,    1, Weak1),"&
 "2 (BC_6, BIDIR1,     bidir,    X,    3,    0, Z)," &
 "3 (BC_2, *,          control,  0)," &
 "4 (BC_1, *,          control,  0)," &
 "5 (BC_1, BIDIR3,     input,    X)," &
 "5 (BC_1, BIDIR2,     output3,  X,    7,    1, Z)," &
 "6 (BC_1, BIDIR2,     input,    X)," &
 "6 (BC_1, BIDIR3,     output3,  X,    4,    0, Z)," &
 "7 (BC_1, *,          control,  1)," &
 "8 (BC_1, IN2,        input,    X)," &
 "9 (BC_1, IN1,        input,    X)," &
 "9 (BC_1, OUT1,       output3,  X,    0,    0, Z)";
```

Cell 0 is simply a control cell between the system logic and the enable for signal OUT1. Cells 4 and 7 are similar.

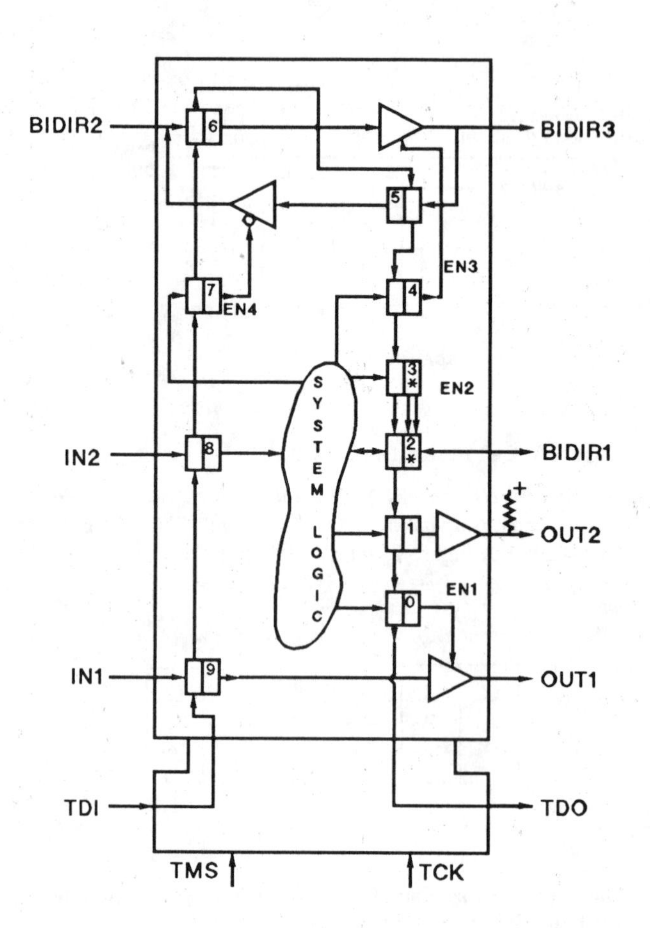

Figure 4. A device illustrating several merged cell situations.

Notice the **safe** bits are assigned to cause the associated drivers to disable. Cell 3 is the control for the reversible cell (*f10-22d*[1]) used on the bidirectional signal BIDIR1.

Cell 1 is a 2-state output data cell. Note that it has the three extra fields indicating that it controls its own open-collector asymetrical driver. Placing a '1' in cell 1 will disable OUT2 by putting it into the 'Weak1' state.

Cell 2 is the reversible cell of figure *f10-22d*[1]. This cell serves as an input if the control cell has turned off the output driver, meaning cell 3 produces a '0'. This cell serves as the data for the driver if the output is enabled. It cannot serve both functions. This is a drawback during test, since the value of BIDIR2 cannot be observed while the driver is turned on. A board level fault could not be seen by this device. Note that the structures for BIDIR2 and BIDIR3 (or *f10-21*[1]) would allow observation of the driver, thus allowing a simple consistancy check.

Cell 5 (and similarly for cells 6 and 9) has merged behavior; it serves as the input receiver for BIDIR3 and as the data source for BIDIR2. It has two lines of description in the Boundary Register definition as a result. The first gives its behavior as an input cell while the second describes its characteristics as an output cell. Note that cell BC_1 used in this capacity must support both *input* and *output3* functions. This is reflected in the definition of BC_1 (see appendix B) where both functions are seen to exist for all instructions.

This example is extreme in dwelling on odd cases. Most device implementations will be quite simple and routine, as the example in Appendix A, the Texas Instruments 74BCT8374[10], illustrates.

PACKAGE DESCRIPTION

The package that describes the Std 1149.1-1990 information needed for BSDL is given in appendix B. This package cannot be modified without changing BSDL itself.

There may be occasion for users to define their own packages for use in conjunction with the 1149.1 package. This is the way to add user-specific Boundary Cell definitions. By placing these in a package, they may be referenced by many entities. While it is possible to place an entire cell description in a package, it is standard practice to place the actual cell description in a *package body* (described next) associated with the package. All that then remains in the user-defined package is the names of the cells. For example, say a user wants to define two new cells for reference in a boundary-scan description. Here is what the package would look like:

```
package My_New_Cells is

    constant MNC_1 : CELL_INFO;  -- My new cell 1
    constant MNC_2 : CELL_INFO;  -- My new cell 2

end My_New_Cells;
```

Of course, to reference these cells, a **use** statement must appear in an entity description that references *My_New_Cells.all*, and the cell names must appear in the BOUNDARY_CELLS attribute string. The definition of these *deferred constants* goes into the related package body.

PACKAGE BODIES FOR DEFINING BOUNDARY CELLS

Now it is time to discuss the description of Boundary Register cells. We have already skimmed this subject in examining the description of the Boundary Register itself, and, we have benefitted from *intrinsic* cell definitions provided by the 1149.1 package body.

What are the important aspects of a cell we need to describe for BSDL to meet its statement of scope? In looking at the variety of possible cell designs given in the standard and the long list of rules governing these designs, this might seem to be a daunting task. It turns out that all of the cells shown in the standard (excepting *f10-22d*[1]) could be modeled as shown in Figure 5, *for the purposes of BSDL scope.*

For the case of cell *f10-22d*[1], its reversible nature can be represented as if it were two cells; one that is left-to-right and the other that is right-to-left, each modeled by Figure 5. The one to choose is defined by the value of the controlling cell. When enabled to drive, the cell works left-to-right and vice versa.

In BSDL, any cell consists of a Parallel Input (PI), a Parallel Output (PO), CAP and UPD flip-flops and connections to/from CAP flip-flops of other cells. Note, the UPD flip-flop may not exist in certain input cell configuration as allowed by the standard. The CAP flip-flop has eight choices of data source as shown. In looking at any particular cell design, usually only two or three of these choices are actually implemented. The '0' and '1' constants may be

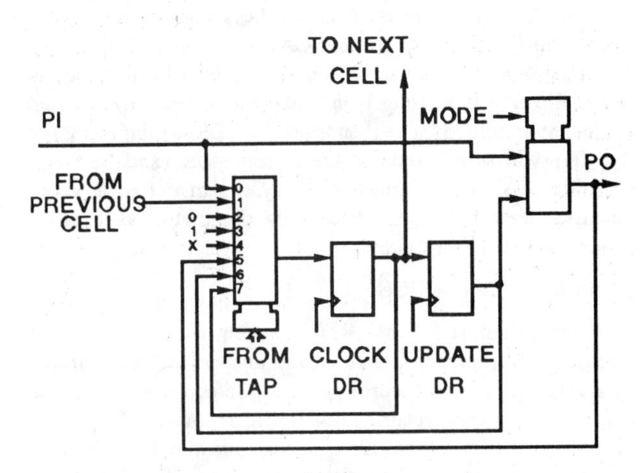

Figure 5. A general BSDL model of a Boundary Register Cell.

loaded into the CAP in certain situations. For example, an output cell design during EXTEST may load a constant into the CAP. The 'X' value denotes a don't-know or don't-care about what is loaded. An example of this is an output cell design during RUNBIST that loads a Linear Feedback Signature bit into the CAP. BSDL alone is not capable of simulating what this value could be. Also, proprietary information about a cell design may be hidden by "X-ing" out the activities of an instruction.

Context is another important factor in analysing Boundary Cells. What is the cell **function**? When a cell is an input cell, then PI **must** be connected to a device pin and PO **must** be connected to the system logic (ignore cell merging here). Now, add the context of the effective instruction. If EXTEST is in effect, then CAP **must** receive PI data. What we are defining here is a *triple* of data:

<function> *<instruction>* *<CAP data source>*

A cell description is a collection of these triples in the form of a VHDL array of records. Each triple tells us a piece of a cell behavior; for a given cell **function**, while a certain **instruction** is in effect, what **data** is loaded into the CAP when passing through the *Capture-DR* state. Since the CAP flip-flop data is what is eventually seen when scanning out data, it is important to know what the CAPs will contain. This data is simple to derive. One simply fixes the cell **function**, and then for each **instruction** supported, traces the **data** flowing to the CAP flip-flop.

What about other details? For example, input cell *f10-10*[1] in the standard produces a '1' on PO while EXTEST is in effect. BSDL does not model this because, during EXTEST on an input, we are looking *outward* from the device, not *inward*. Essentially, we do not care what is being fed to PO during EXTEST on input pins. The device designer *did* which is why the '1' is being injected; probably to satisfy some requirement internal to the device. During and after an EXTEST operation, the 1149.1 standard does not specify the what the state of the system logic will be so there is no point in trying to describe inputs to the system logic during EXTEST. There are similar arguments about what it is necessary to model during INTEST, SAMPLE and RUNBIST.

Many details are prescribed by the standard itself. The UPD flip-flops always get the CAP data transfered during the *Update_DR* state, so we need not describe this. If an UPD flip-flop is missing from a design, it can only be used as an input cell. If it still supports INTEST, then the CAP flip-flop will supply data to the system logic (and data ripple due to shifting is guaranteed by design not to matter), or, the input has been specifically designated as a **clock function** as the rules allow.

Defining a Boundary Cell

A cell is defined as a VHDL constant. It is an array of records with the range of the array unspecified, but implicit from the number of records given in the constant definition. Each field of each record must be filled. A standard practice is that these are filled using positional association rather than named association, as shown, to simplify development of non-VHDL based applications.

We give an example of a Boundary Cell C_Ex_1 that supports EXTEST, SAMPLE and INTEST. It loads a '1' into the CAP during EXTEST if the cell is used as an output or control function. The cell may be used as a simple input function. During INTEST as an input, it reloads the CAP with the data value that was shifted into the cell. Its description is:

```
constant C_Ex_1 : CELL_INFO :=
  ( (Output2, Extest, One),  (Output3, Extest, One),
    (Output2, Sample, PI),   (Output3, Sample, PI),
    (Output2, Intest, PI),   (Output3, Intest, PI),
    (Control, Extest, One),  (Input, Extest, PI),
    (Control, Sample, PI),   (Input, Sample, PI),
    (Control, Intest, PI),   (Input, Intest, CAP) );
```

The values allowed for **function** are the same as shown in Table 2 with the exception that *bidir* is replaced by two functions; *bidir_in* and *bidir_out*. A reversible cell such as *f10-22d*[1] is described only with these functions, with both required for every supported instruction as in cell BC_6 of Appendix B. The control cell, when enabling or disabling the cell as a driver, chooses between the *bidir_out* or *bidir_in* functions respectively. This is the only function with this complication.

The values allowed for an instruction are EXTEST, INTEST, SAMPLE and RUNBIST. Others such as BYPASS have no effect on the Boundary Cells. The values allowed for CAP data are *PI, PO, UPD, CAP, ZERO, ONE,* and *X*.

OTHER BSDL FUNCTIONS

There are some other features in BSDL, some of which we defered in previous discussion.

Instruction Usage

Generally, BSDL is a means for describing static design parameters of an 1149.1 implementation. However, the standard contains two instructions with details of operation that are not statically defined. These are RUNBIST and INTEST. The **instruction_usage** attribute gives additional information about the operation of an instruction. While targetted at the two standard instructions, it could be used to document details about a user-defined instruction as well. The types of information needed are; register identification, result identification and clocking information. This information is placed in string <*UsageString*> with syntax given in Appendix C. Here are examples for describing RUNBIST and INTEST and a user-defined instruction MYBIST:

```
attribute INSTRUCTION_USAGE of My_IC:entity is
  "Runbist (registers Boundary, Signature;" &
          " shift Signature;" &
          " result 0011010110000100;" &
          " clock TCK in Run_Test_Idle;" &
          " length 4000 cycles)," &
  "Intest  (clock SYSCLK shifted)," &
  "MyBist  (registers Seed, Boundary, Bypass;" &
          " initialize Seed 001110101;" &
          " shift Bypass;" &
          " result 1;" &
          " clock SYSCLK in Run_Test_Idle;" &
          " length 125.0e-3 seconds)";
```

The RUNBIST usage shows two registers used. Note, the standard states that only the Boundary Register may be initialized for test operation. A second register *Signature* is also used, and will be placed between TDI and TDO for shifting. When the test is completed, the *result* shifted out from *Signature* should match the given pattern (length must match length of *Signature* register), where the rightmost bit is closest to TDO. The test is run by clocking TCK 4000 times while in state *Run_Test_Idle*.

The INTEST usage tells us that clocking is accomplished by shifting the clock states to signal SYSCLK. This implies a cell structure for input signal SYSCLK that supports INTEST. If this cell had been a **clock** function rather than **input**, then the description would read (clock SYSCLK) and we would have to supply the clocking externally.

The MYBIST usage tells us that **registers** *Seed, Boundary*, and *Bypass* are involved in the test and that *Seed* requires initialization to a pattern. This will have to be done using another instruction since MYBIST places the Bypass register between TDI and TDO. Software could look in the **register_access** attribute for such an instruction. When the test is done, the Bypass register should contain the **result** '1'. Clocking is done with the TAP in *Run_Test_Idle*, with SYSCLK freerunning for 125 milliseconds.

Design Warnings

A device designer may know of situations where the system usage of a device can be subverted via the Boundary-Scan feature to cause circuit problems. As a simple example, a device may have dynamic system logic which requires clocking to maintain its state. Thus, clocking must be maintained when bringing the device out of system mode and into test mode for INTEST. The **design_warning** attribute can be assigned a string message to alert future consumers to the potential for problems. For example:

```
attribute DESIGN_WARNING of My_IC:entity is
      "Dynamic device, " &
      "maintain clocking for INTEST."
```

This warning is for application specific display purposes only. It is a textual message with no specified syntax and is not intended for software analysis.

CONCLUSION

BSDL is an extensible language for defining the basic testability features of a device implemented under the IEEE 1149.1-1990 standard. It is specifically designed for describing the numerous options that may be exercised in such implementations, in a way useful for humans and computers. It is also a *subset and standard practice* of IEEE Std 1076-1987 VHDL and as such may be contained within a larger VHDL description of a device used for modeling or simulation. An added benefit is that a number of compliance violations in a design may be discovered either in attempting to code the device features, or, in semantic checks possible during analysis.

Integrated circuit vendors have been reluctant[2] to embed user accessible testability features within their devices, and are now responding to market pressures for it. The 1149.1 standard makes it much easier to add testability in a prescribed way. However, without a simple, complete, and automated way of describing implementations, these vendors rightfully fear that new support difficulties will result. The concept of a standardized description offers them a way of transfering the support burden to the proper segments of the industry, most notably, the ATE vendors. These same ATE vendors will benefit from the assurance that the descriptions they receive are complete, accurate, and uniform across the IC vendors.

Very recently, a new interest has been expressed for BSDL. ASIC vendors could use a BSDL description of a device in conjunction with the description of its system logic to *automatically* add the Boundary-Scan logic during layout. This offers 1149.1 testability to their customers who may be unfamiliar with the details of the standard and, of course, the BSDL description is available immediately.

The advantages of the 1149.1 standard can be more widely enjoyed if there is some commonality in the description of Boundary-Scan devices across tasks and disciplines. We believe BSDL fills this need.

ACKNOWLEDGEMENTS

A large number of companies on several continents have been involved in drafting the 1149.1 standard. Many of these same companies have materially contributed to the development of BSDL in time and travel commitments. In particular we would like to mention AT&T, Bennetts Associates, British Aerospace, British Telecom, DEC, Electronic Tools, ElektronikCentralen, Ericsson, the ESPRIT Consortium, GenRad, Harris, IBM, ICL, Intel, Logic Automation, Marconi, Mentor Graphics, Motorola, NCR, Philips, Siemens, Teradyne, Texas Instruments, Thomson-CSF, and Unisys.

Hewlett-Packard sponsored a worldwide effort to gain consensus and gather comments. Meetings and presentations were held in Amsterdam, Bobligen, Boston, Carmel, Chicago, Dallas, Denver, Karlsruhe, London, Loveland, New York, Osaka, Paris, Philadelphia, Rome, Stanford, Tokyo, Vail and Zurich.

Special thanks go to the chairs of the IEEE 1149.1 working group, Colin Maunder and Rod Tulloss, who responded to scores of transmissions. Larry Saunders, chair of the IEEE Design Automation Standards Subcommittee was instrumental in the development of the VHDL subset.

The authors communicated with many people to great benefit. We wish to thank all of them, and in particular, Elmer Arment, Bill Armstrong, LaNae Avra, Keith Baker, Dave Ballew, Raymond Balzer, James Beausang, Bill Bell, Ben Bennetts, Leon Bentley, Harry Bleeker, Bill Bruce, Mike Bullock, Bill Den Beste, Bulent Dervisoglu, John Deshayes, Gary Dudeck, Lee Fleming, Peter Fleming, Michael Gallup, Vassilios Gerousis, Grady Giles, Luke Girard, Peter Hansen, Vance Harwood, Jay Hiserote, Najmi Jarwala, Doug Kostlan, Dirk van de Lagemaat, William Lattin, Johann Maierhofer, Ralph Marlett, Ken Mason, Mark Mathieu, Don McClean, Ed McCluskey, Randy Morgan, Carsten Mortensen, Roberto Mottola, Math Muris, Paul Ocampo, Dieter Ohnesorge, Anwar Osseyran, Michel Parot, Alain Plassart, Ken Posse, Jeff Rearick, Gordon Robinson, Rick Robinson, Martin Roche, Derek Roskell, Kevin Schofield, Dave Schuler, Rene Segers, Jay Stepleton, Mark Swanson, John Sweeney, Toshio Tamamura, Michael Tchou, Jake Thomas, Hai Vo-Ba, Rolf Wagner, Allen Warren, Ron Waxman, Lee Whetsel, Harry Whittemore, Tom Williams, Akira Yamagiwa, Chi Yau, and Mike Yeager.

REFERENCES

1. IEEE Standard 1149.1-1990, "IEEE Standard Test Access Port and Boundary-Scan Architecture," *IEEE Standards Board, 345 East 47th Street, New York, NY 10017,* May, 1990
2. Parker, K. P., "Testability: Barriers to Acceptance," *IEEE Design and Test of Computers, vol. 3,* October 1986, pp. 11-15
3. Parker, K. P., "The Impact of Boundary-Scan on Board Test," *IEEE Design and Test of Computers, vol. 6,* August 1989, pp. 18-30
4. Maunder, C. and F. Beenker, "Boundary-Scan: A Framework for Structured Design-for-Test", *Proc. Int'l Test Conference,* 1987, pp. 714-723
5. Hansen, P., "The Impact of Boundary-Scan on Board Test Strategies", *Proc. ATE&I Conference East, pp. 35-40,* Boston, June 1989
6. Private Communications: The authors have benefited from over 300 communications in person, by phone, facsimile, mail and E-mail with many individuals. See the acknowledgements.
7. IEEE Standard 1076-1987, "IEEE Standard VHDL Language Reference", *IEEE Standards Board, 345 East 47th Street, New York, NY 10017,* March, 1988
8. Backus, J. W., "The Syntax and Semantics of the Proposed International Algebraic Language of the Zurich ACM-GAMM Conference", *Proc. Intnl. Conf. on Information Processing,* UNESCO, 1959, pp 125-132
9. van de Lagemaat, D., "Testing Multiple Power Connections with Boundary-Scan", *Proc. 1st European Test Conference,* Paris, April 1989, pp. 127-130
10. Texas Instruments Data Sheet (Preliminary) SN54BCT8374, SN74BCT8374 Boundary-Scan Device with Octal D-Type Flip-Flop, Texas Instruments Inc, Dallas Tx. 1988

[Appendices start on the following page.]

Appendix A, An Example: This example is the Texas Instruments 74BCT8374 Octal D-Type Flip-Flop[10] (see Figure 6). This device has an unusually rich set of user defined instructions.

```
entity ttl74bct8374 is
   generic (PHYSICAL_PIN_MAP : string := "UNDEFINED");

   port (CLK:in bit;  Q:out bit_vector(1 to 8);  D:in bit_vector(1 to 8); GND, VCC:linkage bit;
            OC_NEG:in bit;  TDO:out bit;  TMS, TDI, TCK:in bit);
   use STD_1149_1_1990.all;     -- Get Std 1149.1-1990 attributes and definitions

   attribute PIN_MAP of ttl74bct8374 : entity is PHYSICAL_PIN_MAP;

   constant DW_PACKAGE : PIN_MAP_STRING := "CLK:1,  Q:(2,3,4,5,7,8,9,10),  D:(23,22,21,20,19,17,16,15)," &
                                           "GND:6,  VCC:18,  OC_NEG:24,  TDO:11,  TMS:12, TCK:13,  TDI:14";

   constant FK_PACKAGE : PIN_MAP_STRING := "CLK:9,  Q:(10,11,12,13,16,17,18,19),  D:(6,5,4,3,2,27,26,25)," &
                                           "GND:14,  VCC:28,  OC_NEG:7,  TDO:20,  TMS:21,  TCK:23,  TDI:24";

   attribute TAP_SCAN_IN    of TDI : signal is true;
   attribute TAP_SCAN_MODE  of TMS : signal is true;
   attribute TAP_SCAN_OUT of TDO : signal is true;
   attribute TAP_SCAN_CLOCK of TCK : signal is (20.0e6, BOTH);

   attribute INSTRUCTION_LENGTH of ttl74bct8374 : entity is 8;

   attribute INSTRUCTION_OPCODE of ttl74bct8374 : entity is
     "BYPASS    (11111111, 10001000, 00000101, 10000100, 00000001)," &
     "EXTEST    (00000000, 10000000)," &
     "SAMPLE    (00000010, 10000010)," &
     "INTEST    (00000011, 10000011)," &
     "TRIBYP    (00000110, 10000110)," &     -- Boundary Hi-Z
     "SETBYP    (00000111, 10000111)," &     -- Boundary 1/0
     "RUNT      (00001001, 10001001)," &     -- Boundary run test
     "READBN    (00001010, 10001010)," &     -- Boundary read normal
     "READBT    (00001011, 10001011)," &     -- Boundary read test
     "CELLTST   (00001100, 10001100)," &     -- Boundary selftest normal
     "TOPHIP    (00001101, 10001101)," &     -- Boundary toggle out test
     "SCANCN    (00001110, 10001110)," &     -- BCR Scan normal
     "SCANCT    (00001111, 10001111)," &     -- BCR Scan test

   attribute INSTRUCTION_CAPTURE of ttl74bct8374 : entity is "01010101";
   attribute INSTRUCTION_DISABLE of ttl74bct8374 : entity is "TRIBYP";

   attribute REGISTER_ACCESS of ttl74bct8374 : entity is
     "BOUNDARY (READBN, READBT, CELLTST)," &
     "BYPASS (TOPHIP, SETBYP, RUNT, TRIBYP)," &
     "BCR[2] (SCANCN, SCANCT)";   -- 2-bit Boundary Control Register

   attribute BOUNDARY_CELLS of ttl74bct8374 : entity is "BC_1";
   attribute BOUNDARY_LENGTH of ttl74bct8374 : entity is 18;

   attribute BOUNDARY_REGISTER of ttl74bct8374 : entity is
   -- num  cell     port     function  safe [ccell disval rslt]
     "17   (BC_1,  CLK,     input,    X)," &
     "16   (BC_1,  OC_NEG,  input,    X)," &  -- Merged Input/Control
     "16   (BC_1,  *,       control,  0)," &  -- Merged Input/Control
     "15   (BC_1,  D(1),    input,    X)," &
     "14   (BC_1,  D(2),    input,    X)," &
     "13   (BC_1,  D(3),    input,    X)," &
     "12   (BC_1,  D(4),    input,    X)," &
     "11   (BC_1,  D(5),    input,    X)," &
     "10   (BC_1,  D(6),    input,    X)," &
     "9    (BC_1,  D(7),    input,    X)," &
     "8    (BC_1,  D(8),    input,    X)," &
     "7    (BC_1,  Q(1),    output3,  X,   16,   0,  Z)," &
     "6    (BC_1,  Q(2),    output3,  X,   16,   0,  Z)," &
     "5    (BC_1,  Q(3),    output3,  X,   16,   0,  Z)," &
     "4    (BC_1,  Q(4),    output3,  X,   16,   0,  Z)," &
     "3    (BC_1,  Q(5),    output3,  X,   16,   0,  Z)," &
     "2    (BC_1,  Q(6),    output3,  X,   16,   0,  Z)," &
     "1    (BC_1,  Q(7),    output3,  X,   16,   0,  Z)," &  -- outputs controlled from cell 16 set to 0 are Hi-Z.
     "0    (BC_1,  Q(8),    output3,  X,   16,   0,  Z)";     -- cell 16 has a merged function, both input and control.
end ttl74bct8374;
```

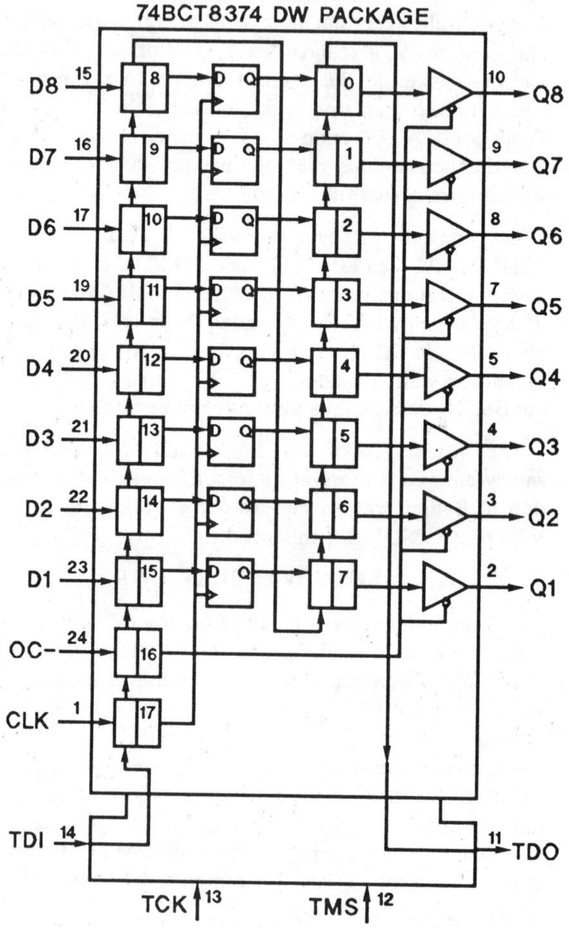

74BCT8374 DW PACKAGE

Figure 6

Appendix B, IEEE Std 1149.1-1990 VHDL Package/Package Body definition.

This is the definition of the VHDL package and supporting package body for IEEE Std 1149.1-1990 attributes, types, subtypes, and constants of BSDL.

```
package STD_1149_1_1990 is

-- Give pin mapping declarations
attribute PIN_MAP : string;
subtype PIN_MAP_STRING is string;

-- Give TAP control declarations
type CLOCK_LEVEL is (LOW, BOTH);
type CLOCK_INFO  is record
  FREQ : real;
  LEVEL: CLOCK_LEVEL;
end record;

attribute  TAP_SCAN_IN     : boolean;
attribute  TAP_SCAN_OUT    : boolean;
attribute  TAP_SCAN_CLOCK  : CLOCK_INFO;
attribute  TAP_SCAN_MODE   : boolean;
attribute  TAP_SCAN_RESET  : boolean;

-- Give instruction register declarations
attribute  INSTRUCTION_LENGTH : integer;
attribute  INSTRUCTION_OPCODE : string;
attribute  INSTRUCTION_CAPTURE : string;
attribute  INSTRUCTION_DISABLE : string;
attribute  INSTRUCTION_PRIVATE : string;
attribute  INSTRUCTION_USAGE : string;

-- Give ID and USER code declarations
type ID_BITS is ('0', '1', 'x', 'X');
type ID_STRING is array (31 downto 0) of ID_BITS;
attribute IDCODE_REGISTER   :  ID_STRING;
attribute USERCODE_REGISTER  :  ID_STRING;

-- Give register declarations
attribute REGISTER_ACCESS : string;

-- Give boundary cell declarations
type BSCAN_INST is (EXTEST, SAMPLE, INTEST,
                    RUNBIST);
type CELL_TYPE is (INPUT, INTERNAL, CLOCK,
                   CONTROL, CONTROLR, OUTPUT2,
                   OUTPUT3, BIDIR_IN, BIDIR_OUT);
type CAP_DATA is (PI, PO, UPD, CAP, X, ZERO, ONE);
type CELL_DATA is record
  CT : CELL_TYPE;
  I   : BSCAN_INST;
  CD : CAP_DATA;
end record;
type CELL_INFO is array of CELL_DATA;

-- Boundary Cell defered constants (see package body)

constant BC_1  : CELL_INFO;
constant BC_2  : CELL_INFO;
constant BC_3  : CELL_INFO;
constant BC_4  : CELL_INFO;
constant BC_5  : CELL_INFO;
constant BC_6  : CELL_INFO;

-- Boundary Register declarations

attribute BOUNDARY_CELLS : string;
attribute BOUNDARY_LENGTH : integer;
attribute BOUNDARY_REGISTER : string;

-- Miscellaneous
attribute DESIGN_WARNING : string;

end STD_1149_1_1990;  -- End of 1149.1-1990 Package
```

```
package body STD_1149_1_1990 is   -- Standard Boundary Cells

-- Description for f10-12, f10-16, f10-18c, f10-18d, f10-21c

constant BC_1 : CELL_INFO :=
 ((INPUT, EXTEST,  PI),         (OUTPUT2, EXTEST,   PI),
  (INPUT, SAMPLE,  PI),         (OUTPUT2, SAMPLE,   PI),
  (INPUT, INTEST,  PI),         (OUTPUT2, INTEST,  PI),
  (INPUT, RUNBIST, PI),         (OUTPUT2, RUNBIST, PI),
  (OUTPUT3, EXTEST,  PI),   (INTERNAL, EXTEST,  PI),
  (OUTPUT3, SAMPLE,  PI),   (INTERNAL, SAMPLE,  PI),
  (OUTPUT3, INTEST,   PI),   (INTERNAL, INTEST,  PI),
  (OUTPUT3, RUNBIST, PI),   (INTERNAL, RUNBIST, PI),
  (CONTROL, EXTEST,  PI),   (CONTROLR, EXTEST,  PI),
  (CONTROL, SAMPLE,  PI),   (CONTROLR, SAMPLE,  PI),
  (CONTROL, INTEST,   PI),   (CONTROLR, INTEST,  PI),
  (CONTROL, RUNBIST, PI),   (CONTROLR, RUNBIST, PI) );

-- Description for f10-8, f10-17, f10-19c, f10-19d, f10-22c

constant BC_2 : CELL_INFO :=
 ((INPUT, EXTEST,  PI),         (OUTPUT2, EXTEST,   UPD),
  (INPUT, SAMPLE,  PI),         (OUTPUT2, SAMPLE,   PI),
  (INPUT, INTEST,   UPD),  -- Intest on output2 not supported
  (INPUT, RUNBIST, UPD),        (OUTPUT2, RUNBIST, UPD),
  (OUTPUT3, EXTEST,  UPD), (INTERNAL, EXTEST,  PI),
  (OUTPUT3, SAMPLE,  PI),    (INTERNAL, SAMPLE,  PI),
  (OUTPUT3, INTEST,   PI),    (INTERNAL, INTEST,  UPD),
  (OUTPUT3, RUNBIST, PI),    (INTERNAL, RUNBIST, UPD),
  (CONTROL, EXTEST,  UPD), (CONTROLR, EXTEST, UPD),
  (CONTROL, SAMPLE,  PI),    (CONTROLR, SAMPLE,  PI),
  (CONTROL, INTEST,   PI),    (CONTROLR, INTEST,  PI),
  (CONTROL, RUNBIST, PI),    (CONTROLR, RUNBIST, PI) );

-- Description for f10-9

constant BC_3 : CELL_INFO :=
 ((INPUT, EXTEST,  PI),         (INTERNAL, EXTEST, PI),
  (INPUT, SAMPLE,  PI),         (INTERNAL, SAMPLE, PI),
  (INPUT, INTEST,  PI),         (INTERNAL, INTEST,  PI),
  (INPUT, RUNBIST, PI),         (INTERNAL, RUNBIST, PI) );

-- Description for f10-10, f10-11

constant BC_4 : CELL_INFO :=
 ((INPUT, EXTEST,  PI),  -- Intest on input not supported
  (INPUT, SAMPLE,  PI),  -- Runbist on input not supported
  (CLOCK, EXTEST,  PI),   (INTERNAL, EXTEST,  PI),
  (CLOCK, SAMPLE,  PI),   (INTERNAL, SAMPLE,  PI),
  (CLOCK, INTEST,   PI),   (INTERNAL, INTEST,   PI),
  (CLOCK, RUNBIST, PI),   (INTERNAL, RUNBIST, PI) );

-- Description for f10-20c, a combined Input/Control

constant BC_5 : CELL_INFO :=
 ((INPUT, EXTEST,  PI),         (CONTROL, EXTEST,   PI),
  (INPUT, SAMPLE,  PI),         (CONTROL, SAMPLE,   PI),
  (INPUT, INTEST,   UPD),       (CONTROL, INTEST,  UPD),
  (INPUT, RUNBIST, PI),         (CONTROL, RUNDIST,  PI) );

-- Description for f10-22d, a reversible cell

constant BC_6 : CELL_INFO :=
 ((BIDIR_IN, EXTEST,  PI),      (BIDIR_OUT, EXTEST, UPD),
  (BIDIR_IN, SAMPLE,  PI),      (BIDIR_OUT, SAMPLE, PI),
  (BIDIR_IN, INTEST,   UPD),  (BIDIR_OUT, INTEST,   PI),
  (BIDIR_IN, RUNBIST, UPD),  (BIDIR_OUT, RUNBIST, PI) );

end STD_1149_1_1990;  -- End of 1149.1-1990 Package Body
```

Appendix C, BSDL Syntax Specification

The BNF syntax descriptions are shown in this appendix. The items described are those contained within VHDL strings and as such, are not part of VHDL syntax. Syntactic items are shown in italics, surrounded by '<' and '>' characters. Keywords such as 'Extest' or 'Output3' are in normal font. Boldface items are BSDL terminals such as **INTEGER, VHDL IDENTIFIER,** or other description. The symbol **NULL** is the empty expansion. All BSDL elements contained within VHDL strings are treated as single, contiguous strings even though they may be expressed as the concatenation of smaller strings. All concatenations should be removed during lexicographical analysis. An asterisk in the leftmost column marks the start of an BNF expression promised in the text of this paper.

* *<MapString>* ::= *<PinMapping>* | *<MapString>* , *<PinMapping>*

<PinMapping> ::= *<PortName>* : *<PhysicalPinDesc>*

<PortName> ::= **VHDL IDENTIFIER**

<PhysicalPinDesc> ::= *<PhysicalPin>* | (*<PhysicalPinList>*)

<PhysicalPinList> ::= *<PhysicalPin>* | *<PhysicalPinList>* , *<PhysicalPin>*

<PhysicalPin> ::= **INTEGER** | **VHDL IDENTIFIER**

* *<OpcodeTable>* ::= *<OpcodeDesc>* | *<OpcodeTable>* , *<OpcodeDesc>*

<OpcodeDesc> ::= *<OpcodeName>* (*<PatternList>*)

<PatternList> ::= *<Pattern>* | *<PatternList>* , *<Pattern>*

* *<Pattern>* ::= **BINARY STRING**

* *<UsageString>* ::= *<UsageDesc>* | *<UsageString>* , *<UsageDesc>*

<UsageDesc> ::= *<OpcodeName>* (*<UsageList>*)

* *<OpcodeName>* ::= Extest | Sample | Intest | Runbist | **VHDL IDENTIFIER**

<UsageList> ::= *<Usage>* | *<UsageList>* ; *<Usage>*

<Usage> ::= *<RegisterDecl>* | *<InitializeDecl>* | *<ShiftDecl>* | *<ResultDecl>* | *<ClockDecl>* | *<LengthDecl>*

<RegisterDecl> ::= Registers *<RegisterList>*

<RegisterList> ::= *<Register>* | *<RegisterList>* , *<Register>*

<Register> ::= **VHDL IDENTIFIER**

<InitializeDecl> ::= Initialize *<Register>* *<Pattern>*

<ShiftDecl> ::= Shift *<Register>*

<ResultDecl> ::= Result *<Pattern>*

<LengthDecl> ::= Length *<LengthSpec>*

<LengthSpec> ::= **INTEGER** cycles | **REAL** seconds

<ClockDecl> ::= Clock **VHDL IDENTIFIER** *<ClockSpec>*

<ClockSpec> ::= in *<TapState>* | shifted | **NULL**

<TapState> ::= Run_Test_Idle

* *<RegisterString>* ::= *<RegisterAssoc>* | *<RegisterString>* , *<RegisterAssoc>*

<RegisterAssoc> ::= *<Register>* (*<OpcodeList>*)

* *<OpcodeList>* ::= *<OpcodeName>* | *<OpcodeList>* , *<OpcodeName>*

* *<CellList>* ::= *<CellName>* | *<CellList>* , *<CellName>*

<CellName> ::= **VHDL IDENTIFIER**

* *<CellTable>* ::= *<CellEntry>* | *<CellTable>* , *<CellEntry>*

<CellEntry> ::= *<CellNumber>* (*<CellInfo>*)

<CellInfo> ::= *<CellSpec>* | *<CellSpec>* , *<DisableSpec>*

<CellSpec> ::= *<CellID>* , *<PortID>* , *<Function>* , *<SafeValue>*

<CellNumber> ::= **INTEGER**

<CellID> ::= **VHDL IDENTIFIER**

<PortID> ::= *<PortName>* | *

<Function> ::= Input | Output2 | Output3 | Control | Controlr | Internal | Clock | Bidir

<SafeValue> ::= 0 | 1 | X

<DisableSpec> ::= *<DisableCell>* , *<DisableVal>* , *<DisableResult>*

<DisableCell> ::= *<CellNumber>*

<DisableVal> ::= 0 | 1

<DisableResult> ::= Z | Weak0 | Weak1

Functional Test and Diagnosis: A Proposed JTAG Sample Mode Scan Tester

Mark F. Lefebvre

Digital Equipment Corporation
100 Minuteman Road
Andover, Massachusetts 01810

ABSTRACT

Emerging trends in physical interconnect technologies have made many of the conventional functional test and diagnosis tools difficult, if not impossible, to utilize in today's manufacture and test processes. The IEEE Standard 1149.1 boundary scan implementation provides the internal access required for analyzing nodal test data. This paper describes a JTAG Sample Mode Scan Tester being developed for diagnosis of at-speed failures in modules .

INTRODUCTION

Advances in physical interconnect technology, made necessary to meet increasing speed and packaging density requirements, are making physical access to the internal networks of a module (populated printed circuit board) increasingly difficult and in many cases, impossible. The traditional method of testing a module for functional defects has been with a functional tester and an edge-connector type fixture. Diagnosis has been typically performed through a combination of limited bed-of-nails access and a hand-held probe, which may or may not have been guided under program control.

This method has worked well for many years on through-hole modules utilizing 100 mil pitch component leads, or with modules of limited complexity. However, with the advent of surface mount technology and high pin count components with a lead pitch of 25 mils or less, the bed-of-nails and guided probe approaches to test have become impractical without the addition of test pads.

Recognizing the limitations of physical access to the internal networks and device leads, product designers are beginning to use boundary scan latches as a testability feature in product designs. Boundary scan latches allow for the capture of electrical stimulus and response data without the loading caused by physically probing the MUT (Module Under Test). Specifications defining a standard implementation for boundary scan have been developed by an industry sponsored committee called the Joint Test Action Group (JTAG) in the IEEE Standard 1149.1, herein referred to as JTAG. It is assumed that the reader is familiar with the IEEE standard. The details of the JTAG boundary scan implementation are provided in IEEE Standard 1149.1 [1].

This paper describes a Sample Mode Scan Tester that is currently being developed for the purpose of diagnosing at-speed functional faults on modules that incorporate the JTAG testability standard. The term "at-speed functional test" refers to the process of sampling response data at MUT speed. However, the JTAG Sample Mode process allow for the data to be shifted to the tester at a much slower speed. This is significant, as it provides a means of performing functional test without the need for the tester to keep pace with product clock speeds.

The tester, with the appropriate software tools, can also be used as a data acquisition system. This capability also facilitates the debug of engineering prototypes, similar to that of a logic analyzer.

PROBLEM STATEMENT

Functional Test Trends

The traditional functional test process can be categorized under one of two scenarios. The first concerns the use of traditional functional automatic test equipment (ATE) and involves stimulus being applied from the tester to the MUT through the edge

Reprinted from *IEEE Proceedings 1990 International Test Conference*, pages 294-303.

connector. The MUT response (ACQUISITION data) to the stimulus is captured by the tester, again through the edge connector. The results are then compared against a known good data base (EXPECT data), and a determination is made on whether the MUT passed or failed the test. Diagnosis is achieved by comparing the EXPECT data with the ACQUISITION data via a guided probe algorithm. In some applications, a fault dictionary is used in place of, or in addition to, the guided probe.

The second scenario is the use of product to test product. In this application, the MUT is plugged into a known good system box, and stimulus is applied via disk- or ROM-based diagnostics. Diagnosis is achieved through a combination of custom diagnostic routines, program directed probing, and the use of electronic instrumentation such as a logic analyzer or an oscilloscope. Fault diagnosis at such a test station is a very complicated process and requires an experienced technician or engineer. In both scenarios, diagnosis would be further complicated without any means of probing the MUT.

Because of the aforementioned physical access restrictions due to emerging module technologies, these traditional functional test methods have begun to break down. Without physical access, guided probe methodologies are no longer feasible. Similarly, the use of test pads for interfacing test instrumentation such as logic analyzers to the MUT is also limited. Due to these restrictions, it is clear that alternative methods of accessing MUT nodal test data are required. Sampling data via boundary scan latches is one such method.

Application Requirements

There are several application requirements that must be addressed to perform Sample Mode testing. The overall objective is to sample *deterministic* nodal test data in a *repeatable* fashion.

- Deterministic - the ability to sample predictable data (i.e. sample cycle *n*).
- Repeatable - given a program that samples cycle *n*, the ability to sample the same data each time the test program is executed.

In order to meet these objectives a number of provisions must be made, both from the tester and the MUT perspectives. The first involves synchronizing the execution of the test sequence to the operation of the tester. To achieve synchronization, the Sample Mode process must be triggered by some event on the MUT that is synchronous with the MUT clock.

The second requirement involves the execution of a repeatable test sequence on the MUT. This will allow test data to be captured in multiple executions of the test sequence and requires a provision for the test sequence to be initiated asynchronously by the tester.

The third requirement is that the tester must have physical access to the JTAG Test Access Port (TAP) interface in order to control the operation of the TAP, and to transfer data between the tester and the boundary scan devices. For the purposes of this application, the JTAG signal pins, in addition to any signals required for synchronization, are brought out to the MUT edge connector.

Finally, in order to effectively diagnose MUT failures to the failing component, the majority of MUT networks must be accessible via a boundary scan latch. As more MUT networks become connected to boundary scan latches, the level of diagnostic resolution increases accordingly [2].

Synchronization Requirements

A means of synchronizing the execution of the MUT test sequence must be developed that will allow deterministic sampling of test data while the MUT is operating at system speed. This usually requires the tester to initiate the test sequence and the tester to trigger off some event that is synchronous with the execution of this test sequence [3].

Once synchronization is achieved, any cycle of test data may be sampled by delaying the JTAG capture sequence such that it aligns with the desired cycle of the test sequence. This process may be repeated under tester control in order that multiple "snapshots" of test data are sampled and ultimately analyzed for diagnostic purposes.

JTAG SAMPLE MODE OPERATION

JTAG Overview

Boundary scan latches allow for the sampling of electronic stimulus and response data without impeding MUT functional performance. The set of boundary scan register latches can be considered as a very wide parallel load, serial shift register. The parallel inputs to the register are physically connected to the device I/O, thereby providing access to the internal networks of the MUT. At the module level, each boundary scan device is daisy-chained to form a scan chain comprised of the individual devices..

The 4 TAP signals (TDI, TDO, TCK, and TMS) are accessible to the Sample Mode Scan Tester via the module edge connector. Figure 1 illustrates how individual JTAG-compatible components can be connected at the module level and brought to the edge connector. Note that the scan output TDO from one device is connected to scan input TDI of the next device in the scan chain. TMS and TCK are connected in parallel to each device in this implementation. A brief description of the TAP signals is also given below.

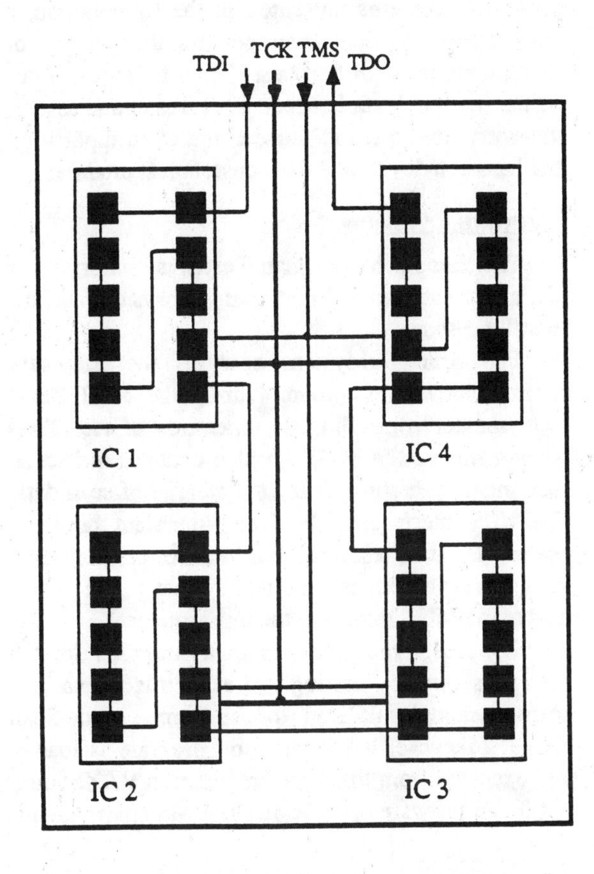

Figure 1 Module-level JTAG Scenario

- **Test Data Input (TDI):** TDI is the serial input to the JTAG device by which test or instruction data are loaded.
- **Test Data Output (TDO):** TDO is the serial output of the JTAG device by which test or instruction data are shifted from a given device.
- **Test Mode Select (TMS):** The logic state of TMS controls and distinguishes the functionality of the TAP controller. The value of TMS is clocked into the TAP controller on the rising edge of TCK.
- **Test Clock (TCK):** TCK is a dedicated test clock input that is normally free-running. The frequency of TCK will determine the speed at which we shift test data from the MUT to the tester.

The JTAG state diagram is shown below in Figure 2. Note that the functionality of the state machine is controlled by manipulating TMS and TCK. The values of TMS are shown.

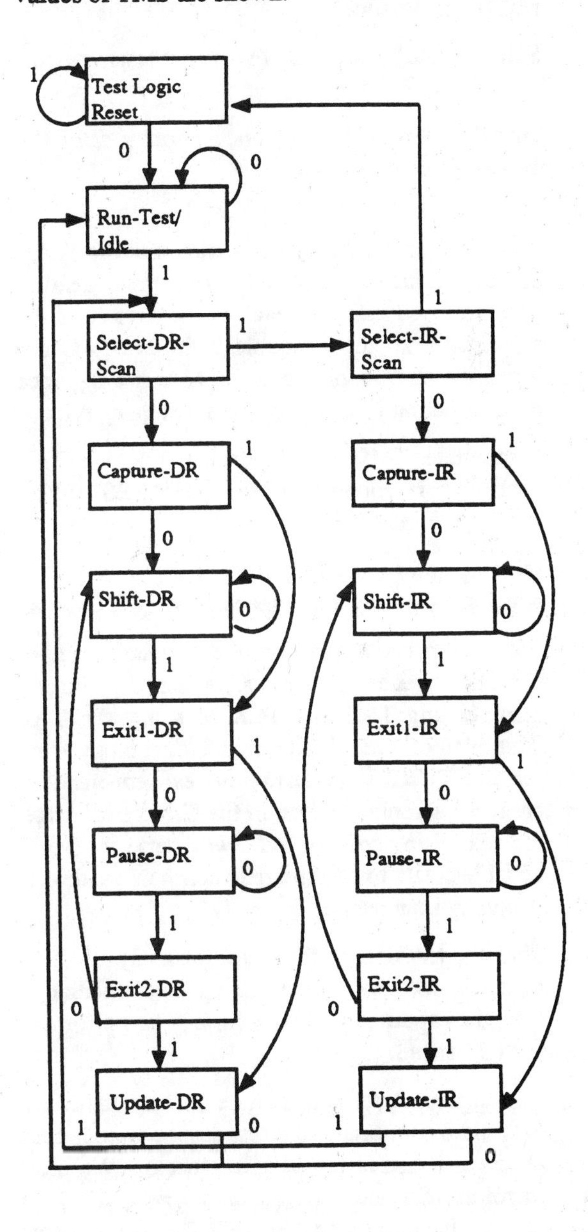

Figure 2 : JTAG TAP State Diagram

JTAG Sample Mode Operational Description

The JTAG Sample Mode sequence is a 3-step process involving the execution of a test, the capture of test data, and the transfer of that test data to the Sample Mode Scan Tester.

TEST - Execute test sequence at MUT speed. The test sequence could be either diagnostics or self-test.

CAPTURE - Sample test data without affecting MUT performance.

SHIFT - Shift nodal data to tester at tester speed.

In JTAG terms, this could be algorithmically described as follows:

1. Load the Sample Mode instruction into the JTAG device Instruction Registers. A Sample instruction must be loaded for each of the devices in the scan chain, so for a module with six JTAG devices, the instruction would be 48 bits long since the Instruction Register has eight bits per device.
2. Manipulate TMS and TCK such that the TAP controller is looping in the RUN-TEST/IDLE state (see Figure 2).
3. Initiate the execution of the MUT test sequence which is operating at the speed of the MUT clock.
4. Capture the nodal response of the desired cycle of the test sequence. This is accomplished by further manipulating TMS and TCK such that the TAP controller of each JTAG device transitions from the CAPTURE state during the test sequence cycle to be sampled. Once in the CAPTURE state, the next rising edge of TCK will cause the MUT nodal test data to be captured at each of the device boundary scan latches.
5. Shift the test data to the tester for analysis. This shift step does *not* have to take place immediately after the capture, nor does it have to be performed at MUT speed.
6. Repeat steps 4 and 5 as required. Since the MUT clock is free-running, the test sequence continues to execute during the SHIFT process. We must therefore take into account the elapsed MUT cycles occurring during the SHIFT process before we perform additional samples.

In order to control the JTAG boundary scan latches for Sample Mode operation, the appropriate JTAG protocol must be programmed by the Sample Mode Scan Tester. This process is presented later in greater detail relative to an actual module that is utilized as a test case.

TESTER DESCRIPTION

In response to the JTAG approach to the testability problems presented in the Introduction, a test system has been developed for the purpose of overcoming these difficulties. What follows is a description of the Sample Mode Scan Tester and associated software tools, and a discussion of an application that has been developed for a custom test module.

Hardware Overview

The Sample Mode Scan Tester is comprised of two major subsystems, the Scan Subsystem and the Host Processor.

The Scan Subsystem contains the necessary control hardware for manipulating the MUT interface, and for controlling the operation of the JTAG components on the MUT. It also contains the actual scan memory hardware for the storage of scan data. The MUT interfaces to the Scan Subsystem through a custom designed test head that includes the necessary logic for synchronizing the test system to the operation of the MUT logic and the test sequence.

The Host Processor acts as a controller for the scan subsystem by hosting and executing the various software modules utilized for the Sample Mode Scan Tester. Likewise, the application program is loaded and executed from the Host Processor, a VAXStation 3500. All software operates in the VMS environment.

Software Overview

The Sample Mode Scan Tester software package provides a comprehensive suite of tools that supports the entire spectrum of the test process, from test program generation to the graphic display of diagnostic data. These tools are integrated into a menu-driven platform that serves as a front end to the test engineer or manufacturing technician. The list of the software modules developed for the Sample Mode Scan Tester includes the following:

- Pattern Converter
- Pattern Editor
- Learn Module

- Tester Control Module
- Boundary Scan Interconnect Test Generation Module
- Waveform Display Module
- Diagnostic Module

Pattern Convertor

The test stimuli have been simulated using a proprietary simulator. The resulting response data must be converted into the format required by the Sample Mode Scan Tester.

The function of the Pattern Converter is to translate these simulation-generated response patterns (called EXPECT patterns) into the correct binary format used by the Scan Subsystem. The pattern converter reads in multiple types of simulator-generated patterns, translates the patterns into an intermediate binary format, and then renders the patterns into the specified tester format. Specific output software modules can be written to tailor the output format to other testers.

Pattern Editor

The Pattern Editor is a tool used for editing and manipulating the EXPECT pattern database. The Editor also has programmable software triggering capability which allows the user to search and trigger on the data based on a sequence of events, as specified by the user. The data can then be displayed or manipulated under program control.

For example, the test engineer may wish to mask indeterminate ACQUISITION data during a scan operation. The Pattern Editor would allow the engineer to specify a sequence of data for the Pattern Editor to "watch for" during the scan operation. When the Editor sees this sequence, the specified bits in the EXPECT database would then be masked by the Editor. Other functions include the ability to edit the database by commands specified through the Waveform Display module.

Learn Module

Two methods of data generation and testing are being developed for the Sample Mode Scan Tester. The first involves generating EXPECT data via simulation and converting this data to the tester format as mentioned above. The second involves "learning" the EXPECT data from a known-good MUT. The latter scenario requires the tester to sample the nodal test data via the JTAG interface much like the normal operation of the tester. However, instead of performing a test on the ACQUISITION data, a nodal database will be constructed that will serve as the EXPECT database.

The Learn Module is an extension of the functions in the Pattern Editor. Its purpose is to compare multiple databases generated from sampling and operate on that data (change states or mask data) based on certain conditions.

The function of this "programmable" Pattern Editor is to compare multiple databases generated from sampling, and then to operate on that data based on conditions defined by the user. The resulting database will serve as the EXPECT data for future tests of that MUT.

Tester Control Program

The function of the Tester Control Program is to interface to the tester hardware, control the testing functions, and allow links to other tools that comprise the tester tools suite. The Tester Control Program is responsible for controlling all system functions such as synchronization, JTAG protocol and data acquisition. It also serves as a front end for test program generation, fault diagnosis and any other applications for the tester.

Interconnect Test Generation Module

In addition to Sample Mode testing, the tester also has the capability to perform boundary scan interconnect testing. A test generation process has been developed to provide test patterns for the purpose of detecting MUT interconnect failures such as shorts and opens. An output module adds the JTAG scan protocol in addition to translating the test patterns from simulator output into tester format.

Waveform Display Module

For the purposes of debug and fault diagnosis, a waveform display tool is being developed for the Sample Mode Scan Tester. The display will have similar functions to those of a logic analyzer. The tool will draw a waveform from the test data that has been sampled from the MUT. MUT failures will be highlighted on the display.

Diagnostic Module

The tester hardware currently performs a real-time hardware compare of the EXPECT data and the ACQUISITION data sampled from the MUT during the execution of the test program. The system passes to the Host Processor the test cycle number and bit(s) which do not match. The Diagnostic Module utilizes this information together with the MUT CAD information to determine the earliest failing cycle. The Diagnostic Module then isolates the fault to a single component and gives a level of confidence of the diagnosis.

MUT DESCRIPTION

The MUT is an internally developed test module designed and fabricated to demonstrate the prototype capabilities of the Sample Mode Scan Tester. The module has been simulated and the resultant EXPECT data has been converted into tester format. A block diagram of the MUT is shown in Figure 3.

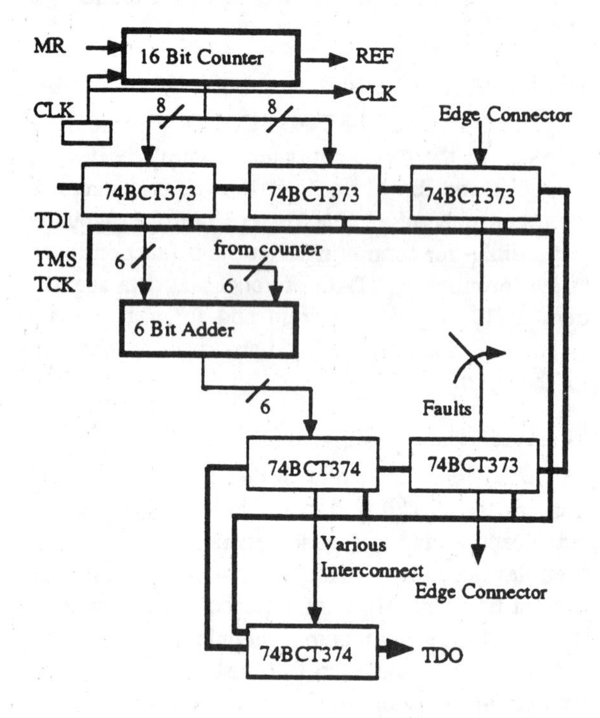

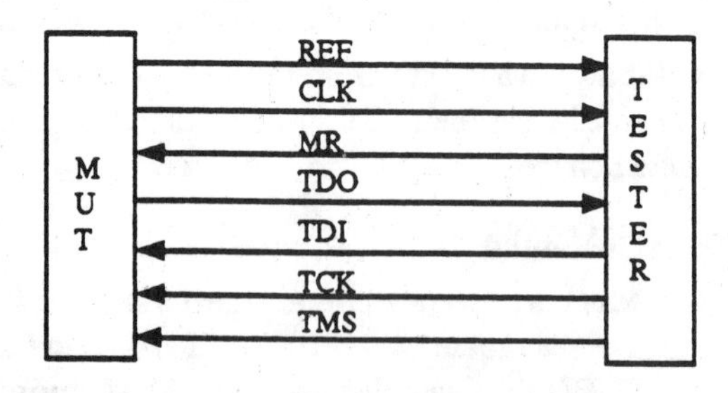

Figure 3 : MUT Block Diagram

The counter's parallel output serves as the stimulus to the module. Therefore, a new "test pattern" is applied to the module for every counter cycle. Since the counter is synchronous with the clock, a test pattern exists for every clock cycle. The test sequence is initiated by the signal MR (Master Reset), which triggers the counter. When the counters count to terminal count, the signal REF is asserted, indicating the completion of the test sequence. When the test sequence is looping, this provision allows for synchronization of the Sample Mode Scan Tester to the test sequence.

The boundary scan ring is comprised of 6 TI SCOPE™ Octals, each having 18 bits of scan data. With this configuration, the boundary scan ring is 108 bits long.

The interface between the tester and the module is also shown in Figure 3. The following is a description of the signal names.

REF: This signal will serve to synchronize the tester to the execution of the test sequence. REF is a trigger signal that is asserted at the completion of the test sequence, synchronously with the system clock.

CLK: This is the system clock and is asserted on the rising edge. The frequency of CLK is 50 MHz

MR: This signal is a Master Reset which originates from the tester. MR allows the tester to asynchronously restart the test sequence.

TDO: Test Data Out is the JTAG serial output of the module under test.

TDI: Test Data In is the JTAG serial input to the module under test.

TCK: Test Clock is the clock used to synchronize the JTAG operations. The speed of TCK is 25 MHz and is provided by the tester.

TMS: Test Mode Select is used by the tester to control the JTAG state machine.

The logic necessary to ensure accurate and programmable synchronization of the Sample Mode process is implemented on an interface between the tester hardware and the MUT.

TEST METHODOLOGY

Overview

Programmable sampling is achieved by programming the Sample Mode Scan Tester to manipulate the JTAG state machine of each of the JTAG devices. Controlling the Sample Mode process involves looping the tester while the TAP Controller is in the

RUN-TEST/IDLE state, then manipulating TMS and TCK to arrive at the CAPTURE state within the appropriate MUT test cycle to be sampled. Refer to the JTAG state diagram in Figure 2.

The specific provisions of this application are as follows:

- **Initial Capture:** Determines when the initial CAPTURE occurs, based on the delays associated with the MUT interface, the tester, and the amount of MUT clock cycles elapsing between the trigger signal and the MUT cycle being tested. This delay will determine the time between REF and the first CAPTURE.
- **Capture Interval:** Determines the interval between successive CAPTURES by accounting for SHIFT overhead. At a minimum, this provision must take into account the number of MUT test cycles that have elapsed during the shift process. In other words, since the MUT clock is still operating, and since the test sequence is still executing unimpeded by the Sample Mode process, the tester must take the elapsed MUT time into account when determining the next cycle to be sampled. This provision allows multiple samples of boundary scan data to be captured during a given pass through the test sequence.
- **Multiple Pass Sampling:** Allows multiple passes of the test sequence to be executed. This is accomplished by programming the tester to restart the test sequence after a given sample, or by having the test sequence continuously loop in a free-running mode. The latter simply requires the tester to track the test cycles, which occurs by default via the synchronization process. This process is illustrated below in Figure 4.

This process provides the capability to sample any cycle of test data during any pass of the test sequence. Subsequently, this sampling procedure is repeated until all desired cycles are sampled, or until the scan tester pattern memory is filled.

For cases where the MUT clock is operating at the same frequency as the JTAG test clock, there will be a one-to-one correspondence between MUT cycles and the tester cycles. If, however, these clocks do not operate at the same frequency, one must account for the difference when performing Sample Mode testing. It is recommended that TCK be programmed to be an integer divisor of the MUT clock. For instance, if the MUT clock is operating at 50 MHz, the test clock should be programmed at 25, 10, or 5 MHz.

Figure 4 : Multiple Pass Sampling

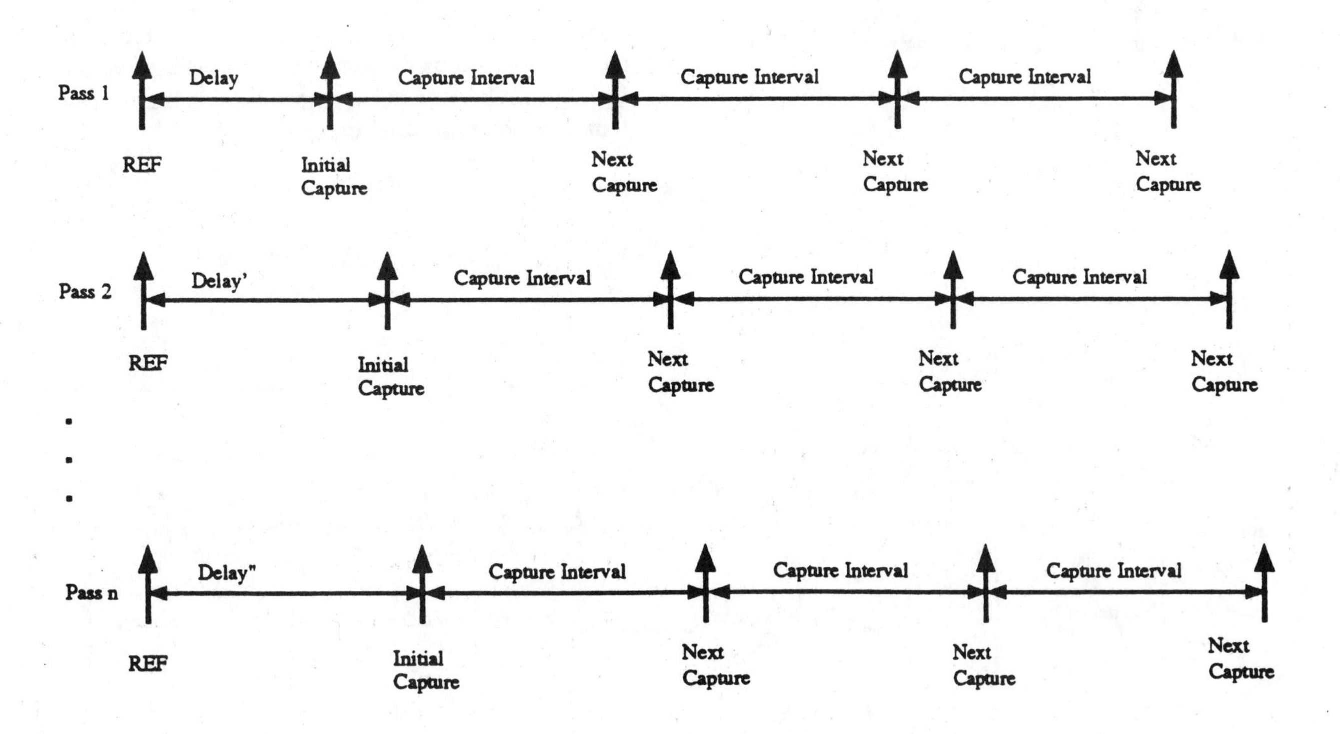

Tester Setup

Prior to loading and executing the JTAG protocol to perform Sample Mode testing, the appropriate tester parameters must be assigned. These parameters include power supply levels, logic levels and timing values, and the assignment of tester scan resources to the appropriate MUT signals.

JTAG Setup

Once the appropriate tester parameters have been assigned, the boundary scan devices must be loaded with the Sample instruction. Assuming the TAP Controller is initially in the Run-Test/Idle state, the sequence for setting up the state machine to perform Sample Mode testing is as follows [4]:

1. Select-IR-Scan - This will select the Instruction Register of each boundary scan device in order that the Sample instruction can be loaded. When TMS is held low in this state, a scan sequence for the Instruction Registers is initiated.
2. Shift-IR - While in this state, the tester will shift the Sample instruction into the Instruction Register via TDI. Since there are 6 boundary scan devices on the MUT, the instruction is 48-bits long (6 x 8 bits).
3. Update-IR - During this state, the Instruction Register contents become a valid instruction.
4. Return to the Run-Test/Idle state for the appropriate time specified by the delay sequence.

In this sequence, the Sample instruction would assume the binary value of 10000010 for each of the 6 devices in the boundary scan chain.

Sample Procedure

Now that the JTAG devices have been set up to capture nodal test data, the tester initiates the MUT test sequence by asserting MR (Master Reset). Again, the test patterns are simply the output of the counter circuit. When the tester sees the trigger signal (REF) asserted, it interprets the next clock cycle as cycle 1 of the test sequence.

All samples will be relative to REF. For example, if we wish to capture data from the 4th cycle of the test sequence, we must program the tester such that the INITIAL CAPTURE occurs after 4 MUT clock cycles, relative to REF.

For successive captures, we have to account for the 5 (EXIT1-DR, UPDATE-DR, RUN-TEST/IDLE, SELECT-DR-SCAN, and CAPTURE-DR) TAP state transitions when calculating the amount of tester delay to perform the next sample. This TAP overhead is added to the time elapsed while shifting data to the tester, and must also be accounted for when specifying the next cycle of the test sequence to be sampled. This value is the CAPTURE INTERVAL.

Using the previous example and keeping in mind that the scan chain is 108 bits long for our MUT, since we have captured data from cycle 4, the next potential cycle to be sampled would be cycle 117 (cycle 4 + 108 cycles of shifting + 5 JTAG state transitions). This sequence is illustrated in Figure 5. In situations where the test clock, TCK, and the system clock are not operating at the same frequency, this CAPTURE INTERVAL would account for the difference.

For each ensuing pass of the test sequence, the tester delays each sample by one cycle in order to create a contiguous database. Such a database would facilitate fault diagnosis by providing a means of determining the earliest failing test cycle.

Figure 5 : Sample Mode Example

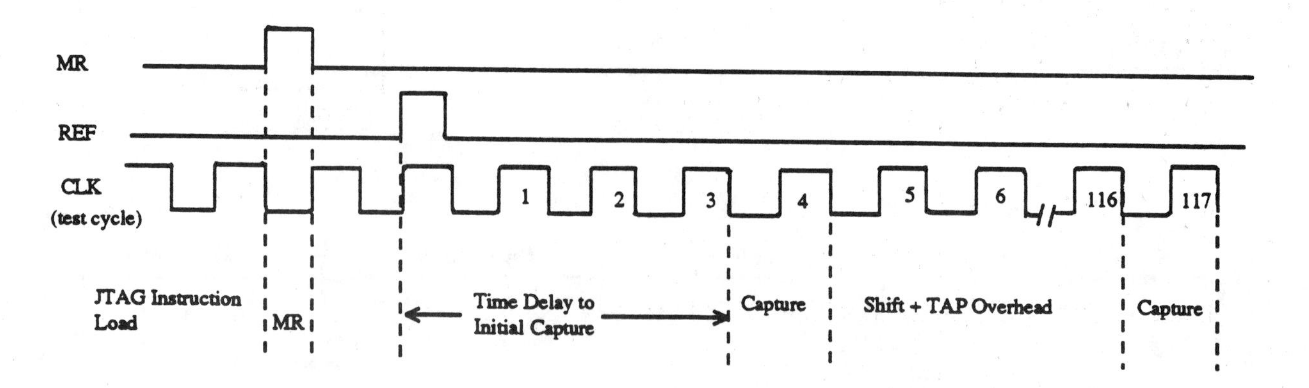

Using the previous example, the resulting test database would resemble the structure shown in Figure 6. Other software tools, such as the Waveform Display or the Diagnostic Module, could then access this database for diagnostic or display purposes. It should be noted that the Sample Mode Scan Tester is not restricted to this pattern sequencing format.

RESULTS

The Sample Mode Scan Tester prototype system is fully capable of sampling nodal test data via the JTAG boundary scan protocol. Using an internally developed test module as a test case, applications have been developed that successfully sample the response data from the MUT, and diagnose MUT failures to the failing scan bit(s) in the scan chain.

A Diagnostic Module, which is currently being developed, will analyze this ACQUISITION data in addition to the corresponding MUT CAD data to further isolate the failure to the failing device. Other tools, such as the Waveform Module, display the ACQUISITION data for further analysis.

CONCLUSIONS

Initial results indicate that Sample Mode testing is a viable means of diagnosing module faults in cases where lack of physical access prevents traditional functional test methods from being used. Using boundary scan latches at device boundaries, and connecting these devices to bring the resulting scan chain to the module edge-connector, provides the internal observation points necessary to diagnose functional test failures.

There are, however, certain limitations with this application. For instance, the MUT must contain the signals and logic necessary to ensure synchronization of the MUT test sequence and the operation of the tester. Also, the relationship of the MUT clock and the test clock will determine how effective the Sample Mode process is for a given application. Assume, for instance, that the test clock is operating at half the speed of the system clock. Since there would be two system cycles for each tester cycle, this relationship would then require that the test clock be variable in order to sample data from both system cycles.

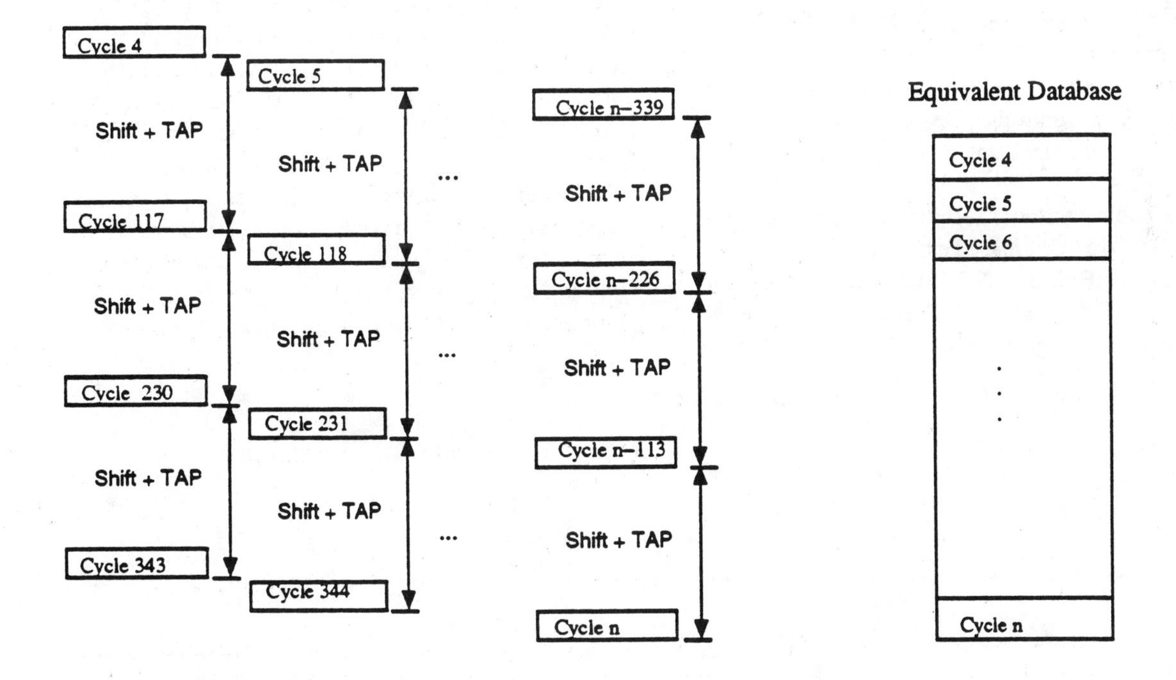

Figure 6 : Example of a Sample Mode Test Database

Finally, unless the MUT has a large percentage of devices incorporating the JTAG standard, diagnostic resolution will be lacking, thereby limiting the usefulness of the Sample Mode application.

The Sample Mode Scan Tester has demonstrated that a low cost test system can be utilized to sample data from modules operating at much higher speeds. This is a significant achievement, as it reduces the need for test equipment to keep pace with product speeds.

ACKNOWLEDGMENTS

The author gratefully acknowledges Lorraine Zambre, Phil McKinley, Al Cossette, John Sweeney, Dave Florcik, Hook Wong and Alex Sokolovsky for their dedication and contributions leading to the development of the Sample Mode Scan Tester.

REFERENCES

[1] IEEE Standard 1149.1-1990: IEEE Standard Test Access Port and Boundary Scan Architecture, 1990.

[2] Sweeney, John, "Testability Implemented in the VAX 6000 Model 400 Computer," *International Test Conference Proceedings*, September, 1990.

[3] Vining, Sue, "Tradeoff Decisions Made for a P1149.1 Controller Design, " *International Test Conference Proceedings*, August 1989, pp. 47-54.

[4] Dahbura, Anton T., Uyar, M., and Yau, Chi W., "An Optimal Test Sequence for the JTAG/ IEEE P1149.1 Test Access Port Controller," *International Test Conference Proceedings*, August 1989, pp. 55-62.

INDEX

Colin M. Maunder

Colin Maunder is an Engineering Adviser at the British Telecommunications Research Laboratories, Martlesham Heath, Ipswich, U.K. His work includes design-for-test consultancy on chip and system development projects as well as the development of new techniques for the maintenance and repair of electronic equipment.

Mr Maunder has been involved in research into test generation and design-for-testability since 1976. Between 1979 and 1981, he worked on the development of the CAMELOT testability assessment program. Later, between 1983 and 1985, he contributed to the development of HITEST -- the first commercial knowledge-based test generation system. More recently, he has been the architect of the design-for-test features included in several integrated circuits designed by British Telecom.

In 1986, Mr Maunder became a member of the Joint Test Action Group (JTAG) and helped draft JTAG's first technical proposal. Later he became chair of JTAG's Technical Subcommittee and, on transfer of work to the IEEE, he became chair of the IEEE P1149.1 Working Group. He is also a charter member of the IEEE Computer Society Test Technology Technical Committee's Testability Bus Standards Steering Committee.

Mr Maunder has lectured on design-fortestability and test generation on many public training courses, both in Europe and in North America. He is a member of the organizing committee for the European Design-for-Test Workshop and of the organizing and programme committees for the European Test Conference.

Mr Maunder received a BSc in Physics from Imperial College, London, in 1973. He is a Chartered Engineer, a member of the IEE (U.K.) and a Senior Member of the IEEE.

Rodham E. Tulloss

Rodham Tulloss is a Distinguished Member of Technical Staff at AT&T Bell Laboratories, Engineering Research Center, Princeton, NJ. He was a Supervisor for 11 years, in which position he initiated and led the development of research into fault simulation, test data translation, automated test generation, built-in self-test, and boundary-scan. At present, he is involved in standards development, in the introduction of built-in self-test and boundary-scan into AT&T designs, and in studies in support of new test technologies.

Dr Tulloss was the first North American member of the Joint Test Action Group (JTAG). He founded the North American JTAG interest group and served as co-chair and, more recently, vice-chair of the IEEE P1149.1 Working Group. He has played a significant role in educating the engineering community about boundary-scan, for example as technical consultant to the 1989 IEEE Satellite Seminar on the topic.

Dr Tulloss is a charter member of the IEEE Computer Society Test Technology Technical Committee's Testability Bus Standards Steering Committee. Between 1981 and 1983, he was co-editor of the *TTTC Newsletter* and played a major role in the expansion of its content which eventually led to the creation of *IEEE Design and Test of Computers* magazine. He is a consultant to the program to develop transatlantic television educational programmes in electronic engineering.

Dr. Tulloss received his PhD from the University of California, Berkeley, in 1971. He also holds an MS in Mathematics from the University of California and a BS in Mathematics from Union College, Schenectady, New York. He is a Senior Member of the IEEE and, outside of engineering, is a prize-winning poet and a recognized expert on fungi of the genus *Amanita*.

Offices of the IEEE Computer Society

Headquarters Office
1730 Massachusetts Avenue, N.W.
Washington, DC 20036-1903
Phone: (202) 371-0101 — Fax: (202) 728-9614

Publications Office
P.O. Box 3014
10662 Los Vaqueros Circle
Los Alamitos, CA 90720-1264
Membership and General Information: (714) 821-8380
Publication Orders: (800) 272-6657 — Fax: (714) 821-4010

European Office
13, avenue de l'Aquilon
B-1200 Brussels, BELGIUM
Phone: 32-2-770-21-98 — Fax: 32-3-770-85-05

Asian Office
Ooshima Building
2-19-1 Minami-Aoyama, Minato-ku
Tokyo 107, JAPAN
Phone: 81-3-408-3118 — Fax: 81-3-408-3553

IEEE Computer Society Press Publications

Monographs: A monograph is an authored book consisting of 100-percent original material.
Tutorials: A tutorial is a collection of original materials prepared by the editors, and reprints of the best articles published in a subject area. Tutorials must contain at least five percent of original material (although we recommend 15 to 20 percent of original material).
Reprint collections: A reprint collection contains reprints (divided into sections) with a preface, table of contents, and section introductions discussing the reprints and why they were selected. Collections contain less than five percent of original material.
Technology series: Each technology series is a brief reprint collection — approximately 126-136 pages and containing 12 to 13 papers, each paper focusing on a subset of a specific discipline, such as networks, architecture, software, or robotics.
Submission of proposals: For guidelines on preparing CS Press books, write thc Editorial Dircctor, IEEE Computer Society Press, PO Box 3014, 10662 Los Vaqueros Circle, Los Alamitos, CA 90720-1264, or telephone (714) 821-8380.

Purpose

The IEEE Computer Society advances the theory and practice of computer science and engineering, promotes the exchange of technical information among 100,000 members worldwide, and provides a wide range of services to members and nonmembers.

Membership

All members receive the acclaimed monthly magazine *Computer*, discounts, and opportunities to serve (all activities are led by volunteer members). Membership is open to all IEEE members, affiliate society members, and others seriously interested in the computer field.

Publications and Activities

***Computer* magazine:** An authoritative, easy-to-read magazine containing tutorials and in-depth articles on topics across the computer field, plus news, conference reports, book reviews, calendars, calls for papers, interviews, and new products.

Periodicals: The society publishes six magazines and five research transactions. For more details, refer to our membership application or request information as noted above.

Conference proceedings, tutorial texts, and standards documents: The IEEE Computer Society Press publishes more than 100 titles every year.

Standards working groups: Over 100 of these groups produce IEEE standards used throughout the industrial world.

Technical committees: Over 30 TCs publish newsletters, provide interaction with peers in specialty areas, and directly influence standards, conferences, and education.

Conferences/Education: The society holds about 100 conferences each year and sponsors many educational activities, including computing science accreditation.

Chapters: Regular and student chapters worldwide provide the opportunity to interact with colleagues, hear technical experts, and serve the local professional community.

Other IEEE Computer Society Press Texts

Monographs

Analyzing Computer Architecture
Written by J.C. Huck and M.J. Flynn
(ISBN 0-8186-8857-2); 206 pages

Desktop Publishing for the Writer: Designing, Writing, Developing
Written by Richard Ziegfeld and John Tarp
(ISBN 0-8186-8840-8); 380 pages

Integrating Design and Test: Using CAE Tools for ATE Programming
Written by K.P. Parker
(ISBN 0-8186-8788-6 (case)); 160 pages

JSP and JSD: The Jackson Approach to Software Development (Second Edition)
Written by J.R. Cameron
(ISBN 0-8186-8858-0); 560 pages

National Computer Policies
Written by Ben G. Matley and Thomas A. McDannold
(ISBN 0-8186-8784-3); 192 pages

Physical Level Interfaces and Protocols
Written by Uyless Black
(ISBN 0-8186-8824-6); approximately 272 pages

Protecting Your Proprietary Rights in the Computer and High Technology Industries
Written by Tobey B. Marzouk, Esq.
(ISBN 0-8186-8754-1); 224 pages

Tutorials

Advanced Computer Architecture
Edited by D.P. Agrawal
(ISBN 0-8186-0667-3); 400 pages

Advanced Microprocessors and High-Level Language Computer Architectures
Edited by V. Milutinovic
(ISBN 0-8186-0623-1); 608 pages

Advances in Distributed System Reliability
Edited by Suresh Rai and Dharma P. Agrawal
(ISBN 0-8186-8907-2); 352 pages

Computer Architecture
Edited by D.D. Gajski, V.M. Milutinovic, H. Siegel, and B.P. Furht
(ISBN 0-8186-0704-1); 602 pages

Computer Communications: Architectures, Protocols, and Standards (Second Edition)
Edited by William Stallings
(ISBN 0-8186-0790-4); 448 pages

Computer Graphics (2nd Edition)
Edited by J.C. Beatty and K.S. Booth
(ISBN 0-8186-0425-5); 576 pages

Computer Graphics Hardware: Image Generation and Display
Edited by H.K. Reghbati and A.Y.C. Lee
(ISBN 0-8186-0753-X); 384 pages

Computer Graphics: Image Synthesis
Edited by Kenneth Joy, Max Nelson, Charles Grant, and Lansing Hatfield
(ISBN 0-8186-8854-8); 384

Computer and Network Security
Edited by M.D. Abrams and H.J. Podell
(ISBN 0-8186-0756-4); 448 pages

Computer Networks (4th Edition)
Edited by M.D. Abrams and I.W. Cotton
(ISBN 0-8186-0568-5); 512 pages

Computer Text Recognition and Error Correction
Edited by S.N. Srihari
(ISBN 0-8186-0579-0); 364 pages

Computers for Artificial Intelligence Applications
Edited by B. Wah and G.-J. Li
(ISBN 0-8186-0706-8); 656 pages

Database Management
Edited by J.A. Larson
(ISBN 0-8186-0714-9); 448 pages

Digital Image Processing and Analysis: Volume 1: Digital Image Processing
Edited by R. Chellappa and A.A. Sawchuk
(ISBN 0-8186-0665-7); 736 pages

Digital Image Processing and Analysis: Volume 2: Digital Image Analysis
Edited by R. Chellappa and A.A. Sawchuk
(ISBN 0-8186-0666-5); 670 pages

Digital Private Branch Exchanges (PBXs)
Edited by E.R. Coover
(ISBN 0-8186-0829-3); 400 pages

Distributed Computing Network Reliability
Edited by Suresh Rai and Dharma P. Agrawal
(ISBN 0-8186-8908-0); 368 pages

Distributed Control (2nd Edition)
Edited by R.E. Larson, P.L. McEntire, and J.G. O'Reilly
(ISBN 0-8186-0451-4); 382 pages

Distributed Database Management
Edited by J.A. Larson and S. Rahimi
(ISBN 0-8186-0575-8); 580 pages

Distributed-Software Enginee..ng
Edited by S.M. Shatz and J.-P. Wang
(ISBN 0-8186-8856-4); 304 pages

DSP-Based Testing of Analog and Mixed-Signal Circuits
Edited by M. Mahoney
(ISBN 0-8186-0785-8); 272 pages

Fault-Tolerant Computing
Edited by V.P. nelson and B.D. Carroll
(ISBN 0-8186-0677-0 (paper) 0-8186-8667-4 (case)); 432 pages

Gallium Arsenide Computer Design
Edited by V.M. Milutinovic and D.A. Fura
(ISBN 0-8186-0795-5); 368 pages

Human Factors in Software Development (2nd Edition)
Edited by B. Curtis
(ISBN 0-8186-0577-4); 736 pages

Integrated Services Digital Networks (ISDN) (Second Edition)
Edited by W. Stallings
(ISBN 0-8186-0823-4); 404 pages

For Further Information:

IEEE Computer Society, 10662 Los Vaqueros Circle, P.O. Box 3014,
Los Alamitos, CA 90720-1264

IEEE Computer Society, 13, Avenue de l'Aquilon, 2,
B-1200 Brussels, BELGIUM

IEEE Computer Society,
Ooshima Building, 2-19-1 Minami-Aoyama,
Minato-ku, Tokyo 107, JAPAN

Interconnection Networks for Parallel and Distributed Processing
Edited by C.-L. Wu and T.-Y. Feng
(ISBN 0-8186-0574-X); 500 pages

Local Network Equipment
Edited by H.A. Freeman and K.J. Thurber
(ISBN 0-8186-0605-3); 384 pages

Local Network Technology (3rd Edition)
Edited by W. Stallings
(ISBN 0-8186-0825-0); 512 pages

Microprogramming and Firmware Engineering
Edited by V. Milutinovic
(ISBN 0-8186-0839-0); 416 pages

Modeling and Control of Automated Manufacturing Systems
Edited by A.A. Desrochers
(ISBN 0-8186-8916-1); 384 pages

Modern Design and Analysis of Discrete-Event Computer Simulations
Edited by E.J. Dudewicz and Z. Karian
(ISBN 0-8186-0597-9); 486 pages

New Paradigms for Software Development
Edited by William Agresti
(ISBN 0-8186-0707-6); 304 pages

Object-Oriented Computing--Volume 1: Concepts
Edited by Gerald E. Peterson
(ISBN 0-8186-0821-8); 214 pages

Object-Oriented Computing--Volume 2: Implementations
Edited by Gerald E. Peterson
(ISBN 0-8186-082108); 214 pages

Office Automation Systems (Second Edition)
Edited by H.A. Freemand and K.J. Thurber
(ISBN 0-8186-0822-6); 324 pages

Parallel Architectures for Database Systems
Edited by A. R. Hurson, L.L. Miller, and S.H. Pakzad
(ISBN 0-8186-8838-6); 478 pages

Programming Productivity: Issues for the Eighties (Second Edition)
Edited by C. Jones
(ISBN 0-8186-0681-9); 472 pages

Recent Advances in Distributed Database Management
Edited by C. Mohan
(ISBN 0-8186-0571-5); 500 pages

Reduced Instruction Set Computers (Second Edition)
Edited by W. Stallings
(ISBN 0-8186-8943-9); 448 pages

Reliable Distributed System Software
Edited by J.A. Stankovic
(ISBN 0-8186-0570-7); 400 pages

Robotics Tutorial (2nd Edition)
Edited by C.S. G. Lee, R.C. Gonzalez, and K.S. Fu
(ISBN 0-8186-0658-4); 630 pages

Software Design Techniques (4th Edition)
Edited by P. Freeman and A.I. Wasserman
(ISBN 0-8186-0514-0); 736 pages

Software Engineering Project Management
Edited by R. Thayer
(ISBN 0-8186-0751-3); 512 pages

Software Maintenance
Edited by G. Parikh and N. Zvegintzov
(ISBN 0-8186-0002-0); 360 pages

Software Management (3rd Edition)
Edited by D.J. Reifer
(ISBN 0-8186-0678-9); 526 pages

Software-Oriented Computer Architecture
Edited by E. Fernandez and T. Lang
(ISBN 0-8186-0708-4); 376 pages

Software Reusability
Edited by Peter Freeman
(ISBN 0-8186-0750-5); 304 pages

Software Risk Management
Edited by B.W. Boehm
(ISBN 0-8186-8906-4); 508 pages

Standards, Guidelines, and Examples on System and Software Requirements Engineering
Edited by Merlin Dorfman and Richard H. Thayer
(ISBN 0-8186-8922-6); 626 pages

System and Software Requirements Engineering
Edited by Richard H. Thayer and Merlin Dorfman
(ISBN 0-8186-8921-8); 740 pages

Test Generation for VLSI Chips
Edited by V.D. Agrawal and S.C. Seth
(ISBN 0-8186-8786-X); 416 pages

VSLI Technologies: Through the 80s and Beyond
Edited by D.J. McGreivy and K.A. Pickar
(ISBN 0-8186-0424-7); 346 pages

VLSI Testing and Validation Techniques
Edited by H. Reghbati
(ISBN 0-8186-0668-1); 616 pages

Reprint Collections

Dataflow and Reduction Architectures
Edited by S.S. Thakkar
(ISBN 0-8186-0759-9); 460 pages

Expert Systems: Software Methodology
Edited by Peter Raeth
(ISBN 0-8186-8904-8); 476 pages

Logic Design for Testability
Edited by C.C. Timoc
(ISBN 0-8186-0573-1); 324 pages

Microprocessors and Microcomputers (3rd Edition)
Edited by J.T. Cain
(ISBN 0-8186-0585-5); 386 pages

Software (3rd Edition)
Edited by M.V. Zelkowitz
(ISBN 0-8186-0789-0); 440 pages

VLSI Technologies and Computer Graphics
Edited by H. Fuchs
(ISBN 0-8186-0491-3); 490 pages

Artifical Neural Networks Technology Series

Artificial Neural Networks: Concept Learning
Edited by J. Diederich
(ISBN 0-8186-2015-3); 140 pages

Artificial Neural Networks: Electronic Implementation
Edited by Nelson Morgan
(ISBN 0-8186-2029-3); approximately 192 pages

Artificial Neural Networks: Theoretical Concepts
Edited by V. Vemuri
(ISBN 0-8186-0855-2); 160 pages

Software Technology Series

Computer-Aided Software Engineering (CASE)
Edited by E.J. Chikofsky
(ISBN 0-8186-1970-8); 110 pages

Communications Technology Series

Multicast Communication in Distributed Systems
Edited by Mustaque Ahamad
(ISBN 0-8186-1970-8); 110 pages

Robotic Technology Series

Multirobot Systems
Edited by Rajiv Mehrotra and Murali R. Varanasi
(ISBN 0-8186-1977-5); 122 pages